Lecture Notes in Earth Sciences

Edited by Somdev Bhattacharji, Gerald M. Friedman, Horst J. Neugebauer and Adolf Seilacher

15

Volker H. Jacobshagen (Ed.)

The Atlas System of Morocco

Studies on its Geodynamic Evolution

Springer-Verlag

Berlin Heidelberg New York London Paris Tokyo

Editor

Prof. Dr. Volker H. Jacobshagen
Freie Universität Berlin
FB Geowissenschaften, Institut für Geologie
Altensteinstrasse 34a, D-1000 Berlin 33, FRG

ISBN 3-540-19086-4 Springer-Verlag Berlin Heidelberg New York
ISBN 0-387-19086-4 Springer-Verlag New York Berlin Heidelberg

Library of Congress Cataloging-in-Publication Data. Geodynamic evolution of the Atlas system/ Morocco / Volker H. Jacobshagen, ed. p. cm.–(Lecture notes in earth sciences; 15) "One of the sources of this book was a conference on the geodynamic aspects of the Atlas system, organized in Berlin in 1986" by the Forschungsgruppe "Mobilität aktiver Kontinentalränder"–Pref.
ISBN 0-387-19086-4 (U.S.)
1. Geology–Morocco–Atlas Mountains. 2. Geology–Morocco–Anti-Atlas Mountains. 3. Geodynamics. I. Jacobshagen, Volker. II. Series.
QE339.M8G36 1988 556.4'4–dc 19 88-12329

Printed in Germany

Printing and binding: Druckhaus Beltz, Hemsbach/Bergstr.
2132/3140-543210

PREFACE

The aim of this volume is to reflect the current state of geoscientific activity focused on the geodynamic evolution of the Atlas system and to discuss new results and ideas. The volume provides a selection of papers on the geological history, structural development, and geophysical data of Morocco. It was not possible to cover all areas of geoscientific interest, however, we hope to shed some light on the major geodynamic problems.

One of the sources of this book was a conference on the geodynamic aspects of the Atlas system, organized in Berlin in 1986 by the Forschergruppe "Mobilität aktiver Kontinentalränder". The financial support for this project is being provided by the Deutsche Forschungsgemeinschaft and by the Freie Universität and the Technische Universität of Berlin (West). Furthermore we are very glad to mention that we have received a great number of publications by authors from different countries who did not take part in the Berlin conference.

It should be noted that, at present, a lot of international geoscientific activity is concerned with Morocco. This is not only due to the fascinating complexity of Moroccan geology and the beautiful outcroups, that may provide the key to many general problems, but also testifies to the liberal and prescient science politics of the Moroccan authorities, most of all the Ministery of Energy and Mining in Rabat. Therefore, on behalf of all the authors from abroad, I take the opportunity to express our deepest gratitude to our host country, especially to the leaders of the Division de la Gêologie, the general director Dr. M. Bensaid and the directors M. Dahmani and M. Hammouda. Thanks are, furthermore, due to Springer Verlag (Berlin-Heidelberg-New York) for offering the opportunity to publish a comprehensive volume and especially to Dr. Wolfgang Engel for his patience and helpful advices.

V. Jacobshagen

C O N T E N T S

Introduction

GEODYNAMIC EVOLUTION OF THE ATLAS SYSTEM, MOROCCO: AN INTRODUCTION

Volker Jacobshagen
Institut für Geologie, Freie Universität Berlin,
Altensteinstr. 34 A, D-1000 Berlin 33

The Atlas System

The northwestern margin of Africa, between the Atlantic Ocean and the Lesser Syrtis, is characterized by high mountain ranges, the elevation of which surmounts 3.000 m in the western part. Jbel Toubkal (4.165 m) in the High Atlas is the highest mountain of North Africa. These ranges have been synchronously uplifted since the Miocene. They differ, however, both in structure and geological history. The Maghrebides consisting of the Rif and Tell Atlas along the Mediterranean coast are a typical Alpidic orogen with complex nappe structures. The ranges in their southern foreland, called the Middle and High Atlas, are anorogenic intracratonic mountain belts separated by elevated plateaus. The eastern prolongation of the High Atlas is formed by the Saharan and Tunisian Atlas. The Anti-Atlas, however, which lies south of the High Atlas in Morocco, originated by simple updoming of the Precambrian basement of the Sahara craton. As all these ranges are geographical and geological related to each other they have been named the Atlas system by M. RICHTER (1970).

Thus, we find a mobile zone in the southern foreland of the Maghrebide orogen, which in southern Morocco extends up to 500 km into the African plate. As the southern border of this zone diverges more and more from the Maghrebides to the southwest, we may assume that its geological evolution cannot be solely deduced from the Mediterranean orogeneses, but seems to be more complex.

Geotectonic subdivision of Morocco

The Moroccan segment of the Atlas system is relatively well known. For an overview, the publications of MICHARD (1976), CHOUBERT and FAURE-MURET (1962, 1971, 1973) and FAURE-MURET and CHOUBERT (1971) are recommended. The geotectonic subdivision of the Moroccan Atlas system is shown in fig. 1 and is briefly described here:

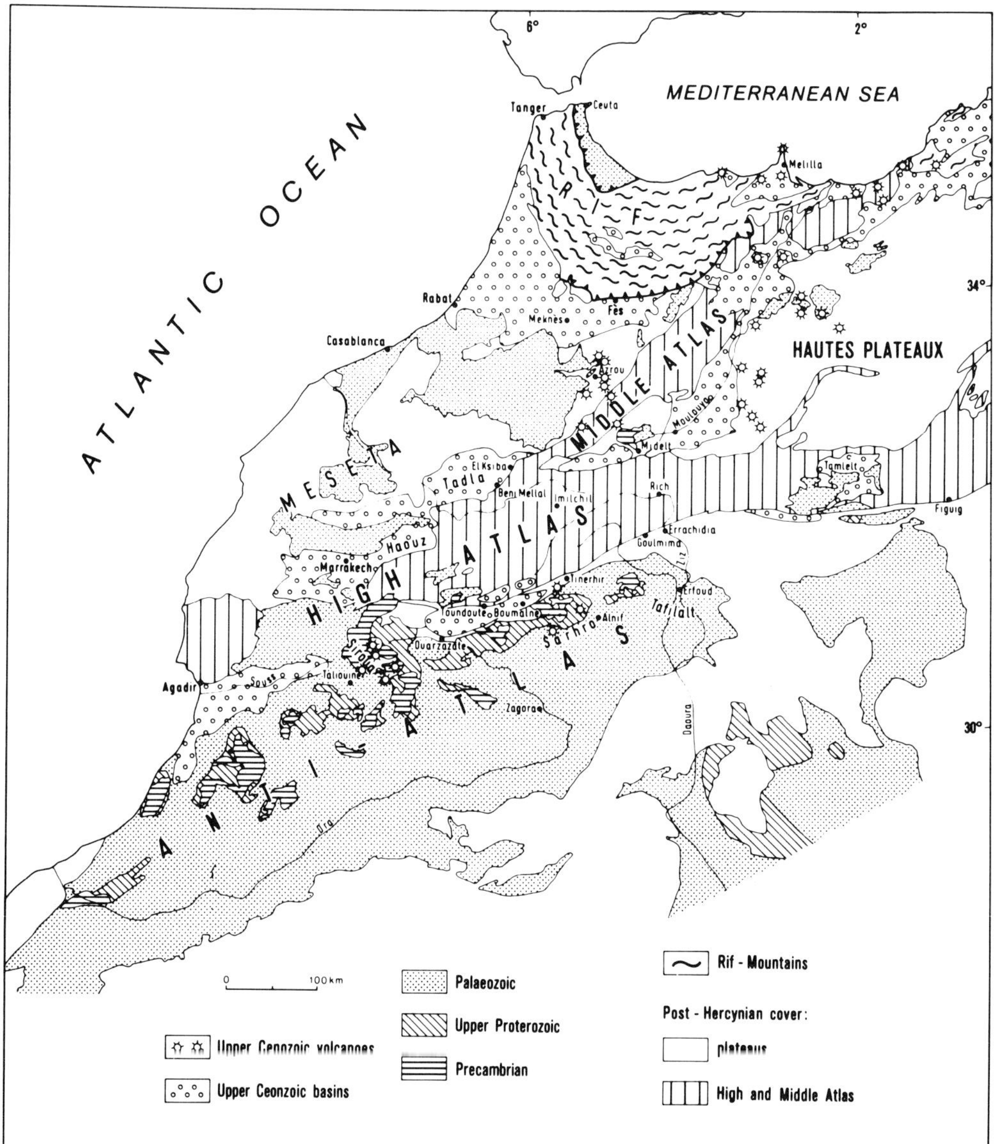

Fig. 1: Geotectonic sketch-map of the Atlas system of Morocco, based on MICHARD (1976)

The narrow arc of the Rif orogen in the north is overthrust towards the S on an Upper Cenozoic marginal trough. To the SW, the orogenic belt is bordered by the hills and plains of the Moroccan Meseta. Its Variscan basement is exposed in large areas from under a cover of flat-lying Mesozoic and Palaeogene sediments. To the SE, the Meseta and the Rif are bordered by the High and the Middle Atlas. Both ranges developed from early Mesozoic rift grabens which originated in correlation to the opening of the Western Tethys and of the North Atlantic Ocean.

Rifting was accompanied by mafic volcanism, which culminated in the Upper Triassic/Liassic and in the Dogger. Some intrusions, however, are of Cretaceous and even Palaeogene age. The structure of the High and Middle Atlas generally correspond to each other, both showing a basement of Palaeozoic or Precambrian rocks that is covered by thick Triassic and Jurassic sediments. Prevailing structures are steep faults, monoclines and strike-slip faults. In places, especially in the central and eastern High Atlas, overthrusts, box folds and even overturned folds exist. However, essential criteria of orogens such as: ophiolites, nappes, flysches, regional metamorphism and granitoids are not developed. Both ranges are accompanied by a few small Cenozoic basins. The Middle Atlas forks off from High Atlas in a northeastern direction, enclosing the so-called High Plateaus of Oran. The Variscan basement of these plateaus is covered by Mesozoic sediments, which are in the southwest overlain by the Cenozoic of the Haute Moulouya.

The High Atlas is separated from the Anti-Atlas by the Souss and the Ouarzazate basins ("sillon préafricaine"). As suggested by RUSSO and RUSSO (1934), many geologists believe that the High Atlas is bordered to the S by an important fault zone, known as the South Atlas Marginal Fault zone ("accident sud-atlasien"). Although several authors doubt that this fault really exists, others have already introduced it to plate tectonic concepts. Some of them even believe it to be the northwestern boundary of the African plate (e.g. DEWEY et al., 1973).

The Anti-Atlas was formed by wide-spanned updoming of marginal parts of the Sahara craton. Precambrian rocks form the core of major domes ("boutonnières") in the axial zone of the range. Normally, this basement is covered by poorly deformed sediments of Uppermost Proterozoic or Lower Palaeozoic age. In the Tafilalt region (Southeast Morocco), the Anti-Atlas dips to the E beneath undeformed Cretaceous sediments of the Sahara platform.

Pliocene or even Quaternary volcanoes of alkaline character are widespread in Morocco, with the exception of the High Atlas. Their geodynamic meaning is not yet clearly understood.

The structural development of Morocco south of the Rif and probably also the Mesozoic magmatism were preordained by a pattern of major faults in the pre-Mesozoic basement, the main directions of which are NE and ENE (DUBOURDIEU 1962, MATTAUER et al. 1972) and, in addition, NNW and ESE. According to these authors, the faults are of late Palaeozoic origin. In the central Anti-Atlas, however, the same pattern originated as soon as the end of the Panafrican cycle (HEINITZ 1984).

Aspects of the geodynamic evolution

Our present geoscientific knowledge of the four ranges of the Atlas system differs widely. The Maghrebides have been intensely investigated for many decades. Usually they have been considered within the context of the alpine orogens of the Mediterranean region. To describe their structure and geologic history with some detail would fill a separate volume. The present book aims, however, mainly to shed some light on the development of the intracontinental Atlas ranges, and to define the general relations between them. Most of the authors have interpreted their geodynamic evolution with respect to major plate tectonic events of the Mesozoic and Cenozoic periods. Under this aspect the Rif orogeny was only briefly reviewed, here, in connection with the alpine development of the Western Mediterranean region by DURAND-DELGA and OLIVIER.

The geodynamic history of the Atlas system started very early. Within the Precambrian cores of the Anti-Atlas, traces of several orogenies are preserved (see WALLBRECHER, this vol.). Recent studies have been mainly focussed on the Panafrican cycle for which plate tectonic models have been proposed by LEBLANC (1981), SCHERMERHORN et al. (1986), and BRABERS (this vol.). But also many new results concerning the Late Proterozoic and Lower Palaeozoic cover of the Anti-Atlas have been produced, which are referred to by BUGGISCH and FLÜGEL, SDZUY and GEYER and by BUGGISCH and SIEGERT in the first part of this book.

The very important Caledonian-Hercynian history, which is documented in the basement of the High and Middle Atlas and of the Moroccan Meseta, has recently been discussed during a symposium "Morocco and

Palaeozoic Orogenesis (I.G.C.P. Project No. 27, The Caledonide Orogen), held in Rabat in 1983. This epoch is, therefore, neglected in the present volume.

A great many of recent investigations have been concentrated on post-Palaeozoic stratigraphy, facies and structural evolution of the Western and Central High Atlas mountains, but our knowledge of the Middle Atlas is at present relatively poor. As both ranges play a key role in deciphering the Mesozoic and Cenozoic evolution of the Atlas system, the majority of contributions to this volume deal with these topics. Seismologic, geoelectric and magnetotelluric studies on a section across both ranges by SCHWARZ and WIGGER are of special importance for the understanding of geodynamic processes. The neotectonic history of the Atlas system is clearly reflected by the sedimentary fillings of the Upper Cenozoic basins adjacent to the Atlas ranges as demonstrated mainly by the contributions of GÖRLER et al. and FRAISSINET et al., but also by geomorphologic features.

Recent geotectonic hypotheses interpreted the evolution of the intracontinental Atlas ranges with respect to plate tectonic configurations and processes (MICHARD et al. 1975, MATTAUER et al. 1977, COURBOULEIX et al. 1981, STETS and WURSTER 1981). All of them referred to the fundamental fractures of the pre-Mesozoic basement. It is generally agreed that these fractures have been repeatedly reactivated since the Triassic, while the orientation of the continental stress-fields changed in time, due to the relative movements of the African and the European plates. MATTAUER et al. (1977) were of the opinion that the post-Palaeozoic development reflects two major geotectonic processes of changing influence: the opening of the Atlantic and the collisions of the West Mediterranean regions. Deviating from this view, STETS and WURSTER (1981) proposed a geotectonic model for the High Atlas which is related only to the Atlantic opening and excluded Mediterranean influences. JACOBSHAGEN, GÖRLER and GIESE (this vol.) support MATTAUER's view in reference to new geological and geophysical data, they also suggest a mechanism to transmit compressional forces from the active northern margin over several hundred kilometers into the African plate.

Acknowledgements: The manuscript would not have been realized without the help of W. Jung, G. Lindemann, D. Reich and C. Di Stefano which is gratefully acknowledged.

REFERENCES

CHOUBERT, G. and FAURE-MURET, A. (1962): Evolution du domaine atlasique marocain depuis le temps paléozoique.- Livre-mém. P. Fallot, Mém. h. ser. Soc. géol. France, 1: 447-527, Paris.

-- (1971): Afrique occidentale: socle précambrien et zones mobiles. Anti-Atlas (Maroc).- in UNESCO, Tectonique d'Afrique, Sci. Terre, 6: 163-175, Paris.

-- (1973): Moroccan Rif.- in A.M. SPENCER (ed.): Mesozoic - Cenozoic orogenic belts: 37-46, Geol. Soc. London.

COURBOULEIX, S., DELPONT, G. and DESTEUCQ, CH. (1981): Un grand décrochement est-ouest au nord du Maroc à l'origine des structures plissées atlasiques. Arguments géologiques et expérimentaux.- Bull. Soc. géol. France, (VII), 23: 33-43, Paris.

DEWEY, J.F., PITMAN, W.C., RYAN, W.B. and BONNIN, J. (1973): Plate tectonics and the evolution of the alpine system. Bull. geol. Soc. Amer., 84: 3137-3180.

DUBOURDIEU, G. (1962): Dynamique wegenérienne de l'Afrique du Nord.- Livre Mém. P. Fallot, Mém. h. ser. Soc. géol. France, 1: 627-644, Paris.

FAURE-MURET, A. and CHOUBERT, G. (1971): Le Maroc. Domaine rifain et atlasique.- in UNESCO, Tectonique de l'Afrique, Sci. Terre, 6: 17-46, Paris.

HEINITZ, W. (1984): Die Deformation des sedimentären Deckgebirges im zentralen Anti-Atlas (Südmarokko).- Berliner geowiss. Abh., (A), 55: 1-84, Berlin.

LEBLANC, M. (1981): Ophiolites précambriennes et gîtes arseniés de cobalt (Bou Azzer - Maroc).- Notes Mém. Serv. géol. Maroc, 280: 1-306, Rabat.

MATTAUER, M., PROUST, F. and TAPPONIER, P. (1972): Major strike-slip faults of Late Hercynian age in Morocco.- Nature, 237: 160-162.

MATTAUER, M., TAPPONIER, P. and PROUST, F. (1977): Sur les mécanismes de formation des chaînes intracontinentales. L'exemple des chaînes atlasiques du Maroc.- Bull. Soc. géol. France, (VII), 19: 521-526, Paris.

MICHARD, A. (1976): Eléments de géologie marocaine.- Notes Mém. Serv. géol., 252: 1-408, Rabat.

MICHARD, A., WESTPHAL, M., BOSSERT, A. and HAMZEH, R. (1975): Tectonique de blocs dans le socle atlaso-mésétien du Maroc: Une nouvelle interprétation des données géologiques et paléomagnétiques.- Earth Planet. Sci. Letters, 24: 363-368, Amsterdam.

RICHTER, M. (1970): Das Atlas-System in Nordafrika.- Z.dt.geol. Ges., 120 (1968): 68-84, Hannover.

RUSSO, P. and RUSSO, L. (1934): Le grand accident sud-atlasien.- Bull. Soc. géol. France, (V), 4: 375-384, Paris.

SCHERMERHORN, L.J.G., WALLBRECHER, E. and HUCH, K.M. (1986): Der Subduktionskomplex, Granitoplutonismus und Schertektonik im Grundgebirge des Sirwa-Doms (Anti-Atlas, Marokko).- Berliner geowiss. Abh., (A), 66: 301-332, Berlin.

STETS, J., and WURSTER, P. (1981): Zur Strukturgeschichte des Hohen Atlas in Marokko.- Geol. Rdsch., 70: 801-841, Stuttgart.

Anti-Atlas

THE ANTI-ATLAS SYSTEM: AN OVERVIEW

Eckard Wallbrecher
Institut für Geologie und Paläontologie
Universität Graz
Heinrichstr. 26, A-8010 Graz

General Situation

The Anti-Atlas System, which lies between the High Atlas and the Sahara (Fig. 1) is a part of the pericratonic zone of the West Sahara Craton (MICHARD 1976). In the Alpine age, which was the time of the orogenesis of the High Atlas, this system acted as a stable crust. At this time, it was transgressed only by the flat Cenoman-Turon sea. It has an Upper Precambrian to Lower Paleozoic cover which was partly folded during Variscan orogeny.

The typical structures of the Anti-Atlas are the numerous, often deeply eroded basement domes which break through the covering rocks as "boutonnières"(button holes). These structures are lined up in a chain which runs from WSW towards ENE and thus follows the NW margin of the West Sahara Craton.

The Anti-Atlas, including the eastern branches Jbel Saghro and Jbel Ougnate, forms the "l'èpine dorsale", with peaks of 2500 m to 2700 m height in Jbel Saghro and 3300 m in Jbel Sirwa. The latter peak, however, is formed by a Ponto-Pliocene volcanic complex which lies on a crystalline uplift.

The Anti-Atlas is separated from the High Atlas by young intramontane molasse basins: the Souss Basin in the west and the Warzazat and Er-Rachidia Basins in the east.

To the south, the Anti-Atlas borders on the Jbel Bani chain which is built up by the cuestas of a flat southerly dipping Cambro-Silurian series. Further to the south follows the Dra Plain and the late Paleozoic layers of Jbel Ouarkziz, which dip under the Meso- and Cenozoic Hamadas of the West Sahara.

The Basin of Tindouf forms a syneclise on the West Sahara Craton with a sedimentary filling of more than 10,000 m depth.

The Boutonnières

The crystalline domes of the Anti-Atlas form the northern border between the West Sahara Craton and the Paleozoic and Cenozoic mountains

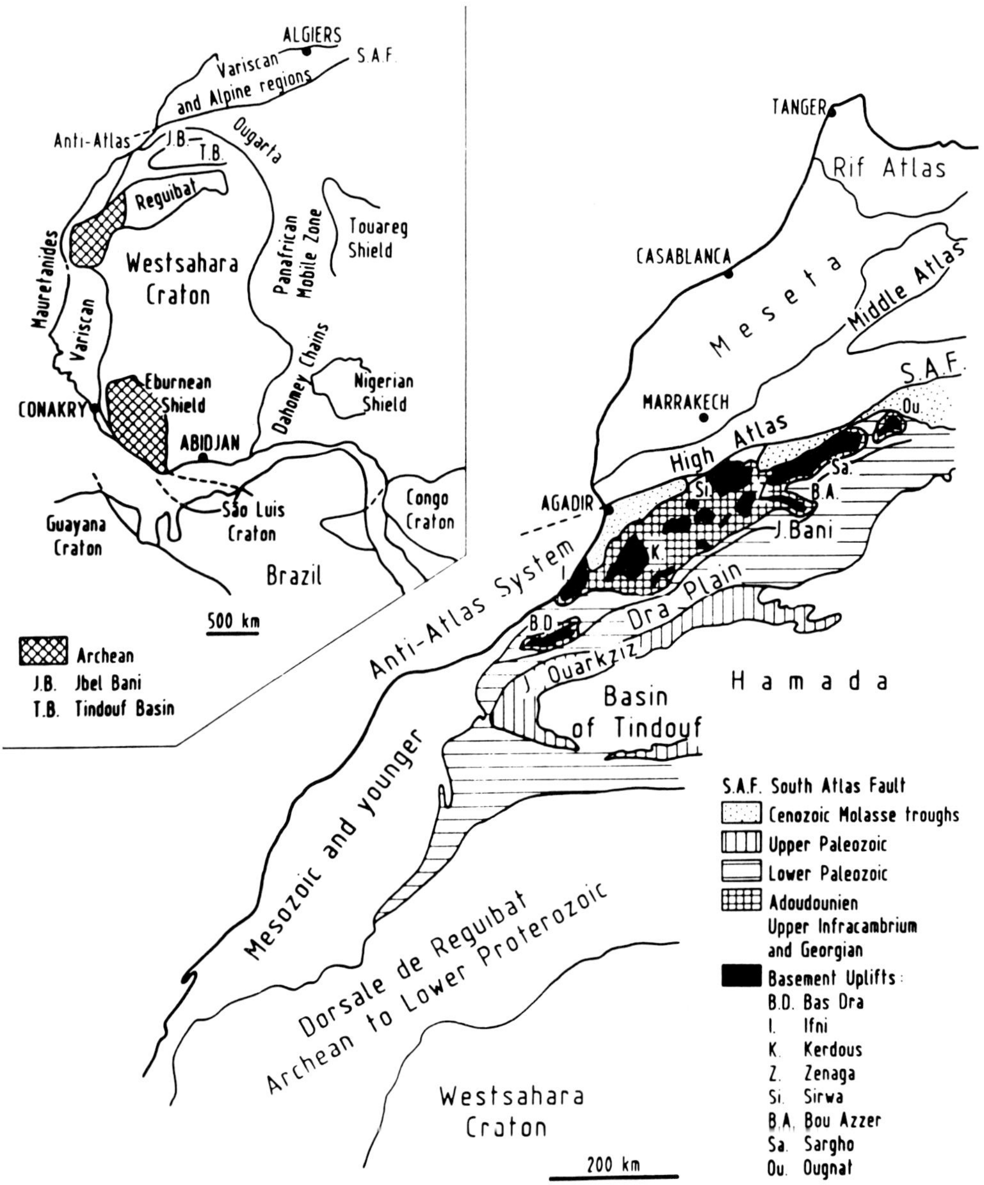

Figure 1: Main structure of the northwestern margin of the West Sahara Craton

of North Africa. CHOUBERT (1963) differentiated six orogenic cycles, among them

1) Precambrium O
This includes, for example, the augengneisses of Jbel Ouiharem (Boutonnière de Kerdous) and the gneisses of Oued Assemlil (Boutonnière de El Graara). According to CHOUBERT, these series belong to the "Zagorides".
2) Precambrium I
This includes the micaschists of the Kerdous Series and the migmatites of the Zénaga Series. CHOUBERT calls these series the "Berberides".
3) Precambrium II
To this belong epi- and pyroclastic layers and greenstones of Bou Azzer, Jbel Sirwa, and Jbel Saghro. CHOUBERT calls these series "Marrocanides".
4) Precambrium III
This includes the Warzazat Series, which is of late Precambrian to Infracambrian age. It is considered to be a molasse of the last (Panafrican) orogenesis. It forms the base of the non-metamorphic covering layers.

Radiometric Dating
New rubidium/strontium dating, carried out by CHARLOT (1982), resulted in a separation between a metamorphic and granitic foundation and a volcano-sedimentary cover.
Foundation
The foundation is formed by the chain of boutonnières and runs parallel to the Dorsale de Lèo in the south and to the Dorsale de Reguibat in the central part of the craton (Fig. 1). Only the western margin, which begins at the western end of the Ivory Coast and leaves the African continent somewhere near Rio de Oro, is of Archean age. All of the crystalline domes of the Anti-Atlas are east of this line, and, therefore, outside of the Archean core of the craton.
A second metamorphic event, dated at 1.80 - 1.65 x 10^9 years (Taznakht granite), took place as a thermic event which produced the metamorphosis of the Azguemerzi granite. The same age has also been attributed to the Dorsale de Requibat (VACHETTE et al. 1973).
CHARLOT has pointed out that these events took place at about the same time as the Hudsonian orogeny in North America, the orogeny of the Svekofennides in Scandinavia, and the Laxfordian orogeny in Scotland. Thus, the Anti-Atlas might be a link between the West African Craton and North America or Northern Europe.

According to CHARLOT, the Precambrium II unit is probably Panafrican, that is, about 700 - 600 x 10^6 years old. CHARLOT showed, using U/Pb dating on zircons, that no new thermic event has rejuvenated these rocks.

The Cover

The lowest of the covering layers is the Warzazat Series. In Jbel Sirwa and Jbel Saghro this series consists of ignimbrites, rhyolites, and conglomerates. Granites and volcanites have been dated at 604 and 560 x 10^6 years (CLAUER 1976, CLAUER & LEBLANC 1977, CHARLOT 1982).

Above the Warzazat Series follows the Adoudounien Series, sometimes with an angular unconformity. It is considered to be of late Infracambrian to Georgian age. The Adoudounien Series consists of a base of conglomerates followed by carbonates, marls, sandstones, and, finally, carbonates again.

Volcanites which are embedded in the lower carbonate layers at Bou Azzer have been dated at 534 x 10^6 years. Later thermic events are not known. In the Boutonnière de Kerdous and in the Boutonnière d'Ifni, however, Variscan closing ages for biotite of 325 x 10^6 and 260 x 10^6 years, respectively, have been measured. Variscan elements of importance can only be found south of the Antiatlas chain in the monoclines of Jbel Bani and Jbel Ouarkziz. They show NE-SW striking structures on the southern slope of the Anti-Atlas in the SW and NW-SE striking structures on the slope of the Ougarta chain in the SE.

Permian to early Mesozoic extensional tectonic events, caused by the opening of the Atlantic, contributed long doleritic dyke systems running SW-NE. The longest of these systems shows its southernmost outcrop in the Jbel Ouarkziz and can be followed for more than 400 km as it cuts through the Bou Azzer structure and ends in the northeastern part of Jbel Saghro. The youngest geodynamic activity seems to have been a Tertiary-Quaternary elevation of the central part of the Anti-Atlas and an important Ponto-Pliocene volcanism of Jbel Sirwa and eastern Jbel Saghro.

REFERENCES

CHARLOT, R. (1982): Caractérisations des événements éburnéens et panafricains dans l'Anti-Atlas marocain. Apport de la méthode géochronologique Rb-Sr. - Notes et mémoires du Service Géologique du Maroc 313:106 p., Rabat.

CHOUBERT, G. (1963): Histoire géologique du Précambrien de l'Anti-Atlas. (Tome 1). Notes et mémoires du Service Géologique du Maroc 162:352 p., Rabat.

CLAUER, N. (1976): Géochimie isotopique du strontium des milieux sédimentaires. Application à la géochronologie de la couverture du craton ouest-africain. Thèse Doct. Etat, Univ. Strasbourg.

CLAUER, N. & LEBLANC, U. (1977): Implications stratigraphiques d'une ètude gèochronologique Rb-Sr sur métasédiments précambriens de Bou Azzer (Anti-Atlas, Maroc). Notes et mémoires du Service Géologique du Maroc 268: 7-22, Rabat.

MICHARD, A. (1976): Eléments de Géologie Marocaine. Notes et mémoires du Service Géologique du Maroc 252: 408 p., Rabat.

VACHETTE, M; SOUGY, J.; CARON, J.P.; MARCHAND, J.; SIMON, B.; TEMPIER, C. (1973): Ages radiométriques Rb/Sr de 2000 à 1700 M.A. de séries métamorphiques et granites intrusifs précambriens dans la partie N et NE de la dorsale Reguibat (Mauritanie septentrionale). C. R. 7e Colloque Géologie Africaine, Florence.

A DUCTILE SHEAR ZONE IN THE PANAFRICAN BASEMENT ON THE NORTHWESTERN MARGIN OF THE WEST AFRICAN CRATON (SIRWA DOME, CENTRAL ANTI-ATLAS

Eckard Wallbrecher
Institut für Geologie und Paläontologie
Universität Graz
Heinrichstr. 26, A-8010 Graz

ABSTRACT

A wrench-fault type shear zone of at least 6 km width was found in the Precambrian basement of the Jbel Sirwa (Central Anti-Atlas, Morocco). It is explained by lateral slip of an (unknown) northern continental mass along the northern margin of the West African Craton. The deformation in the shear zone increases from south to north, producing folds with axes that steepen continuously northwards until they are perpendicular to a (dextral ?) horizontal shear stress. The grain fabric suffered mylonitization that increased in a northern direction, and that, in granitic parent rocks, culminated in superplastic deformation with prism and grain boundary gliding of the quartzes. Pseudotachylitic veins indicate seismic events after the plastic softening of the fabric. This must be explained by either an elevation of the whole complex towards the end of shearing, or by occasionally very high strain rates. The age of the shear zone is late Panafrican (Early Paleozoic).

RESUME

Dans le socle précambrien du Jbel Sirwa (Anti-Atlas central, Maroc), une zone de cisaillement d'une largeur minimale de 6 km a produit une faille verticale de décrochement. On l'explique comme le résultat du mouvement latéral d'une masse continentale (inconnue) le long de la bordure septentrionale du craton ouest-africain.

La déformation dans cette zone s'intensifie du sud au nord, engendrant des plis, dont les axes se redressent vers le nord jusqu'à être perpendiculaires à une contrainte horizontale (?) dextrale. De même, le degré de mylonitisation augmente du sud au nord. Il s'intensifie au centre des roches granitiques pour atteindre le degré de déformations superplastiques avec glissement sur les joints des grains et sur les surfaces prismatiques des minéraux de quartz. Des veines pseudotachylitiques indiquent l'existence d'événements sismiques après la déformation plastique. Ceci est explicable soit par une élévation du complex entier vers la fin du cisaillement soit par des déformations occasionelles très rapides. L'âge de la tectonique est tardi-panafricain (Paléozoique inférieur).

INTRODUCTION

The crystalline dome of the northeastern Jbel Sirwa belongs to a chain of Precambrian basement uplifts, arranged in SW-NE direction, parallel to the northwestern margin of the West African Craton (Fig. 1).

First mapping of this area, the geographic position of which is indicated by the arrow in Fig. 1, was carried out by CHOUBERT (1963). Nonmetamorphic sediments in the south of the area were investigated by JEANNETTE & TISSERANT (1977). A comparison of this uplift with the structure of Bou Azzer was carried out by LEBLANC (1981), who also described tectonic observations and distinguised two tectonic phases.

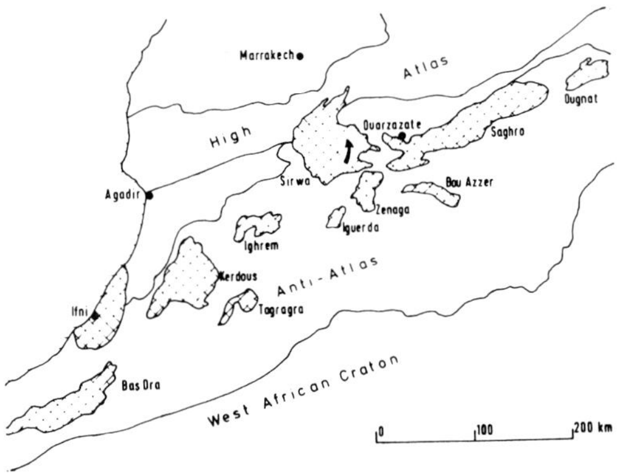

Figure 1: Precambrian uplifts of the Antiatlas and position of the research area in the Sirwa Dome

From these works and from our own observations (SCHERMERHORN et al., 1986) the following sequences have been separated. These are, from south to north (Fig. 2)

1) The Lmakhzan Formation, a foreland sequence of epi- and pyroclastic sediments, containing basaltic dykes and sills.

2) The Tachoukacht Gneiss Formation, a sequence of intermediate gneisses, amphibolites and blastomylonites.

3) The Tourtit Ophiolite Complex, an imbricated and tectonically deformed metamorphic ophiolite sequence of ultrabasic, basic and, to a small extent, acid rocks, containing large lenses of the highly deformed Tourtit Granite-Mylonite.

4) The sedimentary and volcanic cover which forms the Warzazat Group, a sequence which consists mostly of conglomerates, ignimbrites, and rhyolites.

5) Posttectonic granitic intrusions and the N-S striking Daroufarnou Dykes of porphyric granite.

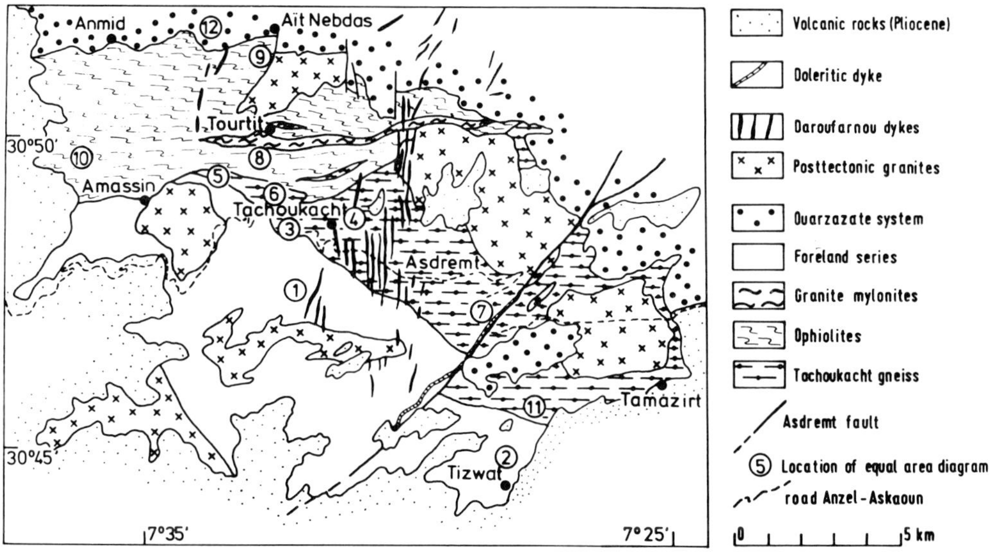

Figure 2: Tectonic and lithological units (SCHERMERHORN et al., 1986, simplified). Numbers indicate the position of the fabric diagrams.

The petrography and geochemistry of these sequences and their geotectonic implications have been described in detail by SCHERMERHORN et al. (1986). These authors describe the geotectonic evolution of the sequences in the following steps:

1) Opening of an ocean between the West African Craton and an unknown continent north of it.

2) Initiation of a south-dipping subduction zone.

3) Development of a volcanic arc (the Tachoukacht Gneiss Formation), flanked by a fore-arc (the Tourtit Ophiolite Complex) and a back-arc (the Lamakhzan Formation) basin.

4) Continent collision, which thrusted the northern continent under the fore-arc ophiolites, under the island arc, and under the margin of the West African Craton. Metamorphism and imbrication are generated in the ophiolites and in the Tachoukacht Gneisses (island arc).

5) Lateral slip along the thus developed geosuture which gives rise to a 6 km wide shear zone which will be described in this paper.

TECTONICAL SETTING

The general tectonical style of the area (Fig. 2) is rather simple. Generally horizontal or only weakly inclined southern foreland sediments are separated by a steeply northward dipping and E-W striking upthrust from the albite-gneisses of the Tachoukacht Gneiss Formation

in the north. A similar thrust separates this formation from the Tourtit Ophiolite Complex further north. This complex, which is strongly imbricated, contains the E-W striking layer of the Tourtit Granite-Mylonite which has a thickness of 100 to 300 m.
In the northern part of the area (generally parallel to the Assif Amsalan Valley), the ophiolites are overlain by a younger horizontal cover of late Precambrian to Infracambrian age which did not suffer the imbrication. All these units are cut by the N-S striking Daroufarnou Dikes.
Numerous posttectonic granite intrusions are also found in all of these units. A penetrative foliation pattern develops continuously from south to north and culminates in the ophiolites producing a pronounced vortex structure and extreme shear deformation in the Tourtit Granite-Mylonite.

Macroscopic Tectonic Observations

1) The Lmakhzan Formation

The layers of the foreland sediments generally dip towards NE at angles of about 30 to 60 degrees (Diagr. 1)+. In spite of some local folding in the area of Tizwat, a unique fold pattern is not discernable (degree of great circle distribution = 21%)++. Up to now, shear zones with brittle strike separation could be observed in these series at three places. These are, from south to north
a) in the Assif Tamazirt, 1 km north of Tizwat (sheet Taghdout)+++. Here we find NNW-SSE striking vertical layers of thin-bedded siltitos which are deformed to kink bands (Fig. 3), indicating dextral slip. The orientation of

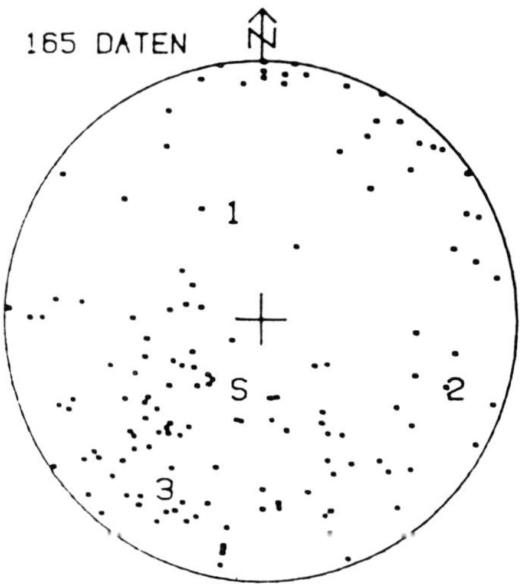

Diagram 1

+) The position of places of fabric measurements for the diagrams are marked by the numbers in Fig. 2.

++) All parameters describing quantitatively preferred orientation and fabric shape are derived from WALLBRECHER (1979). Numbers 1 to 3 are the eigenvectors of the orientation tensor; s is the center of gravity.

+++) All map names refer to the "Carte du Maroc", scale 1:50.000.

the bedding planes and of corrugation lineations in these beds are plotted in Diagramm 2 a) and b)[+].
b) near Bou Wals (trail from Souq at Tlata to Azib-n-Iriri, sheet Tachoukacht). Here, there is a sequence of vertical SW-NE striking shear planes with distances in the dm range between which some rocks of 1 to 2 m width were dextrally rotated.

Figure 3: Kink bands in siltites of the Lmakhzan Formation, north of Tizwat.

c) in the Assif Lmakhzan, 300 m south of the car trail (sheet Tachoukacht). Here we find a lens of serpentinite which is imbricated in the sedimentary layers of the Lmakhzan Formation and shows a dextral movement along kink bands.

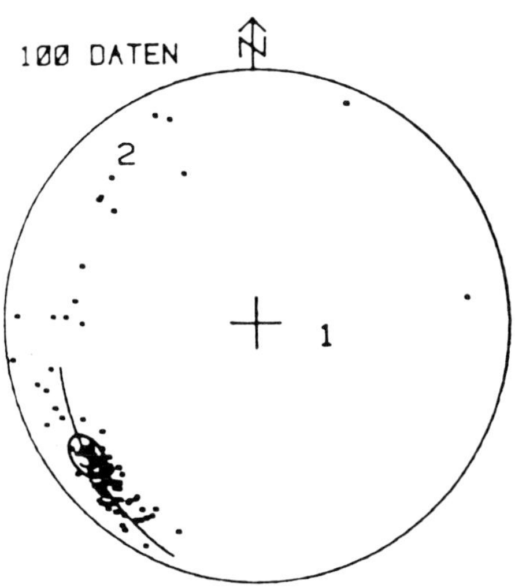

Diagramm 2a

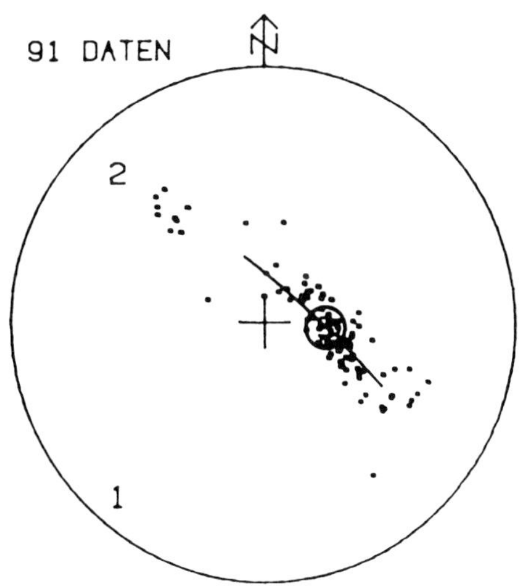

Diagramm 2b

+) In all diagramms inner small circles are the cones of confidence for a level of 99% and the outer small circles are the spherical aperture, a deviation measure (WALLBRECHER, 1979).

2) The Tachoukacht-Gneisses

Steep upthrusts of these gneisses onto the southern foreland was observed in the Assif Lmakhzan, in the Assif Tamazirt, and at the Tizi Izgaran near Amassin (Fig. 4).

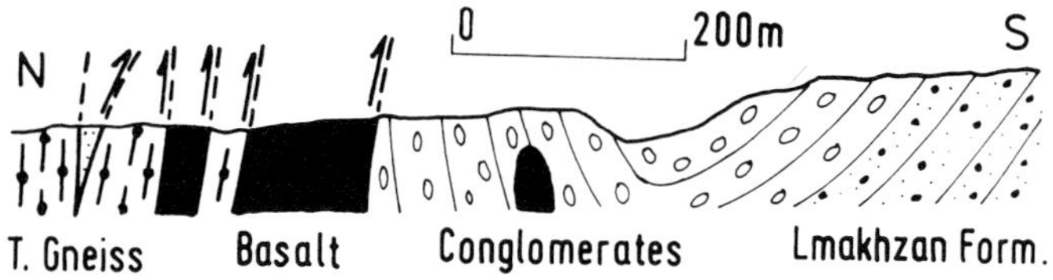

Figure 4: Profile at Tizi Izgaran

At their southern margin, the Tachoukacht Gneisses begin with a fold pattern which is sketched in Fig. 5. The fold axes are flat and arranged in E-W direction . The folds are strongly asymetric and have long vertical or steeply southward dipping limbs. The resulting enveloping surface dips steeply northward. This enveloping surface and the axial plane surfaces of the folds indicate a southward tectonic movement.

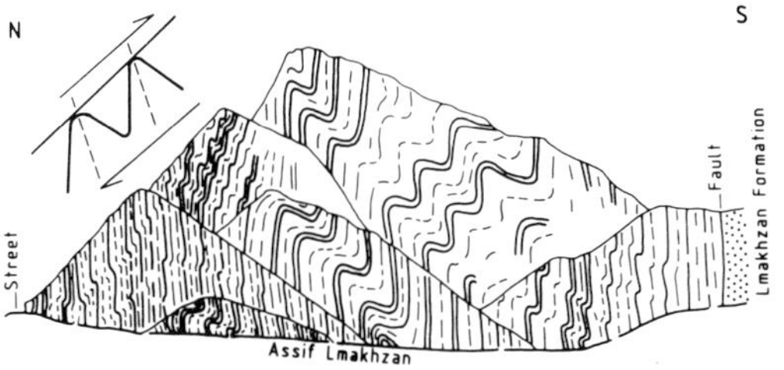

Figure 5: Fold pattern of the Tachoukacht Gneisses, Assif Lmakhzan, south of the car trail

In equal area diagrams, the fold axes of cm to dm range form a flatly eastward-dipping cluster with a center of gravity at 97°/27° (direction of dip and angle of dip, Diagram 3a). The foliation strikes E-W and shows a center of gravity at 174°/66° and a beginning transition from the cluster distribution to a girdle distribution (Diagram 3b). Since eigenvector 1, which belongs to the

Figure 6: Fold with flat axis, Tachoukacht Gneiss, south of the car trail

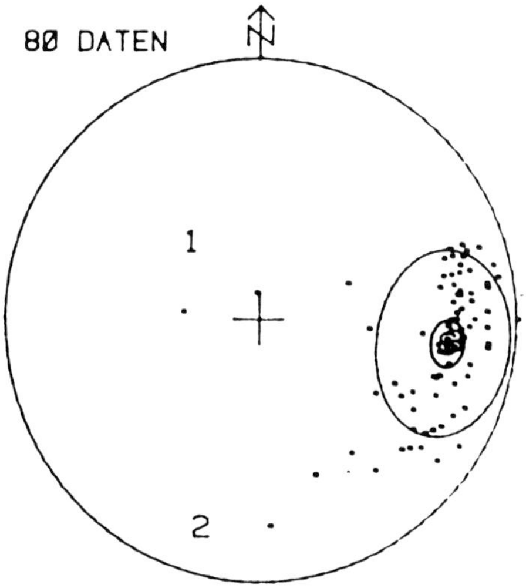

Diagram 3a

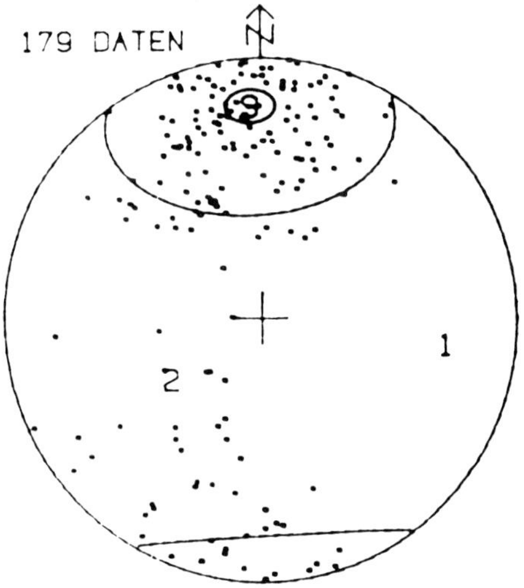

Diagram 3b

smallest eigenvalue of the orientation tensor (SCHEIDEGGER, 1965), coincides with the center of gravity of the fold axes cluster, the foliation is arranged tautozonally to the minor fold axes. This fold pattern, which is established in the Lmakhzan Valley in a zone of about 200 m width, is followed to the north, probably with a floating transition, by a zone in which the center of gravity of foliation poles is approximately the same (Diagram 4a). However, the axis of tautozonality of the foliation is much steeper (67°/43°) and again coincides with an also steeper cluster of the minor fold axes (71°/49°, Diagram 4b).

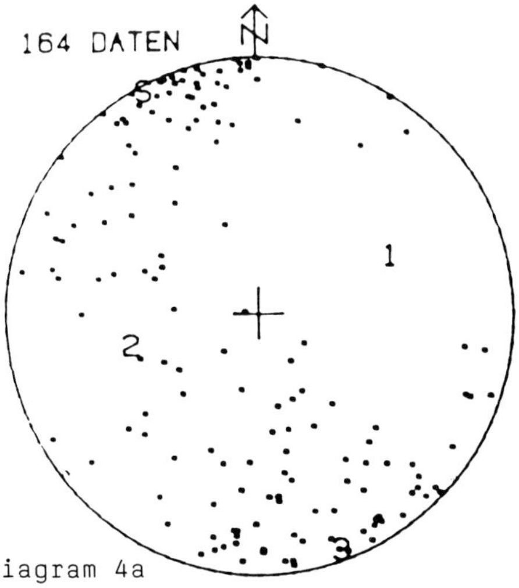

Diagram 4a

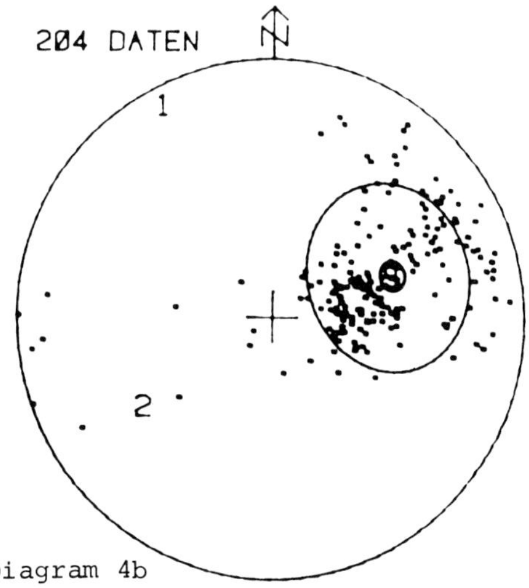

Diagram 4b

An even steeper arrangement of the fold axes in the style of vortex structures with beginning transition to a girdle distribution parallel to the foliation plane and a stronger cluster pattern (great circle distribution = 40%) of the foliation poles is demonstrated by Diagrams 5a and b, measured near Tichkiwin.

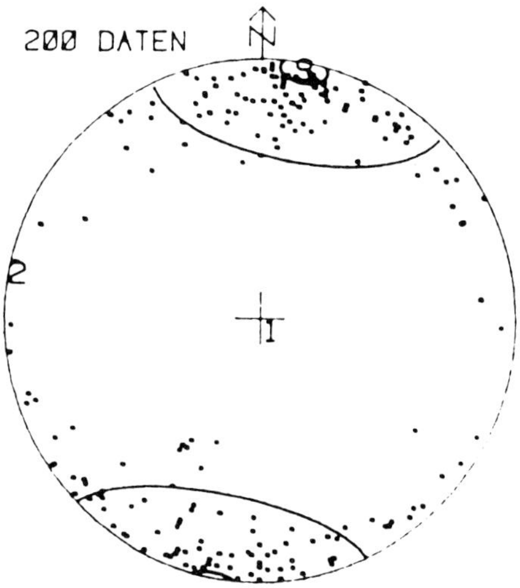

Diagram 5a

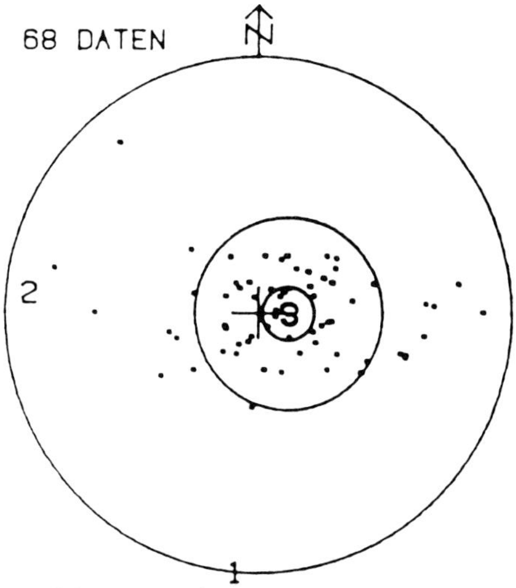

Diagram 5b

As we approach the contact of the ophiolites, this pattern continuously turns into the arrangement of foliation and fold axes shown in Diagram 6a and b (Lmakhzan Valley north of the car trail). The foliation shows almost the same picture as before, but the fold axes now

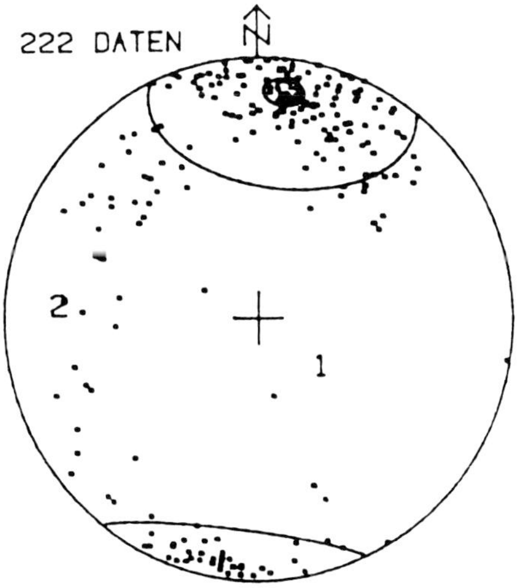

Diagram 6a

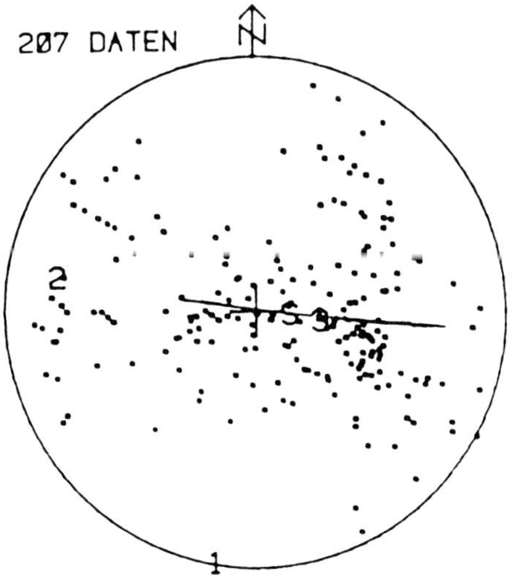

Diagram 6b

are scattered to a girdle distribution in the plane of foliation (degree of great circle distribution = 66%, circular aperture = 99%).
A similar setting of foliation and fold axes was measured at the Asdrem Pass (Diagram 7a and b). The flatter arrangement of both foliation and fold axes fabrics must probably be explained by a tilting of

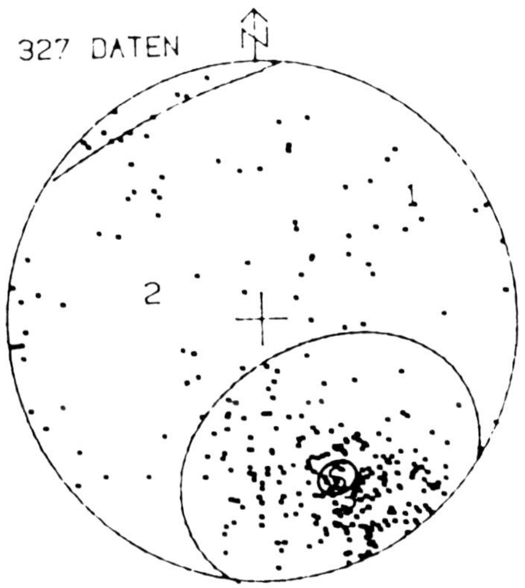

Diagram 7 a

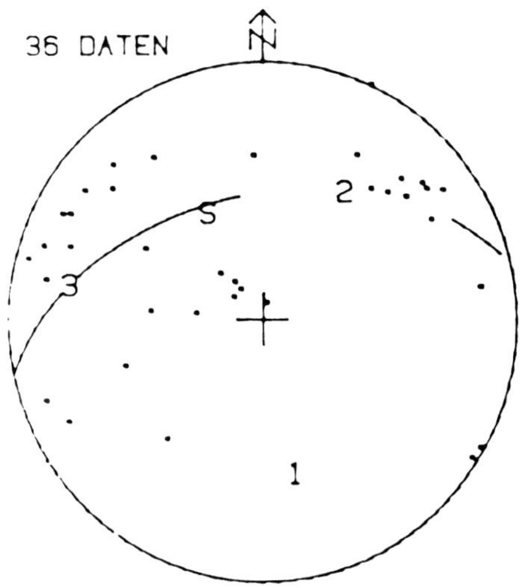

Diagram 7 b

the Asdrem bloc along the younger Asdrem fault (CHOUBERT, 1963) which raised the high plateau of Asdrem upwards in relation to the plain of Tamazirt.
A true vortex style of fold axes was observed E of the Asdrem Pass in the Assif Tamazirt, NNE of Tizwat, sheet Taghdout (Diagram 8a and b). This extremely southern position of vortex structures might be explained by a dextral bloc movement along the young Asdrem fault.

3) The Tourtit Ophiolite Complex and the Tourtit Granite-Mylonite
The ophiolite sequence shows a very strong schistosity and recrystallization parallel to the schistosity planes. Only the cores of basic cumulates resisted the schistosity. The schistosity planes are deformed by extreme vortex structures with vertical fold axes (Fig. 7). The vergency of most folds is dextral.
Diagrams 9a and b show the arrangement of schistosity and fold axes in the southern,and Diagrams 10a and b in the northern part of the ophiolites. Diagram 11 contains fold measures from the region west of Amassin. The foliation planes of the Tourtit Granite-Mylonite also strike E-W. This granite suffered the same tectonical treatment as did the surrounding ophiolites, which proves that the granitic intrusion must be older than the shear tectonic event.

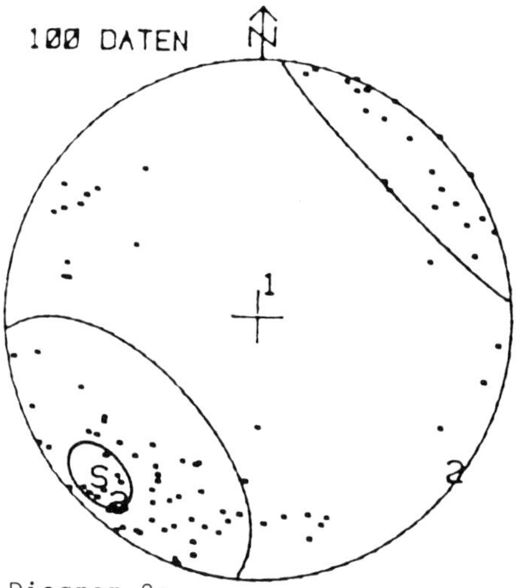

Diagram 8a

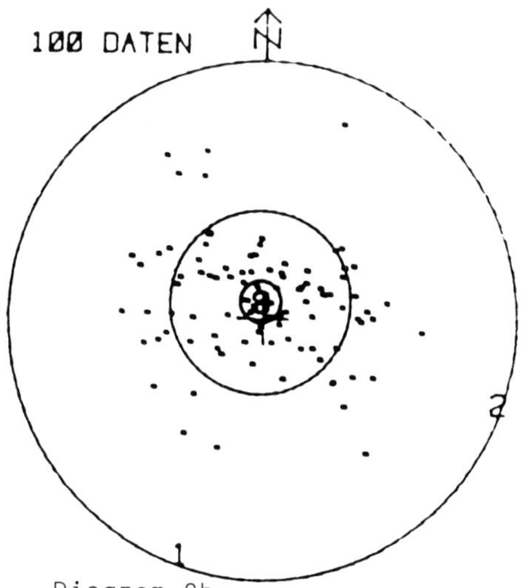

Diagram 8b

Figure 7: Vertical folds in serpentinites, Assif Tasrwin south of Tourtit

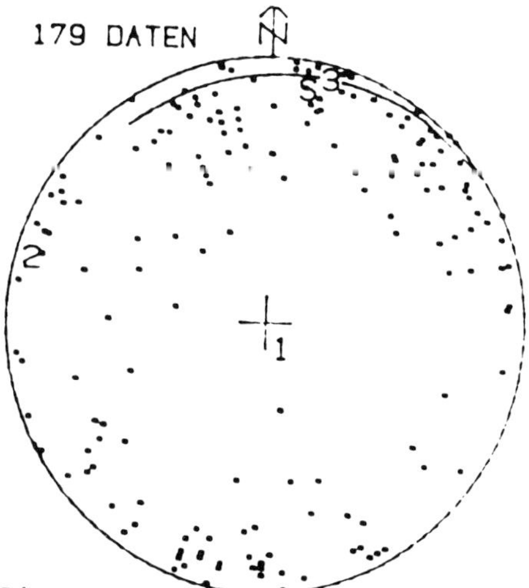

Diagram 9 a

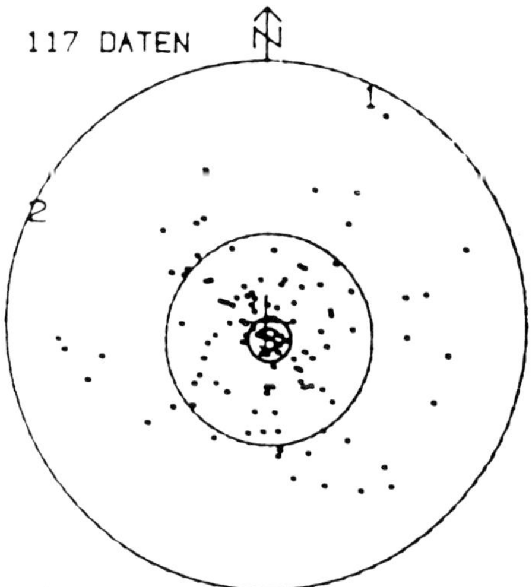

Diagram 9b

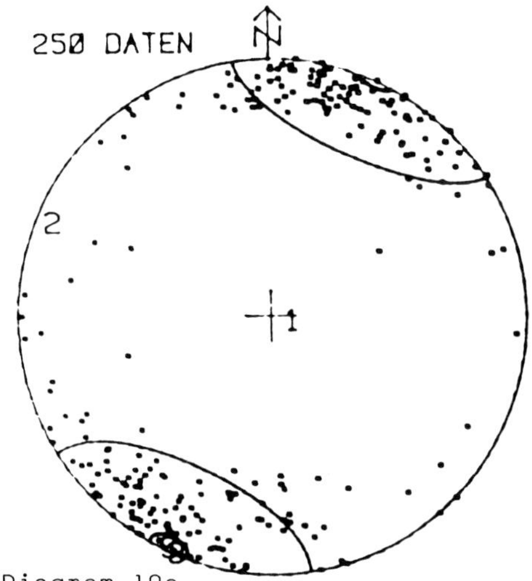

Diagram 10a

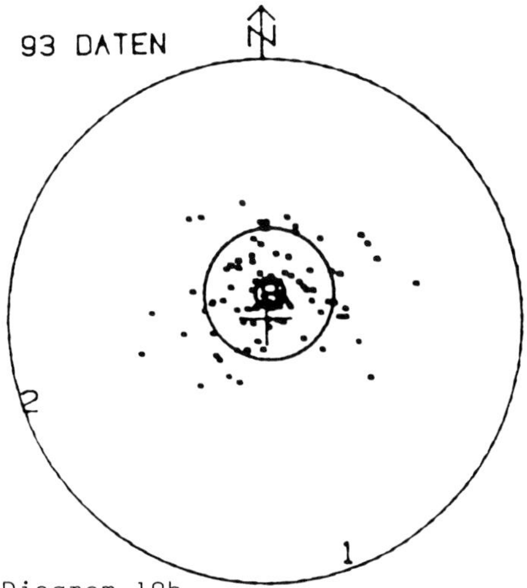

Diagram 10b

4) The Warzazat Group

The younger cover which lies at the northern margin on top of the crystalline uplift shows the same tectonical deformation patterns as the basement, yet in a less pronounced and less penetrative form (Diagram 12). The distance between schistosity planes is greater, vortex structures are not found so

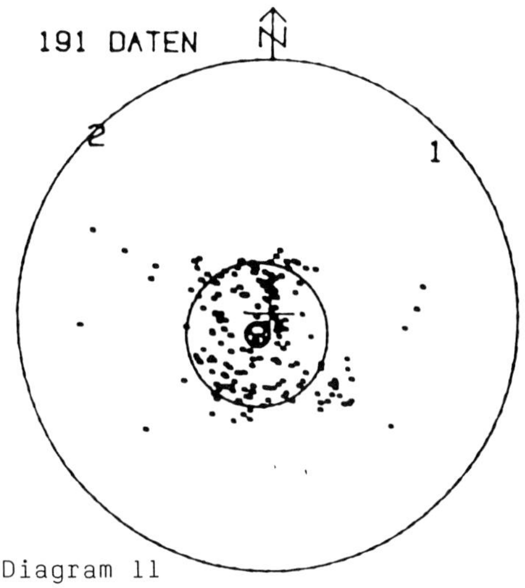

Diagram 11

often, and are concentrated in certain zones.

In the conglomerates, a dissection of the pebbles could be observed in some outcrops (Figure 8).

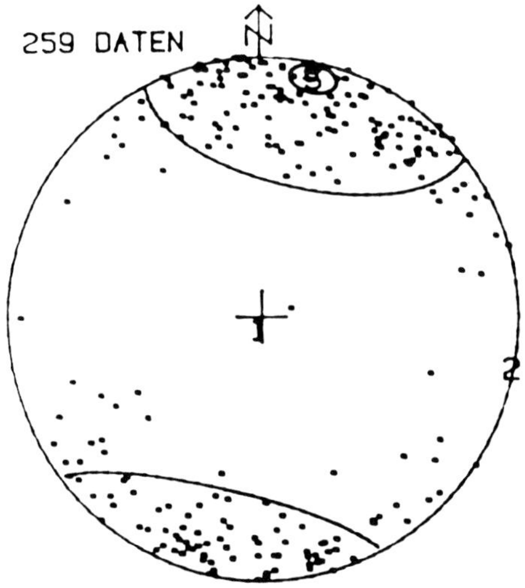

Diagram 12

Figure 8: Sheared pebbles in the conglomerates of the Warzazat Group south of Ait Nebdas

Macroscopic tectonic observations: summary and conclusions

The two most important statistical parameters, the degree of preferred orientation, $R_{\%}$, and the degree of great circle distribution, $G_{\%}$, were derived from the distributions of fold axes in each of the Diagrams 1 to 11. These two parameters were plotted against each other in Figure 9, as proposed by BEER (1981). This plot very clearly demonstrates the development beginning with a tectonical style of very narrow clusters of flat fold axes in the south (Diagram 3), passing through an elongation of the fold axes distribution along a girdle in a zone of transition in which the axes lie with various angles of dip in the foliation plane (Diagrams 4 to 7), and finally ending in vertical folds which again show narrow cluster distributions (Diagrams 8 to 11). To explain this tectonic pattern we presume a system of lateral shear stresses which moved a northern bloc along the margin of the West African Craton thus causing a (dextral?) wrench fault zone.

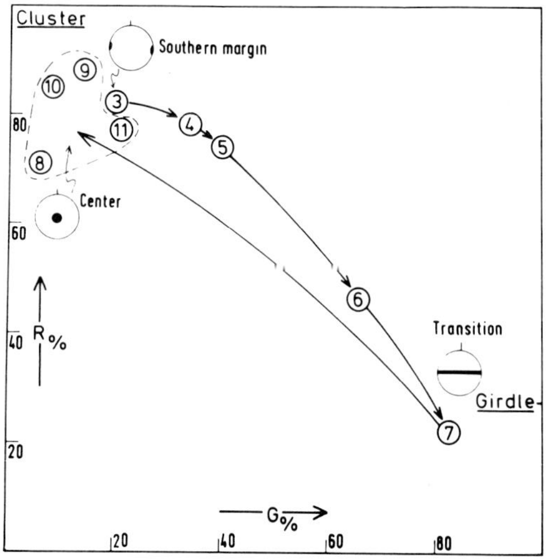

Figure 9: $R_{\%}$ - $G_{\%}$, measures of fold axes, Diagrams 1 to 11

The Tachoukacht Gneisses and the Tourtit Ophiolites, which already suffered upthrusting onto the southern foreland and imbrication during continent collision (SCHERMERHORN et al., 1986), were hereby transformed into an E-W striking mobile zone of at least 6 km width. The center of this zone developed a distinct vortex pattern of folds, while the more marginal southern parts of the zone were less strongly overprinted and thus, to a certain degree, could preserve the older tectonic pattern of southern directed planes of upthrusting and imbrication. The steep thrust and foliation planes which generated in the gneisses and in the serpentinites then served as shear planes for lateral displacement. At the end of the shear deformation the stress field was rebuilt, and an extension in E-W direction (in connection with the opening of the Atlantic) caused the system of N-S striking dykes of the Daroufarnou porphyry granites.

THE GRAIN FABRIC

1) The Tachoukacht Gneiss

At the microscopic scale, mylonitization superimposes all older fabric patterns. It occurs as a grain reduction parallel to the foliation. Rests of the former fabric can only be observed in a small rim at the southern margin of the Tachoukacht gneisses close to the upthrust onto the foreland. In this small zone we find a gneissic to schistose fabric which originally had the paragenesis quartz + plagioclase + biotite + garnet + amphibolite. At first, brittle deformation caused cataclasis; the garnets reacted to this stress pattern by external rotation (Fig. 10) and fracturing. The old quartz and feldspar grains primarily formed mortar textures (Figure 11). This deformation went together with a diaphtoresis, which was indicated in the new grains by the paragenesis quartz + albite + epidote + chlorite.

Towards the north, the deformation rapidly became more plastic, and the Tachoukacht Gneisses turned into ductile protomylites

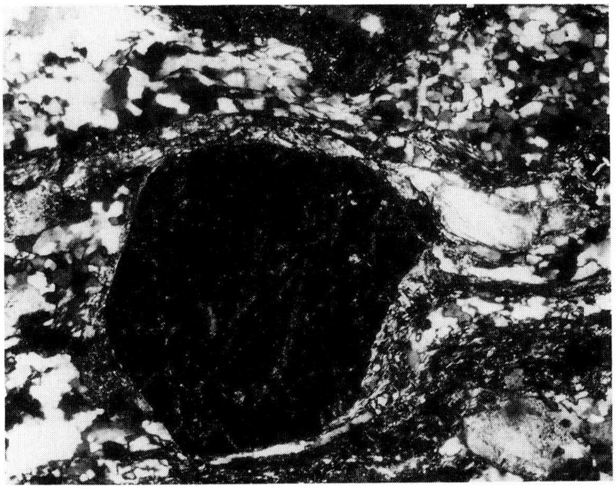

Figure 10: Externally rotated garnet clast with asymmetric pressure shadows (Tachoukacht Gneiss, 60 x, Nic. X)

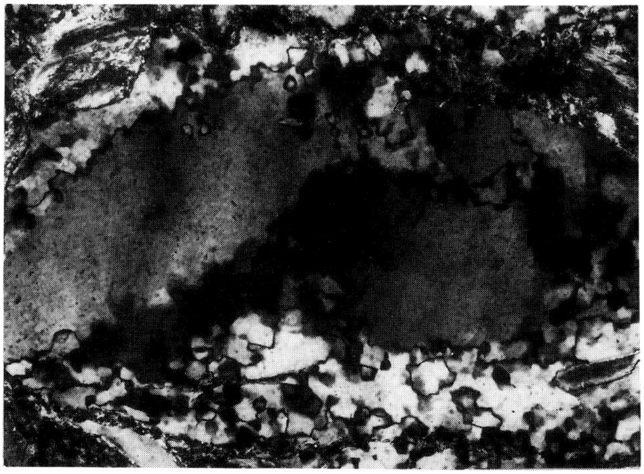

Figure 11: Mortar quartz structures in Tachoukacht Gneiss, 30 x, Nic. X)

and blastomylonites (sensu HIGGINS, 1971). At the beginning, plastic deformation was restricted to single shear bands with characteristically elongated quartzes which show low angle grain boundaries. Between these shear bands there is an old grain assemblage of quartz and feldspar clasts. Recovery patterns of the quartz fabric can also be observed. They form mosaics of isometric grains with characteristic triple junctions and exist in the same sample together with the genetically preceding mortar structures and with the following stadium of recrystallization with lobate grain boundaries. This strain heterogeneity must be caused by the distribution of rigid feldspar paleoclasts (PRICE, 1978). In areas with unusually strong folding, i.e., near measuring point 4, there is an intensive synkinematic albite blastesis. The blasts are packed very densely, reach a size of about 5 mm and obviously went through an external rotation (Fig. 12).

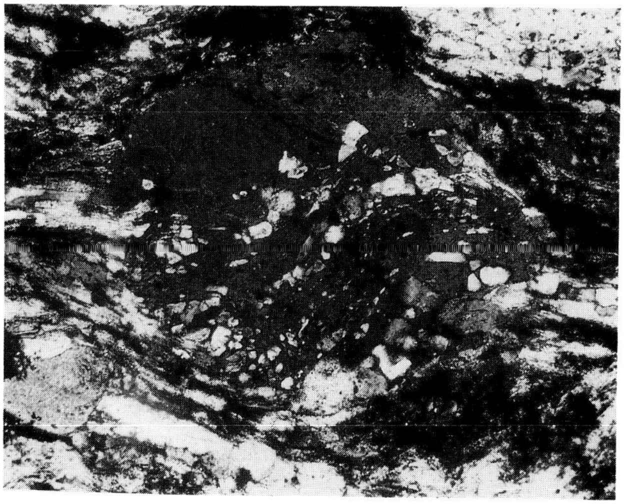

Figure 12: Synkinematic albite blast (Tachoukacht Gneiss, 80 x, Nic X)

2) The Tourtit Ophiolite Complex and the Tourtit Granite-Mylonite

In the ultrabasic and basic rocks of this complex, there is an extended metamorphic recrystallization which follows the foliation planes and which covers a probably preexisting cataclasis or mylonitization. In contrast, the leucocratic migmatite schlieren, which are arranged parallel to the foliation of the serpentinites and which have a thickness of some centimeters to a few meters, are characterized by a strong ductile deformation and strain homogeneity. In these bands of quartz-feldspar rocks most of the strike-slip movement must have taken place under the conditions of strain softening and superplasticity (WHITE et al., 1980). The fabric homogeneity and superplastic deformation culminate in the lens of Tourtit Granite.
The genuine granitoid grain fabric first developed into cataclasites, remnants of which survived in some nests within the Tourtit Granite. Plastic deformation turned these cataclasites into mylonites and ultramylonites with very fine subgrain aggregates of quartz and neoblasts of feldspar. Only a few feldspar clasts still exist. These show a marginal disintegration into subgrains and a schlieren-like arrangement of very small new grains in the pressure shadows of the clasts. The quartz grains are strongly elongated with an average ratio of axes of >5:1. Only low angle grain boundaries are visible.
An intercalation of layers of elongated new quartz grains with a mylonitic fabric of very small feldspar and quartz aggregates (Fig. 13) can be observed. Mylonites of this type are generated at high temperatures and relatively low stresses and strain rates (BOULLIER & GUEGUEN, 1975). According to these authors the ratio T/Tm between the temperature T at the time of deformation and the melting temperature Tm of the mineral phase (feldspar) must be >0.5. In the entire area investigated, the geometric pattern of the grain fabric is characterized

Figure 13: Intercalation of elongated quartzes and feldspars with strong grain reduction (Tourtit Granite-Mylonite, 13 x,Nic.X)

by the perpendicular relationship of the stretching lineation, which is manifested by elongation of quartz and by feldspar lineation rods, to the fold axes of the vortex structures in the neighboring serpentinites.

This means that the serpentinites reacted to the shear stress only by folding and forming of B-tectonites (SANDER, 1950) in a "rolling the dough" deformation (LISTER & PRICE, 1978) with b-axes vertical to the shear stress, whereas most of the shear movement which can be considered as a simple shear deformation must have been concentrated in the relatively small leucocratic migmatite bands and in the Tourtit Granite.

In the Tourtit Granite-Mylonites, 1-2 mm wide dark veins were found which contain rounded grains of parent rock. They show sharp, sinuated contacts to the host rock (Fig. 14). Even though we have only optical microscopic data up to now, and there is no indication of glassy material in these bands, the observations point strongly to pseudotachylitic bands with their characteristic injection veins (cf. SIBSON, 1975; ALLEN, 1979; PASSCHIER, 1982). Pseudotachylites are explained as former rock melts generated by rapid melting which is caused by friction fusion along seismically active fault zones (ALLEN, 1979; ERMANOVICS et al., 1972; FRANCIS; 1972; GROCOTT, 1977; SIBSON, 1975,1977; WATTS & WILLIAMS, 1979).

Figure 14: Pseudotachylite vein (Tourtit Granite-Mylonite, 10 x, Nic. //)

The fact that the injection veins are undeformed and cut more or less vertically through the mylonite bands clearly shows that the pseudotachylites were generated after the mylonitization. This observation does not conform to the development expected at a shear zone. This should begin with brittle behaviour during which pseudotachylites may develop and should be followed by a ductile reaction which can be supported by a reduction of the effective normal stress caused by hydration along a fault zone.

These observations may be explained in two different ways:
a) The entire ductilely deformed zone was elevated into a higher crustal level with transition to brittle behaviour. In experiments TULLIS and YUND (1972) showed that this behaviour of rock material is mostly temperature controlled, with a transition at about 400° C.
b) An aseismic ductile deformation and plastic softening of the fabric which took place in the deeper crust at a very low strain rate was interrupted by seismic events in greater depth.

Quartz-c-axes patterns

Method

The quartz-c-axes distribution patterns were registered quantitatively in two separate ways. On the one hand, 200 grains of a thin section were measured with the usual universal stage method in sections of defined orientation. Using a small computer program (WALLBRECHER, 1986), the thin sections were then rotated into the orientation which they would have had in the field. In this orientation the quartz-c-axes were plotted into an equal area projection and their statistical parameters were calculated.
Parallel to this method an integral photometric measuring method (PRICE, 1978) was used in order to obtain a quick overview of the distributions and to register the heterogeneity of the strain fields caused by the distribution of rigid clasts. The method may be compared to the axial distribution analyses of SANDER (1950). In the field of view, which is fixed by the optical system of the microscope, the intensity of monochromatic light transmitted after passing the polarizer, the thin section, a gypsum plate of 550 nm, and the analyzer is measured by a photodiode fixed on the phototube of the microscope. The microscope stage is turned through 360°, and intensity is measured at each degree of turning.
In order to carry out these measurements the apparatus of PRICE (1980) was modified as shown in Figure 15. Movement of the stage, measurements of the intensities and storage of data are controlled by a computer built for this purpose by the Technisches Büro C. H. HILGERS, Königswinter, Germany. The intensities were plotted as rose diagrams for each field of view. These diagrams thus represent the distributions of quarz-c-

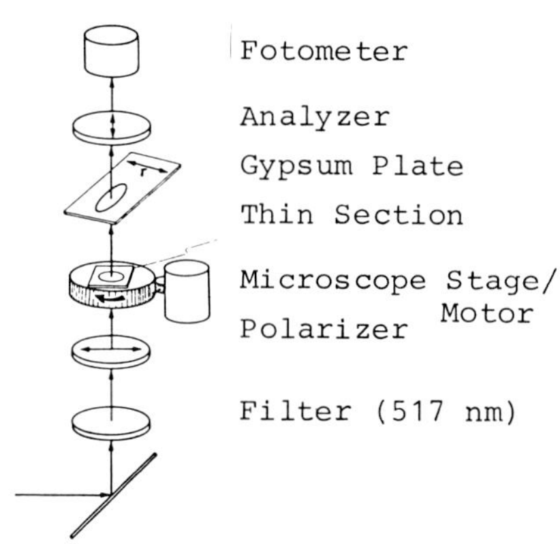

Figure 15: Device for integral measurement of quartz-c-axes azimuths

axes azimuths in the thin section. The method is quite quick (approximately one minute per rose diagram) and has the advantage that the distribution of quarz-c-axes in the individual domains of the fabric can be analysed.

Results of the universal stage measurements

Diagrams 13 and 14 show the distributions of quartz-c-axes in two samples of the Tachoukacht gneisses. In addition, the mylonitic foliation (shear plane sf) of the samples and the axes of the finite

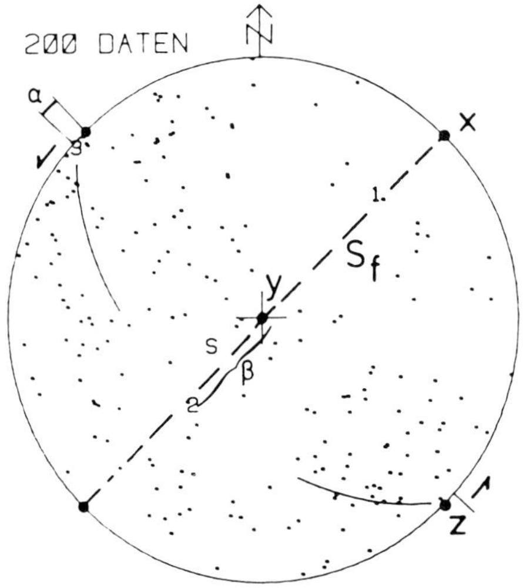

Diagram 13

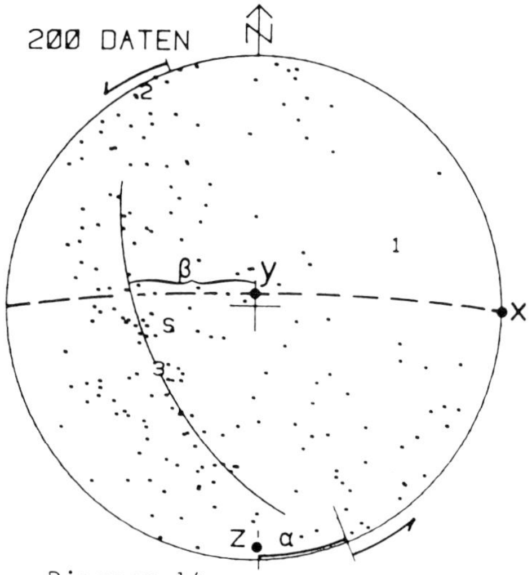

Diagram 14

strain ellipsoid (X>Y>Z) are plotted into the diagrams. The Y-axis is the direction of axes of minor folds at the place where the sample was taken.

Both diagrams show a partial girdle distribution (solid lines) with a goodness-of-fit to a great circle distribution of $G_{\%}$ = 61% and $G_{\%}$ = 89%, respectively.

This means that they most closely fit the model "quartzite I with axial elongation" of LISTER et al. (1978). Theoretically, this fabric type should form a girdle around the X-axis. LISTER et al. (1978) explain this distribution by lattice gliding with both basal slip along (0001) and slip along the rhomboedric planes $\{01\bar{1}1\}$. Since the girdle distributions are not around X, gliding must have taken place on other lattice planes, too.

Following the method of SIMPSON (1981), the position of the girdle distribution to the orientation of the finite strain ellipsoid is marked by the angles α (angle between Z and the direction of strike of the great circle) and β (angle between Y and the great circle, measured in (XY)), in order to determine the obliquity of the girdle distribution compared with the (YZ) plane (LAURENT & ETCHECOPAR, 1976).

The divergence angle (35° and 38°, respectively) from the theoretically expected girdle in (YZ) might be explained by a preexisting preferred orientation on which the elongation in X was superimposed. This means that the true deformation path cannot be explained by simple shear alone (SIMPSON, 1981). The older preferred orientation could have been generated during the period of imbrication tectonics which preceded the shear tectonic period. The declination (4° and 20° sinistrally, respectively) could be explained by a later rotation around X which took place in a very late stage and in which the stress became so weak that only the mobile quartz fabric was influenced and forced to arrange itself between the unmobile clasts of feldspar and mica (LISTER, 1977; VOLLBRECHT, 1981). This explanation is also in accordance with the observation of LISLE (1985) that the strength of a quartz fabric is dependant on the content of other minerals in the sample.

A totally different situation was found in the Tourtit Granite-Mylonite. Here we found cluster distributions with $R_{\%} = 31\%$ and $G_{\%} = 48\%$ (Diagram 15) and more pronounced clusters with $R_{\%} = 64\%$ and $G_{\%} = 46\%$ (Diagram 16). In order to interpret these parameters correctly we have to consider that the measure $G_{\%}$ is especially sensitive to tautozonality in general and less sensitive to equal distribution along a great circle (WALLBRECHER, 1986). The preferred orientation of quartz-c-axes, represented by eigenvector $\vec{\lambda}_3$ is, in Diagram 15,

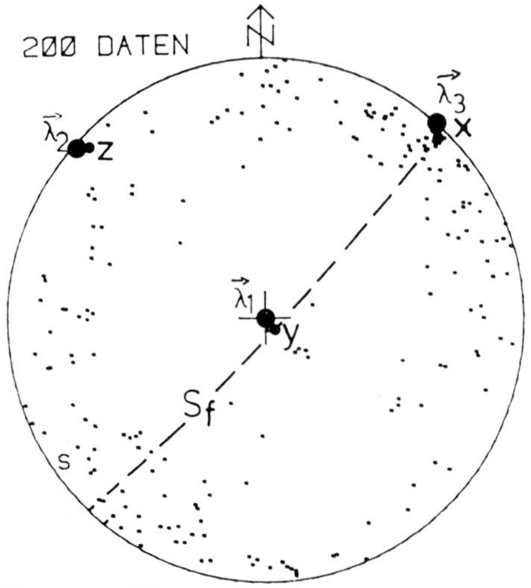

Diagram 15

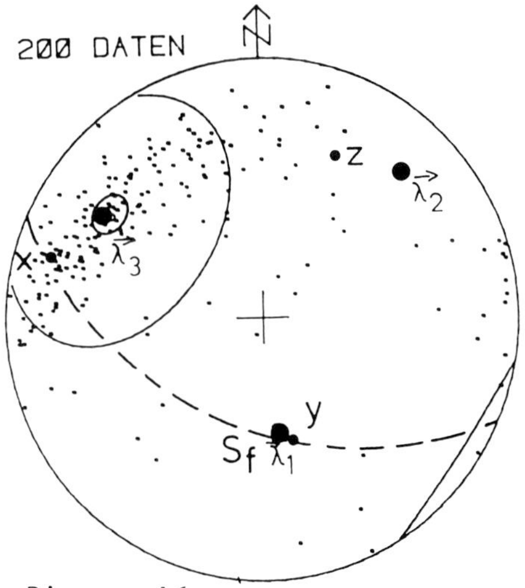

Diagram 16

parallel to the X-axis of the finite strain ellipsoid, and in Diagram 16, at least close to the X-axes. The orientation ellipsoid, represented by the eigenvectors $\vec{\lambda}_1, \vec{\lambda}_2, \vec{\lambda}_3$ (large dots), is parallel to the strain ellipsoid with $\vec{\lambda}_1 \parallel Y$, $\vec{\lambda}_2 \parallel Z$, $\vec{\lambda}_3 \parallel X$. The elongation of the quartz-c-axes distribution is around the Y-Axis in these cases. As far as we know, a clustering of quartz-c-axes parallel to the direction of maximal elongation (X) has only been described recently by MAINPRICE et al. (1986) from migmatites of the Ojén nappe of southern Spain, which were deformed at high temperature (750° C), and from another sample from sheared granitic veins in the Vosges Massif. The authors derive a temperature of 650° C to 750° C and a hydrostatic pressure of P_{H_2O} = 350 MPA for the second sample. In experiments, BLACIC (1975) and LINKER et al. (1984) found a transition from basal to prism gliding with increasing temperature or decreasing strain rate which is controlled by the water content of the crystals. Since the same pattern is seen in the Tourtit Granite-Mylonites, we can assume that temperatures were reached which were, at least, high enough to start prism slip. It is also possible that grain boundary gliding took place. This would mean that the Tourtit Granite-Mylonites are SP mylonites (superplastic sensu BOULLIER & GUEGUEN, 1975). The differing orientation of the quartz-c-axes and oblique girdle distributions in the Tachoukacht gneisses, on the one side, and partial girdles around Y with a center in X in the Tourtit Granite-Mylonites, on the other, is made especially clear by synoptic diagrams in which the mylonite foliation is rotated in an E-W plane and the b-lineation of the fold pattern into a vertical position (Diagrams 17 and 18).

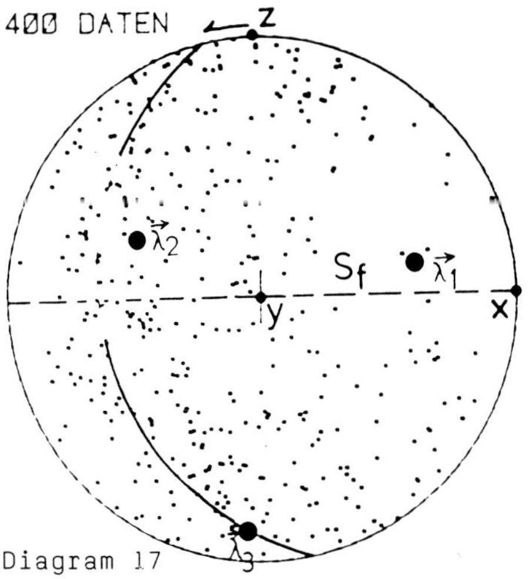

Diagram 17

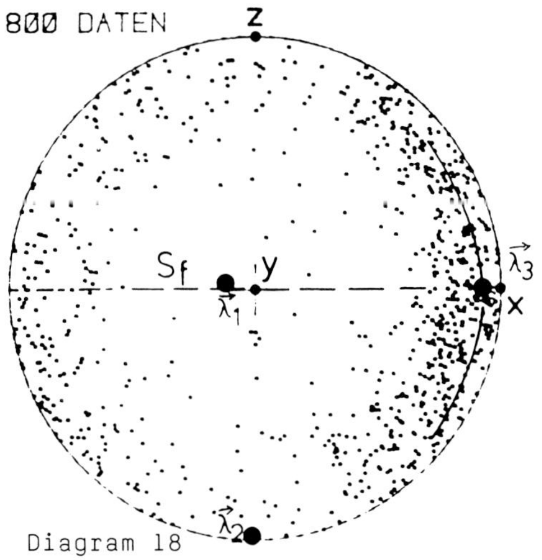

Diagram 18

Results of the photometric measures

The same thin sections in which the quartz-c-axes were measured with the universal stage were also measured with the photometric method (15 domains per section). The light transmission spectra were plotted as rose diagrams (Fig. 16 and 17). In these plots the orientation of the finite strain ellipsoid is the same as in Diagrams 17 and 18.

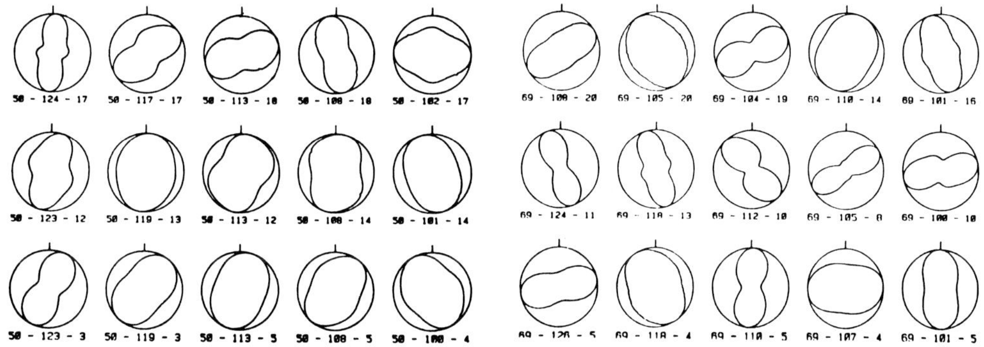

Figure 16: Rose diagrams of integrally measured azimuths of quartz-c-axes, Tachoukacht Gneiss.

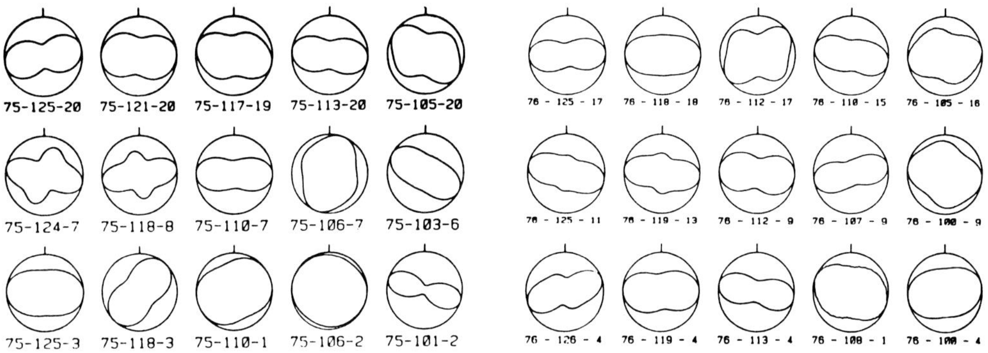

Figure 17: Rose diagrams of integrally measured azimuths of quartz-c-axes, Tourtit Granite-Mylonite.

In each domain, the Tachoukacht gneisses show distributions which can be explained by oblique girdles (Fig. 16). The orientations themselves differ from each other. This points strongly to strain heterogeneity due to a high content of clasts of feldspar, mica, and garnet.

In contrast, the orientation of quartz-c-axes distributions in the Tourtit Granite-Mylonites is almost parallel to the mylonitic foliation (Fig. 17). This proves much higher strain-homogeneity and plastic softening of the fabric.

GENERAL CONCLUSIONS

In the Sirwa area of the Antiatlas, a segment of the collision zone between the West African Craton and a northern continental mass is exposed (Fig. 18).

A first phase of deformation and metamorphism was caused by imbrication tectonics. This caused a steep E-W striking schistosity and imbrication of rock sequences. Fold axes were flat and stroke E-W. The enveloping surfaces of these folds dip steeply to the north.

With increasing lateral shear, a wrench fault system developed which caused a broad shear zone of at least 6 km width. The strike slip shear stresses caused different strain patterns in the three lithological units involved. Cataclasites to protomylonites were generated in the marginal gneisses. The mechanism of initiation of ductile deformation in these rocks was mainly basal slip in the quartz fabric and external rotation of the clastic minerals. The serpentinites reacted only by forming vortex structures vertical to the shear stress. Numerous E-W arranged migmatite veins of quartz-feldspar composition and the elongated band of Tourtit granite in the center of the zone transported most of the shear movement. This was made possible by plastic to superplastic simple shear deformation of the quartz fabric. High temperatures probably caused a dehydration of the serpentinites, so we have to assume a very high water pressure. This, together with almost anatectic conditions and very high flow stress, activated prism gliding and grain boundary mechanisms in the quartz fabric. This geometrical strain softening reinforced the shear movement in these zones.

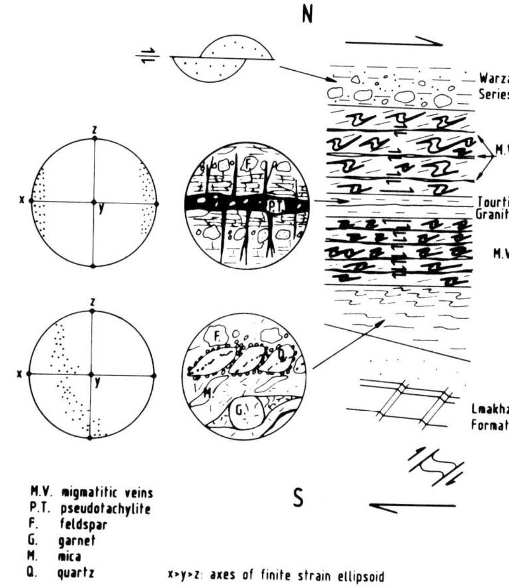

Figure 18: Synoptic sketch of structure patterns of the shear zone in the Sirwa area.

In the entire shear zone, these movements added up to a considerable lateral dislocation of the northern continental block. Finally, the plastic softening of the material might have reached a point at which slip movements with seismic velocities became possible. These formed pseudotachylite bands in the superplastically deformed migmatitic bands. The geographic range of this Panafrican collision zone and of the ductile shear zone is not yet known.

REFERENCES

ALLEN HA (1979): Mechanism of frictional fusion in fault zones. J. Struct. Geol. 1: 231-293

BEER W (1981): Die strukturelle Entwicklung der Metamorphite des Bayerischen Waldes. Diss. Univ. Göttingen, 186 pp

BLACIC JD (1975): Plastic deformation mechanism in quartz: The effect of water. Tectonophysics 27:271-294

BOULLIER AM; GUEGUEN Y (1975): Sp-mylonites: Origin of some mylonites by superplastic flow. Contributions Mineral. Petrol. 50:93-104

CHOUBERT G (1963): Histoire géologique du Précambrien de l'Anti-Atlas. Notes et mém. du service géol. du Maroc 162:352 pp

ERMANOVICS JF; HELMSTAEDT H; PLANT AG (1972): An occurence of Archean pseudotachylite from southeastern Manitoba. Canadian J. Earth Sci. 9:257-265

FRANCIS PW (1972): The pseudotachylite problem. Comments on Earth Sci. (Geophys.) 3:35-53

GROCOTT J (1977): The relationship between Precambrian shear belts and modern fault systems. J. Geol. Soc. London 133:257-261

HIGGINS MW (1971): Cataclastic rocks. U. S. Geol. Surv. Prof. Pap. 687:97 pp

JEANETTE D; TISSERANT D (1977): Les épisodes tectoniques et intrusifs du Précambrien supérieur de l'Anti-Atlas occidental. Estudios geol. 33:315-326

LAURENT P; ETCHECOPAR A (1976): Mise en évidence à l'aide de la fabrique du quartz d'un cisallement simple à déversement ouest dans le Massif de Dora Maira (Alpes occidentales). Bull. Soc. Géol. France (Série 7) 18:1387-1393

LEBLANC M (1981): Ophiolites précambriens et gîtes arséniés de cobalt (Bou Azzer, Maroc). Notes et mém. Serv. Géol. du Maroc 280:306 pp

LINKER MF; KIRBY SH; ORD A; CHRISTIE JM (1984): Effects of compression direction on the plasticity and rheology of hydrolytically weakened synthetic quartz at atmospheric pressure. J. Geophys. Res. 89:4241-4255

LISLE R (1985): The effect of composition and strain on quartz fabric intensity in pebbles from a deformed conglomerate. Geol. Rundsch. 74:657-663

LISTER GS (1977): Discussion: Crossed girdle c-axes fabrics in quartzites plastically deformed by plane strain and progressive simple shear. Tectonophysics 39:51-54

LISTER GS; PETERSON MS; HOBBS BE (1978): The simulation of fabric development in plastic deformation and its application to quartzite: The model. Tectonophysics 45:107-158

LISTER GS; PRICE GP (1978): Fabric development in a quartz-feldspar mylonite. Tectonophysics 49:37-78

MAINPRICE D; BOUCHEZ JL; BLUMENFELD P; TUBIA JM (1986): Dominant c slip in naturally deformed quartz: Implications for dramatic plastic softening at high temperature. Geology 14:819-822

PASSCHIER CW (1982): Pseudotachylite and the development of ultra-mylonite bands in the Saint-Barthélmy massif, French Pyrenees. J. Struct. Geol. 4:69-79

PRICE GP (1978): Study of heterogenous fabric and texture within a quartz-feldspar mylonite using the photometric method. Geol. Soc. Am. Bull. 89:1359-1372

---- (1980): The analysis of quartz c-axes fabrics by the photometric method. J. of Geol. 88:181-195

SANDER B (1950): Einführung in die Gefügekunde der geologischen Körper. 409 pp, Springer, Wien

SCHEIDEGGER AE (1965): On the statistics of the orientation of bedding planes, grain axes, and similar sedimentological data. U. S. Surv. Prof. Pap. 525C:C164-C167

SCHERMERHORN LJG; WALLBRECHER E; HUCH M (1986): Der Subduktionskomplex, Granitplutonismus und Schertektonik im Grundgebirge des Sirwa-Doms (Anti-Atlas, Marokko). Berliner Geowiss. Abh. (A) 66:301-322

SIBSON RH (1975): Generation of pseudotachylite by ancient seismic faulting. Geophys. J. R. Astron. Soc. 43:775-794

SIMPSON C (1981): Ductile shear zones: A mechanism of rock deformation in the orthogneisses of the Maggia Nappe, Ticino, Switzerland. Diss. ETH Zürich, 265 pp

TULLIS J; YUND RA (1977): Experimental deformation on dry Westerly granite. J. Geophys. Res. 82:5705-5718

VOLLBRECHT A (1981): Tektogenetische Entwicklung der Münchberger Gneismasse (Quarzgefüge-Untersuchungen und Mikrothermometrie an Flüssigkeitseinschlüssen). Göttinger Arb. Geol. Paläont. 24:122 pp

WALLBRECHER E (1979): Methoden zum quantitativen Vergleich von Regelungsgraden und -formen strukturgeologischer Datenmengen mit Hilfe von Vektorstatistik und Eigenwert-Analyse. N. Jb. Geol. Paläont. Abh. 159:113-149

---- (1986): Tektonische und gefügeanalytische Arbeitsweisen: Grafische, technische und statistische Verfahren. 240 pp, Ferdinand Enke, Stuttgart

WATTS MJ; WILLIAMS GD (1979): Fault rocks as indicators of progressive shear deformation in the Guigamp region, Brittany. J. Struct. Geol. 1:323-332

WHITE SH; BURROW SE; CARRERAS J; SHAW ND; HUMPHREY FJ (1980): On mylonites in ductile shear zones. J. Struct. Geol. 2:175-188

TECTONOTHERMAL EVOLUTION OF THE LATE PANAFRICAN OROGENY IN THE CENTRAL ANTI-ATLAS (SOUTHERN MOROCCO)

Yannis Bassias
Institut für Geologie der Freien Universität Berlin,
Altensteinstr. 34 A, D-1000 Berlin 33

Eckard Wallbrecher
Institut für Geologie und Paläontologie, Universität Graz
Heinrichstr. 26, A-8010 Graz

Alexander Willgallis
Institut für Mineralogie der Freien Universität Berlin
Takustr. 6, D-1000 Berlin 33

ABSTRACT

Medium grade gneisses and metapelites from the northeastern Sirwa Mountains show complex retrograde pressure-temperature paths related to a lateral slip of a northern continental mass along the northern margin of the West African Craton. Early uplift and erosion were interrupted by heating and subsequent mylonitization. An older pre-mylonitic garnet generation is weakly zoned and pervasively chloritized, whereas younger garnets show growth over older plagioclases, inclusion trails and textural unconformities. Muscovites from rocks with both garnet and biotite are syn- to post-mylonitic. Those from rocks which bear only garnets are pre- and post-mylonitic. The strong mylonitization was caused by an east-west shear zone which developed during the lateral movement of the continental masses and seems to appear at the end of an isobaric path at about 8 kb in which temperatures decreased from 600 to 500 ^{o}C, while the pressure $P_{tot} - P_{H_2O}$ was maintained between 2 and 4 kb.

RESUME

Dans la partie nord-est de la montagne de Sirwa, des gneiss et des méta-pélites de moyen degré enregistrent des trajectoires retrogrades complexes. Le soulèvement et l'érosion précoce, à la suite d'un événement prograde, sont interrompus par un échauffement et une mylonitisation liés à un mouvement latéral entre une masse continentale située au nord et la marge nord du craton ouest-africain. Des grénats anté-mylonitiques sont zonés et intensivement chloritisés. Des grénats syn- à post-mylonitiques se cristallisent sur des anciens plagioclases, ils contiennent des inclusions et montrent des discontinuités dans leur texture. Des muscovites provenant des roches à grènat et biotite sont syn- à post-mylonitiques. Celles provenant des roches à grénat sans biotite sont anté- et post-mylonitiques. La mylonitisation intense est liée a une zone de cisaillement E-W, qui se produit, probablement, à la fin d'une période de conditions isobares à 8 kbar environ pour des températures comprises entre 600 et 500°C et pour $P_{tot} - P_{H_2O}$ maintenu au moins entre 2 et 4 kb.

INTRODUCTION

The Anti-Atlas mountain chain in Southern Morocco consists of a Precambrian basement covered by Paleozoic layers. The crystalline basement crops out in several windows, which were formed by uplift and erosion. They are arranged in a SW-NE direction, parallel to the northwestern margin of the West African Craton (see WALLBRECHER 1987a, this vol.)

Three other papers in this volume (WALLBRECHER 1987a, b and BRABERS 1987) present a review of older and recent works on the geological setting and geodynamic context of the Precambrian basement of the Anti-Atlas.

We carried out an analysis of the metamorphic conditions under which the generation of gneisses, metapelites, and amphibolites in the basement exposed in northeastern Jbel Sirwa (Fig. 1) took place. This analysis includes pressure-temperature-deformation paths derived from geothermometric-geobarometric calibrations.

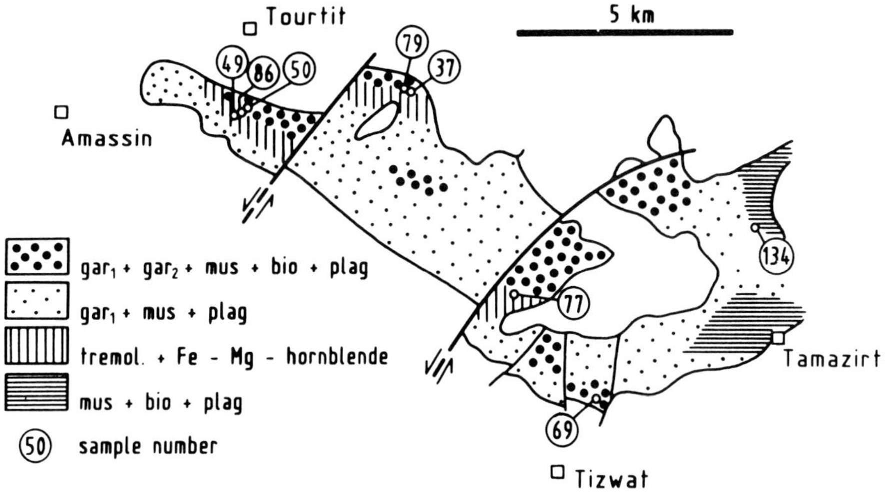

Fig. 1 The Tachoukacht Gneiss Formation of Jbel Sirwa (for location and geological setting see WALLBRECHER 1987b, Fig. 1, this vol.)

GEOLOGICAL SETTING

According to SCHERMERHORN et al. (1986) the studied rocks are part of the Tachoukacht Gneiss Formation, which was thrust between the Lmakhzan Formation, an epi- and pyroclastic sedimentary foreland sequence which contains basaltic dykes and sills, to the south, and the Tourtit Ophiolite Complex, an imbricated, deformed and metamorphised ophiolite sequence which consists essentially of ultrabasic and basic rocks, containing lenses of highly deformed mylonitic granite, to the north.

These authors came to the conclusion that the metamorphism and imbrication that affected the ophiolite and gneiss sequences were caused by a continental collision with an underthrusting northern continent.

TECTONIC SETTING

An analysis of the fabric and the tectonic evolution of the Tachoukacht Gneisses is given by WALLBRECHER (1987b, this vol.). Three main points can be summarized as follows:

- Imbrication tectonics caused a steep E-W striking schistosity, a fold pattern with flat E-W fold axes, and a repetition of rock sequences.

- A lateral shearing caused a wide shear zone with high temperature and low strain rate, associated with SP-mylonites in anatexites, within the ophiolite sequence.
- Subsequent elevation caused re-establishment of a brittle regime.

PETROGRAPHIC SETTING

From samples collected from the entire area of exposure of the gneisses, an important trend from south to north in the lithology was observed in each section studied. This trend is characterized by increasing blastomylonitic texture, increasing quartz content and the appearance of 'fresher' metamorphic assemblages, ranging from the garnet + muscovite to the garnet + biotite zone.

Muscovites, biotites, garnets, amphiboles, chlorites, epidotes, chloritoids and plagioclases were analyzed by microprobe. Representative diagrams of their chemical features are shown below.

MUSCOVITE

Muscovite is present in almost all metapelites and gneisses. Texturally, it is an uncommon part of the matrix. It crystallizes essentially during and after the mylonitization. Intergrowth with biotite is rare; intergrowth with chlorite is common and dates a latest syn- to post-mylonitization time of low grade facies. From the Si-Al-(Fe_{tot} + Mg) proportions, a weak variation from the ideal muscovite towards the phengite composition is observed. A trend towards substitution of the octahedral sites in the sense of the Tschermak substitution is observed on the Si - Al^{IV}/Al^{VI} - diagram (Fig. 2). The

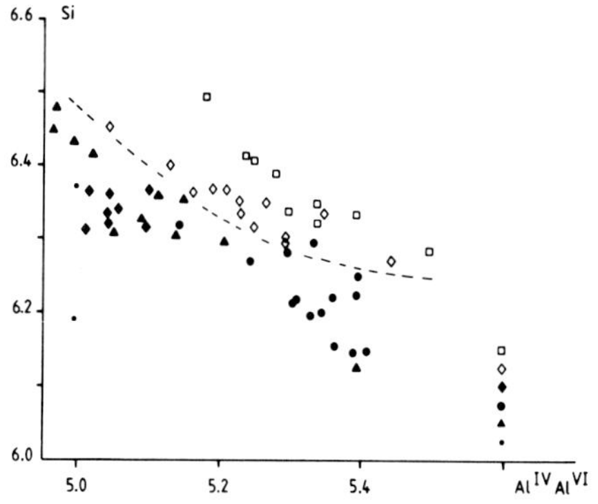

Fig. 2
Si/(Al^{IV}/Al^{VI}) diagram for muscovites. Open symbols: garnet-biotite-muscovite-assemblages (Samples: squares 69, diamonds 77). Filled symbols: garnet-muscovite-assemblages (Samples: diamonds 50, large dots 86, triangle 49). Small dots sample 134 (biotite-muscovite assemblage).

tetrahedral Si varies from 6.13 to 6.59 atoms per formula unit. These data lie at the muscovite-phengite join. However, the phengite component is small, and a ferri-muscovite component is possible.

Most of the analyses show minor amounts of iron and magnesium and a deficiency in the alkali content. As can be seen from Fig. 2, the influence of the ferri-muscovite content weakens with increasing 'grade'. However, projected data from each grade plot parallel to the phengite join. Normalized atomic formulae show significant amounts of ferric iron; there is sometimes more ferric than ferrous iron. The data show, although within the limits of analytical uncertainty, that muscovite with a 2:1 layer composition has minor inter- and intrasample variations of composition depending on metamorphic grade and/or metamorphic assemblage; the composition varies no more than in different spots of the same sample. An exception is the variation in the XII sites. Solid solutions of paragonite in muscovite range up to 16 mol% for the rims and up to 13 mol% for the centers of grains (Fig. 3). The amount of BaO was as high as 0.76%. In contrast, Ca plays only a minor role in substitutions in the XII sites.

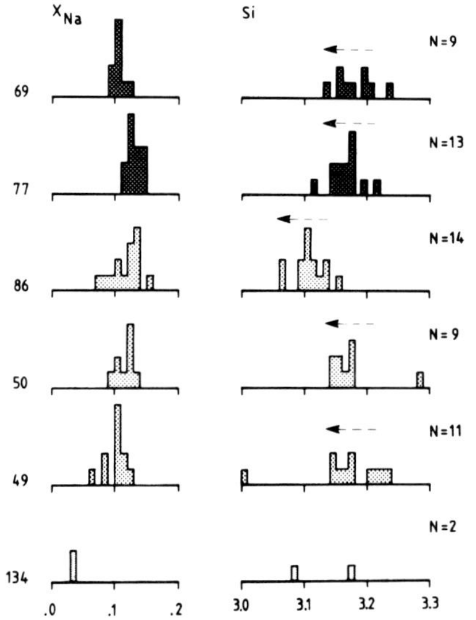

Fig. 3
X_{Na} and Si^{4+} variations in muscovites. Graphic densities refer to assemblages as in Fig. 2. Arrows indicate compositional variations from older to younger muscovites.

BIOTITE

Biotites are mostly pleochroic and iron-bearing. They are intermediate between the Mg and Fe end members. The total number of ions in the octahedral sites remains below 6.0, indicating a deviation from trioctahedral stoichiometry. This implies substitution of the type $(Mg,Fe)^{2+}$-Al^{3+} in octahedral sites and/or Fe^{3+}-Al^{3+} in tetrahedral sites. A deviation towards a 5% to 20% solution of dioctahedral mica in biotite is probable. Variations in X_{Mg} are, at a minimum, between 0.43 and 0.5 for matrix biotites and for biotites in contact with garnet. For the latter the Al^{VI} varies from 0.43 to 0.98 and is, on the average, higher than that of the biotite ± muscovite 'grade'.

Titanium amounts in the garnet + biotite 'grade' are limited to 1.5 weight % TiO_2 or 0.18 atoms per six octahedral sites. According to GUIDOTTI (1984), these values are consistent with the greenschist grade for metapelites and metabasites. However, titanium is not involved in substitutions (Fig. 4). A positive correlation with X_{Mg} has been suggested. There is an Al-Ti substitution for the biotite + muscovite 'grade'. Sodium biotites (Na=0.01 - 0.078) have higher X_{Mg}, implying a probable exchange between Fe and Na at octahedral sites. Because the total of Al and Si is always higher than 8, the presence of Fe^{3+} at the tetrahedral sites is improbable. The approximate ratio Fe^{2+}/Fe^{3+}, ca. 13.67 for pelites and semi-pelites (GUIDOTTI 1984), is confirmed by the normalized atomic formulae.

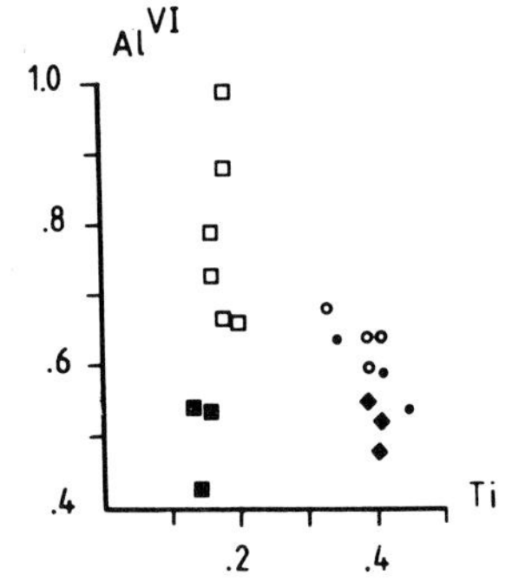

Fig. 4
Al^{VI}/Ti diagram for biotites.
Symbols are the same as in Fig. 2.
Open symbols indicate composition of the cores.
Filled symbols indicate rim composition.
Diamonds are younger biotites.

Potassium is the predominant cation at XII sites. Barium varies from 0.01 to 0.07 atoms and, in the presence of K-feldspar, characterizes high grade (TRACY & ROBINSON, 1976). A slight positive correlation between XII totals and X_{Mg} was also observed.

GARNET

The garnets show polymetamorphic patterns. Two main generations can be distinguished. The first generation garnets are small, subautomorphic, without inclusions and pervasively chloritized. In the matrix they occur associated with high X_{an} plagioclase (anorthite 30-40 mol%) and brown to brownish green biotite. These garnets show a fair increase in Mg from the core to the rim and have flat patterns for Fe, Mn and Ca. They are rotated and their main internal fractures are oriented subvertically to the mylonitization foliation. Garnets of the second generation are large and strongly zoned with inclusions of plagioclase, epidote and quartz. They grew in two stages, the earlier of which is manifested by:

1) 'pushing aside' forms and displaced brown-green biotite flakes.
2) cores of garnets surrounded by brown-green biotite as a result of an intermediate prograde reaction, followed by final retrograde reaction and rim crystallization.
3) garnet crystallizing over plagioclase with high X_{an} (0.35-0.40).

Although matrix muscovite seems to be associated with these garnets, it was never found in contact with them. The latest stage of crystallization concerns garnets with almost identical chemical composition. However these have higher grossular and spessartine content, and they crystallize on the matrix together with large brown-green biotite flakes. The presence of both these stages indicates polymetamorphic or complex retrograde patterns. Such patterns are generally attributed to either diffusion during prograde metamorphism or to retrograde re-equilibration. There is another special feature that has been rarely observed: homogenous garnet 'stringers' filling mylonitization fractures. These garnets are not fractured and may be syn- to post-mylonitic.

AMPHIBOLE

Two kinds of amphiboles have been found in the amphibolites intercalated in the gneisses: actinolite and hornblende. Actinolites are green and have a brownish Mg-hornblende core. They crystallize in rocks rich in rotated epidote (allanite ?) and they coexist with younger fine grained epidote and with low X_{an} plagioclase. Their X_{Mg} varies from 0.56 to 0.72.

Hornblendes are pleochroic from blue-green to grass-green and show, from core to rim, reversal patterns according to the Si-X_{Mg} projection. These reversals are materialized in the join between Mg- and Fe-hornblende (Fig. 5). When both amphiboles are plotted, they define a Fe-hornblende → actinolite trend which, presumably, is a retrograde one.

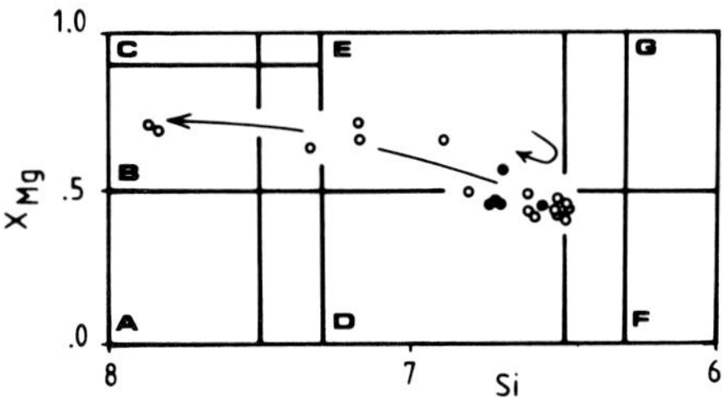

Fig. 5
X_{Mg}/ Al^{VI}-diagram for amphiboles
Open circles: sample 37, filled circles: sample 79
A= ferro-actinolite, B= actinolite, C= tremolite, D= ferro-hornblende, E= Mg-hornblende, F= ferrotschermakite, G=tschermakite

EPIDOTE

Epidote appears in gneisses as well as in amphibolites. The pre-mylonitic generation in the gneisses is rich in Si (38.9 weight % SiO_2) and in Fe^{2+} for low amounts of Al^{VI} (2.43 atoms per formula unit). In amphibolites the grains have subhedral, cross fractured and rotated cores which are overgrown by fine grained rims of iron-poor and Ca-rich epidote. Younger post-mylonitic epidote fills large fractures. It is related to the pumpellyite-prehnite facies. This epidote is frequently observed near the NE-SW faults that affected the studied area (Fig. 1).

CHLORITE

Analyses of chlorites are plotted in Fig. 6. The composition varies between ripidolite and pycnochlorite (Si=5.3 to 6.2), with a small range of Fe/(Fe+Mg). High amounts of MnO (0.50 to 0.84 weight %) were detected in younger chlorites. In co-existing muscovite-chlorite pairs, the amount of MnO varies around 0.75 wt%.

CHLORITOID

Two different generations of chloritoid have been found: the first one is pleochroic and forms large dark green lustrous crystals with straight regular sides and rounded angles. It is strongly tectonized and dates a pre-mylonitic crystallization. The second generation is post-mylonitic with small crystals growing in all directions, often in 'rosettes'.

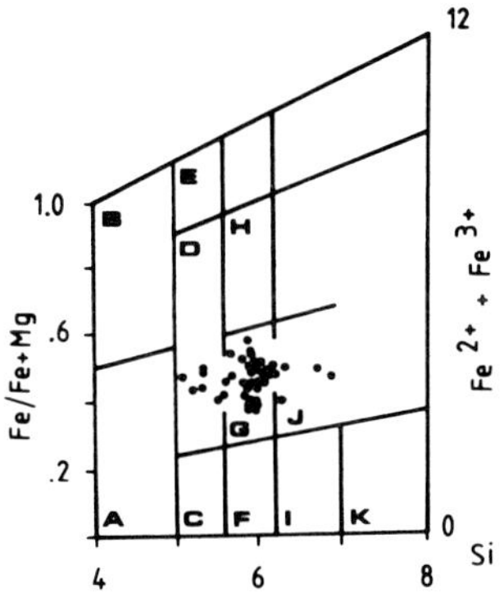

Fig. 6
Fe/(FeMg-(Fe^{2+}+Fe^{3+})-diagram for chlorites (HEY 1954): A= corundophilite, B= pseudo-thuringite, C=sheridanite, D= ripidolite, E= daphnite, F= chlinochlore, G= pycno-chlorite, H= brunsvigite, I= penninite, J= diabantite, K= talc-chlorite.

FELDSPAR

Plagioclase is the dominant mineral for most of the studied rocks. K-feldspar is not abundant but exists in almost every studied thin section.

Plagioclase forms porphyroblasts or microporphyroblastic associations. Very often the large porphyroblasts have inclusions of quartz which define previous trails as those of the second generation of garnets. Very fine-grained plagioclase has also been observed. The anorthite content of the plagioclase in the garnet assemblages varies from 44 to 19% and from core to rim no reversal patterns have been found. Large variations within a single crystal have not been observed which implies low to medium grade conditions (BUCHER et al.1983). Latest pure albite has also been observed. Rocks from the biotite + muscovite assemblages show plagioclase with an anorthite content varying from 20% to 0%.

TURMALINE

Turmaline is a common accessory mineral with blue-grey homogenous colour and few inclusions. In sample 80 an internal rim was observed in some crystals which could imply a probably detrital origine.

QUARTZ

Quartz is ubiquitous in the assemblages studied. All mineral assemblages are assumed to have equilibrated with it.

METHODS OF PRESSURE-TEMPERATURE CALIBRATIONS

Temperature dependent Fe-Mg partitioning has been used on coexisting garnet-biotite and garnet-muscovite. Serious changes in the K_D of each calibration are caused by the Fe^{3+} which is commonly present in micas. For this purpose normalized atomic formulae have been used according to LAIRD & ALBEE (1981) for the estimation of Fe^{2+}. For garnet-biotite pairs the calibration of HODGES et al. (1981), GOLDMAN & ALBEE (1977) and PERCHUK (1970) have been applied. According to the relevant literature the first calibrations have an analytical error of 20°C. The larger scatter that can appear within a narrow metamorphic zone may be related with the scatter between the Mg-Fe exchange temperature and the thermal peak from sample to sample. The second calibration obtains temperatures of the lower limit and overcorrects garnet with high X_{Ca}. The calibration of HODGES & SPEAR (1982) accounts for non-ideality in garnet solid solution. However, it is supposed that at the low amphibolite facies, between 450 and 600°C, the spessartine component has an ideal mixing behaviour with the grossular-pyrope component. HOINKES (1986) proposed a correction of the grossular component in the lower amphibolite facies. In the case of our data this does not greatly affect the estimated temperatures.

For garnet-muscovite pairs the calibration of GREEN & HELLMAN (1982) was used. The calibration for pelitic and basaltic rocks with high CaO content was used in order to border the variations in temperature. An interpolated temperature has also been estimated according to the Mg value (MgO/(MgO+FeO) mol.prop.) of the concerned muscovite. However, this interpolation involves possible errors if the non-ideality of the exchange is not linear and suggests that these interpolated temperatures cannot be taken into account as absolute values. Another uncertainty is caused by the presence of high $X_{Spessartine}$ which could lead temperatures of the second generation of garnets to higher ranges.

Pressure estimates have been based on the equilibria

$$3 \text{ anorthite} = \text{grossular} + 2 \text{ Al-silicate}$$

according to GHENT (1976). The grossular activity (γ_{gros}) was calculated from the expression of GANGULY & KENNEDY (1974) which takes the spessartine component into account. For an anorthite content lower than 0.50, γ_{an} is given by ORVILLE (1972) as $\log_{10} \gamma_{an} = 0.1060$.

Pressure and temperature for garnet-muscovite and plagioclase were derived by simultaneously solving the garnet-muscovite geothermometer and the garnet-plagioclase geobarometer. Water fugacities were calculated according to the equilibrium

paragonite (in musco.)+ quartz = albite (in plag.)+ Al-silicate + H_2O,

according to GHENT (1976). γ_{par} was estimated from the expression of CHENEY & GUIDOTTI (1979) which is only a function of the temperature. The parameters A, B and C for plagioclase barometry and muscovite water fugacities were chosen according to the 5 kb - temperature intersection for garnet-biotite pairs, and according to the pressure - temperature intersection for garnet-muscovite pairs. As barycenters for garnet-biotite pairs the middle values of three pressure-temperature values derived from the temperature calibration were used. For garnet-muscovite pairs barycenters were deduced from the interpolated temperature according to the Mg value of the muscovite.

Pressure-temperature paths were deduced from the rectangles of error which derive from separate calibration for each pair of minerals. Total pressure and water fugacities must be considered as maximum and minimum, respectively, since no aluminium-silicate was observed in the thin sections. Application of the correction of HODGES et al. (1982) increases all temperatures obtained by other calibrations derived from natural (Ca + Mn) bearing garnets by ca. 80°C and all pressures by 1 kb. Thus another probable path could be parallel to that proposed in Fig. 7a and 7b for relatively lower pressures and temperatures. However, the choice of the path does not affect the conclusions in the proposed geodynamic scheme since part of the retrograde metamorphism was almost isobaric.

CONDITIONS OF METAMORPHISM

On the basis of existing experimental studies of phase equilibria, the following estimates can be made: The upper stability of epidote for epidote bearing rocks (curve 1, Fig. 7) is restricted to lower pressures and temperatures according to the stability of muscovite + quartz (curve 2, Fig. 7). The upper stability of chloritoid (curve 3, Fig. 7) restricts the temperature even more. The stability of Mg-chlorite + muscovite is given by the curve 4. Although the products of the assemblage are not all found (phlogopite + kyanite is missing) the reactants make up part of the most common assemblage in the studied area. Minimum temperatures of stability for muscovites (paragonite is not associated) were estimated from the K/(K+Na) ratio (EUGSTER et al. 1972) and may be higher than $400^{o}C$ without any pressure correction. The presence of K-feldspars and the absence of prograde muscovite in samples from the garnet + biotite assemblages indicates that the breakdown of earlier muscovites is proable, and partial melting occured prior to the mylonitization. This hypothesis is also supported by the very small variations of X_{Na} in muscovites, compared with that of the garnet + muscovites assemblages. Additionally, small melanosomatic horizons which contain rounded isolated plagioclases (highly albitic) have been found in six localities (Fig. 1). These horizons are always in contact with amphibolites and seem to separate the northern border of the garnet$_1$ + muscovite quartzitic gneisses from the southern border of the garnet$_1$ + garnet$_2$ + muscovite $\pm$ biotite gneisses.

The p-t- path from samples with garnet$_2$ + muscovite shows a complete pressure retrograde - temperature retrograde evolution in the stability

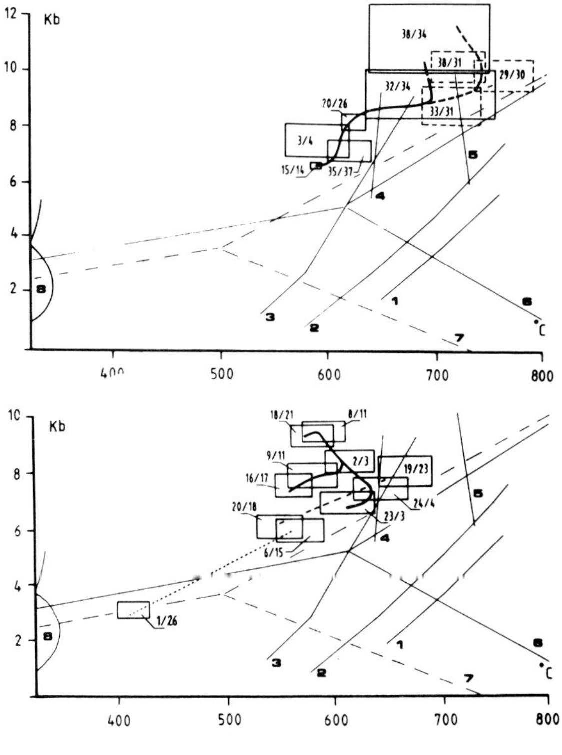

Fig. 7
Pressure-temperature stability fields for the analyses samples.
a) garnet-biotite-muscovite-plagioclase-assemblages (sample 69) ----= probable equilibria
b) garnet-muscovite-plagioclase (samples 50, 86, 49)

——= sample 50, ----= sample 86, ...= sample 49

1: Upper stability of epidote (LIOU 1973). 2: Muscovite + quartz=andalusite+sanidine+H_2O (CHATTERJEE & JOHANNES 1974). 3: Chloritoid=Al-Fe-anthophyllite+staurolite+hercynite (GRIEVE & FAWCETT 1974). 4:Mg-chlorite+muscovite=phlogopite +kyanite+quartz+H_2O (BIRD & FAWCETT 1973). 5: Upper stability of muscovite. 6: Al-silicate polymorphs (RICHARDSON et al. 1969). 7: Al-silicate polymorphs (HOLDAWAY 1971). 8: Field of prehnite-pumpellyite facies for basalt andesite systems.

field of muscovite and thus can well explain the large variations in X_{Na} of muscovite (Fig. 3). The p-t-time path from samples with garnet$_2$ + biotite + muscovite shows a similar evolution which enters into the stability field of the muscovite during the isobaric part of the retrograde history, long after the heating.

Biotite in the immediate vicinity of garnets is altered to a green colour and has a low X_{Mg} of ca. 0.45. Away from garnets, the biotite is commonly brown to brown-green and has a X_{Mg} varying from 0.50 to 0.57. Microprobe analyses show that the brown ones are enriched and the green ones are depleted in Al^{VI}. However, the Ti-content is retained at very low levels (0.14 to 0.18 atoms per formula unit) for both biotites.

Garnets are totally depleted of Ti. The green biotite grew either directly from the breakdown of $garnet_1$, or the $garnet_1$ reacted to form chlorite in the earliest stages of the retrograde event. This is also supported by the observed strong chloritization prior to mylonitization. Thus, this garnet reacted later to form green biotite which is depleted of Ti. THOMPSON et al. (1977) have used intermediate prograde consuming reactions which could represent tectonic events or polymetamorphism. ZEN (1981) proposed for assemblages similar to those of the studied area a reaction in which chloritoid plays an important role, as follows:

low-Ca-garnet + high-an-plagioclase + biotite + chloritoid + H_2O + quartz = high-Ca-garnet + high-ab-plagioclase + muscovite + chlorite

In the case of our study, this reaction can explain relatively well the evolution before and after the thermal peak (Fig. 7), the large variations of the grossular content of garnets of the second generation, as well as the decreasing anorthite content in plagioclase, the appearance of muscovite, and the younger Mn-rich chlorite.

In the Mn/Fe - Mg/Fe diagram (Fig. 8) garnets of the second generation are plotted from different profiles and show, from core to rim, an encreasing Mn/Fe-ratio which is independent of their Ca-variations. This final enrichment of the garnet in Mn could well explain the presence of younger Mn-rich chlorites.

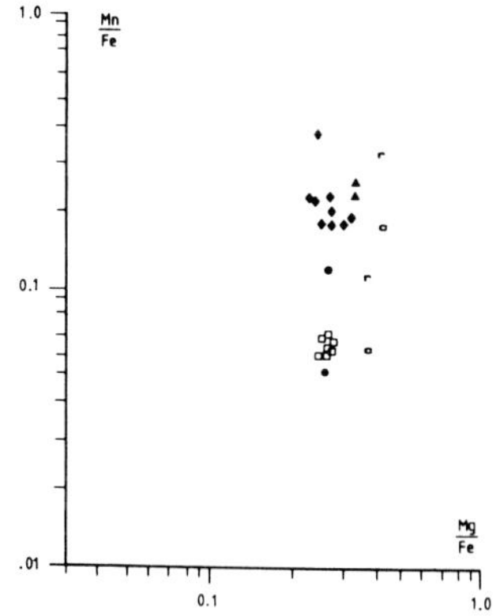

Fig. 8
(Mg/Fe)/(Mn/Fe)-diagram for garnets. Mn/Fe increases constantly from core to rim. Symbols are the same as in Fig. 2

GEODYNAMIC EVOLUTION

A normal retrograde path, convex towards the temperature, is the classic evolution of a metamorphic field after uplift and erosion (ENGLAND & RICHARDSON 1978). However, the retrograde history of the Tachoukacht Gneisses and metapelites (Fig. 9) shows other characteristics which can be grouped in two stages as follows:

- An increase in temperature seems to be manifested in both garnet-biotite and garnet-muscovite 'grade' with peaks of about 750° and 650° for pressures of 10 kb and 8 kb, respectively.
- this thermal effect is not only a result of uplift and erosion because it is followed by an almost isobaric decrease of temperature between 680°C and 620°C for the garnet-biotite and the garnet-muscovite 'grade', respectively.

Field observations and mapping show that the thermal effect could be related to the intrusive granites situated on the norhern border of the gneisses and meta-pelites (see WALLBRECHER 1987b, Fig. 2, this vol.). The presence of migmatites, high molar volume of water in the muscovites of the early retrograde stage and the absence of retrograde large garnets

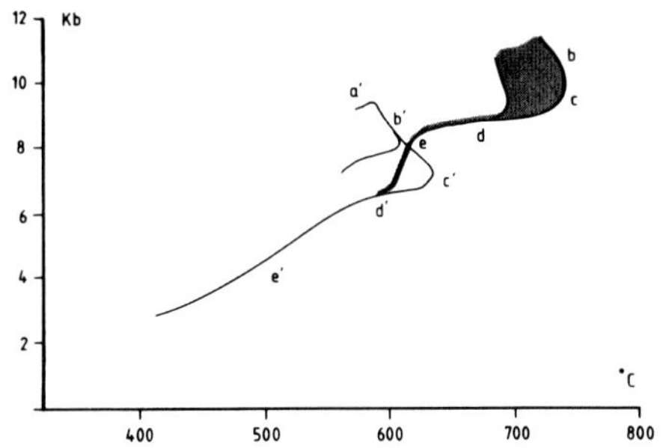

Fig. 9
p-t-deformation paths for garnet-biotite-muscovite-plagioclase (heavy symbols) and garnet-muscovite-plagioclase. a+a'=prograde paths (first garnet-generation). b+b' = prograde temperature paths (first garnet generation). c+c' = thermal peak metamorphism after uplift, erosion, and intrusion related with migmatization of the neighbouring ophiolites. Early stage garnets of the second generation. d+d' = isobaric retrograde paths, mylonitization and late stage garnets of the second generation. e+e' = isothermal-like path for garnet + biotite + muscovite + plagioclase assemblages, moderate to high geothermal gradient for the garnet + muscovite + plagioclase assemblages.

on the southern part of the studied profiles reinforce the proposed model of a thermal effect caused by the presence of a hot intrusive body.

The subsequent important decrease in temperature at almost constant pressure is illustrated in the pressure-temperature diagram (Fig. 7) as a concave-towards-the-temperature path. This shape is translated as a period of important tectonic activity (mylonitization). The isobaric feature, for important decrease of temperature, may also mean that the duration of this period was restricted in time. Field observations and mineralogical data suggest also a short duration since pseudotachylitic features were observed frequently (WALLBRECHER 1987b, this vol.) and because small variations in the X_{Mg} of chlorites, muscovites and biotites as well as in the X_{Mg} of bulk rock imply rates of re-equilibration not slower than those of the destabilization of the paragenses.

The X_{molar} volume of H_2O during the retrograde isobaric path (Fig. 9) is retained at high levels between 0.90 and 0.60 and probably has favoured the strain energy. The P_{total} $-P_{water}$ is between 2 and 4 kb for the same period and increases towards the end of the isobaric path for decreasing temperature. The partial pressure of water during mylonitization was relatively low at 3 kb, since the earlier higher grade must have removed most of the available water. During mylonitization the rocks were uplifted to mid-crustal level.

Although the retrograde evolution can be relatively well defined, no information about the prograde event (or events) is available. Also the starting point of the retrograde metamorphism of Fig. 7 cannot be proposed as the initial one. The granitic intrusion near the northern border affected both the gneisses and the meta-pelites and the Tourtit Ophiolite Complex (BASSIAS & WALLBRECHER 1987). Thus they are younger than the collision which transported the Tourtit Series and the Tachoukacht Series. It seems that these intrusions followed an uplift and erosion period, probably a rapid one, which created shorter and easier paths for the upraising hot material but also, antithetically, created advantages for its cooling.

ACKNOWLEDGEMENT

We would like to thank the Deutsche Forschungsgemeinschaft for financial support for field excursions in April 1986 and Oktober 1986, as well as the Institute for Mineralogy of the Freie Universität Berlin, especially Ms. Siegmann, for the ARL electron micorprobe analyses.

The work was carried out while one of the authors (Y.B.) held Research Fellowship from the Alexander von Humboldt Stiftung. We are grateful to this organization for generous support.

We thank Prof.Dr. C. TRIBOULET, Laboratoire de Pétrographie Métamorphique, Université Pierre et Marie Curie, Paris France for critically reading the manuscript.

Finally we would like to thank the Service Géologique du Maroc for the friendly cooperation and logistic help without which this project could not have been carried out.

REFERENCES

BASSIAS I. & WALLBRECHER E. (1987): Pressure-temperature path of late Proterozoic gneisses (Central Anti-Atlas, Morocco).- EUG 4th Biennial Meeting, 13-16 April, Strasbourg, 7 (2-3): 290.

BIRD GW. & FAWCETT JJ. (1973): Stability relations of Mg-chlorite-Muscovite and Quartz between 5 and 10 kb water pressure.-J. Petrol. 14: 415-428.

BRABERS P. (1987): A plate tectonic model for the Panafrican orogeny in the Anti-Atlas, Morocco.- (this volume).

BUCHER K., FRANK E. & FREY M. (1983): A model for the progressive regional metamorphism of margarite-bearing rocks in the Central Alps.- Am.J.Sci., 283: 370-395.

CHATTERJEE N.D. & JOHANNES W. (1974): Thermal stability and standard thermodynamic properties of synthetic $2M_1$-muscovite, $KAl_2(AlSi_3O_{10}(OH)_2)$.- Contr. Mineralogy and Petrology, 48: 89-114.

CHENEY J.T. & GUIDOTTI C.V. (1979): Muscovite + plagioclase equilibria in sillimanite + quartz bearing metapelites, Puzzle Mountain area, northwest Maine.- Am.J.Sci., 279: 411-434.

ENGLAND P.C. & RICHARDSON S.W. (1977): The influence of erosion upon the mineral facies of rocks from different metamorphic environments.- Geol. Soc. London J., 134: 201-213.

EUGSTER H.P., ALBEE A.L., BENCE A.E., THOMPSON J.B. JR. & WALDBAUM D.R. (1972): The two-phase region and excess mixing properties of paragonite-muscovite crystalline solutions.- J. Petrol., 13: 147-179.

GANGULY J. & KENNEDY G.C. (1974): The energetics of natural garnet solid solution I. Mixing of the aluminosilicate end-members.- Contr. Mineral. Petrol., 48: 137-148.

GHENT E.D. (1976): Plagioclase-garnet-Al_2SiO_5-quartz: a potential geobarometer-geothermometer.- Am. Mineralogist, 61: 710-714.

GOLDMAN D.S. & ALBEE A.L. (1977): Correlation of Mg/Fe partitioning between garnet and biotite with $^{18}O/^{16}O$ partitioning between quartz and magnetite.- Am.J.Sci., 277: 750-767.

GREEN T.H. & HELLMAN P.L. (1982): Fe-Mg partitioning between coexisting garnet and phengite at high pressure and comments on a garnet-phengite geothermometer.- Lithos, 15: 253-266.

GRIEVE R.A.F. & FAWCETT J.J. (1974): The stability of chloritoid below 10 kb P_{H_2O}.-Journal of Petrology, 15: 113-139.

GUIDOTTI C.V. (1984): Micas in metamorphic rocks.- in: S.W. BAILEY (ed.), Reviews in Mineralogy, Mineral.Soc.Amer., 13: 357-456.

HEY M.H. (1954): A new review of chlorites.- Min. Mag., 30: 277.

HODGES K.V. & SPEAR F.S. (1982): Geothermometry, geobarometry and the Al_2SiO_5 triple point at Mt. Moosilauke, New Hampshire.- Am. Mineral., 67: 1118-1134.

HOINKES G. (1986): Effect of grossular-content in garnet on the partitioning of Fe and Mg between garnet and biotite.- Contr. Mineral. Petrol., 92: 393-399.

HOLDAWAY M.J. (1971): Stability of andalusite and the aluminum silicate phase diagram.-Am.Jour.Sci., 271: 97-131.

LAIRD I. & ALBEE A.L. (1981): High pressure metamorphism in mafic schists from Northern Vermont.- Am.J.Sci., 281: 97-126.

LIOU J.G. (1973): Synthesis and stability relations of epidote, $Ca_2Al_2FeSi_3O_{12}(OH)$.-Journal of Petrology, 14: 383-419.

ORVILLE P.M. (1972): Plagioclase cation exchange equilibria with aqueous chloride solution: Results at 700°C and 2000 bars in the presence of quartz.- Am.J.Sci., 272: 234-272.

PERCHUK L.L. (1970): Equilibrium of biotite with garnet in metamorphic rocks.- Geochem.Int. 1970: 157-179.

RICHARDSON S.W., GILBERT M.C. & BELL P.M. (1969): Experimental determination of kyanite-andalusite and andalusite-silimanite equilibria: the aluminosilicate triple point. Am.J.Sci. 259-272.

SCHERMERHORN L.J.G., WALLBRECHER E. & HUCH K.M. (1986): Der Subduktionskomplex, Granitplutionismus und Schertektonik im Grundgebirge des Sirwa-Doms (Anti-Atlas, Marokko).- Berliner geowiss. Abh. (A). 66: 301-322.

TRACY R.J. & ROBINSON P. (1976): Garnet composition and zoning in the determination of temperature and pressure of metamorphism, central Massachusetts.- Am.Mineral., 61:)762-775.

THOMPSON A.B., TRACY R.J., LYTTLE P. & THOMPSON J.B. (1977): Prograde reaction histories deduced from compositional zonation and mineral inclusions in garnet from the Gassetts schist, Vermont.- Am.J.Sci., 277: 1152-1167.

WALLBRECHER E. (1987 a): The Anti-Atlas System, an overview.- (this vol.)

WALLBRECHER E. (1987 b): A ductile shear-zone in the Panafrican basement on the northwestern margin of the West African craton (Sirwa Dome, Central Anti-Atlas).- (this volume).

ZEN E. (1981): Metamorphic mineral assemblages of slightly calcic pelitic rocks in and around the Taconcic Allochtone, Southwestern Massachusetts and adjacent Connecticut and New York.- Geol. Surv. Prof. Papers 1113, Washington.

APPENDIX

SAMPLE 69	c	r	c	r	c	r	c	r		
SPOT	1	3	20	15	29	28	32	33	35	38
X_{al}	0,685	0,697	0,700	0,688	0,668	0,670	0,669	0,674	0,670	0,664
X_{pyr}	0,171	0,157	0,158	0,172	0,172	0,163	0,174	0,168	0,169	0,171
X_{gr}	0,101	0,102	0,098	0,098	0,114	0,109	0,110	0,109	0,103	0,114
X_{sp}	0,042	0,043	0,046	0,043	0,046	0,048	0,048	0,050	0,517	0,051
X_{Mg}	0,200	0,202	0,184	0,197	0,205	0,193	0,206	0,200	0,202	0,205

	86		50							49	
SAMPLE	c	r	c	r	c	r	r	r		c	r
SPOT	19	20	8	9	24	23	2	16	18	6	1
X_{al}	0,629	0,635	0,588	0,588	0,559	0,520	0,590	0,581	0,591	0,555	0,538
X_{pyr}	0,157	0,164	0,152	0,149	0,152	0,136	0,147	0,158	0,150	0,162	0,159
X_{gr}	0,173	0,124	0,142	0,139	0,159	0,154	0,139	0,138	0,140	0,162	0,167
X_{sp}	0,042	0,077	0,118	0,124	0,130	0,189	0,124	0,122	0,119	0,121	0,136
X_{Mg}	0,189	0,196	0,203	0,198	0,214	0,208	0,195	0,209	0,203	0,226	0,228

TABLE 1: Garnet mole fractions for used cores and rims. X_{al} = Fe/(Fe+Mg+Ca+Mn), X_{pr} = Mg/(Fe+Mg+Ca+Mn), X_{gr} = Ca/(Fe+Mg+Ca+Mn), X_{sp} = Mn/(Fe+Mg+Ca+Mn), al = almandine, py = pyrope, gr= grossular, sp= spessartine, X_{Mg} = Mg/(Fe+Mg).

SAMPLE	50					49		86		69	
SPCT	3	4	11	17	21	15	26	18	23	30	31
X_{Mg} mu	0,37	0,36	0,38	0,35	0,36	0,34	0,34	0,35	0,33	0,54	0,49
X_{mg} mu	0,45	0,43	0,45	0,41	0,40	0,34	0,41	0,42	0,40	0,61	0,60

SAMPLE	69					
SPOT	2	4	14	34	36	37
X_{Mg} bi	0,45	0,53	0,57	0,48	0,55	0,55

TABLE 2: X_{Mg} of biotites X_{Mg} and X_{mg} = MgO/(MgO+FeO) (molar proportions of muscovites used for calibrations).

SAMPLE	SPOTS gr - mica	$T \pm \Delta T$[°C]		X_{gr}	X_{an}	$P \pm \Delta P_b$		log γ par $\pm$ Δlog γ par		$fH_2O \pm \Delta fH_2O$		XH_2O	$P_{tot}-P_{H_2O_b}$	P_{H_2O}/P_{tot}
69	38 34 bi	690	60	0,114	0,359	11232	1131	-		-		-	-	-
	38 31 m	720	27	0,114	0,395	10632	909	-		-		-	-	-
	29 30 m	757	29	0,114	0,395	9952	566	-		-		-	-	-
	33 31 m	708	26	0,109	0,395	9350	873	-		-		-	-	-
	32 34 bi	698	57	0,109	0,250	9772	1180	-		-		-	-	-
	20 26 bi	620	8	0,098	0,251	8241	158	0,838	0,004	6151	571	0,78	2085	0,70
	3 4 bi	593	36	0,102	0,200	7706	754	0,853	0,019	4556	1940	0,72	3149	0,57
	35 37 bi	619	19	0,109	0,200	7274	380	0,820	0,009	4834	1132	0,80	2430	0,66
	15 14 bi	588	2	0,096	0,200	6851	29	0,835	0,000	3167	64	0,67	3483	0,48
86	20 18 m	547	21	0,124	0,200	6015	396	0,829		1776		0,55	4175	0,30
	19 23 m	667	25	0,173	0,200	8096	474	-		-		-	-	-
50	18 21 m	585	21	0,140	0,437	9223	400	0,851		7126		0,78	1996	0,78
	8 11 m	592	22	0,121	0,153	9364	187	0,830		8700		0,89	663	0,93
	2 3 m	617	23	0,139	0,257	8244	431	0,802		6408		0,83	1818	0,78
	24 4 m	643	25	0,159	0,178	7366	457	0,767		5958		0,93	1311	0,82
	23 3 m	593	23	0,155	0,178	6830	430	0,780		4387		0,83	2413	0,65
	9 11 m	582	21	0,139	0,257	7734	416	0,804		4469		0,72	3234	0,58
	16 17 m	555	8	0,138	0,237	7410	410	0,873		3086		0,59	4226	0,42
49	6 15 m	566	21	0,162	0,192	5805	425	0,841		1195		0,37	3721	0,35
	1 26 m	413	15	0,192	0,192	3020	301	0,923		158		0,28	2690	0,06

TABLE 3: Pressure-temperature, water fugacity (fH_2O) and molar volume of water (XH_2O) data in garnet $\pm$ biotite = muscovite + plagioclase assemblages.

— indicates that muscovites are not in equilibrium. Water fugacity values for samples 86, 50, 49 derive from pressures and temperatures which correspond to the X_{mg} value of used muscovites.

A PLATE TECTONIC MODEL FOR THE PANAFRICAN OROGENY IN THE ANTI-ATLAS, MOROCCO

Peteralv M. Brabers *
Fysico-Chemische Geologie, Katholieke Universiteit Leuven
Celestijnenlaan 200C, 3030 Heverlee, Belgium

ABSTRACT

Combined field observations and petrological evidence from the Taznakht block, in the Anti-Atlas, Morocco, have led to a new stratigraphic classification of the Precambrian, applicable in the other Precambrian blocks in the Anti-Atlas and High Atlas as well.

Lithostratigraphic and geochronological correlations made between Precambrian formations in the different blocks, have allowed to propose a plate tectonic model explaining the Panafrican Orogeny in the Atlas by the opening and closing of an ensialic marginal basin associated with southward subduction of an oceanic plate below the northern edge of the West-African craton.

RESUME

Une combinaison d'observations de terrain et de données pétrographiques de la boutonnière de Taznakht a mené à une classification nouvelle de la stratigraphie précambrienne, qui est valable aussi dans les autres boutonnières de l'Anti-Atlas et du Haut-Atlas.

Des corrélations lithostratigraphiques et géochronologiques entre les formations précambriennes des différentes boutonnières nous ont permis de proposer un modèle géodynamique expliquant l'orogénèse panafricaine par l'ouverture et fermeture (obduction) d'un bassin marginal ensialique contemporaine de subduction d'une plaque océanique vers le sud au-dessous de la bordure septentrionale du craton ouest-africain.

Introduction

Geographically, the Anti-Atlas is a WSW-ENE trending mountain chain in South Morocco, bordered in the north by the High Atlas and in the south by the Moroccan Sahara. The geology is marked by a number of Precambrian blocks or "boutonnières", surrounded by a Palaeozoic sedimentary cover (Fig. 1). From west to east the main blocks are: the Lower-Dra, Ifni and Kerdous blocks; the Jbal Sirwa block with it's northern extension into the High-Atlas; the Taznakht, Bou-Azzer and the Jbal Saghro blocks. An important feature of the Anti-Atlas is the "accident majeur de l'Anti-Atlas", a major east-west trending fault zone crosscutting it (fig. 1).

* present address: Geosurvey, Blijde Inkomststraat 31, B3000 Leuven

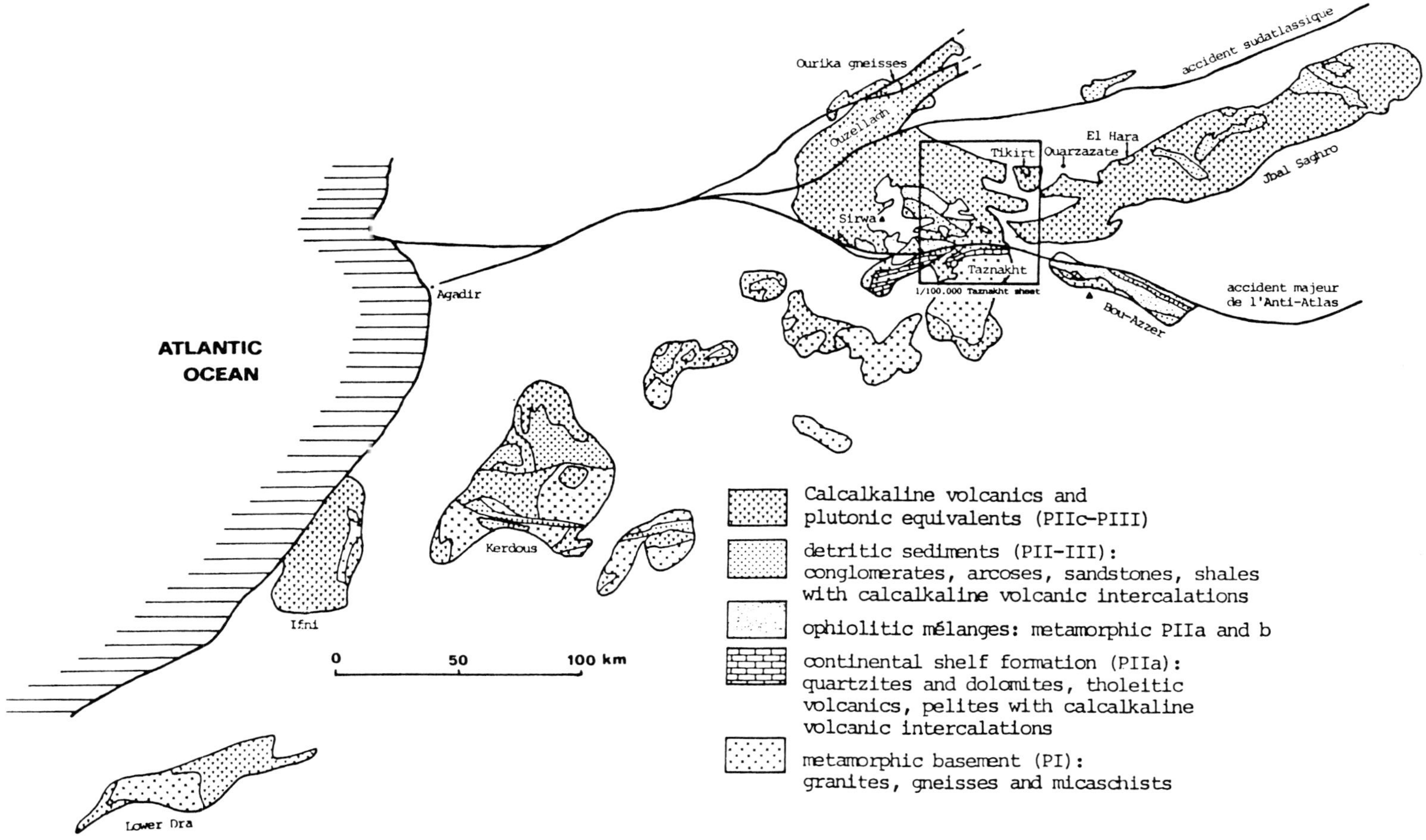

Figure 1: The Precambrian blocks in the Anti-Atlas.

It may be represented by the ash flow tuffs of Jbal Amsemsa northwest of the Zenaga plain (Fig. 2). The Jbal Amsemsa tuffs are unconformably overlain by volcanics of the Ouarzazate series (Upper PIII) and are lithologically quite distinct from the Sirwa-Saghro series (Lower PIII) exposed 400 m north of them. Their orientation is the same as the pretectonic continental shelf quartzites (PIIa) in the same area, dipping 30 degrees northwest. The absence of detritic intercalations in the Jbal Amsemsa tuffs, shows that these tuffs probably were deposited in a period of negligeable erosion in a mountainless environment. Such an environment can only have existed before the onset of the Panafrican Orogeny. During the main tectonic phase at the end of the PII, however, the Anti-Atlas probably underwent an important uplift associated with intensive erosion. Post-tectonic formations, therefore, are necessarily marked by abundant detritic sediments, while the Jbal Amsemsa tuffs, devoid of sedimentary intercalations, probably have been erupted pretectonically. Moreover, the similar orientation of the tuffs compared to the pretectonic continental shelf quartzites may indicate a pretectonic origin as well.

3. The gneisses of Tamazight, Imghlay, Oued Assemlil, Tiffadine, Ourika and El Hara (fig. 1) always have been assigned to the PI or even the PO (Choubert 1963) because of their high metamorphic grade. The absence of the Taznakht and Azguemerzi granites and the presence of ultramafics associated with the gneisses, show that they are quite different from the Eburnean basement (PI) as it is known in the Zenaga plain. A detailed petrographical study shows that they represent the metamorphic equivalents of the continental shelf sediments and volcanics (Brabers 1986).

4. The quartzites of Jbal Mimount (near Taghdout) and most PII-III quartzites in the Anti-Atlas in general, have always been assigned to the PII (Choubert 1963). This mistake is understandable because of the lithological similarities between the PII-III quartzites and the PIIa continental shelf quartzites. The misidentification has been caused by not recognizing the PIIa - PII-III unconformity at the PII type-locality near Taghdout described by Choubert and Termier (1951). The Taghdout profile, in fact, is not suitable at all for defining the PII because the upper part of the PIIa is hidden below the unconformity and the lower (quartzitic) part is cut off by a reverse fault separating

This paper is based mainly on results from a detailed study in the northern part of the Taznakht block (fig. 2).

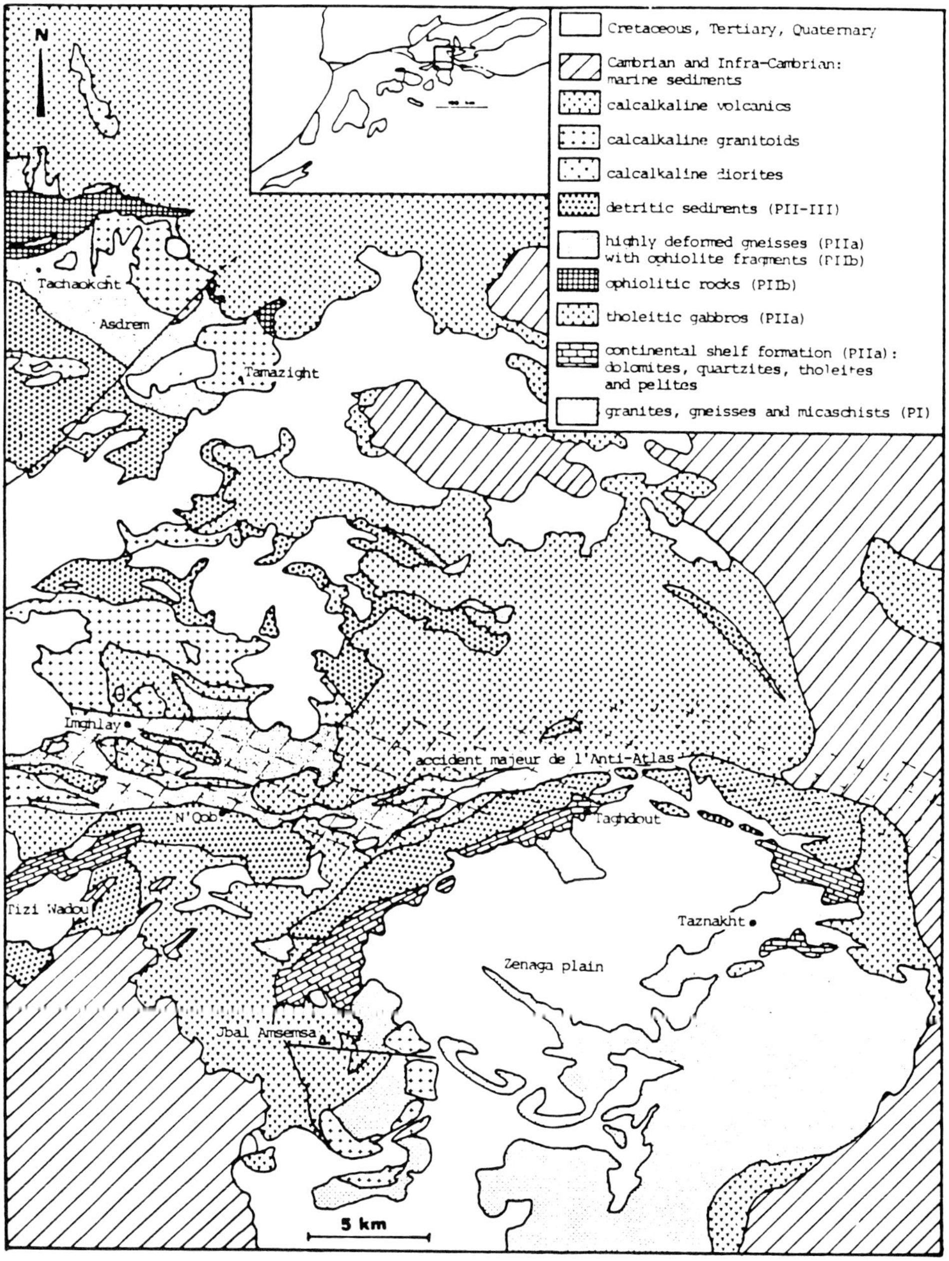

Figure 2: The Taznakht block.

1. Stratigraphy

1.1 New classification

A new classification of the Precambrian in the Anti-Atlas is proposed in fig. 3.

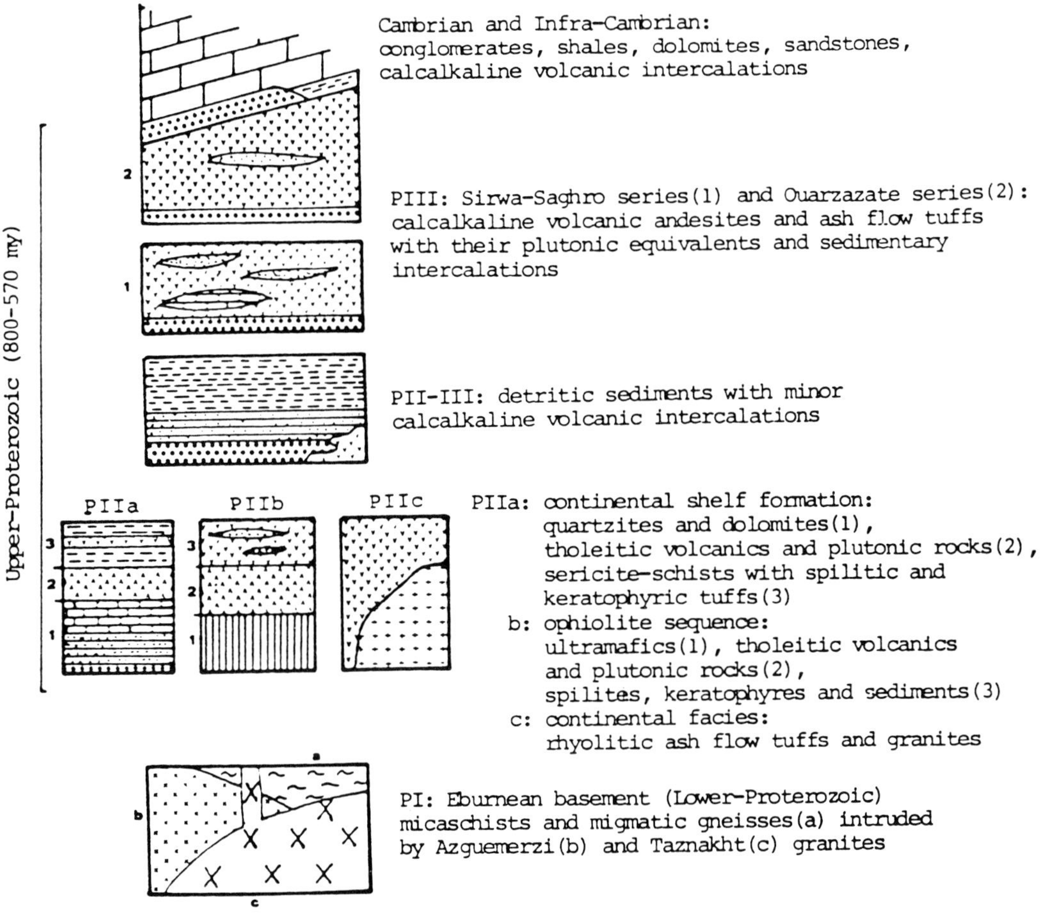

Fig. 3: Stratigraphy of the Precambrian in the Anti-Atlas.

The metamorphic Eburnean basement (Precambrian I) is unconformably covered by formations of the Precambrian II in which three facies may be distinguished: a continental shelf facies (PIIa), an oceanic (ophiolitic) facies (PIIb) and a continental (volcanic) facies (PIIc). North of the accident majeur de l'Anti-Atlas the continental shelf sediments have been gneissified due to overthrusting of ophiolites from the oceanic facies. Quartzites, dolomites, tholeitic volcanics, tholeitic dolerites, sericite-schists have been transformed into metaquartzites, scapolite gneisses, fine-grained amphibolites and chlorite-schists, amphibolite lenses, garnet-bearing gneisses. The ophiolitic formation is highly disrupted and metamorphosed. It includes pre- and syntectonic trondhjemites and quartz-diorites associated with tholeitic metagabbros and a volcano-sedimentary sequence consisting of spilites, keratophyres and sediments. Ophiolite fragments are found scattered all over the gneissified continental shelf formation.

The next level in the stratigraphic column is a detritic sedimentary sequence (PII-III) unconformably covering older formations. This sequence is related to east-west and northwest-southeast trending reverse faults that have set vertical or even overturned the sediments. The sediments probably have been generated by erosion of blocks uplifted along reverse faults.

The Precambrian III consists of a huge bimodal calcalkaline volcanic complex in which a folded sequence may be recognized (the Sirwa-Saghro series, Choubert 1963) unconformably overlain by the Ouarzazate series (Choubert 1963). A marine sequence of Cambrian to Infra-Cambrian age shows a minor angular unconformity with the Ouarzazate series.

1.2 Comparison with other stratigraphic classifications

The different Precambrian formations have been assigned a number and placed into the different classification schemes in table I. The newly proposed classification scheme shows some essential differences with Choubert's (1963) and Leblanc's (1981) classification schemes.

1. The meta-arkoses covering the Taznakht and Azguemerzi granites in the Zenaga plain have in the past been misinterpreted as ectinites (orthoschists) belonging to the PI at the same stratigraphic level as the PI micaschists. However, they consist of the erosion products of the basement granites, gneisses and micaschists and they didn't undergo any metamorphic influence of the granite intrusions. Therefore, they must be younger than the granites and they are, in fact, unconformably covering them.

2. The existence of pretectonic calcalkaline volcanism in a continental facies (PIIc) has never been recognized in the past.

Choubert 1963 based on data from the Central Anti-Atlas	Leblanc 1981 based on data from the Bou-Azzer block	New classification based on data from the Taznakht block
Adoudounian	Adoudounian	Adoudounian
PIII: Ouarzazate series(11)	PIII: Ouarzazate series(11)	PIII: Upper-PIII: Ouarzazate series(11) Lower-PIII: Sirwa-Saghro series(10)
PII-III: Sirwa-Saghro series(10)	PII supérieur: Tidiline Formation - metavolcanics(10) - detritic sediments(9)	PII-III: Jbal Mimount quartzites(9)
PII: Taghdout formation including dolomites(4) and Jbal Mimount quartzites(9)	PII inférieur PII2: - volcano-sedimentary sequence of Ousdrat(7) - ophiolite sequence of Bou-Azzer(6) PII1: Tachdamt formation(4)	PIIc: Jbal Amsemsa ash flow tuffs(8) PIIb: Tachaokcht formation - calcalkaline volcanics(7) - ophiolitic rocks(6) PIIa: - Ophiolitic mélanges: Tamazight and Imghlay gneisses(5) - Continental shelf formation: Taghdout dolomites, Tizi Wadou andesites and shales(4), metaärkoses covering Taznakht and Azguemerzi granites(4)
PI - amphibolites and ultramafics of Bou-Azzer and Tachaokcht(6) - Tamazight and Imghlay gneisses(5) - Azguemerzi and Taznakht granites(2) - ectinites(3) and Timjicht micaschists(1) PO: Oued Assemlil gneisses(5)	PI: Oued Assemlil gneisses(5)	PI: - Azguemerzi and Taznakht granites(2) - micaschists(1)

Table I: A comparison between different stratigraphic classifications for the Precambrian in the Anti-Atlas.

a southern uplifted block from a northern downfaulted block. The Jbal Mimount quartzites undoubtedly are the erosion products of the uplifted block that have been set vertical due to dragfolding associated with reverse faulting (see section 2.3). Figure 4 explains our view on the Taghdout profile.

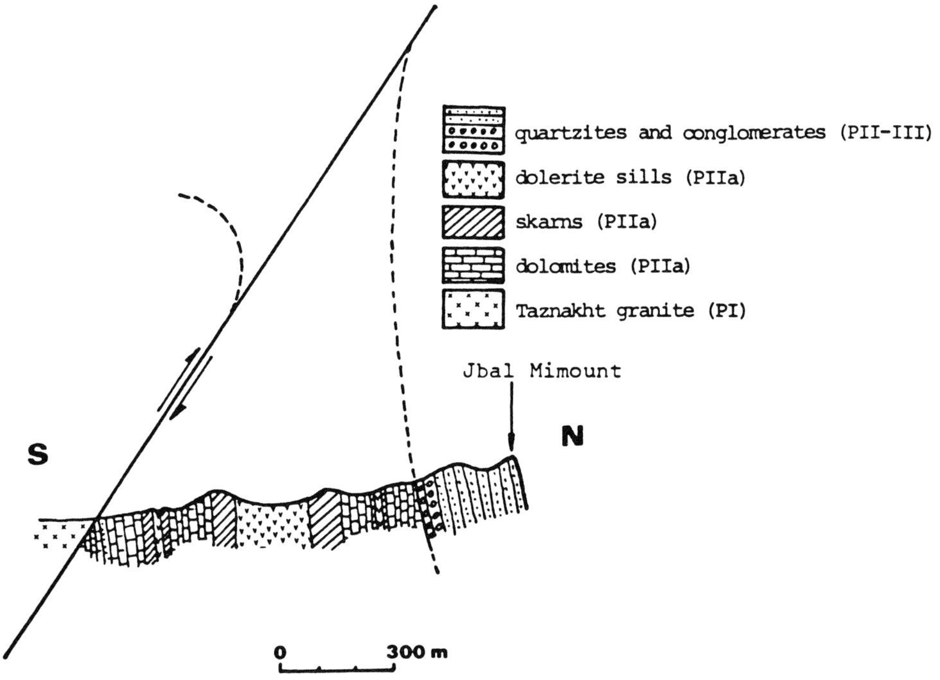

Fig. 4: The Taghdout profile.

The original misinterpretation of this profile has led to a general confusion concerning quartzites all over the Anti-Atlas. Useful criteria to distinguish between PIIa and PII-III quartzites are:

1) PII-III quartzites are associated with and bounded by reverse faults. As a consequence, they usually have been set vertical or even overturned. This, generally, is not the case with PIIa quartzites, unless these are accidentally situated near such a reverse fault.

2) PIIa quartzites are intercalated or overlain by dolomites, while PII-III quartzites are overlain by shales. The Jbal Mimount quartzites can be traced about 10 kilometers westward where

shales are found covering them. They, therefore, undoubtedly should be assigned to the PII-III.

5. The folded volcanosedimentary Sirwa-Saghro series was classified with the detritic PII-III series by Leblanc and Choubert. Arguments for this may be:

1) the lithological similarities between both series: both contain calcalkaline volcanics as well as detritic sediments

2) the similar tectonic history both have undergone: the reverse faulting during the PII-III, may still have been active a while during the Lower PIII, because in some cases the volcanosedimentary series has been set vertical close to some reverse faults.

Nevertheless, we prefer to classify the Sirwa-Saghro series separately as Lower PIII, because:

1) There exists a clear unconformity between the Sirwa-Saghro series and the detritic PII-III series. This unconformity is well exposed at Imdghar-n-Izdar north of the Zenaga plain and 3 kilometers northeast of Tamezzarra northwest of the Zenaga plain (see 1/50,000 Taghdout sheet).

2) There are important lithological differences between both series: calcalkaline volcanics occur abundantly in the Sirwa-Saghro series, while in the detritic PII-III series, they are rather insignificant.

3) The detritic PII-III series is always associated with reverse faults while the Sirwa-Saghro series may occur elsewhere as well.

2. Tectonic evolution

The Precambrian formations in the Anti-Atlas have been submitted to numerous tectonic movements during the Panafrican, the Hercynian and Alpine orogenies. Reactivation of older fault systems often makes it difficult or impossible to assign a correct age to every single fault.

2.1 Thrusting

The first important Panafrican tectonic movement is southward thrusting during ophiolite obduction. Most tectonic features of this phase are lost due to the overprint of intensive strike-slip faulting during the second tectonic phase. What remains is: 1) a clear foliation in the ophiolite related rocks; 2) a regional metamorphism reaching lower amphibolite facies conditions in the Tamazight area, decreasing in metamorphic grade southward in the Imghlay area and disappearing south of the accident majeur; 3) the smaller amphibolite lenses south of the larger ophiolite fragments, highly deformed by the overriding ophiolite sheet and 4) a single 40 to 50 degrees northward dipping thrust fault in the accident majeur north of N'Qob (Fig. 5).

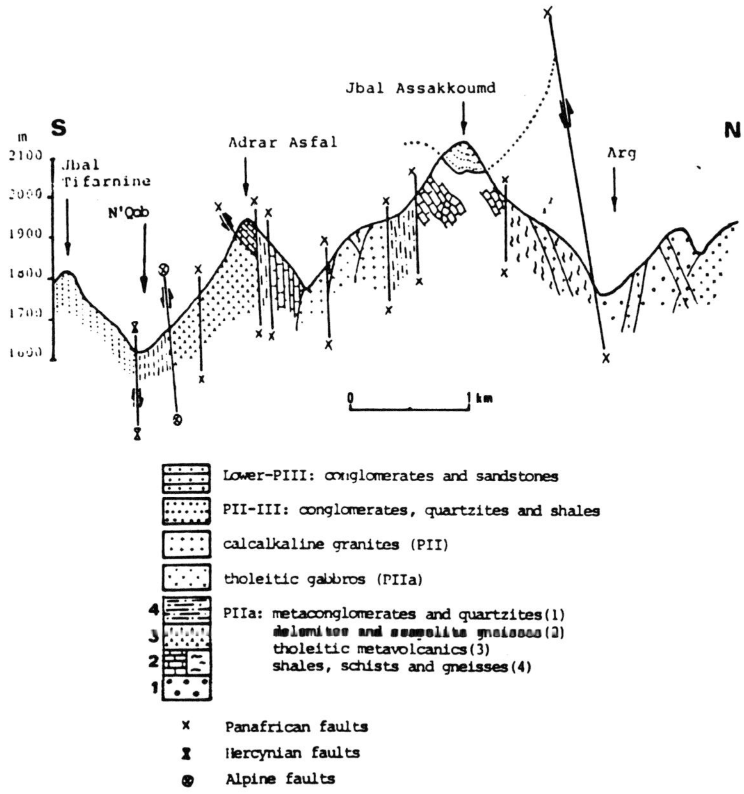

Fig. 5: The "accident majeur de l'Anti-Atlas" near N'Qob.

2.2 Strike-slip

The PI and PII formations usually are highly deformed by intensive strike-slip movements along east-west trending fault zones. The sense of movement can be determined from a fault crosscutting the Jbal Amsemsa ash flow tuffs: this is a left-lateral fault (Fig. 2).

The most pronounced effect of this tectonic phase is that most of the formations affected by it, have been rotated vertically. Another feature is the greenschist retromorphism associated with it, which has remobilized the rare earth elements to such extent that it obscured the original primary lithogeochemical patterns (Brabers 1986).

An example of a typical strike-slip fault zone is the accident majeur de l'Anti-Atlas near N'Qob, where most faults appear to be strike-slip faults of Panafrican age (Fig. 5).

2.3 Reverse faulting

As mentioned before, the PII-III formations usually are bounded by synsedimentary reverse faults that have set vertical or even overturned the stratification. These faults are not necessarily seen exactly at the contact between the PII-III sediments and their substratum; they may be seen in the sediments as well as in the substratum, depending on the erosion level (Fig. 6).

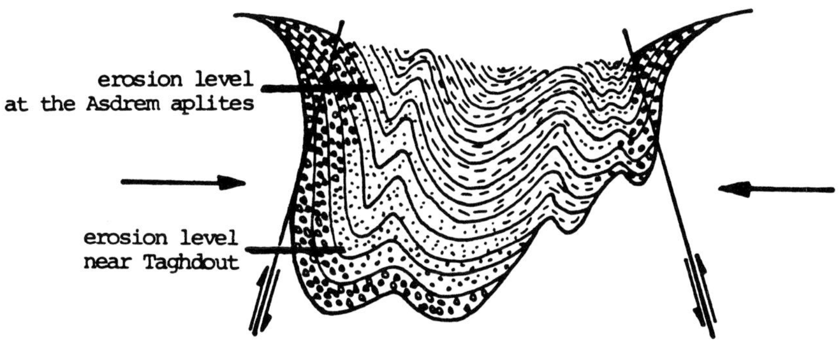

Fig. 6: Synsedimentary PII-III reverse faulting.

In the Taghdout profile, for example, the reverse fault representing the southern limit of the Jbal Mimount quartzites (PII-III) is situated in the substratum (Fig. 5).

The other configuration, with the reverse fault situated in the PII-III sediments, is found on the Asdrem plateau south of Tachaokcht (fig. 2). A number of north-south striking granitic aplites are crosscutting the PIIa gneisses north of the PII-III formation. The aplites can be traced 50 to 80 m southward into the PII-III sandstones and shales before disappearing abruptly.

The sudden disappearance of these aplites in the PII-III sediments may be explained in terms of differential erosion associated

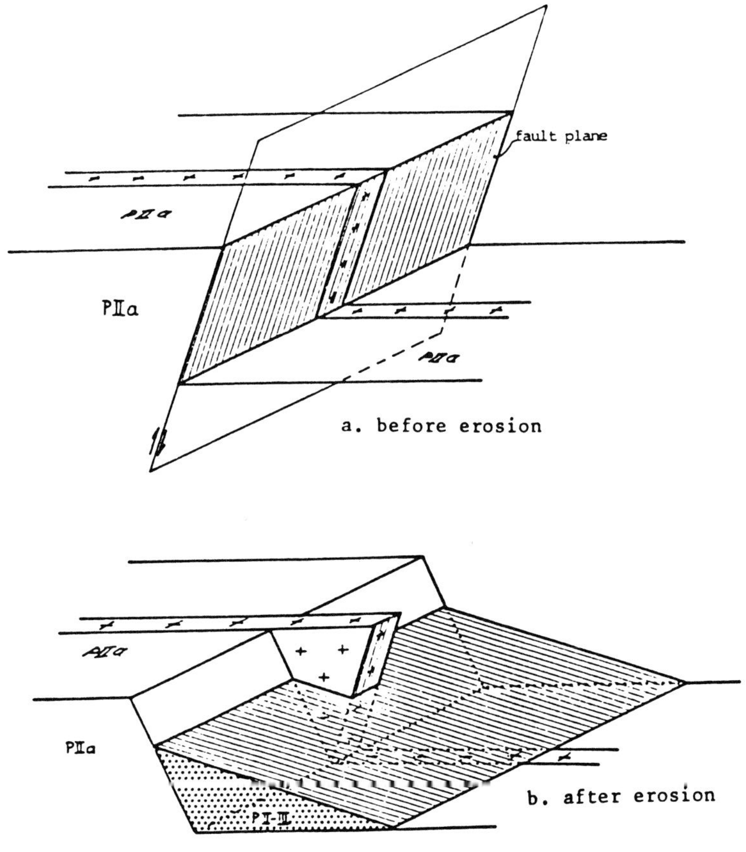

Fig. 7: PII-III faulting and sedimentation at the Asdrem aplites.

with reverse faulting (Fig. 7). The aplitic dikes are more resistant to erosion than the surrounding gneisses and stick out in relief. When, during reverse faulting, the northern block was uplifted, erosion started attacking gneisses and aplites in the uplifted block. The aplites, however, being more resistant to erosion than the gneisses, were eroded much more slowly than the gneisses and after a certain period of sedimentation, were surrounded by the PII-III sediments. Thus, although the aplitic dikes at the first sight give the impression of having intruded the PII-III formation, they are in fact older than the sediments. The reverse fault can be situated south of the southern edge of the aplites (Fig. 7).

2.4 Folding

Contrary to the subhorizontal Upper-PIII volcanics, the Lower PIII volcanics as well as the PII-III sediments are folded in east-west trending folds. Depending on the competence of the rocks, the style of folding may vary from tight isoclinal folds to extended gentle folds.

2.5 Hercynian tectonics

The Hercynian tectonics are marked by vertical movements alongside the "accident majeur de l'Anti-Atlas". The northern block was downfaulted and the Palaeozoic sedimentary strata covering the uplifted southern block slipped downward onto the downfaulted block. Décollement structures showing folded Palaeozoic formations on top of unfolded Precambrian basement rocks, are found northeast of the Zenaga plain.

2.6 Alpine tectonics

During the Alpine orogeny, which in fact, is still active, northern and northwestern blocks were uplifted relative to southern and southeastern blocks. Examples of Alpine faults are seen at Tizi Asdrem, where the Asdrem plateau is separated from the Tamazight plain by such a fault (Fig. 2) and in the "accident majeur" near N'Qob (Fig. 5). Both faults have displaced Late Tertiary (or Early Quaternary) phonolites of Jbal Sirwa.

3. Geochronology

The abundance of geochronological data in the Anti-Atlas and High-Atlas allows correlations to be made between the different Precambrian blocks. A reinterpretation of stratigraphic and geochronological data leads to the conclusion that the new stratigraphic classification set up for the Taznakht block is valid in all other Precambrian blocks in the Anti-Atlas and High-Atlas as well.

Geochronological data and correlations between the different blocks are discussed elsewhere (Brabers 1986).

4. Plate tectonic model (Fig. 8)

The stratigraphic, tectonic and petrographic data discussed in the former sections may now be translated into plate tectonic terms.

1) Rifting

At the onset of the PII, subsidence may have occurred in the Anti-Atlas, resulting in the deposition of PIIa quartzites and dolomites. The subsequent tholeitic volcanism could have marked the opening of an ensialic back-arc basin associated with southward dipping subduction at the northern edge of the West-African craton. The subduction zone and the volcanic arc associated with it are not exposed and may be situated north of the Atlas.

2) Expansion

As expansion proceeded, oceanic crust was formed in the marginal basin and a continental shelf probably existed on which the PIIa pelites were deposited. The marginal basin expansion may be seen as a result of seaward migration of the subduction trench due to the fact that more oceanic lithosphere was consumed in the trench than the amount produced at the midocean ridge. As discussed by several authors (Vlaar and Wortel 1976, Molnar and Atwater 1978, England and Wortel 1980), such a situation is likely to occur where old oceanic lithosphere is subducted easily, fast and steeply dipping.

3) Compression

As the oceanic lithosphere at the trench becomes younger, the subduction process will slow down and the subduction angle will decrease. After a certain period of time then, a situation will be reached that as much oceanic lithosphere is consumed as the amount produced at the midocean ridge. As a consequence, lateral pressure on the marginal basin increases and expansion in the marginal basin ceases.

Due to the decreasing subduction angle the focus of calcalkaline volcanism is shifted southward over the marginal basin towards the West-African continent. In the marginal basin, this pretectonic subduction volcanism may be represented by the trondhjemites, spilites and keratophyres associated with the ophiolite suite. A mechanism for

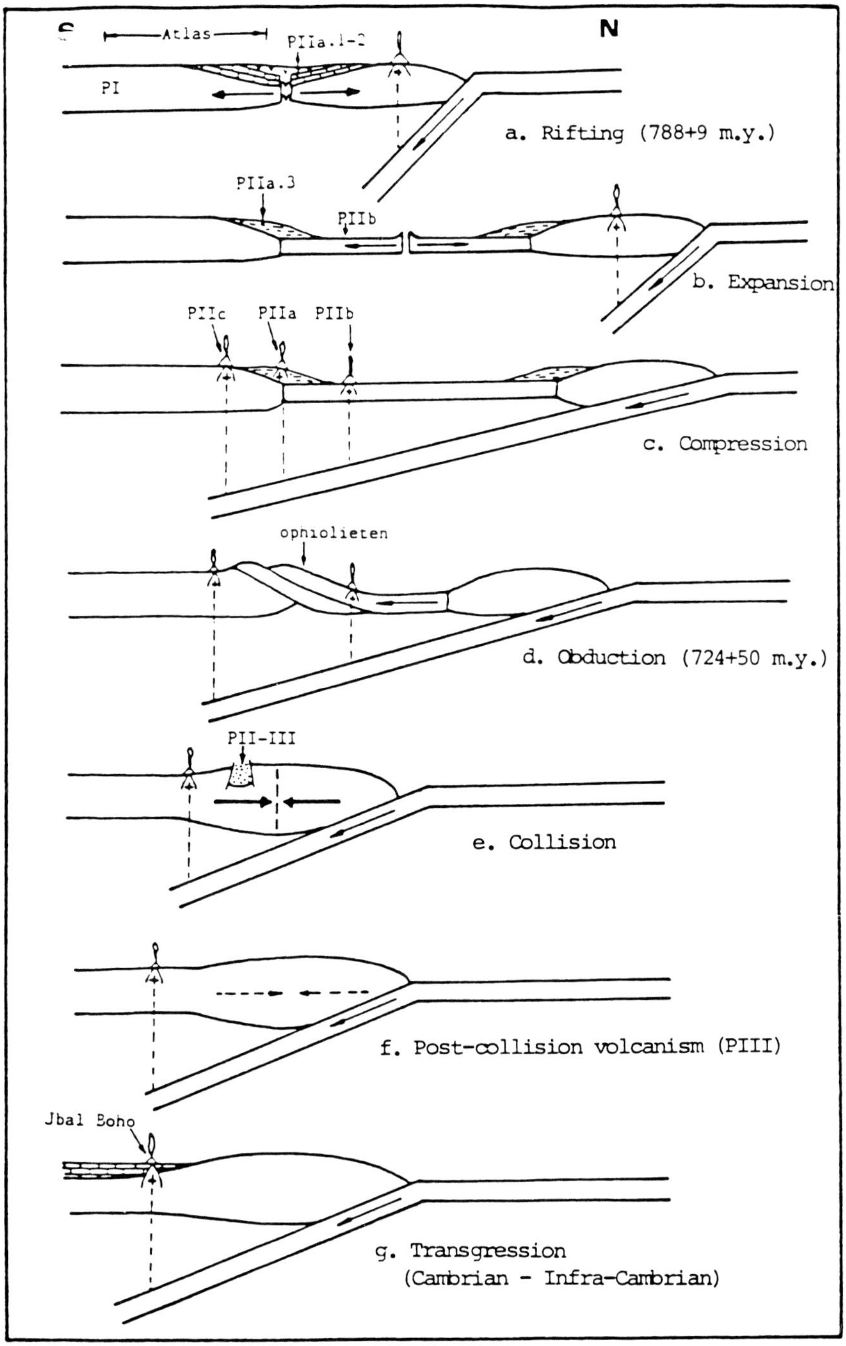

Fig. 8: Plate tectonic evolution of the Atlas during the Panafrican Orogeny

generating trondhjemitic magmas by partial melting of tholeitic gabbro (Arth 1979; Payne and Strong 1979) is supported by the occurrence of nebulitic textures in similar rocks in the Bou-Azzer area (Leblanc 1981). Trondhjemitic magmas, thus, may have been generated by partial melting of tholeitic gabbros in the oceanic crust due to an influx of fluids rising up from the oceanic slab subducted below the marginal basin. The calcalkaline volcanic intercalations in the pelitic series of the continental shelf formation (PIIa) and the pretectonic subaerial ash flow tuffs of Jbal Amsemsa (PIIc) show that the southward shifting subduction volcanism may have reached the continent before obduction started.

4) Obduction

As obduction started shortly after the first appearance of calcalkaline volcanism at the continental margin, it is not unlikely that the obduction process, in fact, was initiated by the weakening of the contact zone between the oceanic and the continental lithosphere due to this magmatic activity. In response to the pressure at the subduction trench, thrust planes could have been generated in the mechanically weak contact zone and the marginal basin ophiolites were thrust over the continental shelf.

The greenschist-amphibolite facies metamorphism marking the continental shelf formation may have been caused by frictional heating and the residual heat of the overriding ophiolite slab (cfr. Woodcock and Robertson 1977).

The maximum north-south distance between the ophiolitic mélanges of El Hara and the Ourika in the north and those of Tiffadine, Imghlay and Oued Assemlil in the south is about 70 km (Fig. 1). This may be considered a minimum for the original marginal basin width.

The strike-slip tectonics active in this period may be interpreted as a result of the tangential component of the oceanic plate movement relative to the continental plate. The normal component of this movement was absorbed by subduction of the oceanic plate and obduction of the marginal basin, while the tangential component was absorbed by strike-slip movements in the area weakened by the calcalkaline magmatism. Granitic magmas thus may have served as a lubricant for strike-slip movements while thrusting was lubricated by trondhjemitic magmas.

5) Collision (PII-III)

Obduction of the marginal basin probably continued untill collision occurred between the northern continental fragment and the West-African continent. Whereas before collision the normal component of the relative plate movement was absorbed by obduction and subduction, after collision the entire normal component had to be absorbed by subduction alone. As a consequence, pressure on the continental margin may have increased and the collision zone was uplifted. This may explain the reverse faulting and high sedimentation rates during the PII-III.

6) Post-collision volcanism (PIII)

After collision, subduction of the oceanic lithosphere continued just as before and the calcalkaline volcanism associated with it as well. The main difference with the pre-collision volcanics is the abundance of detritic intercalations due to the intensive erosion in the uplifted collision zone.

As for the disappearance of strike-slip tectonics in this period, one can argue that the thickening of the continental crust during collision may have meant a higher resistance against strike-slip movements. As a consequence, the focus of strike-slip tectonics may have shifted elsewhere towards weaker zones.

The unconformity between the Lower-PIII and Upper-PIII is not explained by this plate tectonic model. A possible explanation probably has to be sought outside Morocco.

7) Marine transgression

In the Late Precambrian, a marine transgression came over the Atlas and the Adoudounian sediments were deposited. In the northern areas of the Anti-Atlas and High Atlas, the first marine sediments were deposited later during the Middle Cambrian.

The numerous calcalkaline volcanic intercalations in the marine formations (Boudda et Choubert 1972, Buggisch 1986, Pouit et Jouravski 1960, Proust 1961, Van Looy 1980-1985, Van Guestaine et Van Looy 1983, Viland 1972) show that the subduction zone north of the West-African craton was still active.

Conclusions

1. A new stratigraphic classification is proposed for the Precambrian applicable in all Anti-Atlas and High Atlas blocks. The stratigraphic sequence comprises an Eburnean Precambrian I, three facies in the Precambrian II, a detritic Precambrian PII-III and a calcalkaline volcanic Precambrian III.

2. Criteria are proposed to distinguish between PIIa quartzites associated with dolomites, and PII-III quartzites associated with shales and synsedimentary reverse faulting. These criteria are applied to reinterpret the correct stratigraphic position of different quartzite formations in the Anti-Atlas.

3. In addition to the pretectonic calcalkaline volcanics known in the oceanic facies (PIIb) and the continental shelf facies (PIIa), pretectonic volcanic activity is recognized in a continental facies (PIIc) as well.

4. All gneisses north of the "accident majeur de l'Anti-Atlas", such as the Tamazight and Imghlay gneisses, the Oued Assemlil, Tiffadine, Ourika and El Hara gneisses, are interpreted to be the metamorphic equivalents of the PIIa platform formation, gneissified by ophiolite obduction. They contain mafic and ultramafic fragments of an ophiolite disrupted by Panafrican strike-slip tectonics and, therefore, they may be called an ophiolitic mélange. Their stratigraphic position is thus changed from the Precambrian I, as generally accepted in the past, towards the Precambrian II.

5. For the trondhjemites associated with the Anti-Atlas ophiolites, a mechanism is proposed to explain their genesis by partial melting of tholeitic gabbros involving an influx of fluids from a subducted plate. The trondhjemites as well as the spilites and keratophyres associated with them, are thus interpreted as a manifestation of subduction volcanism in an oceanic facies.

6. The Panafrican orogeny is explained by the creation and closure of a marginal basin associated with southward subduction below the northern edge of the West-African craton. An evolution of expansion towards compression was coupled with southward shifting of subduction volcanism due to a decreasing subduction angle. The closure of the marginal basin resulted in ophiolite obduction and

finally in collision of the northern continental fragment with the West-African continent.

Acknowledgements

This paper has been the subject of a Ph.D. thesis at the Katholieke Universiteit Leuven in Belgium. It would not have been possible without the help of Prof. D.E. Vogel, the thesis promotor, who has been conducting and stimulating geological research work in the Anti-Atlas and High-Atlas for the last 13 years until his untimely decease on May 18th 1986. His enthusiasm and the stimulating discussions with him are highly appreciated by the author.

References

Arth J.G. 1979. Some trace elements in trondhjemites. Their implications to magma genesis and palaeotectonic setting. In: Trondhjemites, dacites and related rocks (F. Barker), pp. 123-132.

Boudda A. et Choubert G. 1972. Sur la limite inférieure du Cambrien au Maroc. C.R.Acad.Sc.Paris, 275, D, 5-8.

Brabers P.M. 1986. Een plaattectonisch model voor de Panafrikaanse orogenese in de Anti-Atlas, Marokko. Ph.D. thesis - Katholieke Universiteit Leuven.

Buggisch W. 1986. The Lower-Cambrian of the Anti-Atlas: facies and palaeogeography. Workshop: Orogeny on continental margins, Berlin 19/2/1986.

Charlot R. 1982. Caractérisation des évènements éburnéens et panafricains dans l'Anti-Atlas marocain. Apport de la méthode géochronologique Rb/Sr. N.M.Serv.Géol.Maroc 313.

Choubert G. et Termier G. 1951. Les calcaires précambriens de Taghdout et leurs organismes problématiques. N.M.Serv.Géol.Maroc 85, 9-33.

Choubert G., Faure-Muret A., Gauthier H., Hindermeyer J., Hollard H. 1960. Carte géologique du flanc nord de l'Anti-Atlas Central dans la région de Ouarzazate (1/100.000).

Choubert G., Faure-Muret A., Morin P., Hollard H. 1961. Carte géologique de la plaine des Zenaga (1/100.000).

Choubert G. 1963. Histoire géologique du précambrien de l'Anti-Atlas. N.M.Serv.Géol.Maroc 162.

Coleman R.G. and Donato M.M. 1979. Oceanic plagiogranite revisited. In: Trondhjemites, dacites and related rocks (F.Barker), pp 149-168.

England P. and Wortel R. 1980. Some consequences of the subduction of young slabs. E.P.S.L. 47, 403-415.

Leblanc M. et Lancelot J.R. 1980. Interprétation géodynamique du domaine panafricain de l'Anti-Atlas (Maroc) à partir de données géologiques et géochronologiques. Can.J.E.Sc. 17, 1, 142-155.

Leblanc M. 1981. Ophiolites précambriennes et gîtes arséniées de cobalt (Bou-Azzer, Maroc). N.M.Serv.Géol. Maroc 280.

Molnar P. and Atwater T. 1978. Interarc spreading and Cordilleran tectonics as alternates related to the age of subducted oceanic lithosphere. E.P.S.L. 41, 330-340.

Pouit G. et Jouravski G. 1960. Le gisement de manganèse de Tizi-n-Isdid (région de l'Ounein, Haut-Atlas). Mines et Géologie 11, 21-30.

Payne J.G. and Strong D.F. 1979. Origin of the Twillingate trondhjemite, North-Central Newfoundland: partial melting in the roots of an island arc. In: Trondhjemites, dacites and related rocks (Barker F.), pp 489-516.

Proust F. 1961. Etude stratigraphique, pétrographique et structurale du Bloc Oriental du Massif Ancien du Haut-Atlas (Maroc). Thèse de doctorat - Université de Montpellier.

Saunders A.D., Tarney J., Stern C.R., Dalziel J.W.D. 1979. Geochemistry of Mesozoic marginal basin floor. Igneous rocks from southern Chili. Geol.Soc.Am.Bull. part I, 90, 37-258.

Vanguestaine M. et Van Looy J. 1983. Acritarches du Cambrien Moyen de la vallée de Tacheddirt (Haut-Atlas, Maroc) dans le cadre d'une nouvelle zonation du Cambrien. Ann.Soc.Géol.Belgique T.106, 69-85.

Van Looy J. 1980. Stratigrafie en paleontologie van de Paleozoische sedimentaire gesteenten van het dal van Tacheddirt (Hoge Atlas, Marokko). Licentiaatsthesis - Katholieke Universiteit Leuven.

Van Looy J. 1985. Het kaartblad Tazenakht 1/100.000 Anti-Atlas, Marokko. Kartering, lithostratigrafie, biostratigrafie van Precambrium tot Tremadoc.

Viland J.C. 1972. Les minéralisations cuprifères de Tanfit. Mines et Géologie 36, 5-32.

Vlaar N.J. and Wortel M.J.R. 1976. Lithospheric aging, instability and subduction. Tectonophysics 32, 331-351.

Woodcock N.H. and Robertson A.H.F. 1977. Origins of some ophiolite-related metamorphic rocks of the "Tethyan" belt. Geol. 5, 373-376.

THE PRECAMBRIAN/CAMBRIAN BOUNDARY IN THE ANTI-ATLAS (MOROCCO) DISCUSSION AND NEW RESULTS

Werner Buggisch and Erik Flügel
Institut für Geologie, Universität Erlangen-Nürnberg, Schloßgarten 5, D-8520 Erlangen
and Institut für Paläontologie, Universität Erlangen-Nürnberg, Loewenichstr. 28, D-8520 Erlangen

Abstract

Controversial theories of the age of the Latest Precambrian and Early Cambrian are discussed based on published radiometric and paleontological data. New fossil findings of calcareous algae (*Kundatia composita* KORDE) in the Série des calcaires inferieurs give evidence of the Early Cambrian age of the Lower Adoudounien, which was formerly correlated with the Latest Precambrian.

Introduction

The basement of the Anti-Atlas consists of Precambrian rock-units with multiple deformation and metamorphism. The stratigraphy of the basement is still under discussion (CHOUBERT 1963; CLAUER et al. 1982; LEBLANC & LANCELOT 1980) and will not be dealt with in this paper.

The Panafrican Orogeny affected the Anti-Atlas during the Late Precambrian. The Precambrian metamorphics are unconformably overlain by sequences which can be interpreted as molasse of the Panafrican Orogeny: the Série d'Ouarzazate in the central Anti-Atlas and the Série d'Anezi in the anticline of Kerdous in the western Anti-Atlas. These sequences are characterized by coarse-grained, partly volcano-detritic sediments and andesitic to rhyolitic volcanic rocks and ignimbrites. The molasse sediments locally grade either conformably or with low angular discordance into the overlying sediments of the 'Adoudounien' which overlap the older Precambrian rocks of the Anti-Atlas.

The ages of the Série d'Ouarzazate and of the Adoudounien are under discussion as well. These units were established by CHOUBERT (1952; 1963) and the lithostratigraphic succession is still generally accepted by most geologists.

Middle Cambrian:	Schistes à Paradoxides Niveau d'Ourmast
Lower Cambrian: (Géorgien)	Grès terminaux Complexe schisteux Série schisto-calcaires Série des calcaires supérieurs
'Infracambrian': (Adoudounien)	Série lie de vin Série des calcaires inférieurs (Série de base)
— — — — — —	— — — — — —
'Precambrian III':	Série d'Ouarzazate

Tab. 1. Lithostratigraphic succession of the Latest Precambrian and Early Cambrian in the Anti-Atlas after CHOUBERT (1952; 1963).

During the Adoudounien and Géorgien the sea prograded with two regressive megacycles from the region of Agadir into the Gulf of Anti-Atlas. Except for a short regression during the Upper Cambrian, a marine environment prevailed in the Anti-Atlas from the Middle Cambrian to the Lower Carboniferous.

Geochemical analyses of lavas and pyroclastics intercalated in the Série des calcaires inférieurs south of Tazenakht indicate an intracontinental volcanism with affinities to the analytical data of the Afra volcanics (GASSNER 1986). Therefore, the transgression of the Adoudounien sea was probably combined with the subsidence of an 'Anti-Atlas-Graben' (aulacogen linked with the Japetus Ocean?).

During the Hercynian Orogeny, the sedimentary sequences of the central Anti-Atlas were hardly affected by deformation; the 'metamorphism' did not exceed the stage of diagenesis. The Lower Paleozoic strata of the western Anti-Atlas, however, show increasing strain and metamorphism (see BUGGISCH, this volume).

Discussion of previous work

Although the lithostratigraphy has been generally accepted, the theories on the stratigraphic correlation of the Série d'Ouarzazate and the Adoudounien diverge considerably. The first biostratigraphical subdivision based on trilobites was established by HUPE (1952). New fossils – archaeocyathids (DEBRÉNNE & DEBRENNE 1978) and trilobites (SDZUY 1978) verified the Lower Cambrian age of the Géorgien (upper part of the Série des calcaires supérieurs to Grès terminaux). The Adoudounien was hence regarded to be older than the Lower Cambrian. Therefore, the Adoudoudien and the Série d'Ouarzazate were assigned to the 'Infracambrian' and to the Precambrian III.

The first radiometric dating of volcanics from the Série d'Ouarzazate and of rhyolites from the High Atlas revealed K/Ar-ages of 315 ma and 215 ma and scattered Rb/Sr whole-rock ages with a mean value of about 420 – 430 ma (JUREY et al. 1974). These radiometric ages were obviously rejuvenated because the strata are conformably overlain by fossiliferous Lower Cambrian sediments.

U/Pb dating of zircons from magmatic rocks – previously assumed to be of 'Late Precambrian' or 'Infracambrian' age – revealed unexpectedly young ages. Geochronological correlations by different authors are summarized in Tab. 2 and Fig. 1; new data from the literature (after 1974) will be discussed below (numbers refer to Fig. 1):

(1) An U/Pb zircon age of 615 ± 12 ma was determined for the granodiorite of Bleida, which corresponds to the time of emplacement (LEBLANC & LANCELOT 1980). In relation to the older Precambrian basement the intrusion is postkinematic. It predates the deformation which affected the Precambrian II3.

(2) CLAUER (1974) and CLAUER & LEBLANC (1977) established a Rb/Sr age of 623 ± 18 ma (recalculated by DUCROT & LANCELOT, 1977) for the fraction < 2μm of metapelites of the Precambrian II3.This age is interpreted as the age of the very low grade to low grade metamorphism during the Panafrican Orogeny.

(3) A Rb/Sr whole-rock age of 618 ± 22 ma was determined for trachytes at the base of the Série d'Ouarzazate (YAZIDI 1976: after CLAUER et al. 1982, recalculated).

(4) The 'ignimbrite 3' of the upper part of the Série d'Ouarzazate revealed an U/Pb zircon age of 563 ± 10 ma. The Rb/Sr whole-rock age of the ignimbrites is about 489 ± 18 ma (MIFDAL & PEUCAT 1985).

(5) The 'ignimbrite complex 5' of the Série d'Ouarzazate yielded an U/Pb zircon age of 586 ± 20 ma (or 574 ± 5 ma). The discordance of the age increases with the decreasing grain size of the analyzed zircons (MIFDAL & PEUCAT1985).

(6) Subvolcanic rocks of the Série d'Ouarzazate were dated by BENZIANE (1974: after CLAUER et al. 1982, recalculated). A Rb/Sr whole-rock age of 573 ± 21 ma was determined.

(7) An U/Pb zircon age of 534 ± 10 ma was established by DUCROT et al. (1976; recalculated by DUCROT & LANCELOT 1977) from the syenite of the Jebel Boho (= Alougoum). The Rb/Sr whole-rock age is on the order of 310 ± 10 ma.

Western Anti-Atlas: Several rocks from the Ifni anticline (dome-like structure = 'boutonnière') range between 660 and 554 ma (JEANETTE & TISSERANT 1977). CHARLOT (1976) reported a Rb/Sr whole-rock age of 530 ± 10 ma (not recalculated) in sampled from the postkinematic Tafraoute granite of the Kerdous anticline. BONHOMME &

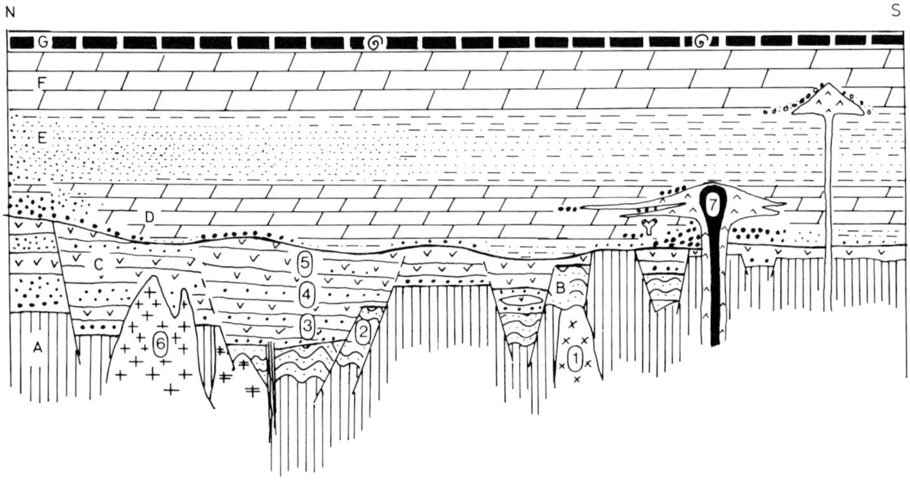

Fig. 1. Schematic section across the central Anti-Atlas during the Lower Cambrian (not to scale). Units A to C according LEBLANC & LANCELOT (1980).

	Series after CHOUBERT 1952;1963		BOUDDA et al. 1979	BOUKHARI et al. 1979	DESTOMBES et al. 1985
G	calcaires de base	Etage d'Amouslek	Amouslekien	Géorgien	slate and limestone Formation
				570 - 600 m.a.	
F	calcaires supérieurs	Adoudounien (= Infracambrian)	Tioutien	Taliouinien	Upper limestone Fm.
E	lie de vin		Taliouinien	650 - 680 m.a.	lie de vin Fm.
D	calcaires inférieurs		Adoudounien	Adoudounien	lower limestone and dolomite Formation
				800 - 900 m.a.	
C	Série d'Ouarzazate	Precambrian III	Precambrian III	Precambrian III	Precambrian III
				1000 m.a.	
B	Pre-cambrian II	Precambrian II	Precambrian II3	Precambrian II3	
				1450 m.a.	
A	older Precambrian basement				

Tab. 2. Subdivision and correlation of the Uppermost Precambrian and Early Cambrian according to several authors (A to G compare with Fig. 1).

HASSENFORDER (1985) analyzed the fraction < 2μm from the pelites of the Precambrian and the Adoudounien. Rb/Sr ages are scattered between 371 ± 8 and 364 ± 24 ma; K/Ar ages range from 357 ± 9 to 291 ± 7 ma.

CHOUBERT et al. always questioned the credibility of the radiometric data (CHOUBERT & FAURE-MURET 1971; 1977; CHOUBERT et al. 1979; BOUDDA et al. 1979). They favored an earlier age for the Precambrian, the Série d'Ouarzazate and the Adoudounien (Tab. 2) based on compared successions and paleoclimatologic and paleontologic arguments.

Paleontologic arguments

SDZUY (1978) described olenellid trilobites (*Eofallotaspis*) from the Série des calcaires supérieurs of the western Anti-Atlas (Tiout). He correlated this fauna with the lowermost Atdabanian of the Siberian platform. In the High Atlas the oldest trilobites were found immediately above the lie de vin facies, but the lithostratigraphic units of the High Atlas and the Anti-Atlas cannot be compared with complete certainty for the reason that they are separated by the Atlas boundary fault. Based on the same trilobites, BOUDDA et al. (1979) correlated the Série lie de vin (Taliouinian) with the 'Riphéen terminal = Vendian'. The Série des calcaires supérieurs was compared with the Tommotian, the 'Niveau de Tiout' containing *Eofallotaspis* was regarded to represent transition beds and the Série schisto-calcaires was correlated with the Atdabanian.

DEBRENNE & DEBRENNE (1978) argued against these correlations on the basis of archaeocyathids found 10m above the first trilobites in the same section (Tiout). They compared the upper Série de calcaires supérieurs with the Atdabanian, but not with the lower Atdabanian. According to these authors the Precambrian/Cambrian boundary ranges lower in the section.

Stromatolites from the Série lie de vin and the Série des calcaires supérieurs were described by SCHMITT (1978; 1979). Based on these stromatolites, SCHMITT correlated the Série lie de vin with the Vendian to Riphean III. This correlation was clearly contradicted by BERTRAND-SAFATI (1981) due to the fact that SCHMITT had neglected all the arguments for a Cambrian dating: the abundant algal bioherms of thrombolites and *Renalcis*- and *Epiphyton*-like microstructures.

According to TIMOVEIEV (in CHOUBERT et al. 1979), acritarchs are evidence of a Middle Riphean age for the Precambrian II3 and an Upper Riphean age for the Série d'Ouarzazate and the Adoudounien.

Based on analyses of amino-acids, PRASHNOVSKY (in CHOUBERT & FAURE-MURET 1977) estimated the age of the Série des calcaires inférieurs at 700 - 800 ma.

Paleoclimatological arguments

The regressive facies of the Série lie de vin was explained by glacio-eustatic sea-level changes in conjunction with the Varanger glaciation (CHOUBERT & FAURE-MURET 1977; BOUKHARI et al. 1979; BOUDDA et al. 1979). MONNINGER (1979: Pl. 17, Fig. 6) even thought he had found pseudomorphs after ice crystals in thin sections of the Série lie de vin (the pseudomorphs are probably due to Ca-sulfates). However, in contrast to the glaciation-model, all the distinctive features indicate a hot and dry climate and a frequently hypersaline environment: casts of halite, stromatolites and red beds.

New data and fossil findings

Complete sections of the Adoudounien were studied E of the road from Tazenakht to Foum Zguid at the SW margin of the El Graara anticline (GASSNER 1986). These sections expose the marginal facies of the Adoudounien which is strongly influenced by volcanism (Fig. 2).

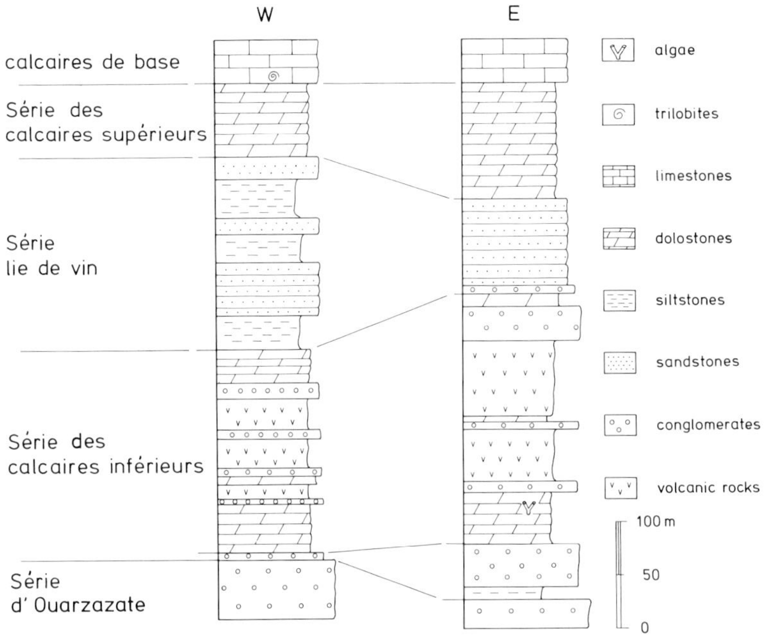

Fig. 2. Columnar sections of the Adoudounien at the SW margin of the Bou Azzer- El Graara anticline. Distance between the two sections about 4 km (after GASSNER 1986).

The Série d'Ouarzazate predominantly consists of rhyolites, ignimbrites, tuffs and conglomerates. The deposition of the Adoudounien begins above a minor unconformity and contains thin-bedded partly dolomitic red mudstones and siltstones. Poorly sorted greywackes are intercalated. They contain volcanic lithoclasts of acid and intermediate composition and marginally resorbed quartz grains. These basal beds are overlain by massive dark conglomerates with clasts of ignimbrites, rhyolites and quartz within a dolomitic matrix.

Typical stromatolitic carbonates were deposited above this first unit of the Série des calcaires inférieurs. Owing to the lava-flows and volcanoclastics, they can be subdivided into a lower and an upper dolomite sequence. Distinct algal structures were found within the lower dolomite sequence of the Série des calcaires inférieurs.

The Série lie de vin is exposed in the 'grès de Tikirt' facies. It is overlain by the Série des calcaires supérieurs containing thrombolites, oolites and early trilobites in the upper part.

Lower Cambrian Calcareous Algae

Distinct algal structures occur in sample X_1 ci.

D e s c r i p t i o n: The algal structures are represented by calcareous micritic areas (maximum dimension 10 mm x 20 mm) which contain tubular elements (Plate 1, Figs. 6,8). The outline of the structures is irregularly lobate. The lumina of the tubes are filled with sparry calcite and surrounded by differentiated, concentric micrite zones which include very small spar-filled cells (Plate 1, Fig. 5).

Longitudinal sections show inflated 'tubes' (Plate 1, Fig. 7); tangential and oblique sections exhibit peripheral spine-like processes which are sometimes truncated (Plate 1, Figs. 3, 5). Some tubes are constricted, some are subdivided by densely spaced, bent and thin micritic partitions.

The following pattern can be recognized within the 'wall' of the tubes (starting with the innermost part):

a) a very thin black and undifferentiatied micrite zone bordering the lumina of the tubes (Plate 1, Fig. 3);

b) a zone (width up to 35 µm) with closely spaced or isolated cross-sections of cells which may occur in one or two layers (Plate 1, Fig. 2). The diameter of the cells is between 12 and 20 µm; larger cells are found in the outer layer;

c) a variably thick, very irregularly bounded zone exhibiting irregular concentric laminations and also isolated cells (diameter up to 15 µm). This zone consists of up to five lamellae which are characterized by alternating thin micritic or pelmicritic layers and sparry layers in between. This marginal zone is gray in transmitted light. Quartz-grains incorporated within this zone indicate a weaker calcification than in the central zones.

The inner diameter of the tubes varies between 100 and 240 µm.

D i s c u s s i o n: The existence of cell structures, differentiated according to size and distribution, as well as the tubular pattern indicates the algal nature of these structures. The algae were sometimes strongly calcified. The first occurrence of calcareous algae is known near the base of the Tommotian (RIDING & VORONOVA, 1984). The Cambrian algal flora is domintated by cyanophyceans and possible chlorophyceans (CHUVASHOV & RIDING 1984) which can be attributed to six morphological groups (RIDING & VORONOVA 1985). A classification of the alga described using these groups is not possible because of the occurrence of characteristics typical of the dendritic, tubiform and the tuberous form groups.

The morphological classification does not embrace all Cambrian algae. This is also true for *Kundatia composita* KORDE, 1973 – a species considered to be a red alga. The criteria of this species correspond to the criteria of the alga described with regard to the differentiation of the thallus, the morphology and distribution of the cells, as well as the size of the cells. The species is characterized by encrusting thalli with chain-like cells arranged along the axis of the thalli (corresponding to the 'tubes').

A g e: *Kundatia composita* KORDE has been described until now only from the Bagradskian Horizon of the Lower Cambrian. The species was found near the river Kiya in the Alatau region, Kuznets basin. Together with *Kundatia* the following species occur: *Kordephyton crinitum* (KORDE), *Epiphyton leptosum, Razumovskia* sp., *Renalcis* sp. and *Siphonomorpha* sp.. According to RIDING & VORONOVA (1984: Fig. 2) the first occurrence of *Kordephyton* in Siberia and Mongolia is Tommotian III, of *Epiphyton* Tommotian II, of *Razumovskia* Middle Atdabanian (perhaps Lower Atdabanian), and of *Renalcis* Nemakitdaldynian (a time interval prior to Tommotian). The Bagradskian horizon is attributed to the higher Lower Cambrian by KORDE (1971: p. 86); it forms the upper unit of the Aldanskian Group. The Bagradskian horizon overlies limestones with archaeocyathids.

Plate 1: Lower Cambrian Algae, Anti-Atlas

Kundatia composita KORDE. East of the road Tazenakht to Foum Zguid, southwestern margin of the El Graara anticline. Sample X_1ci from the lower dolomite sequence of the Série des calcaires inférieurs.

Fig. 1. Cross-section. The wall of the algal tube consists of micrite with closely spaced small cells (arrow). x 42

Fig. 2. Cross-section, exhibiting closely-spaced cells occurring in two layers. Larger cells are concentrated within the outer layer. x 70

Fig. 3. Cross-section with spine-like projections (arrow). x 70

Fig. 4. Several 'tubes' with differentiated walls. Oblique and longitudinal sections of the cells are indicated by arrows. x 42

Fig. 5. Cross-section. Concentrical wall structure and truncated spine-like projections. x 70

Fig. 6. Algal thallus with cross-sections and longitudinal sections of 'tubes', separated by dark, inhomogenous micrite. x 10

Fig. 7. Cross-section and longitudinal section, showing inflated 'tubes' (arrow = x 30

Fig. 8. Algal thallus with irregularly distributed 'tubes' surrounded by concentrical wall structures. x 10

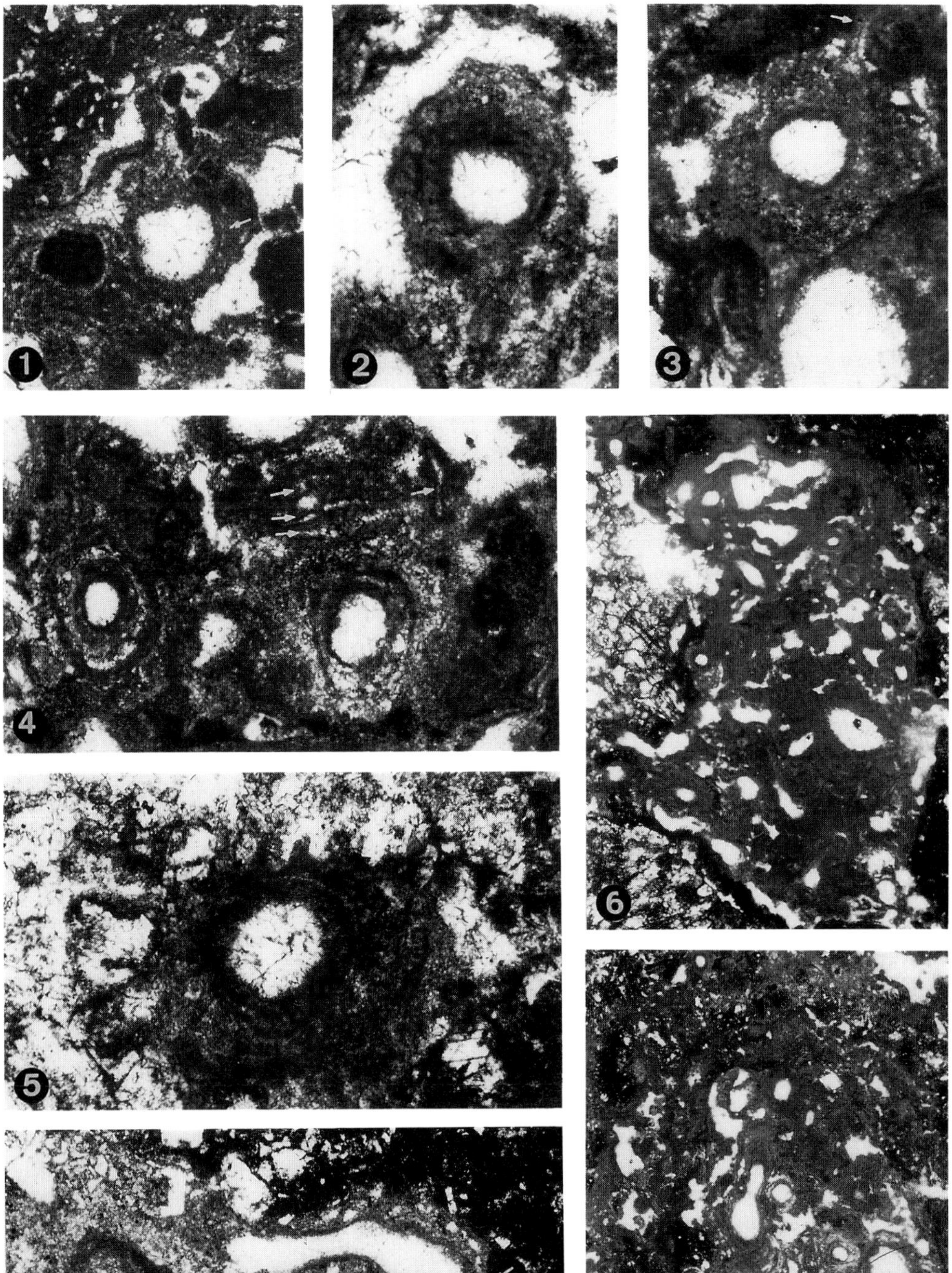
1
2
3
4
5
6
7
8

Algae, most probably identical with *Kundatia*, are also known from the Lower Cambrian of Sardinia: SELG (1985) and BECHSTÄDT et al. (1985) describe algal/archaeocyathid reef limestones from the Matoppa Member of the Nebida Formation in southwestern Sardinia (upper Atdabanian). A re- investigation of the thin-sections shows that these limestones yield archaeocyathids together with algae of the *Kundatia* type, *Epiphyton, Renalcis* and *Razumovskia.*

Conclusions

In summarizing all the radiometric data, the following interpretation appears to be reasonable: The intrusion of the granodiorite of Bleida (615 ± 12 ma) predates the last strong deformation which affected the central Anti-Atlas. This deformation is more or less of the same age as the very low grade to low grade metamorphism which affected the sediments of the Precambrian II3 (623 ± 18 ma). Therefore, the age of this deformation is about 620 ma (602 - 627 ma) which can be correlated with the Panafrican Orogeny of the Pharusian Chains (Ahaggar: ALLEGRE & CABY 1972).

During the postorogenetic crustal extension, the molasse basins of the Ouarzazate and Anezi subsided along synsedimentary faults. These troughs were filled with conglomerates, volcanic and subvolcanic rocks between 618 ± 22 and 563 ± 10 ma. In the western Anti-Atlas intrusions continued up to 530 ± 10 ma (granite of Tafraoute). The syenite of the Jebel Boho (534 ± 10 ma) is intercalated in the Série des calcaires inférieurs (formerly 'Infracambrian'). Therefore, the correct interpretation of the age of this syenite (syngenetic or later intrusion?) is difficult.

Most of the Rb/Sr whole-rock ages were obviously rejuvenated. The rejuvenation in the western Anti-Atlas can be explained by the Hercynian metamorphism (360 - 290 ma). However, the temperature did never exceed the 300^{0} C isotherme in the central Anti-Atlas (see BUGGISCH, this volume). This contradicts the views of LEBLANC & LANCELOT (1980) and ODIN et al. (1983). Probably devitrification and alteration were important factors for the rejuvenation in this area.

The radiometric age of the Precambrian/Cambrian boundary is still under discussion. ODIN et al. (1983) proposed an age of 520 - 540 ma. If the Precambrian/Cambrian boundary is defined at 570 ma, the radiometric data of the Anti-Atlas suggest a correlation of the Adoudounien with the Lower Cambrian.

The occurrence of *Kundatia composita* KORDE in the Série des calcaires inférieurs of the central Anti-Atlas provides evidence of a Lower Cambrian age for the entire Adoudounien. This is in accordance with the U/Pb zircon ages dicusssed above.

References

ALLEGRE, C.J. & CABY, R (1972): Chronologie absolue du Précambrien de l'Ahaggar occidental. – C. R. Acad. Sci., **275-D,** 2095-2098, Paris

BECHSTÄDT, T., BONI, M. & SELG, M. (1985): The Lower Cambrian of SW-Sardinia: From a clastic Tidal Shelf to an Isolated Carbonate Platform, Facies, **12,** 113-140, Pl. 11-13, 5 Figs., Erlangen

BERTRAND-SARFATI; J. (1981): Problème de la limite Précambrien-Cambrien: la section de Tiout (Maroc); les stromatolites et leur biostratigraphie (SCHMITT 1979): critique et observations. – Newsl. Stratigr., **10,** 20-26, Berlin

BONHOMME, M.G. & HASSENFORDER, B. (1985): Le métamorphisme hercynien dans les formations tardi- et post-panafricaines de l'Anti-Atlas occidental (Maroc). Données isotopiques Rb/Sr et K/Ar des fractions fines. – Sci. Geol. Bull., **38,** 175-183, Straßburg

BOUDDA, A., CHOUBERT, G & FAURE-MURET, A. (1979): Essai de stratigraphie de la couverture sédimentaire de l'Anti-Atlas: Adoudounien – Cambrien Inférieur. – Not. Mém. Serv. Geol. Maroc, **271,** 96 pp, Rabat

BOUKHARI, A., CHOUBERT, G., FAURE-MURET, A. & TIMOFEIEV, B. (1979): Sur la microflore des couches terminales du Précambrien III et sur l'âge de l'Adoudounien au Maroc. – C. R. Acad. Sci., **288**, 13-14, Paris

CHARLOT, R. (1976): The Precambrian of the Anti-Atlas (Morocco): a geochronological synthesis. – Precambrian Res., **3**, 273-299, Amsterdam

CHOUBERT, G. (1952): Histoire géologique du domaine de l'Anti-Atlas. – Not. Mém. Serv. Geol. Maroc, **100**, 77-194, Rabat

— (1963): Histoire géologique du Précambrien de l'Anti-Atlas. – Not. Mém. Serv. Geol. Maroc, **162**, 352 pp. ,Rabat

CHOUBERT, G. & FAURE-MURET, A. (1971): Afrique occidental: Socle précambrien et zones mobiles. Anti-Atlas (Maroc). – UNESCO, Tectonique de l'Afrique, Sci. Terre, **6**, 165-175, Paris

— (1977): Difficultés géochronologiques dans le Précambrien de l'Anti-Atlas (Maroc). – C. R. Somm. Soc. Geol. Fr., **2**, 72-74, Paris

CHOUBERT, G., FAURE-MURET, A. & TIMOFEIEV, B. (1979): Premiers resultats de récherche d'Acritarches dans le Précambrien de l'Anti-Atlas. – C. R. Acad. Sci. Paris, **288 D**, 191-194, Paris

CHUVASHOv, B. & RIDING, R. (1984): Principal Floras of Paleozoic Marine Calcareous algae. – Palaeontology, **27**/3, 487-500, 9 Figs., Oxford

CLAUER, N. (1974): Utilisation de la méthode rubidium-strontium pour la datation d'une schistosité de sédiments peu métamorphisés: application au Precambrien II de la boutonnière de Bou Azzer-El Graara (Anti-Atlas, Maroc). – Earth Planet. Sci. Lett., **22**, 404-412, Amsterdam

CLAUER, N. & LEBLANC, M. (1977): Implications stratigraphiques d'une étude géochronologique Rb-Sr sur métasédiments précambriens de Bou Azzer (Anti-Atlas, Maroc). – Not. Mém. Serv. Geol. Maroc, **38**, 7-12, Rabat

CLAUER, N., CABY, R., JEANETTE, D. & TROMPETE, R. (1982): Geochronology of sedimentary and metasedimentary rocks of the West African Craton. – Precambrian Research, **18**, 53-71, Amsterdam

DEBRENNE, F. & DEBRENNE, M. (1978): Archaeocyathid fauna of the lowest fossiliferous levels of Tiout (Lower Cambrian, Southern Morocco). – Geol. Mag., **115**, 101-119, Cambridge

DESTOMBES, J., HOLLARD, H. & WILLEFERT, S. (1985): Lower Paleozoic rocks of West Africa and the western part of central Africa. – In: HOLLAND, C.H. (ed.): Lower Paleozoic rocks of the world, **4**, 91-336, Chichester (Wiley)

DUCROT, J. & LANCELOT, J. (1977): Problème de la limite Précambrien-Cambrien: étude radiochronologique par la méthode U/Pb sur zircons du volcan du Jbel Boho. – Canad. J. Earth Sci., **14**,2771-2777, Ottawa

DUCROT, J., LANCELOT, J.R. & LEBLANC, M. (1976): Age panafricain de la diorite de Bleida (Anti-Atlas, Maroc) et conséquences. – 6e Réun. ann. Sci. Terre, p. 150, Orsay

GASSNER, W. (1986): Das Infra- und Unterkambrium am Südostrand des Präkambriumaufbruches von Bou Azzer/El Graara (Zentraler Anti-Atlas, Marokko). – Unpubl. Dipl. Arb., 92 pp, Erlangen

HUPE, P. (1952): Contribution à l'étude du Cambrien inférieur et du Précambrien III de l'Anti-Atlas marocain. – Not. . Mém. serv. géol. Maroc, **103**, 362 pp, Rabat

JEANETTE, D. & TISSERANT, D (1977): Les épisodes tectoniques et intrusifs du Précambrien Supérieur de l'Anti-Atlas occidental. – Estud. Geol., **33**, 315-326, Madrid

JUREY, A., LANCELOT, J.R., HAMET, J., PROUST, F. & ALLEGRE, C.J. (1974): L'âge des rhyolites du Précambrien III du Haut Atlas et le problème de la limite Précambrien-Cambrien. – 2e Réun. ann. Sci. Terre, P. 230; Nancy

KORDE, K.B. (1971): Vodorosli kembriya. – Akad. nauk SSSR, Trudy paleont. inst., **139**, 349 pp., 49 Pls., 65 Figs., Moskva (Nauka)

LEBLANC, M. & LANCELOT, J.R. (1980): Interpretation géodynamique du domain panafricain (Précambrien terminal) de l'Anti-Atlas (Maroc) à partir de données géologiques et géochronologiques. – Can. J. Earth Sci., **17**, 142-155, Ottawa

MIFDAL, A. & PEUCAT, J.-J. (1985): Datations U-Pb et Rb-Sr du volcanisme acide de l'Anti-Atlas Marocaine et du

MIFDAL, A. & PEUCAT, J.-J. (1985): Datations U-Pb et Rb-Sr du volcanisme acide de l'Anti-Atlas Marocaine et du socle sousjacent dans la région de Ouarzazate. Apport au problème de la limite Précambrien- Cambrien. – Sci. Geol. Bull., **38**, 185-200, Straßburg

MONNINGER, W. (1979): The section of Tiout (Precambrian/Cambrian boundary beds, Anti-Atlas, Morocco): An environmental model. – Arb. Paläont. Inst. Würzburg, **1**, 289 pp., Würzburg

ODIN, G.S., GALE, M.H., AUVRAY, B., BIELSKI, M., DORE, F., LANCELOT, J.R. & PASTEELS, P. (1983): Numerical dating of Precambrian-Cambrian boundary. – Nature, **301**, 21-23, London

RIDING, R. & VORONOVA (1984): Assemblages of calcareous algae near the Precambrian/Cambrian boundary in Siberia and Mongolia. – Geol. Mag., **121**/3, 205-210, 3 Figs., London

— & — (1985): Morphological Groups and Series in Cambrian Calcareous Algae. – In: TOOMEY, D.F. & NITECKI, M.H. (eds.): Paleoalgology. – 56-78, 17 Figs., Berlin-Heidelberg-New York-Tokyo (Springer)

SCHMITT, M. (1978): Stromatolites from the Tiout section. Precambrian- Cambrian boundary beds, Anti-Atlas, Morocco. – Geol. Mag., **115**, 95-100, Cambridge

— (1979): The section of Tiout (Precambrian/Cambrian boundary beds, Anti-Atlas, Morocco): stromatolites and their biostratigraphy. – Unpubl. Thesis, University Würzburg

SELG, M. (1985): Die siliziklastisch-karbonatische Wechsellagerung der unterkambrischen Nebida-Formation (SW-Sardinien): Entstehung einer Karbonat-Plattform. – Unpubl. Thesis, Geowiss. Fak. Univ. Freiburg, 105 pp., 9 Pls., 15 Figs., Freiburg

SDZUY, K. (1978): The Precambrian-Cambrian boundary beds in Morocco (preliminary report). – Geol. Mag., **115**, 83-94, Cambridge

The Base of the Cambrian in Morocco

K. Sdzuy & G. Geyer
Institut für Paläontologie, Universität Würzburg,
Pleicherwall 1, D-8700 Würzburg

Abstract

The paper summarizes problems of datings of Latest Precambrian to Early Cambrian rocks of Morocco. The known palaeontological data are reviewed briefly. Characteristics and correlatability of the earliest trilobites of Morocco are discussed. Additional remarks concern the appearance of hard part animals at the beginning of the Cambrian.

1. Introduction

In Morocco Lower Cambrian rock successions are complete, well exposed, rich in fossils, and there is no break between Precambrian and Cambrian deposits. The historical importance of the Moroccan Cambrian bases mainly on publications of Pierre Hupé. Hupé erected a detailed biostratigraphical scheme based on trilobites (Hupé, 1952), and he published a classical monograph on Lower Cambrian trilobites (Hupé, 1953). He improved the latter in an internal thesis (Hupé, 1959), which never was published (revised biostratigraphical frame in Hupé, 1960). However, his first monograph led to an improved international correlation of the Early Cambrian and its publication can be regarded as the beginning of modern research on Early Cambrian.

Good reviews on the history of investigations on the Moroccan Cambrian are given by Boudda, Choubert & Faure-Muret (1979) and Hollard (1985), respectively. Although Cambrian successions are known from the Moroccan Meseta as well as from Jbilet, the High Atlas Mountains and the Anti-Atlas, the latter became "type" area for lithostratigraphy and for biostratigraphy of the Moroccan Early Cambrian until today. This was induced by good outcrops as well as by huge research activity in this region, above all by Georges Choubert and Pierre Hupé. Choubert published most important articles on lithostratigraphy of the final Precambrian and Early Cambrian (e. g. Choubert, 1942; Choubert in Hupé, 1953; Choubert, 1963). According to Hupé the base of the Cambrian is marked by appearence of archaeocyatha. Trilobites were said to appear somewhat later, but Hupé had good reasons to believe the early trilobites of Morocco being distinctly the oldest known on earth so far. Later, trilobites clearly older than those from Hupé's oldest zone I and provisionally referred to an *Eofallotaspis* zone herein were discovered in a section near Tiout (Sdzuy, 1978), not far from the section of Hupé's Tazemmourt section. Therefore the section of Tiout is the one where most information about the

2b. Calcaires inférieurs

The Calcaires infèrieurs (Adoudounien, Choubert, 1958; vs. Boukhari et al., 1979) are a monotonous complex of about 1000 m thickness. It is essentially composed of shallow water carbonates transgressing discordantly over various Precambrian rocks. The lightly coloured massively bedded dolomites and limestones are locally interrupted by mudstones, pelitic shales and sandstones. The siliceous intercalations are often mineralized by diverse ores.
The formation commences with a so-called Série de base (Choubert, 1952), a mostly siliclastic transgressional sequence of about 100 m thickness (Boudda et al., 1979), comprising dolomites, conglomerates, and first of all a unit with argillites and shales, interrupted by conglomeratic or arenitic-arkosic layers.
Choubert et al. (1950, 1952a, 1952b), Buggisch et al. (1979), and Buggisch & Heinitz (1984) reported stromatolites from that formation, Buggisch & Flügel (this volume) describe a calcareous alga.
A striking volcanic activity is reported from the central Anti-Atlas, were trachytes and andesites are intercalated into the dolomites. The centre of volcanism was found in Jbel Bokho near Alougoum (Choubert, 1952a; Boudda et al., 1979). Volcanoclastic layers may contain ripple marks and soft pebbles (Buggisch et al., 1979).

2c. Série lie de vin

The name 'Lie de vin' was chosen by Choubert (1942) because of the typical purple colour of the shales contained. "Taliwinien" is more or less a synonym (Boudda et al., 1979, Boukhari et al., 1979). Although the rocks composing this formation vary to a large extend, they are mostly continentally-terrigenously influenced, giving indications of the regressive phase of the first post-orogenic cycle (e. g. Choubert in Hupé, 1953). The thickness differs from 300 m in the Warzazate region in the east (Boudda et al., 1979) to about 1000 m south of Taroudant (Monninger, 1979).
Red sandstones and conglomerates dominate close to the basinal shore in the east. The totally detritic succession of preponderant sandstones which advances from east of Warzazate into the central Anti-Atlas and interfingers the middle member of the Série lie de vin is called Grès de Tikirt. Its thickness is up to 225 m south of the Jbel Kissane (Buggisch et al., 1979). Towards the centre of the basin the deposits pass over into fine-grained rocks.
The most typical facies is found in sections on the southern flank of the Anti-Atlas: The succession is largely formed by purple shales; dolomites and limestones are more often intercalated towards the top, especially in the upper third of the sequence (Hollard, 1985). Some thick limestone beds may be used as lithological marker horizons cropping out over large areas (e. g. "Barre de Tata").

The well studied section of Tiout represents cyclic sedimentation, deposited under rather stable conditions (low water energy, shallow water depths, insignificant tidal range, partly high salinity; Monninger, 1979). Trace fossils ("worm burrows") are common. Algal build-ups can be used as facies indicators, in particular columnar stromatolites can be useful index fossils.

2d. Calcaires supérieurs

The deposition of the Calcaires supérieurs was initiated by a new transgression. The formation mainly consists of dark limestones; light grey to pink carbonates (especially finely crystalline dolostones) are less common. Obvious detrital deposits are absent. The known thicknesses vary between 75 m at Bou Azzer (Boudda et al., 1979) and up to 300–400 m in the region of Tiznit.

South of the Jbel Sarhro the facies changes to the so-called Série schisto-calcaire orientale with shales and light limestones.

The formation was deposited under shallow water conditions. Mud cracks and ignimbrite horizons point to local emersion (Buggisch et al., 1979).

Owing to a remarkable facial change in the Tiout section Monninger & Schmitt (1975) called the upper portion of the formation "transition layers". Instead of black micritic limestones dark oolites and more and more marlstones compose the succession. A similar change of facies is described from the region south of the Jbel Kissane (Buggisch et al., 1979). The thickness of this latter member called "calcaires de base" is up to 20 m. Thrombolites occur in the basal limestones; moreover trilobite fragments, ooids and intraclasts are reported from the carbonate beds. Biostromal and domal stromatolites are generally reported from the southern Jbel Sarhro (Buggisch et al., 1979).

Schmitt (1979a, 1979b) described stromatolites and reported trace fossils from the lower part of the Calcaires supérieurs of Tiout. In some beds of the upper part archaeocyatha form local bioherms (Debrenne, 1975, Debrenne, 1976, Debrenne & Debrenne, 1978). Besides this Sdzuy (Sdzuy, 1978; Liñan & Sdzuy, 1978; Sdzuy, 1981) reported trilobite-bearing horizons.

2e. Série schisto-calcaire

The classical Early Cambrian studied by Hupé commences with an alternation of generally dark limestones and light shaly slates, called Série schisto-calcaire (Choubert, 1952b).

A rather typical lithological sequence is exposed at Amouslek (southwest of Taroudant). The section has been described several times (e. g. Choubert in Hupé, 1953; Choubert, 1963). The series of shales is interrupted by limestone beds, the latter largely formed

by archaeocyathan bioherms. The formation may be subdivided into three lithological units which altogether are more than 200 m thick. The similar section of Tazemmourt was of great importance for the investigations of Pierre Hupé.
In the section of Oued Boutergui (region of Tiznit; Berneker & Geyer, in prep.) the sequence has a thickness of about 200 m, with dark limestones largely displacing the basal shales. In the northwestern Anti-Atlas (e. g. northern part of Asif Aouerga) the succession can be formed essentially by limestones containing only thin shale layers (p. p. "Calcaires massifs de Tiznit" sensu Boudda et al., 1979).
Some published estimations of thicknesses in the western Anti-Atlas are possibly exaggerated. According to Choubert (1942) the formation is up to 390 m thick in the Tata area, but this might possibly not comprise the identical lithostratigraphical unit as in current use.
Sedimentation was influenced by varying water depths and perhaps shoreline oscillation under shallow marine shelf conditions. Towards the east the content of detrital elements increases with shallowing water depths (Choubert, 1964).
The sequence often contains abundant shelly faunas. Trilobites and archaeocyatha were described in detail by Hupé (1953) and Debrenne (1958, 1961, 1964, 1973), respectively. The famous archaeocyathan limestones of Sidi Moussa d'Aglou near Tiznit (Bourcart, 1927; Bourcart & Le Villain, 1931; Debrenne, 1964) belong to this formation.

3. Stromatolites and calcareous algae

3a. Occurrences of stromatolites and algae

The only fossils discovered below trilobite bearing beds are algal build-ups. Low algal mats, stromatolites, and thrombolites are widespread, but published palaeontological notes are sparse: Choubert et al. (1950, 1952a, 1952b) reported stromatolites from the Calcaires inférieurs. Stromatolites from the Série lie de vin were described by Schmitt & Monninger, 1977, Schmitt, 1978, Schmitt, 1979a, and Schmitt, 1979b (review in Boudda et al., 1979). Columnar stromatolites are also known from the Calcaires supérieurs of the Tiout section (Schmitt, 1979a; Schmitt, 1979b) and lithologic equivalents of the Lemdad area (Schmitt, 1979b). (Buggisch et al., 1979), Buggisch & Heinitz (1984), and Heinitz (1984) mentioned stromatolites from the Calcaires supérieurs east of the Jbel Sirwa.

A detailed study of the stromatolites of the Tiout section was presented by Schmitt (1979a) giving the following biostratigraphical information:
The lower sequence of the Série lie de vin furnishes a number of horizons with poorly preserved stromatolites. Schmitt (1979a) described a *Linella* cf. *avis* Krylov, 1967 pointing to a late Riphean or Vendian age (Schmitt, 1979a).
From the middle sequence of the Série lie de vin *Tungussia* cf. *inna* Walter, 1972 and

Patomia fsp. indet. were described, indicating most probably Vendian age.
The upper sequence of the Série lie de vin yielded *Nimbophyton fungiforme* Schmitt, 1979, *N.* ? fsp. indet., *Tioutella rupta* Schmitt, 1979, *Igoudinia hemisphaerica* Schmitt, 1979, *Tifounkeia* n. fsp. T, and *Tungussia* f. indet. (besides thrombolites and other unnamed stromatolites).
From the lower to middle portion of the Calcaires supérieurs Schmitt (1979a, 1979b) described *Acaciella* cf. *angepena* Preiss, 1972, which implies an Early Cambrian age. In addition, *Tioutella bouddai* Schmitt, 1979 and some horizons with thrombolites are reported.
From the Lemdad River section stromatolites are only described from trilobite bearing beds (Schmitt 1979a, 1979b; "Ounein region").

A skeletal alga is described by Buggisch & Flügel (this volume) from the Calcaires inférieurs east of the Jbel Sirwa and supposed to indicate Early Cambrian age.
Frequent algae intergrowing with archaeocyatha are *Epiphyton* and *Renalcis*, partly forming pure algal boundstones (e. g. Oued Boutergui, Berneker & Geyer, in prep.). *Girvanella* appears in several localities, *Tolypothrix* and *Skythonema*? were found in archaecyathan limestones of Tiout (Berneker, pers. comm.).

3b. Problems of datings

The datings of Schmitt (1979a, 1979b) were criticized by Semikhatov (1977), Clauer (1980), and Bertrand-Sarfati (1981). Semikhatov mentioned a stromatolite from the lower part of the Série lie de vin of Tiout and stresses the similarity to the Siberian *Parmites victorius* from the Early Cambrian. Especially Bertrand-Sarfati stresses similarities of the forms described by Schmitt with Early Cambrian stromatolites. Associated thrombolites also are thought to indicate Early Cambrian age by Bertrand-Sarfati (1981). On the other hand, it is known that the occurrence of thrombolites is controlled by facies. The biostratigraphical usability of stromatolites depends of course on a sufficiently broad spectrum of forms. The diversity of stromatolites in the Tiout section certainly is not sufficient to date rocks with precision, but there is a good conformity in the succession of the indicated biostratigraphical levels of Schmitt (1979a). In contrast, we find it difficult to accept datings based on the same or even more poor amount of forms. However, in spite of the critique we were informed by Walter (written pers. comm.) that Schmitt's work was indeed carried out carefully.
Clauer (1980) presented a comparison between biostratigraphic datings and isotopic geochronology of the Upper Precambrian of the Anti-Atlas, based mainly on an absolute dating of trachytic volcanism, which implies that the Calcaires inférieurs were deposited after 618±22 m. y. b. p. However, this radiometric age does not strongly contradict the biostratigraphic data given by Schmitt. In addition it should be mentioned that a large

percentage of recent U/Pb datings strongly opposes provable biostratigraphical data basing on trilobites (e. g. Ducrot & Lancelot, 1977, Leblanc & Lancelot, 1980).
Moreover Tucker (1986) published recent carbon isotope analyses on rocks of the Tiout section. Two positive $\delta^{13}C$ excursions (in the upper part of the Calcaires inférieurs and in the lower part of the Calcaires supérieurs) could be traced which are interpreted as increased levels of organic productivity, believed to reflect the evolutionary increase of microplancton close to the Riphean/Vendian boundary and the increased organic matter produced during the rapid metazoan evolution at the very beginning of the Cambrian (Tucker, 1986). Although the significance of some data may be questioned and nothing can be said about the precise positions of boundaries we would like to stress the similarity of the sketched stratigraphy with that of Schmitt (1979a).
Buggisch & Flügel (this volume) report the calcareous alga *Kundatia composita* Korde from the Calcaires inférieurs east of the Jbel Sirwa. The findings are thought to indicate Early Cambrian age. However, *Kundatia composita* was only known from a single area of the Siberian platform before. Its occurrence in Siberia together with algae which are only known from Early Cambrian and much younger strata, gives in any case no evidence for Cambrian age. Moreover, one may doubt a sufficient short time range of a calcareous alga to allow reliable correlation. In addition, there is no prove that the Calcaires inférieurs in the area studied by Buggisch and his collaborators are time equivalents of those in the Tiout section.

4. Cambrian shelly fossils

4a. Occurrence of very early Cambrian fossils in Morocco

The earliest trilobites known to Hupé (1953) were those of the zone of *Fallotaspis tazemmourtensis*, which is well documented only in the Tazemmourt section and the only still earlier Cambrian fossils were Archaeocyatha. About 1970 early trilobites were discovered near the Lemdad river, southern slope of the High Atlas, north of Tafinegoult and in the following years one of us (K.S.) collected trilobites from a large number of horizons. The lowest horizon which yielded *Lemdadella* Sdzuy, 1978, is immediately on top of sediments that strongly resemble the Lie de vin of the Anti-Atlas (see Sdzuy, 1978: fig. 3) and consequently it was thought that *Lemdadella* might be considerably older than the *F. tazemmourtensis* zone. Later it turned out, however, that it probably has the age of that zone.
The section where the in all probability earliest Cambrian fossils of Morocco were found is that near Tiout, northern slope of the central Anti-Atlas, southeast of Taroudant. It was carefully studied by us (see Introduction), the rock succession is typically that described in chapter 2. Regarding the problem of the base of the Cambrian, the Tiout section is the most important one in Morocco.

4b. Tommotian, Atdabanian, and the *Eofallotaspis* zone

The base of the Tommotian stage of Siberia is generally accepted as the base of the Cambrian. The identification of this base in other regions will depend upon the correlation with Siberian successions, and the only dependable method to do this is biostratigraphy. In Siberia the Tommotian is characterized by the absence of trilobites and by the presence of some archaeocyathids and brachiopods, and especially of "small shelly fossils". The latter are skeletal elements of phosphatic material partly of unknown systematic position, partly thought to represent well known phyla like molluscs.

The first undoubtedly Cambrian fossils in the Tiout section are trilobites. None of the Tommotian fossils mentioned above has been found below the first trilobites although the beds were thoroughly investigated with various methods. A few phosphatic fossils were found in association with trilobites but they are not sufficient for correlation. It should be mentioned that geochemical investigations of Monninger have shown that in sediments older than the *Eofallotaspis* zone (see below) phosphate is extremely rare.

In Siberia, the earliest trilobites occur near the base of the Atdabanian stage. Whether this justifies the correlation of the early trilobite bearing horizons of the Tiout section with the Atdabanian will be discussed below. It must be stressed, however, that a correlation on the base of undoubted specific or at least generic identity is not possible, and only a correlation of this type would be sufficiently dependable.

In the Tiout section no fossils doubtlessly representing the zone of *Fallotaspis tazemmourtensis*, the earliest zone known to Hupé, have been found. It can be demonstrated in the classical section of Tazemmourt which is just a few kilometers distant from Tiout that trilobites found near Tiout below the Série schisto-calcaire are older than this zone. Here, the Calcaires supérieurs seem to be very similar to those in the Tiout section. *F. tazemmourtensis* appears about 40 meters above the top of the limestones, so the zone named for it is well documented. Well below the top of the Calcaires supérieurs we found *Hupetina antiqua* Sdzuy, 1978 which in the Tiout section occurs in the earliest trilobite horizon, T 1. We did not measure the Tazemmourt section very exactly, but the distance from the top of the formation seems to be similar to that in the Tiout section.

Trilobites have been found in the upper part of the Calcaires supérieurs of the Tiout section in 17 horizons, T 1 – T 17 (see Sdzuy, 1978: fig. 2). Those trilobites belong to a biostratigraphical unit which in Morocco is not named yet and to which no name in use elsewhere can be applied, owing to difficulties of correlation (see below). In order to facilitate discussions we call this unit the *Eofallotaspis* zone. This name is somewhat provisional because it is probable that more than one zone will be distinguished when the faunas will be completely described.

Fossils of biostratigraphical value from the *Eofallotaspis* zone are trilobites and archaeocyatha, the latter were studied by F. Debrenne. Both groups will be discussed separately below.

4c. Very early Cambrian trilobites and correlation

Hundreds of specimens of trilobites have been collected from the *Eofallotaspis* zone of the Tiout section and all of them belong to two types: 1) Olenellids (*Eofallotaspis* Sdzuy, 1978) and 2) opisthoparians with thick ocular ridges and moderately long eyes. We deal here only with the second, and because *Bigotina* Cobbold, 1935 from Normandy is quite typical for this group we informally call it the *Bigotina* group. Drawings of typical specimens from the Tiout and the Lemdad (see above) sections were published by Sdzuy (1978: figs. 2b, 3), photos of *Bigotina* and of specimens from Tiout by the same author (figs. 13-20).

One problem of the *Bigotina* group is that on one hand there exists great similarity between trilobites we consider to belong to that group, that on the other hand clear differences can be seen, and that it is now known what the significance of similarity and of differences is. It is possible that we are dealing with a closely related group of rather low taxonomic rank (perhaps family group) and that the differences are those of species or at most of genera. But it is also quite possible that the *Bigotina* group is the stem group of opisthoparian trilobites and that phyletic lineages were separated genetically but not yet strongly morphologically and that only later in Cambrian times characters of high ranking taxa became distinct. Repina (1969) who referred trilobites, we would consider as belonging to the *Bigotina* group, to taxa existing in the late Early Cambrian was apparently influenced by the latter philosophy. -- It is obvious that on the taxonomic problems depend others regarding the use of those trilobites as index fossils.

There are still other reasons why the use of very early Cambrian trilobites as index fossils is problematical. It is obvious that the first trilobites must have developed from some unknown ancestors in an unknown region and that from there the early trilobites must have spread over the oceans. If this is true, then the primitive trilobites (perhaps the *Bigotina* group) must have appeared in different regions at different times. There is still another problem: Migration implies factors strongly influencing evolution like geographic isolation of small populations, change of environment, etc. The result must have been a rapid genetic change and therefore it must be expected that primitive trilobites of different regions that are closely related cannot be recognized as such. This, of course, is also true for other fossil groups under similar conditions. -- The use of very early Cambrian fossils as index fossils for long distance correlation is even more problematical, if it is true that changes of environment influenced the regional distribution of life near the base of the Cambrian (see 4e).

4d. Archaeocyatha

Archaeocyatha are common fossils in most of the Lower Cambrian rocks of Morocco. Im-

portant localities with archaeocyathan bioherms from the Série schisto-calcaire are known from north of the Jbel Taïssa (Debrenne, 1958; 1961), from Tazemmourt (Debrenne, 1960), and from most probably equivalent strata of the Lemdad area.
Notes on archaeocyatha from the Calcaires supérieurs were published for the Tazemmourt section (Debrenne, 1960) and for the section of Tiout. From that part of the Tazemmourt section only one species was described (*Coscinocyathus minutus* Debrenne, 1960), but Hupé (1959) and e. g. Boudda et al. (1979) mentioned several horizons bearing archaeocyatha.
At Tiout archaeocyatha appear also in numerous horizons within the *Eofallotaspis* zone (Debrenne, 1975; Debrenne & Debrenne, 1976; Debrenne, 1976; Debrenne & Debrenne, 1978). Seven species are known only from this section, five species were already described from other Moroccan localities; most of them have a wide stratigraphical range. "*Aldanocyathus biohermicus* (Zhuravleva, 1960)?" is similar to a Siberian species from the Atdaban and Lena stages, but like *Dictyocyathus*? sp. very poorly preserved and thus its recognition is doubtful. *Protopharetra circula* var. *colonialis* (Debrenne, 1964) is also known from Sardinia, Spain, and Siberia, but has a long time range, as well. Therefore, the correlation with the Siberian stages has to be based upon genera, not species. Debrenne & Debrenne (1978) regard the archaeocyatha bearing horizons of the Calcaires supérieurs of Tiout as probably not very early Atdabanian. This is founded essentially on the appearance of *Neoloculicyathus*, *Tumulifungia*, and *Gordonifungia*. Of these genera, only *Tumulifungia* was described from more than one region before, but the Moroccan specimens differ largely in morphology from the Siberian species. Therefore biostratigraphical correlations of the *Eofallotaspis* zone based on archaeocyatha appear to be not very reliable in the present state of knowledge.

4e. Early Cambrian radiation event and/or change of environment?

The stratigraphical problems we are confronted with when dealing with the base of the Cambrian are certainly caused or at least influenced by the fact that we are dealing with a time span that is unique in the Phanerozoic: It is the time when animals bearing hard skeletons are documented for the first time, and the fact that their Precambrian ancestors are still unknown creates one of the greatest enigmas of Earth history. They are unknown in spite of a tremendous amount of research on the late Precambrian in the last decades and this led to the assumption of a rapid evolution, of a radiation of animal life, near the beginning of Cambrian time. We will not discuss this, but we want to point out that the mode of appearance of hard-part animals in early Cambrian time cannot be explained by a radiation alone.
The first Cambrian fossils did not really appear at once, but obviously they first appeared in very limited regions and then needed a considerable time to spread over the

early Cambrian seas. Very few sections with Tommotian fossils are known, but there are many in which the existence of Tommotian sediments without fossils may be suspected, and the more we ascend in the Cambrian successions the more widely are fossils spread. It is obvious that in early Cambrian time, there existed vast regions where sedimentation occurred and where either animals with hard parts did not exist or where their rests were not fossilized. The most likely explanation is that in those regions the environment was not favourable for animal life, and it is quite likely that similar environments existed in late Precambrian time.

Monninger's (1979) sedimentological investigations of the Lie de vin and the Calcaires supérieurs of the Tiout section have clearly shown that there was a very long period in which the environment was exceptional: There are various indicators for very shallow water (the most obvious being the many stromatolites), yet there are no signs of water agitation, currents apparently were very slow and there was little influx of terrigenous material. The connection between the basin and the open ocean probably was partly interrupted, there were periods of hypersalinity, and the extensive development of stromatolites makes us suspect that the water was oversaturated with oxygen. It apparently was an environment not favourable to marine animal life.

With the appearance of the first trilobites a slight change of the rock composition occurs, the most notable being the appearance of oolitic limestones. Obviously water agitation and currents were markedly stronger than before, and this probably led to normal marine conditions that were favourable to benthic animals. A change of environment is also indicated by the results of Tucker (1986) who investigated carbon isotopes in the Tiout section.

Stromatolites were enormously widespread in the late Precambrian and it is quite possible that conditions unfavourable to benthic animals were widespread too. If this is true then the correlation of the base of the Cambrian will be even more problematical than for the reasons discussed above. Conditions unfavourable to animals in the Precambrian should be considered as a possible explanation for the puzzling fact that the ancestors of the Cambrian faunas are still unknown. -- There is still another aspect regarding life conditions near the base of the Cambrian and stromatolites: It is an open question when the process of changing the atmosphere of earth to its present state was terminated and there are authors that suppose that this was as late as some time within the Phanerozoic. The stromatolites, however, prove that assimilation by organisms occurred on a large scale over hundreds of millions of years before the beginning of Cambrian times, and we think it is quite impossible to imagine that the change of the atmosphere to its present state was not finished long before this time.

5. Conclusions

The trilobites from the *Eofallotaspis* zone of Tiout (and of Tazemmourt) are the earliest trilobites known in Morocco and their appearence apparently was made possible by a change of environment. For reasons which we have discussed, a long-distance correlation of the *Eofallotaspis* zone is not possible and therefore we cannot decide whether its age is early Atdabanian (we would exclude a younger age) or whether it goes down into the Tommotian. The presence of trilobites certainly is no proof against Tommotian age.

Later Early Cambrian faunas from Morocco and similar ones from Spain (Sdzuy, 1962) can be correlated with Siberian ones and it can be shown that the zones IV and V of Hupé are uppermost Atdabanian. We think it to be probable then that the *Eofallotaspis* zone and the earliest four or five zones of Hupé together comprise a time span that is at least equal to the Atdabanian.

The base of the Cambrian cannot be located with precision in the Tiout section. Evidence from stromatolites points to a position near the base of the Calcaires supérieurs, but it is no firm evidence. A position of the base of the Cambrian far below the Calcaires supérieurs has been suggested repeatedly. We cannot exclude that, but on the other hand we do not think that it is of great importance whether rocks that cannot be well dated are regarded as Vendian or as Cambrian.

6. References

Bertrand-Sarfati, J. (1981): Problème de la limite Précambrien-Cambrien: la section de Tiout (Maroc); les stromatolites et leur biostratigraphie (SCHMITT 1979): critiques et observations. - Newsl. Stratigr., **10**(1): 20-26, 1 Tab.; Berlin, Stuttgart.

Boudda, A. (1968): Zur geologischen Kartierung 1 : 25000 im Gebiet von Ounein (Massif ancien du Haut-Atlas, Marokko). - Masters Thesis, Paleont. Inst. Univ. Würzburg: 52 pp., 35 figs., 2 pl.; Würzburg (unpubl.).

Boudda, A. & Choubert, G. (1972): Sur la limite inférieur du Cambrien au Maroc. - C. R. Acad. Sci. Paris, D **275**: 5-8; Paris.

Boudda, A., Choubert, G. & Faure-Muret, A. (1979): Essai de Stratigraphie de la Couverture sédimentaire de l'Anti-Atlas: Adoudounien - Cambrien Inférieur. - Notes Mém. Serv. géol Maroc, 271: 96 pp., 20 figs.; Rabat.

Boukhari, A., Choubert, G., Faure-Muret, A. & Timofeiev, B. V. (1979): Sur la microflore des couches terminales du Précambrien III et sur l'âge de l'Adoudounien au Maroc. - C. R. Acad. Sci. Paris, **288**: 13-14; Paris.

Bourcart, J. (1927): Découverte du Cambrien à Archaeocyathus dans l'Anti-Atlas marocain. - C. R. somm. Soc. géol. Fr., (1-2): 10-11; Paris.

Bourcart, J. & Le Villain, G. (1928a): Sur le gisement à Archaeocyathus de Sidi Mouça d'Aglou (Anti-Atlas marocain). - C. R. somm. Soc. géol. Fr., (10): 115; Paris.

--- (1928b): Sur la faune des calcaires cambriens de Sidi Mouça d'Aglou près de Tiznit (Sud marocain). - C. R. Acad. Sci. Paris, **187**: 1066-1068; Paris.

--- (1931): La faune des calcaires cambriens de Sidi Mouça d'Aglou (Anti-Atlas marocain). - Notes Mém. Serv. Mines Carte géol. Maroc, 15: 44 pp., 1 tab., pl. I-X; Macon.

Buggisch, W. & Heinitz, W. (1984): Slumpfolds and other early diagenetic deformations in the early Cambrian of the Western and Central Antiatlas (Morocco). - Geol. Rdsch., **73**(2): 809-818, 8 figs., 1 tab.; Stuttgart.

Buggisch, W., Marzela, C. & Hügel, P. (1979): Die fazielle und paläogeographische Entwicklung der infrakambrischen bis ordovizischen Sedimente im Mittleren Antiatlas um Agdz (S-Marokko). - Geol. Rdsch., **68**(1): 195-224, 21 figs., 1 tab.; Stuttgart.

Choubert, G. (1942): La constitution et la puissance de la série primaire de l`Anti-Atlas. - C. R. Acad. Sci. Paris, **215**: 445-447; Paris.

--- (1943): Sur le Géorgien de l`Anti-Atlas. - C. R. Acad. Sci. Paris, **216**: 69-70; Paris.

--- (1952a): Le volcan géorgien de la région d`Alougoum (Anti-Atlas). - C. R. Acad. Sci. Paris, **234**: 350-352; Paris.

--- (1952b): Histoire géologique du domaine de l`Anti-Atlas. In: Choubert, G. & Marçais, J.: Géologie du Maroc. - XIXe Congrès Géol. Intern., Alger 1952, Monogr. rég., Ser. 3, Maroc, 6: 77-194, 16 maps; Alger.

--- (1953): Le Précambrien III et le Géorgien de l`Anti-Atlas. Introduction stratigraphique. - In: Hupé, P.: Contribution à l`étude du Cambrien inférieur et du Précambrien III de l`Anti-Atlas marocain. Notes Mém. Serv. géol. Maroc, 103: 17-39, 4 figs.; Bagnolet/Seine ("1952").

--- (1958): L'Adoudounien et le Precambrien III dans l'Anti-Atlas. - In: Les Relations entre Précambrien et Cambrien. Problèmes des séries intermédiaires. Coll. intern. Centre Nat. Rech. Sci., Paris, 1957, **LXXVI**: 143-162, 2 figs., 1 tab.; Paris.

--- (1963): Histoire géologique de Précambrien de l`Anti-Atlas. Tome 1. - Notes Mém. Serv. géol. Maroc, 162: 352 pp., 33 figs., 81 Phot., 5 maps, 7 colour maps; Rabat.

--- (1964): Nouvelles précisions sur la stratigraphie des formations calcaires et schisto-calcaires du Cambrien inférieur de l`Anti-Atlas. - C. R. Soc. géol. France, (8): 337-338; Paris.

Choubert, G., Boudda, A & Faure-Muret, A. (1975): Essai de stratigraphie de la couverture sedimentaire de l`Anti-Atlas: Adoudounien - Cambrien. - Notes Mèm. Serv. géol Maroc, 271: 95 pp., 20 figs.; Rabat (1979). & Reunion Lim. Cambrien-Précambrien, Agadir-Rabat, Dok. 1A: 57 pp.; Rabat (unpubl.).

Choubert, G., Dresnay, R. du & Hindermeyer, J. (1950): Sur les calcaires à Collenia de la région de Safsaf-Aïn Chair. - Notes Mém. Serv. géol. Maroc, **3**, 76: 93-103; Rabat.

Choubert, G., Hindermeyer, J. & Hollard, H. (1952a): Sur la présence de Collenia dans le Précambrien III et le Géorgien de l`Anti-Atlas. - C. R. Acad. Sci. Paris, D **234**: 543-545; Paris.

--- (1952b): Note préliminaire sur le Collenia de l`Anti-Atlas. - Notes Mém. Serv. géol. Maroc, 95: 85-102; Rabat.

Clauer, N. (1980): Nouvelles précisions sur la géochronologie isotopique et biostratigraphique du Précambrien de l'Anti-Atlas au Maroc. - C. R. somm. Soc. géol. Fr., (3): 84-87; Paris.

Debrenne, F. (1958): Sur quelques Archaeocyatha du Jebel Taissa (Anti-Atlas occidental). - Notes Serv. géol. Maroc, **16**, 143: 59-73, 2 figs., 2 pl.; Rabat.

--- (1960): Archaeocyatha des lentilles calcaires de Tazemmourt (Anti-Atlas). - Notes Mem. Serv. Géol. Maroc, **18**, 147: 7-26, pl. I-V; Rabat.

--- (1961): Nouvelles données sur la faune d'Archaeocyatha du Jbel Taissa (Anti-Atlas occidental). - Notes Serv. géol. Maroc, **20**, 152: 7-37, pl. I-VI; Rabat.

--- (1964): Archaeocyatha. Contribution à l'étude des faunes cambriennes du Maroc, de Sardaigne et de France. - Notes Mém. Serv. géol. Maroc, 179: 371 pp., 52 pl.; Rabat.

--- [Debrenn] (1976): Drevnejshie arkheotsiaty Marokko. - Izv. Akad. Nauk SSSR, Seriya Geol., 12: 157-159, fig.; Moscow.

Debrenne, F. & Debrenne, M. (1976): Archaeocyathid fauna of the lowest fossiliferous levels of Tiout (Lower Cambrian - Southern Morocco). - Réun. Gr. Trav. Limite Précamb.-Camb. Maroc, Mars-Avril 1976, C. R. activité: 22 pp., pl. I-IV; Rabat (unpubl.).

--- (1978): Archaeocyathid fauna of the lowest fossiliferous levels of Tiout (Lower Cambrian - Southern Morocco). - Geol. Mag., **115**(2): 1101-119, 3 tab., pl. 1-4; London.

Ducrot, J. & Lancelot, J. R. (1977): Problème de la limite Précambrien-Cambrien. Étude radiochronologique par la méthode U-Pb sur zircons du volcan du jbel Boho (Anti-Atlas marocain). - Can. J. Earth Sci., **14**: 2771-2777; Ottawa.

Geyer, G. (1986): Mittelkambrische Mollusken aus Marokko und Spanien. - Senckenbergiana leth., **67**(1/4): 55-118, 6 figs., 1 tab., 6 pl.; Frankfurt a. M.

Heinitz, W. (1984): Die Deformation des sedimentären Deckgebirges im zentralen Anti-Atlas (Südmarokko). - Berliner geowiss. Abh., (A), 55: 84 pp., 64 figs.; Berlin.

Hollard, H. (1985): Precambrian-Cambrian Boundary and Lower Cambrian. - In: Destombes, J., Hollard, H. & Willefert, S.: Lower Palaeozoic Rocks of Morocco. In: Holland, C. H. (ed.): Lower Palaeozoic Rocks of the World, Vol. 4: Lower Palaeozoic of north-western and west central Africa: 101-157, figs. 5-28, pl. 1-4; Chichester, New York, Brisbane, Toronto, Singapore (J. Wiley & Sons).

Hupé, P. (1952): Sur les zones de Trilobites du Cambrien inférieur marocain. - C. R. Acad. Sci. Paris, **235**: 480-481; Paris.

--- (1953): Contribution à l'étude du Cambrien inférieur et du Précambrien III de L'Anti-Atlas marocain. - Notes Mém. Serv. géol. Maroc, 103: 402 pp., 99 figs., 4 tab., pl. I-XXIV; Bagnolet/Seine ("1952").

--- (1959): Nouvelle contribution à l'étude du Cambrien marocain. - Thèse Sci. Fac. Sci. Univ. Paris, 447 pp., 12+5 figs., 3 maps, pl. A-W, 1-10, 10bis, 11-32; Paris (unpubl.).

--- (1960): Sur le Cambrien inférieur du Maroc. - Rep. XXIth int. geol. Congr., Norden 1960, **8**, 8. Late Pre-Cambrian and Cambrian stratigraphy: 75-85, tab.; Kopenhagen.

Liñan, E. & Sdzuy, K. (1978): A trilobite from the Lower Cambrian of Córdoba (Spain) and its stratigraphical significance. - Senckenberg. leth., **59**(4-6): 378-399, 2 figs., 1 pl.; Frankfurt a. M.

Monninger, W. (1979): The section of Tiout (Precambrian/Cambrian boundary beds, Anti-Atlas, Morocco): An environmental model. - Arb. Paläont. Inst. Würzburg, 1: 289 pp., 49 figs., 32 tab., 22 pl.; Würzburg (unpubl.).

Monninger, W. & Schmitt, M. (1975): The section of Tiout: Lithology and algal structures. - Réun. Gr. Trav. Limite Précambrien/Cambrien, Agadir-Rabat, 1975: 8 pp.; Rabat (unpubl.).

Repina, L. N. (1969): Trilobity nizhnego kembriya i srednego kembriya yuga Sibiri (nadsemejstvo Redlichioidea). Chast' II. - Akad. Nauk SSSR, Sib. Otdel., Trudy Inst. Geol. Geofiz., **67**: 107 pp., 52 figs., pl. I-IV; Moscow.

Schmitt, M. (1978): Stromatolites from the Tiout section, Precambrian-Cambrian boundary beds, Anti-Atlas, Morocco. - Geol. Mag., Special issue, **115**(2): 95-99, 3 figs., pl.; Cambridge.

--- (1979a): The Section of Tiout (Precambrian/Cambrian Boundary beds, Anti-Atlas, Morocco): Stromatolites and their Biostratigraphy. - Arb. Paläont. Inst. Würzburg, 2: 188 pp., 44 figs., 23 pl.; Würzburg (unpubl.).

--- (1979b): New stromatolites from the Late Precambrian of the Anti-Atlas and from the Lower Cambrian of the High Atlas, Morocco. - Senckenberg. lethaea, **60**(1/3): 39-49, 7 figs., 2 pl.; Frankfurt a. M.

Schmitt, M. O. & Monninger, W. E. W. (1977): Stromatolites and Thrombolites in Precambrian/Cambrian Boundary Beds of the Anti-Atlas, Morocco: Preliminary Results. - In: Flügel, E. (ed.): Fossil Algae: 80-85, 2 pl.; Berlin, Heidelberg (Springer).

Sdzuy, K. (1962): Trilobiten aus dem Unter-Kambrium der Sierra Morena (S-Spanien). - Senckenberg. leth., **43**(3): 181-229, 4 figs., pl. XVIII-XXIII; Frankfurt a. M.

--- (1978): The Precambrian-Cambrian boundary beds in Morocco (Preliminary Report). - Geol. Mag., **115**(2): 83-94, 6 figs., 1 pl.; Cambridge.

--- (1981): Der Beginn des Phanerozoikums - Paläobiologische und stratigraphische Probleme. - Natur und Museum, **111**(11): 390-399, 20 figs.; Frankfurt a. M.

Semikhatov, M. A. (1977): Méthode paléontologique dans la corrélation du Précambrien supérieur des régions péri-atlantiques de l'Europe, de l'Afrique et de l'Amérique. - Estud. Geologicos, **33**: 365-372; Madrid.

Tucker, M. E. (1986): Carbon isotope excursions in Precambrian/Cambrian boundary beds, Morocco. - Nature, **319**(6048): 48-50, 2 figs.; New York.

PALEOGEOGRAPHY AND FACIES OF THE 'GRES TERMINAUX'
(UPPERMOST LOWER CAMBRIAN, ANTI-ATLAS/MOROCCO)

Werner Buggisch & Rainer Siegert
Institut für Geologie der Universität, Schloßgarten 5, D-8520 Erlangen

Abstract

The facies and paleogeography of the uppermost Lower Cambrian of the Anti-Atlas is described. Five different paleogeographic units were recognized: Fluviatile sandstones in the east, continental to marine facies in the center, marine sediments and volanic rocks in the northwest and interfingering of shallow marine sediments with the 'Delta of Asrir' in the southwest. The Ait Ounein section is separated from the Anti-Atlas by the High Atlas boundary fault.

1. Introduction

Above the first Lower Cambrian regressive megacycle of the Anti-Atlas (Série des calcaires inférieurs to Série lie de vin), a second sedimentary cycle starts with the transgression of the Série des calcaires supérieurs. Abundant

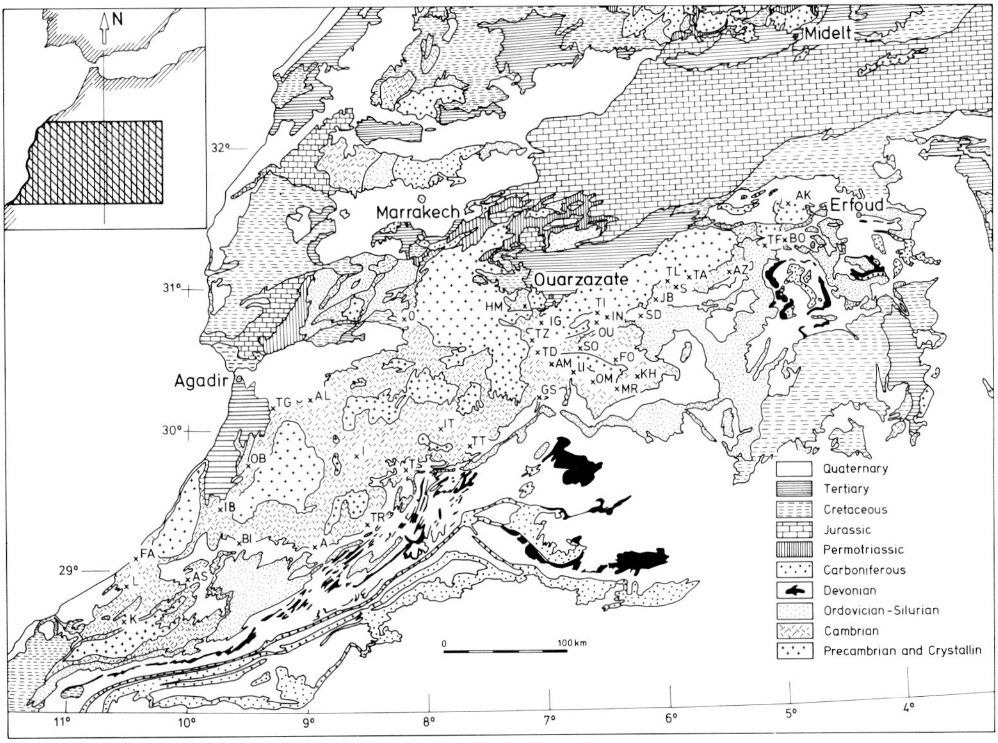

Fig. 1. Map of the Anti-Atlas showing the studied sections.

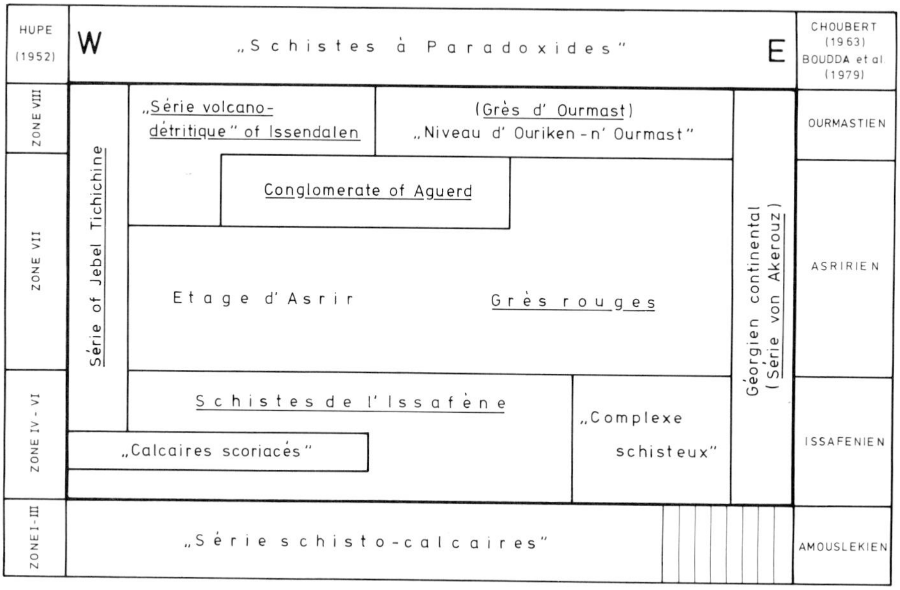

Tab. 1. Stratigraphy of the uppermost Early Cambrian in the Anti-Atlas.

trilobites and archaeocyathid bioherms in these Lower Cambrian sediments are evidence of an open marine environment in the western Anti-Atlas up to the central Anti-Atlas (particularly in the black limestones of the "calcaires de base'). The overlying series (schisto-calcaires, complexe schisteux and grès terminaux) mark the second regression in the Lower Cambrian prior to the Middle Cambrian transgression, which covered most parts of the Anti-Atlas.

This paper describes the lithology and facies of the clastic sediments below the Middle/Upper Cambrian boundary and outlines paleogeographic maps and facies models of the western and central Anti-Atlas. This investigation is based on 40 sections (Fig. 1) and several hundred samples and thin-sections.

2. Previous work

The first survey of the Cambrian in the Anti-Atlas by NELTNER (1938) established the stratigraphic outlines, which are still valid today. Based on trilobites, HUPE(1952) subdivided the Lower Cambrian into different zones and classified the clastic sediments below the Lower/Middle Cambrian boundary as Zone VI to VIII. The term 'grès terminaux' was established by CHOUBERT (1952) for the sandstones of the trilobite Zone VII. In 1963, he named the sedimentary time point of the grès terminaux 'étage d'Asrir', referring to the type locality. The term 'niveau à Micmacca' or 'niveau d'Ourmast' was used for the Lower/Middle Cambrian transition beds. Volcanoclastic intercalations in these sequences were described in the western Anti-Atlas; in the High Atlas they were described by BOUDDA et al. (1974).The biostratigraphy of the Lower to Middle Cambrian boundary beds in the western Anti-Atlas was studied in detail by GEYER (1984).

3. Lithostratigraphy

The clastic sequences below the Lower/Middle Cambrian boundary are not at all uniform. The variety in lithology

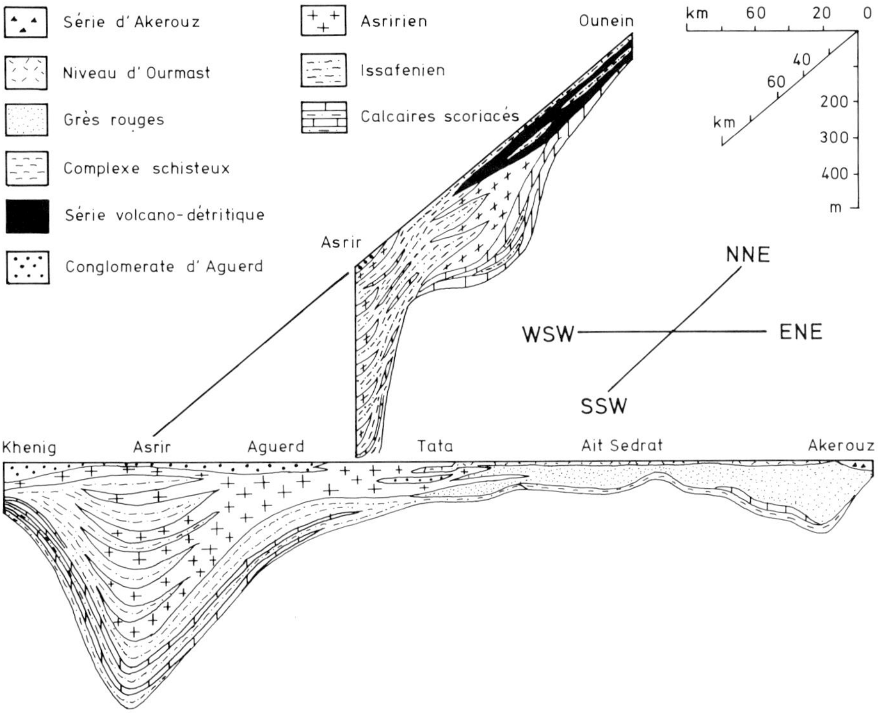

Fig. 2. Sections across the depositional area of the 'grès terminaux' in the Anti-Atlas.

documents a differentiated depositional area with an interfingering of terrestrial and shallow marine sediments.

In order to classify the different units, new terms will be introduced (underlined in Tab. 1) and existing terms will be defined more precisely in this paper. The term 'grès terminaux' is applied as a generic term for sandstones of the trilobite Zone VII.

Five different paleogeographic units were recognized within the trilobite Zone VII (HUPE 1952) in the Anti-Atlas:

(1) The eastern Anti-Atlas (between Foum Zguid/Tazenakht and Erfoud). This area is characterized by fluviatile sandstones.

(2) The central Anti-Atlas (between Tata and Foum Zguid) with the transition from continental to marine facies.

(3) The northwestern Anti-Atlas (region of Tiznit-Taroudant) distinguished by marine sediments and thick volcanic sequences.

(4) The southwestern Anti-Atlas (Tan-Tan, Goulimime, Bou Izakarn) characterized by the interfingering of the 'Delta of Asrir' with shallow marine sediments.

(5) The Ait Ounein section is situated north of the southern Atlas boundary fault and cannot be clearly correlated with the sections of the Anti-Atlas.

3.1. Eastern Anti-Atlas

A 70 m thick sequence of white and grey sandstones alternating with red-violet mudstones is exposed in the easternmost section of the Anti-Atlas about 60 km WNW of Erfoud. This Série d'Akerouz (Géorgien continental) consists of laminated or sometimes cross-bedded sandstones. The bed thickness is about one meter. The grain-size parameters (mean = 0.06-1.0 mm; sorting and skewness) indicate transportation and sedimentation in slowly moving water. Pedogenic broken quartz-grains (Pl. 1, Fig. 1) floating in a carbonate matrix indicate temporary sedimentation above the ground water table under arid conditions (caliche).

The upper Lower Cambrian of the other sections of the eastern Anti-Atlas are characterized by the succession:

grès d´Ourmast
grès rouges
complexe schisteux

Red-violet mudstones, siltstones and thin-bedded fine-grained sandstones are typical of the complexe schisteux. Frequent ripple marks, small-scale cross-bedding, desiccation cracks and casts of halite argue for a depositional environment with occasional water currents, high salinity and abundant desiccation (lagoon to salt marsh, REINECK 1985).

3.1.1. Grès rouges

The grès rouges are the typical 'grès terminaux' in the eastern Anti-Atlas. They consist of light-red, thin-bedded, 20 – 300 cm thick layers of fine-grained sandstones forming up to 15 m thick units. These units wedge out laterally after short distances (maximum width about 300 m). Erosion channels with a few meters in width (point bars), layers

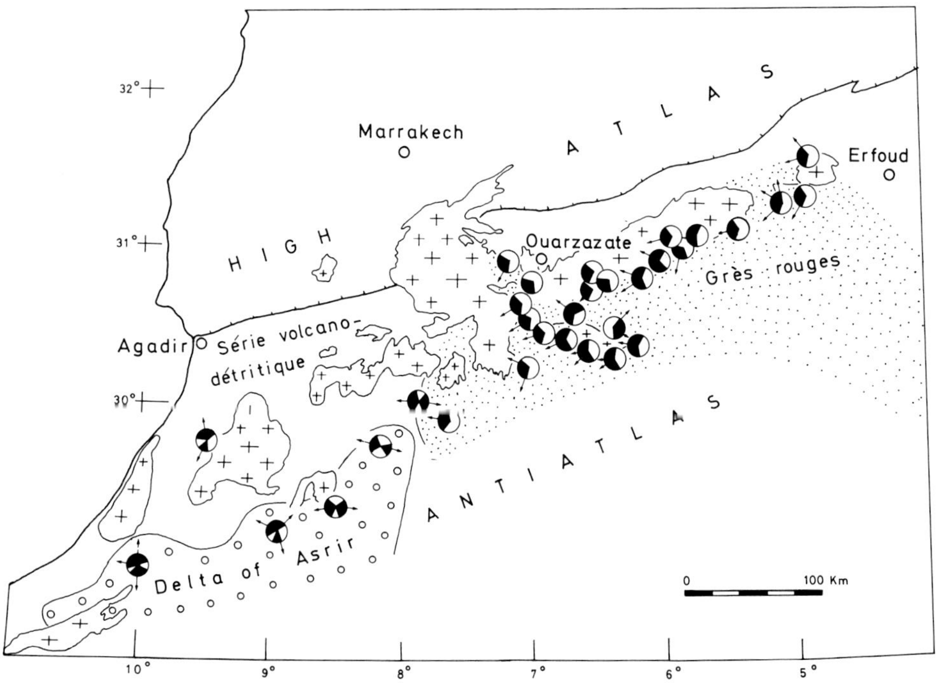

Fig. 3. Paleocurrent pattern according to the interpretation of cross-bedding orientation.

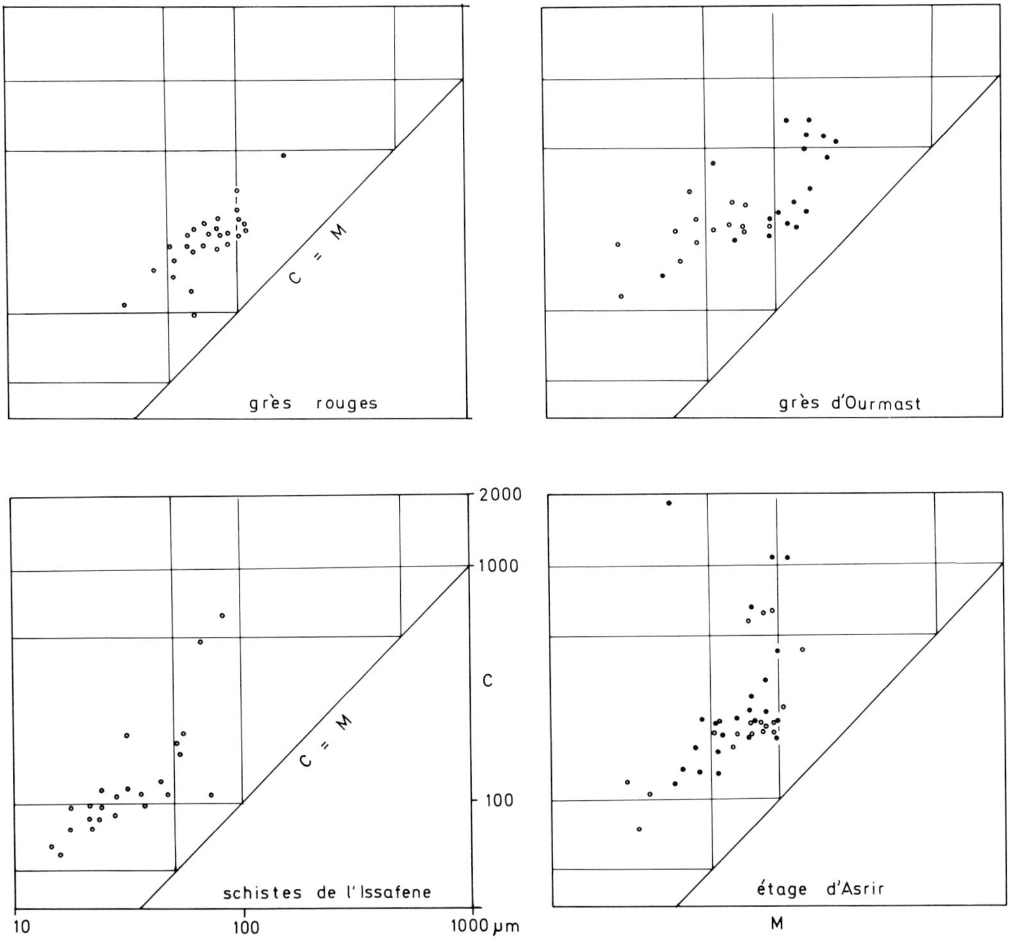

Fig. 4. CM-patterns in the Passega-diagram.

with mud pebbles and asymmetric current ripples with a wave length of a few centimeters up to more than one meter are frequently observed.

Arched trough-shaped cross-bedded sand bodies, several meters wide and about 50 cm thick, are typical of the grès rouges. Statistical measurements of cross-beddings (Fig. 3) indicate a W or WSW directed transport in outcrops at the southern margin of the Jebel Sarhro and W or WNW directed transport in the vicinity of the anticline (boutonnière) of Bou Azzer - El Graara.

The sandstones are fine-grained (Md 0.06-0.07 mm), sorting is good to medium with a predominance of the fine-grained fraction; roundness ranges from rounded to subrounded. The pattern of the CM-diagram ranges from 'uniform' transport to 'graded suspension' (Fig. 4).

According to the diagram of PETTIJOHN et al. (1973) the sandstones of the grès rouges are classified as 'arcosic arenites' or 'subarcoses' (Fig. 5, Pl. 1, Figs. 2,3). The quartz content increases and the matrix content decreases from E to W. About 50 - 60% of the sandstones consist of quartz. Metamorphic quartz grains are unimportant in comparison with the non-metamorphic volcanic quartz with sharp extinction and resorption features which make up the bulk of the grains. At the southern margin of the anticline of the Jebel Sarhro, only 25 to 30% of

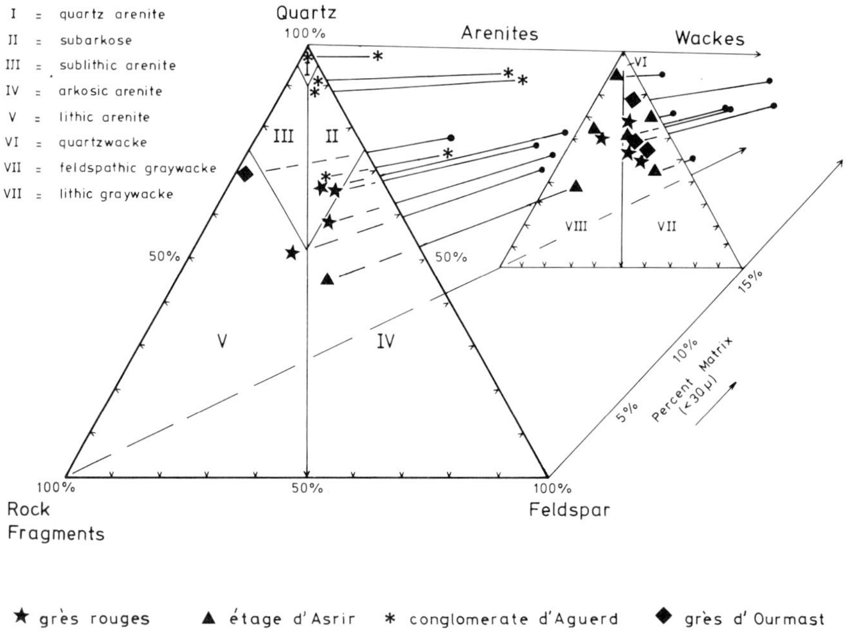

Fig. 5. Classification of the upper Lower Cambrian sandstones.

the quartz grains exhibit undulatory extinction, whilst in the vicinity of the anticline of Bou Azzer - El Graara the amount of undulatory quartz grains is about 40%.

Feldspars (plagioclase, orthoclase and microcline) make up about 20 - 30% of the sandstones. Rock fragments (up to 10%) consist of metamorphics and plutonics.

The grès rouges are cemented by syntaxial overgrowth of detritic quartz and feldspar grains (Pl. 1, Figs. 2, 3).

Two different source areas probably existed in the NE and SE according to the data from current directions and the modal composition of the grès rouges. The sandstones were deposited in braided river systems. The thickness of the sections Tazlaft and Talioulat (Fig. 2) indicates an increased rate of subsidence in this area. The base of these sections consists of bioturbated calcareous arenites; typical cross-bedding is missing. The thick sequences of the eastern Anti-Atlas probably represent lake deposits. Recent braided rivers are usually characterized by the dominance of coarse-grained sand and gravel (REINECK & SINGH 1980), but fine-grained sediments are described as well, for instance from the Brahmaputra and the Mekong river (COLEMAN 1969; MORGAN 1970).

3.1.2. Grès d'Ourmast

Dark or black, medium-grained to coarse-grained sandstones interbedded with thin green-black to dark-violet mudstones occur at the base of the 'niveau d'Ourmast'. This formation is named 'grès d'Ourmast' in this paper. It is overlain by light-green mudstones and coquinas of fossils of the 'brècche à Micmacca'.

The thickness of the grès d'Ourmast varies considerably (Fig. 6). The orientation of cross-bedding suggests a NE - SW directed pattern of usually unimodal paleocurrents ; a NW - SE transport is measured in one section (IG = Ighels 21 Km NE Tazenakht). Asymmetric ripples and mud pebbles are common. Bioturbation and grazing trails can be observed in the upper part of the grès d'Ourmast. These horizons are partly cemented by carbonates and they

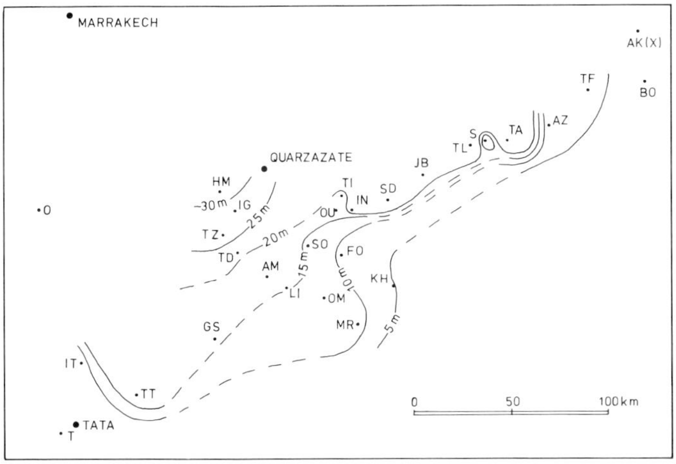

Fig. 6. Thickness of the grès d'Ourmast.

may contain fossil fragments (trilobites, archaeocyathids and echinoderms).

A different facies of the grès d'Ourmast is exposed in the Targant section (TT). Lenticular and flaser bedding and small-scale cross-bedding of mudstones and fine-grained sandstones are attributes of a muddy to sandy tidal flat (REINECK 1972).

The grain-size is not uniform: rolling and traction transport and transport in uniform to graded suspension is documented in CM-diagrams (Fig. 4). Whereas rolling transport dominated in the eastern Anti-Atlas, water energy decreased towards the central Anti-Atlas. In this area, fluvial transport was obviously unimportant in comparison with transport by waves and marine currents.

The grès d'Ourmast are composed of dominating undulate quartz grains, not-undulate quartz, rock fragments (20 %), feldspars (10-20%) and abundant matrix (up to 30%). According to the diagram (Fig. 5), the grès d'Ourmast consist of lithic and arcosic arenites and arcosic wackes.

The depositional environment of the grès d'Ourmast was predominantly fluviatile in the E with increasing marine influence toward the W and toward the top of the sections (beach and delta sediments in the central Anti-Atlas and tidal flat deposits at Targant).

3.2. Central Anti-Atlas

The facies of the grès rouges interfingers with the sediments of the étage d'Asrir in the central Anti-Atlas. A change in color from red to green is combined with the facies change. Cross-bedding measurements give evidence of a SW transport direction in the E, whereas the bimodal paleocurrent patterns in the sections IT and TT are indicative of an estuary or probably tidal-dominated delta environment.

3.3. Southwestern Anti-Atlas

The sedimentation of the southwestern Anti-Atlas is mainly influenced by the 'Delta of Asrir'. The following facies units and formations can be distinguished:

1. Basin sediments	'schistes de l'Issafène'
	'Série de Jebel Tichinchine'
2. Bar and platform sediments	'Calcaires scoriacés'

3. Delta sediments: 'étage d´Asrir'
'conglomerate d´Aguerd'

3.3.1. Schistes de l'Issafène

The up to 350 m thick sequence consists of 25 m thick units of green mudstones and up to 3 m thick limestones. Crystal tuffs occur in some layers . The fine-grained sediments are usually laminated, occasionally cross-bedded and frequently bioturbated. According to the CM-diagram (Fig. 4), the sediment transport is to be classified as 'pelagic' to 'uniformly suspended' (PASSEGA 1964). The mudstones and siltstones are composed of quartz and feldspar grains floating in a matrix of carbonate, chlorite and sericite. The cystal tuffs consist of plagioclase, sanidine, orthoclase and rock fragments. The feldspars are altered to sericite and carbonates.

The reddish brown, massive, meter-thick sandy carbonate units consist of quartz and feldspar containing biosparites with trilobites, echinoderms and archaeocyathids (Pl. 1, Fig. 4). Chlorite and oxychlorite are neomorphic in fossil shells.

The schistes de l´Issafène are open marine basin sediments deposited below the wave base.

3.3.2. Etage d'Asrir

The étage d´Asrir can be correlated with the grès rouges. The up to 70 m thick basal sandstones are overlain by green mudstones in alternation with sandstones. The étage d´Asrir is overlain by the conglomerate of Aguerd or by the Série volcano-détritique.The transition to the Middle Cambrian 'schistes à Paradoxides' is continuous in some sections of the western Anti-Atlas.

While cross-bedding shows bimodal, E-W directed current patterns in the sections IT, T and TZ, no clear direction can be recognized in the western sections. The mean of the grain sizes varies from silt to coarse sand; the sorting is poor to good. All possibilities from rolling transport to uniform suspension are shown in the CM-diagram (Fig. 4). Accordingly, the modal composition is very inhomogeneous.

Corresponding to the variability of the étage d'Asrir, the sediments were deposited in different environments as follows:

S Anti-Atlas:	predominant lagoon
SW Anti-Atlas:	prodelta to delta platform
NW Anti-Atlas:	shelf to shore, pyroclastics

The source area of the 'delta of Asrir' is situated south of the Anti-Atlas where mainly metamorphic rocks and intermediate to basic volcanics were eroded. It is still an open question as to where the source area for the sediments of the northwestern Anti-Atlas was situated. Local emersion combined with erosion evidence the redistribution of sediments in this area.

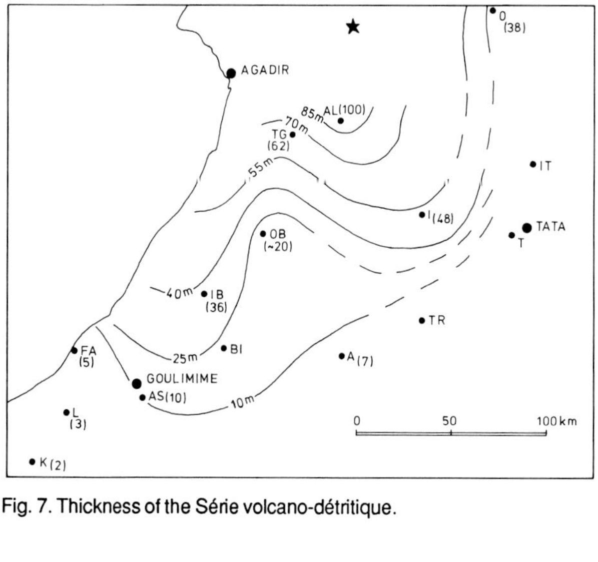

Fig. 7. Thickness of the Série volcano-détritique.

3.3.3. Conglomerate of Aguerd

In the sections Aguerd (A), Asrir (AS), Khenig (K), and Tata (T) the étage d'Asrir is overlain by a 10 - 20 m thick sequence of massive quartzitic coarse sandstones and conglomerates. Planar and through-shaped cross-beds and scour fills indicate NE to NW directed paleocurrents. 65 - 95% of the components consist of quartz (recrystallized metaquartzites, volcanic quartz grains with embayments); feldspars and rock fragments are less abundant (Pl. 1, Fig. 5).

The basal units of the conglomerate d'Aguerd were deposited in a littoral environment evidenced by the occurrence of *Scolithos*. Sedimentary structures and granulometric parameters substantiate a fluviatile environment for the sedimentation of the upper units. Therefore, the conglomerates of Aguerd probably represent the topsets of the Delta of Asrir.

3.4. Northwestern Anti-Atlas and Ait Ounein

Only marine environment is known in the northwestern Anti-Atlas for the late Upper Cambrian and the Middle Cambrian period.

3.4.1. Calcaires scoriacés

The approximately 100 m thick reddish brown limy marls contain a rich fauna. A direct terrigenous influx cannot be observed; the few lithoclasts are of volcanic origin. Possibly, the calcaires scoriacés were deposited on a submarine rise which separated the depositional area of the schistes de l'Issafène from the open marine.

3.4.2. Series of Jebel Tichinchine

The depositional area of the series of the Jebel Tichinchine is situated between the Delta of Asrir and the open marine northwestern Anti-Atlas. Green slaty mudstones with intercalated decimeter-thick limestones alternate with thin fine-grained sandstones. The increasing abundance of crystal tuffs in the upper part of the section is indicative of a correlation with the Série volcano-détritique.

Biosparites and oolites (section BI) confirm a sedimentation in an open marine shallow water at the base of these sections. Other sedimentary structures indicate deposition below the wave base and a reduced environment within the sediment (pyrite). The area of deposition probably represents a partly restricted offshore basin.

3.4.3. Série volcano-détritique

The Lower Cambrian section Amouslek (AL) is well known for its abundance of fossils. The section is situated about 30 km SW of Taroudant at the southern border of the Souss Plain. The Série volcano-détritique d'Jebel Issendalen is exposed at the top of the complete section. This series is known throughout the northwestern Anti-Atlas. Some horizons of volcanics can even be traced south of Goulimime. The thickness amounts to approximately 100 m in the type section, decreasing towards the southwest and east (Fig. 7).

Multiple eruption cycles characterize this series. Each cycle starts with splintery green tuffites, overlain by up to 10 m thick coarse-grained volcanoclastics. The cycle terminates with a 1 cm thick horizon of precipitated iron oxides. The base of the tuffitic layers consists of closely packed pyroclastic material (fragments of devitrified volcanic

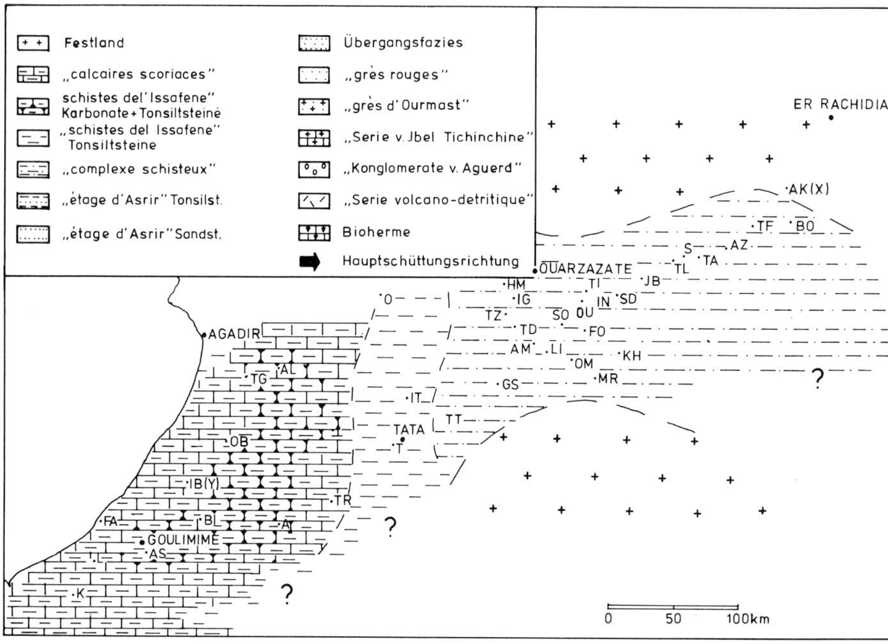

Fig. 8. Paleogeography of the Issafènien.

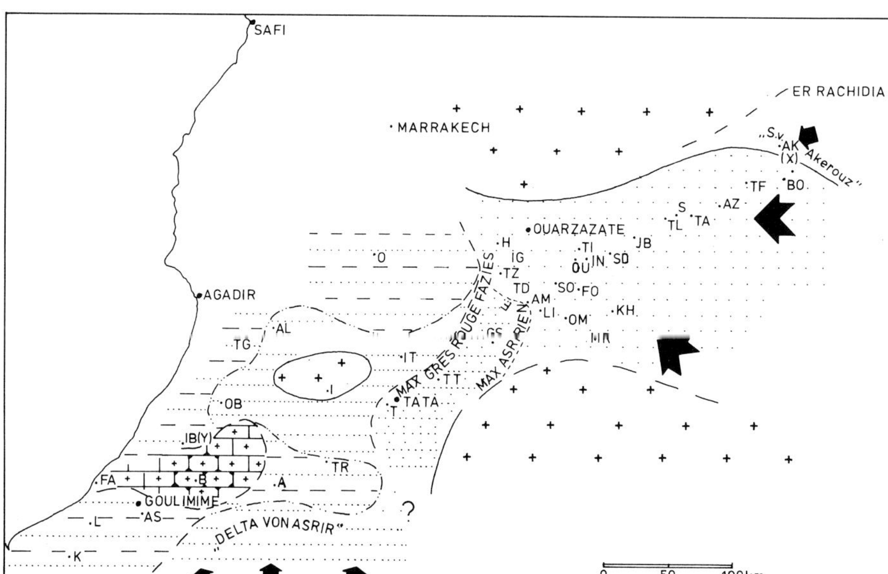

Fig. 9. Paleogeography of the Asririen.

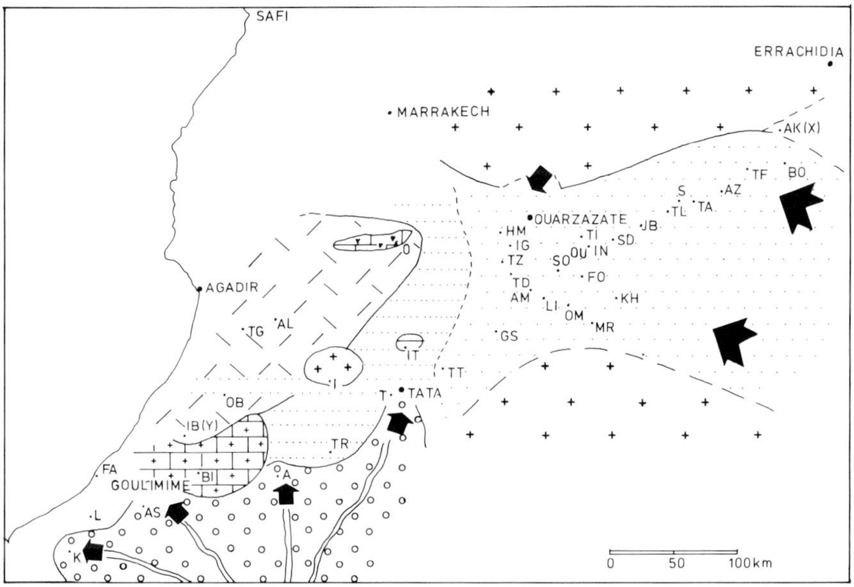

Fig. 10. Paleogeography at the boundary Asririen/Ourmastien.

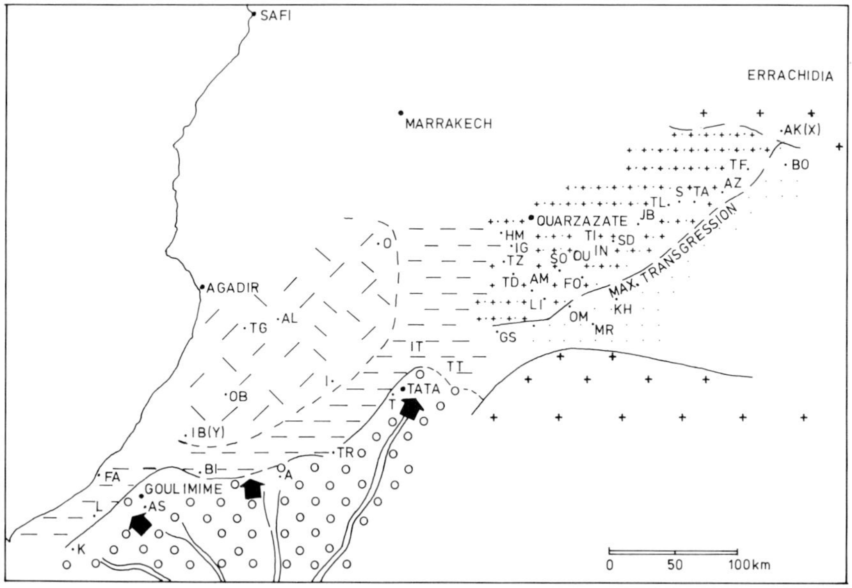

Fig. 11. Paleogeography of the Ourmastien.

glasses). Dispersed chlorite, calcite and ore form the matrix. Non-vulcanogenetic compounds (for instance metaquartzites) become more frequent towards the top. The coarse-grained unit (Pl. 1, Fig 6) is composed of trachytic rock fragments in a tuffaceous matrix.

The Série volcano-détritique was deposited in a shallow marine basin by pyroclastic turbidites and ash-falls. The location of the eruption center cannot be definitely determined; CHOUBERT & FAURE-MURET (1956) supposed that this eruption center was situated about 100 km east of Agadir.

4. Conclusions

4.1. Serie d'Issafène (Trilobite Zone IV-VI)
(Fig. 8)

Three lithologic units can be distinguished: the complexe schisteux in the eastern Anti-Atlas, the schistes de l'Issafène and the calcaires scoriacés in the western Anti-Atlas.

The facies of the reddish mudstones of the complexe schisteux extends over several hundred kilometers without a major change. The sediments were deposited in a flat lagoon or flute plain under aride climate. The unit complexe schisteux might have been deposited diachronously.

The transition into lithology of the schistes de l'Issafène is located east of Tata. Green mudstones, crystal tuffs and intercalated carbonate horizons are predominant west of this boundary. Trilobites indicate an open marine environment for the schistes de l'Issafène. The distinct increase in pyroclastics point to an intensified volcanic activity in the western Anti-Atlas.

The carbonates of the calcaires scoriacés were deposited on top of a wide submarine ridge in the westernmost part of the Anti-Atlas. This ridge probably separated the depostitional area of the schistes de l'Issafène from the open ocean (Japetus ocean?).

4.2. Asririen (Trilobite Zone VII)
(Fig. 9)

The pattern of sedimentation is completely different in the Asririen . Whereas mudstones prevailed in the Série de l'Issafène, the Asririen is characterized by fine-grained detrital sediments which were distributed by several river systems into the Anti-Atlas basin. Two river systems can be traced in the eastern Anti-Atlas. The first exhibits a nearly E-W directed flow regime and the second a dominance of SE-NW sediment transport. Both river systems of the grès rouges represent the type of 'braided river' with slow water flow.

The fluviatile facies of the grès rouges gradually passes into the marine sandstones of the étage d'Asrir in the central Anti-Atlas. Sediments of a tidal-dominated delta are found in the transitional area. During the same period, the Delta of Asrir developed in the southwestern Anti-Atlas prograding towards the marine basin in the north.

The conglomerates of Aguerd represent the littoral to fluviatile topset bed of the Delta of Asrir. They are exposed throughout a width of appr. 300 km at the southern margin of the Anti-Atlas.

The area of Issafène is characterized by local erosion and redepostition of the sediments. The mudstones of the étage d'Asrir were deposited outside of the influx of sand.

The carbonates of the Série d´Jebel Tichinchine were deposited in the former sedimentation area of the calaires scoriacés. The upper Asririen exhibits an intensified volcanic activiy leading to an increased deposition of volcanoclastics.

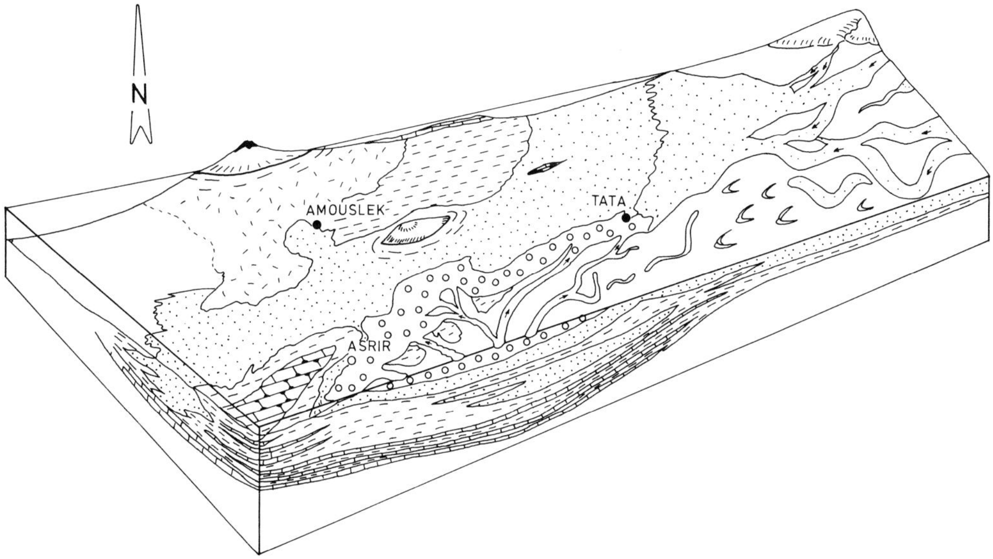

Fig. 12. Model of the upper Lower Cambrian deposition area in the Anti-Atlas.

4.3. Transition Asririen - Ourmastien (Trilobite Zone VII - VIII)
(Fig. 10)

A new transgressive cycle of sedimentation started with the deposition of the grès d'Ourmast. The source area of these series was located in the north, in the region of the present-day High Atlas. The grès d'Ourmast wedge out towards the W. They are replaced by only few meters of mudstones and siltstones.

The volcanic activity increased considerably in the western Anti-Atlas. The up to 100 m thick Série volcano-détritique was formed by eruptions with submarine currents of pyroclastics and a considerable amount of ash-fall.

The Delta of Asrir reached its maximum extent at the Asririen - Ourmastien boundary.

4.4. Ourmastien (Trilobite Zone VIII)
(Fig. 11)

Coquina horizons intercalated within the increasingly more fine-grained sediments of the grès d'Ourmast indicate the transgression of the sea which covered nearly the entire Anti-Atlas - with the exception of some areas in the SE and E. The prograding of the Delta of Asrir terminated. The volcanic activity decreased in the northwestern Anti-Atlas.

The schistes à *Paradoxides* are indicative of the Middle Cambrian transgression which considerably exceeded the former area of sedimentation in the Anti-Atlas.

1
4
2
5
3
6

Plate 1: Lithology of the 'grès terminaux'
(thin sections of Figs. 1 - 3, and 5 with crossed nicols;Figs. 1, 4 - 6: scale = 1 mm, Figs. 2, 3: scale = 0.1 mm)

Fig. 1. Pedogenic broken quartz within a carbonate matrix. Série of Akerouz (sample AK 11).
Fig. 2. Grès rouges. Detrital quartz cemented by secondary quartz overgrowth (sample AM 32).
Fig. 3. Grès rouges. Detrital feldspar with secondary feldspar overgrowth (sample AZ).
Fig. 4. Schistes de l'Issafène. Sandy biosparite with echinoderms and trilobites (sample A 11).
Fig. 5. Conglomerate d'Aguerd. Component of partly recrystallized metaquartzite (sample T 16).
Fig. 6. Série volcano-détritique. Lithic arenite composed of reworked volcanic rock fragments (sample AL 18).

Acknowledgment

The authors are indebted to the authorities of Morocco - particularly to Dr. Bensaid and M. Dahmani of the Ministère de l'Energie et des Mines - for the permission to work in the Anti-Atlas and for support during fieldwork. We gratefully acknowledge financial support from the Deutsche Forschungsgemeinschaft (Bu 312-10/1-2). Furthermore, we wish to thank H. Bogle for the translation and proof reading.

References

BOUDDA, A., CHOUBERT, G. & FAURE-MURET, A. (1974): Note au sujet des 'grès terminaux' du Cambrien inférieur de l'Anti-Atlas et de la transgression du Cambrien moyen. – C. R. somm. Soc. géol. France, **5**, 138-140; Paris

CHOUBERT, G. (1952): Histoire géologique du domaine de l'Anti-Atlas. - Not. Mém. Serv. géol. Maroc, **100** , 77-194; Rabat

— (1963): Histoire géologique du Précambrien de l'Anti-Atlas. - Not. Mém. Serv. géol. Maroc, **162**,352 pp.; Rabat

CHOUBERT, G. & FAURE-MURET, A. (1956): L'activité volcanique de la fin du Géorgien dans l'Anti-Atlas et le Haut Atlas occidentaux. - C. R. Acad. Sci. Paris, **242**, 2735-2738; Paris

COLEMAN, J. M. (1969): Brahmaputra river, channel processes and sedimentation. - Sed. Geol., **3**, 129-239; Amsterdam

GEYER, G. (1984): Biostratigraphische und systematisch- paläontologische Untersuchungen im Grenzbereich Unter-/Mittelkambrium von Marokko. - Diss. Univ. Würzburg, 229 pp.; Würzburg

HUPE, P. (1952): Contribution à l'étage du Cambrien inférieur et du Précambrien III de l'Anti-Atlas marocain. - Not. Mém. Serv. géol Maroc, **103**, 402 pp.; Rabat

MORGAN, J. P. (1970): Depositional processes and products in the deltaic environment. - In: MORGAN, J. P. (ed): Deltaic sedimentation. - SEPM Spec. Paper, **15**, 31-47; Tulsa

NELTNER, L. (1938): Etudes géologiques dans le sud Marocain. - Not. Mém. Serv. géol. Maroc, **42**, 298 pp.; Rabat

PASSEGA, R. (1964): Grain size representation by CM-pattern as a geological tool. - J. Sed. Petrol., **34**, 830-847; Menasha

PETTIJOHN, F. J., POTTER, P. E. & SIEVER, R. (1973): Sand and Sandstones. - 618 pp.; Heidelberg (Springer)

REINECK, H. E. (1972): Tidal flats. - SEPM Spec. Paper, **17**, 146-159; Tulsa

— (1985) Aktuogeologie klastischer Sedimente. - 348 pp.; Frankfurt (Kramer)

REINECK, H. E. & SINGH, I. B. (1980): Depositional sedimentary environments. - 451 pp.; Heidelberg (Springer)

DIAGENESIS AND VERY LOW-GRADE METAMORPHISM OF THE LOWER CAMBRIAN ROCKS IN THE ANTI-ATLAS (MOROCCO)

Werner Buggisch

Institut für Geologie, Universität Erlangen-Nürnberg, Schloßgarten 5, D-8520 Erlangen

Abstract

The diagenesis and the very low-grade metamorphism of the 'grès terminaux' in the Anti-Atlas were studied on the basis of the 'crystallinity' of illite and the deformation of quartz. The illite crystallinity improves from E to W. The quartz cements of the quartzitic sandstones are not deformed in the eastern Anti-Atlas. Undulation of quartz cement is observed in the central Anti-Atlas. The deformation increases towards the WSW where subgrains of quartz are found in quartz veins.

Introduction

The diagenesis and the very low-grade metamorphism (WINKLER 1967) were studied in the strata of the Anti-Atlas on the basis of clay minerals, 'illite crystallinity', authigenic minerals and the grade of quartz deformation. The stratigraphy, paleogeography and facies of the Lower Cambrian strata are discussed by BUGGISCH & FLÜGEL and BUGGISCH & SIEGERT (this volume). The illite crystallinity was measured in one section containing strata from the 'Série des calcaires inférieurs' to the 'grès terminaux' of the central Anti-Atlas, in order to analyse the influence of the sedimentary environment. The quartz deformation and the illite crystallinity were studied in the grès terminaux between Erfoud in the E and Agadir and Tiglit in the W in order to establish the regional pattern of the diagenesis and very low-grade metamorphism.

Phyllosilicates

The clay mineral association of the eastern Anti-Atlas is dominated by illite and mixed layer minerals. Furthermore montmorillonite and kaolinite were detected in the X-ray diffraction records. The stability of kaolinite and its replacement by illite and chlorite depends considerably on the K^+/H^+ and $(Mg^{++} + Fe^{++})/H^+$ concentration ratios (KISCH 1983). Therefore, the disappearance of kaolinite is documented in many deep wells at depths between 2000 m (90^0 C) and more than 5100 m (180^0 C), but not at as great a depth as the appearance of anchizonal illite "crystallinities" (references summarized in KISCH 1983).

An overburden pressure has already resulted from more than 3000 m thick Middle Cambrian and Ordovician sediment in the region south of Ouarzazate. Therefore, the new formation of kaolinite during weathering (Cretaceous or later) must be taken into consideration as well.

Chlorite is the main phyllosilicate in the upper Early Cambrian sediments of the western Anti-Atlas. The chlorites which were frequently altered to oxychlorites replaced detrital micas as well as fossil shells. The replacement of mica by chlorite is known from the diagenesis and the 'early metagenesis' (KARPOVA 1969 after KISCH 1983: Fig. 5–9). The matrix of the clastic sediments in the western Anti-Atlas is composed of authigenic chlorite and illite. Intergrowth of quartz, chlorite and illite/sericite forming 'beards' can be observed in thin sections of rock samples from the southwesternmost Anti-Atlas (Pl. 1, Fig. 6, 7).

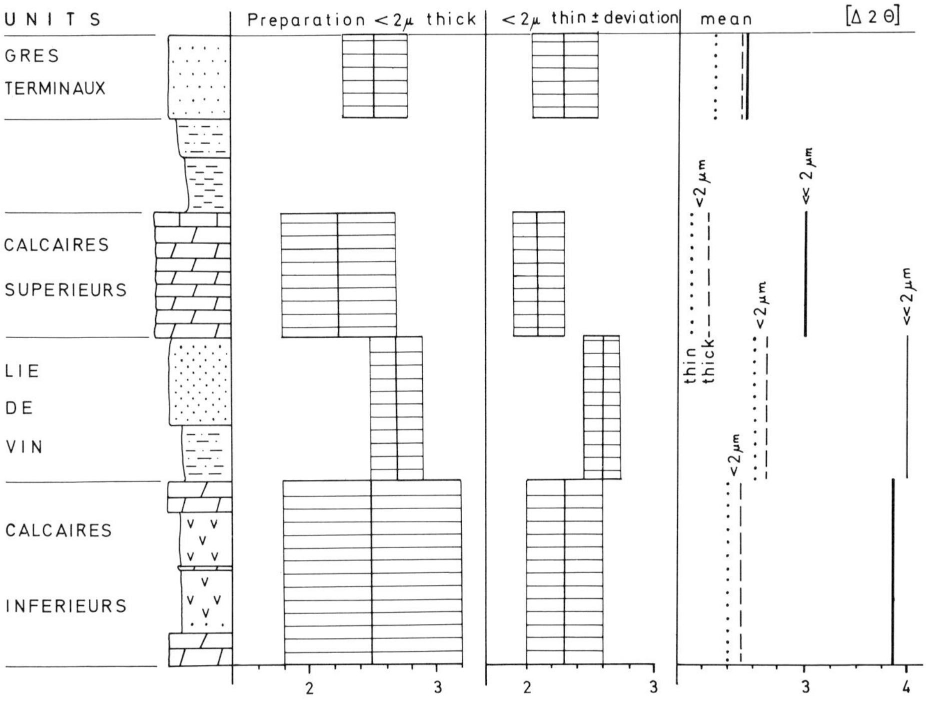

Fig. 1. 'Illite crystallinity' in the Lower Cambrian rocks of the section Ait Mrabt (AM of Fig. 1 in BUGGISCH & SIEGERT, this volume) in the central Anti-Atlas with alternating marine carbonates and mudstones and fluviatile sandstones.

Formation of authigenic minerals

The overburden was sufficient to mobilize quartz by pressure solution. In this way, enough SiO_2 was available for cementation. Authigenic overgrowth of feldspar (plagioclase and orthoclase) is also common in the central and eastern Anti-Atlas.

GASSNER (1986) described the alteration of olivine to chrysotile along fissures in volcanic rocks which are intercalated in the 'Série des calcaires inférieurs' west of the 'boutonnière of Bou Azzer-El Graara'. Chrysotile was formed due to hydrothermal solutions which were probably autometamorphically produced during the cooling of the lava. Chrysotile becomes instable in the greenschist facies. Therefore, later heating never exceeded the temperatures of the very low-grade metamorphism in this area.

'Illite crystallinity'

The sharpness of the illite (001) diffraction peak at 10 Å becomes more pronounced with an increasing degree of burial diagenesis. Therefore, the sharpness of this peak was used as a parameter of the incipient metamorphism by WEAVER (1961). Later, the width at half peak was used as a scale of the 'illite crystallinity' (KUBLER 1967, WEBER 1972). In this paper, the delta 2 theta value of the width at half diffraction peak of the 10 Å peak is measured (CuKα).

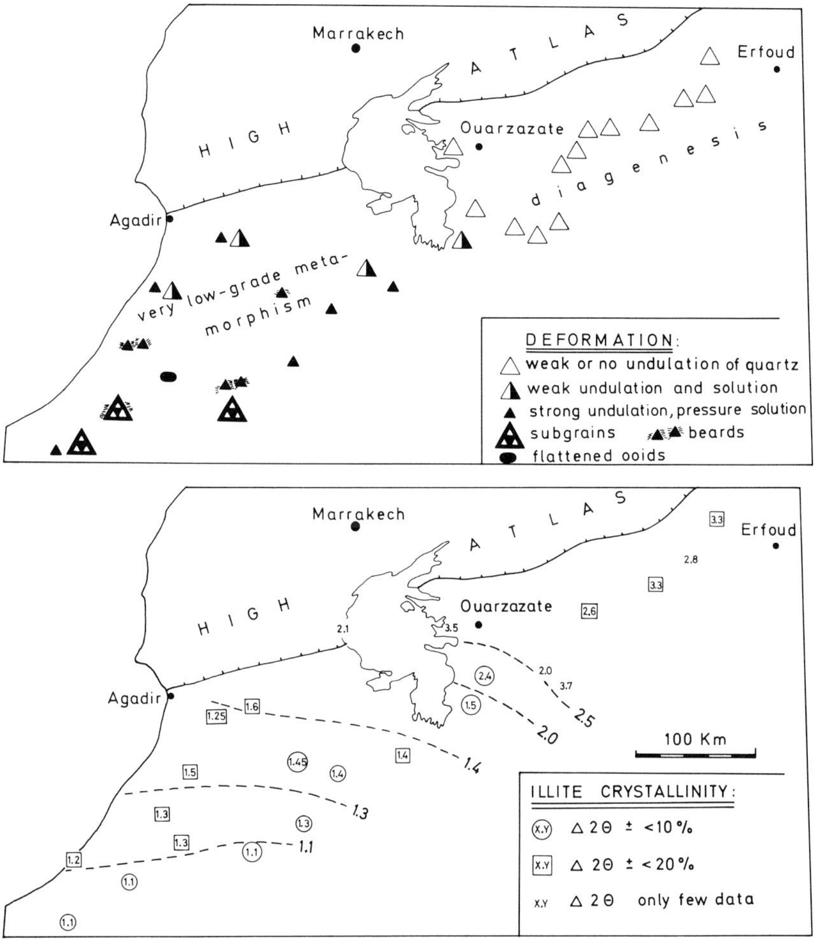

Fig. 2. 'Illite crystallinity' of the fraction < 2 µm and quartz deformation in the upper Early Cambrian 'grès terminaux' of the Anti-Atlas. Illite values measured as delta 2 theta of the (001) peak.

The value of 'illite crystallinity' depends considerably on the experimental methods, e.g. the preparation of samples and X–ray diffraction conditions (KISCH 1980, BUGGISCH 1986).The fractions < 2 µm and << 2 µm were pipetted off from 0.01 n ammonia solutions after suitable settling periods, and sedimented on glass slides. Preparations of varied thickness (TEICHMÜLLER et al. 1979) were used because of the dependence of the peak width on the thickness of the preparations. The samples were measured with a Philips X-ray diffractometer (generator PW 1729: fine focus tube 20 KV, 30 mA; diffractometer PW 1840 with automatically variable divergence-slit: speed 0.005/ 2 theta/min, chart 50 mm/2 theta, range $1x10^3 - 5x10^3$, TC 5 sec, slit 0.2 mm).

The measured peak widths are narrower than those of KUBLER (1967), WEBER (1972) or even KISCH (1980) depending on the thickness of the preparations and diffractometer conditions. The peak width depends on the grain–size of the preparations as well. Broadening of the peaks is observed when very small grain sizes are used. Some detrital mica might still remain present in the fraction <2 µm. The mode of preparation is always a compromise between different sources of errors. Therefore, the value of the 'illite crystallinity' is taken only as a relative parameter in this paper.

The 'illite crystallinity' is poor in the eastern Anti–Atlas and improves in the western Anti-Atlas (Fig. 2). The transition zone is relatively narrow. The best sharpness of the 10 Å peaks is observed in the southwesternmost Anti–Atlas.

The improvement of the illite crystallinity coincides with the transition from fluviatile to marine facies of the sediments. The crystallinity was studied in a section with alternating continental and marine strata (Fig. 1) in order to examine whether the improvement of the illite crystallinity is dependant on the sedimentary environment or on the burial metamorphism. The illite (001) peak of thin < 2μm preparations is a little bit broader in the fluviatile strata of the studied section compared with the marine sediments. The change in the sharpness of the illite peak from E to W is much more significant than the dependency on the environment. Therefore, the improved 'illite crystallinity' is due to a higher overburden or to an increased stress in the western Anti-Atlas.

Deformation and subgrain formation of quartz

The behavior of quartz depends on stress and temperature (VOLL 1960, 1969; MAJORIBANKS 1976; BELL & ETHERIDGE 1976 and ETHERIDGE & WILKIE 1981). The quartz grains of the grès terminaux are affected only by a weak pressure solution in the eastern Anti-Atlas. Only weak or no undulation of volcanic quartz grains or of quartz cement is observed in thin sections from this area.

Undulation of the quartz grains and of the authigenic overgrowth of quartz is common west of the line Tazenakht-Foum Zguid (Fig. 2). This boundary coincides more or less with the area in which the 'illite crystallinity' distinctly improves. Increasing stress is documented by the deformation lamellae of quartz and by the loss of quartz cement due to pressure solution. These phenomena are restricted to competent layers. The strain is buffered by the matrix within the incompetent rocks.

Boundary migration and formation of quartz subgrains are restricted to quartz veins within the grès terminaux of the southwestern Anti-Atlas. Recrystallization of quartz is not observed in the studied thin-sections of the grès terminaux.

Schistosity

A distinct schistosity is not developed in the Cambrian rocks of the Anti-Atlas. Flattening of oolites is observed in the section Bou Izakarn (BI) in the southwestern Anti-Atlas (Pl. 1, Fig. 8). 'Beards' of illite and chlorite and new growth of chlorite and illite in the matrix with distinct orientation are also restricted to the southwestern Anti-Atlas (Pl. 1, Fig. 6, 7).

Plate 1 (Fig. 1 – 7 with crossed nicols)

Fig. 1. Undeformed quartz grains with authigenic overgrowth (sample AM, scale = 0.5 mm)

Fig. 2. Weak undulation of quartz in detritic grains and authigenic overgrowth (sample AM, scale = 0.1 mm)

Fig. 3. Undulation and deformation lamellae in detritic grains and cements (sample K, scale = 0.5 mm)

Fig. 4. Strong pressure solution affecting detritic quartz grains (sample TT, scale = 0.5 mm)

Fig. 5. Formation of subgrains and incipient boundary migration in quartz veins (sample AS, scale = 0.1 mm)

Fig. 6. 'Beards' of chlorite, illite and quartz (sample A, scale = 0.5 mm)

Fig. 7. 'Beards' of chlorite, illite and quartz (sample A, scale = 0.1 mm)

Fig. 8. Flattened oolites (sample BI, scale = 0.5 mm)

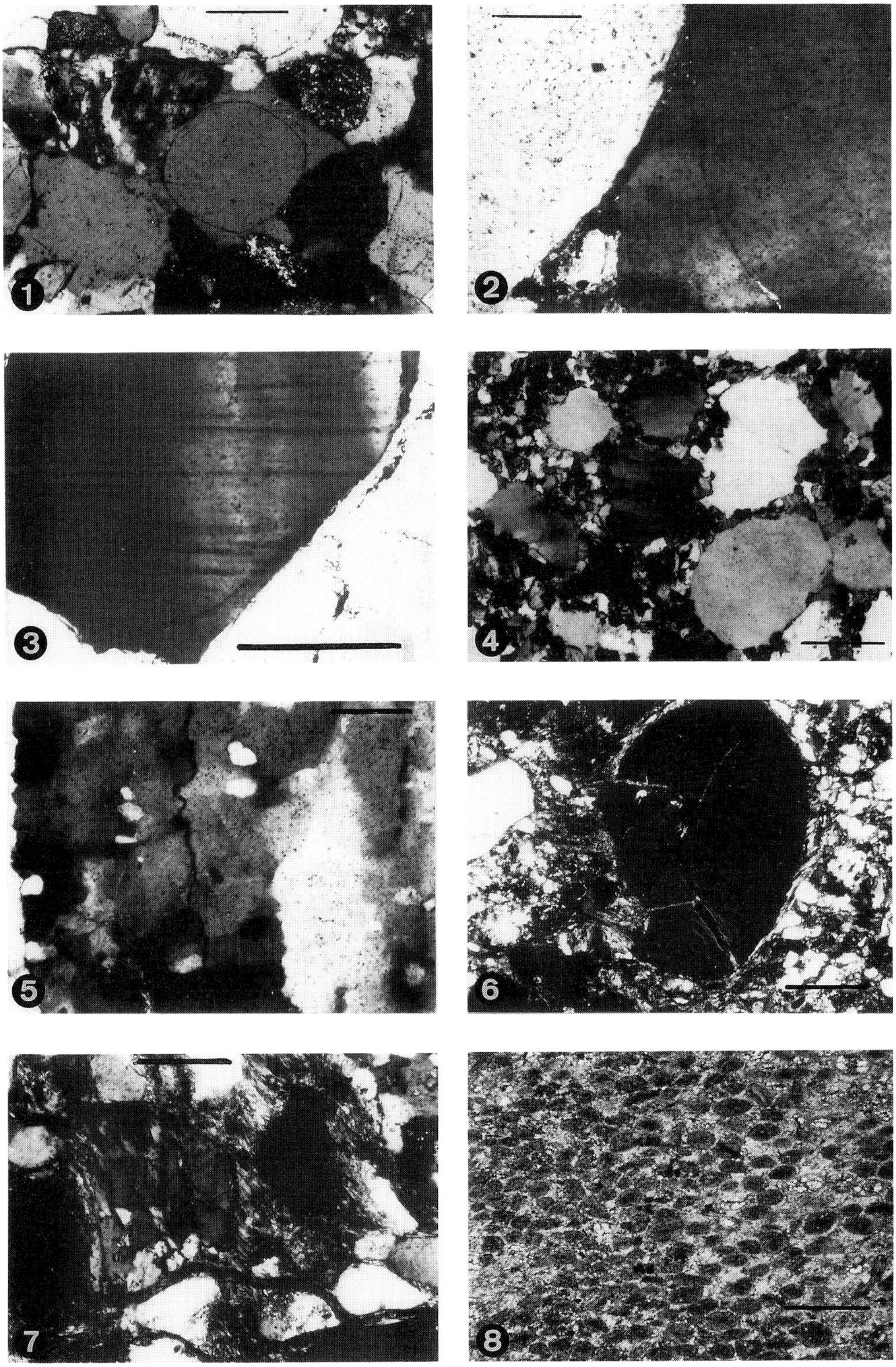
1
2
3
4
5
6
7
8

Conclusion

The eastern Anti-Atlas is characterized by broad illite diffraction peaks and by weak deformation of quartz. The Lower Cambrian sediments still exhibit the diagenesis stage. West of Tazenakht-Foum Zguid, the 'illite crystallinity' improves and the quartz grains and the authigenic quartz overgrowth reveal undularory extinction. The quartz deformation and the 'illite crystallinity' increases towards the southwestern Anti-Atlas.

The boundary of diagenesis and very low-grade metamorphism coincides with the mobility of the socle. The eastern Anti–Atlas is characterized by low rates of subsidence and by the dominance of continental-fluviatile sedimentation during the Lower Cambrian. High rates of subsidence and marine sediments prevailed in the more mobile western Anti-Atlas. While the Paleozoic sediments are only weakly deformed in the eastern Anti-Atlas, a foldbelt was formed during the Hercynian Orogeny in the (south)western Anti-Atlas.

Acknowledgement

The author is indebted to Dr. W. v. Gosen (Erlangen) for many discussions of the results.

References

BELL, T.H. & ETHERIDGE, M.A. (1976): The deformation and recrystallization of quartz in a mylonite zone, central Australia. – Tectonophysics, **32**: 235-267; Amsterdam

BUGGISCH, W. (1986): Diagenese und Anchimetamorphose aufgrund von Conodontenfarbe (CAI) und "Illit-Kristallinität" (IC). Methodische Untersuchungen und Daten zum Oberdevon und Unterkarbon der Dillmulde (Rheinisches Schiefergebirge). – Geol. Jb. Hessen, **114**: 181-200; Wiesbaden

ETHERIDGE, M.A. & WILKIE, J.C. (1981): An assessment of dynamically recrystallized grainsize as paleopiezometer in quartz-bearing mylonite zones. – Tectonophysics, **78**: 475-508; Amsterdam

GASSNER, W. (1986): Das "Infra"- und Unterkambrium am Südostrand des Präkambriumaufbruchs von Bou Azzer / El Graara (zentraler Antiatlas, Marokko). – Unpubl. Dipl. Arb., 92 pp, Erlangen

KISCH, H.J. (1980): Incipient metamorphism of Cambro-Silurian clastic rocks from the Jamtland Supergroup, Central Scandinavian Caledonids, Western Sweden: illite crystallinity and 'vitrinite' reflectance. – J. geol. Soc. London, **137**: 271-288; London

-- (1983): Mineralogy and petrology of burial diagenesis (burial metamorphism) and incipient metamorphism in clastic rocks. – In: LARSEN, G. & CHILINGAR, G.V. (eds): Diagenesis in sediments and sedimentary rocks, 2: – Development in Sedimentology, **25B**: 289-493; Amsterdam (Elsevier)

KUBLER, B. (1967): La cristallinité de l'illite et des zones tout à fait supérieurs du métamorphism. – In: Colloque Etage Tectoniques. A la Baconnière: 105-122; Neuchâtel

MAJORIBANKS, R.W. (1976): The relation between microfabric and strain in a progressively deformed quartzite sequence from central Australia. – Tectonophysics, **32**: 269-293; Amsterdam

TEICHMÜLLER, M., TEICHMÜLLER, R. & WEBER, K. (1979): Inkohlung und Illit-Kristallinität. Vergleichende Untersuchungen im Mesozoikum und Paläozoikum von Westfalen. – Fortschr. Geol. Rheinl. Westf., **27**: 201-276; Krefeld

VOLL, G. (1960): New works on petrofabrics. – Liverpool and Manchester Geol. J., **2**: 503-567

-- (1969): Klastische Mineralien aus den Sedimentserien der Schottischen Highlands und ihr Schicksal bei aufsteigender Regional- und Kontaktmetamorphose. – Habil. Schr., 206 pp, Berlin

WEAVER, C.E. (1961): Clay minerals of the Ouchita structural belt and adjacent foreland. – Univ. Texas Publ. **6120:** 147-160

WEBER, K (1972): Kristallinität des Illits in Tonschiefern und andere Kriterien schwacher Metamorphose im nördlichen Rheinischen Schiefergebirge. – Jb. Geol. Paläont. Abh., **141**: 333-363; Stuttgart

WINKLER, H.G.F. (1967): Die Genese der metamorphen Gesteine. – 2nd ed., 237 pp; Berlin (Springer)

High and Middle Atlas

DEFORMATION OF IGNEOUS DIKES IN AND AROUND THE TICHKA GRANITE, HIGH ATLAS - MOROCCO

Jean-Paul Schaer
Institut de Géologie, Université de Neuchâtel
11, rue E.-Argand, CH-2007 Neuchâtel

In Morocco, north of the Anti-Atlas, as in the rest of North Africa, the non-metamorphic or weakly metamorphic Paleozoic formations are often pierced by small (10-15 km in diameter) granitic intrusions, with well developed metamorphic aureoles. The Tichka granite in the western High Atlas south of Marrakech is one of them and it has received a great deal of attention from geologists and petrologists, which is due to the complexity of the intrusion and to the excellent quality of the outcrops dissected by deep valleys in an arid climate.

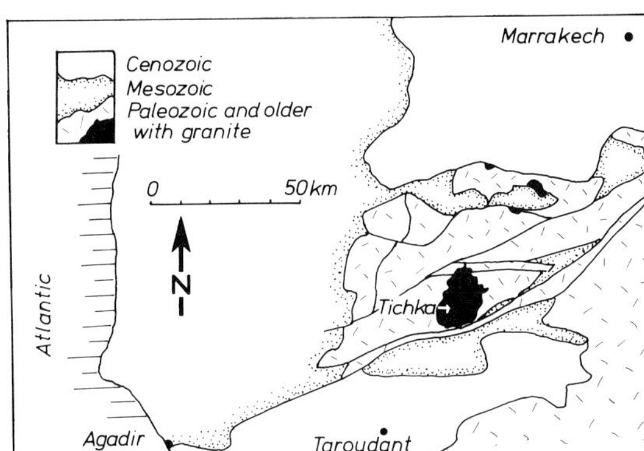

Fig. 1. Generalized geological map of SW Morocco with the Tichka granite

The formations surrounding the Tichka igneous complex are considered to be Cambrian with the possibility of some Infra-Cambrian. These weakly metamorphosed rocks contain graywackes, schists, limestones and considerable volumes of volcano-detrital material (tuffs, lapilli with few lavas) of andesitic to dacitic composition. As is the case in other deformed rock sequences, it is not easy to deduce the original thickness of the total pile or of the different lithological groups, especially as there are many indications of important lateral facies variations. We propose a total thickness of 3-5 km for the Cambrian, the volcanics often exceeding 1 km, whereas the limestone are locally a few hundred m. thick. In harmony with other Hercynian granites in Morocco the Tichka intrusion is considered to be of Carboniferous age. Radiometric age determinations by the Rb/Sr method give 305 and 318 ± 10 m.a. (Termier et al. 1972) The intrusion is remarkable for its petrographical heterogeneity; diorites, granodiorites, and granites are about equal in proportion, whereas gabbros are extremely rare. Contacts between the different igneous rocks are often sharp but they can also be transitional. Throughout the intrusion there are numerous lenses or slabs of metasedimentary inclusions, especially abundant in the south; they have undergone assimilation to a varying degree, but the lithologies are generally still recognizable. They range from several centimeters to more than a kilometer. On the whole, magmatic and metasedimentary inclusions form an interlocked elliptical pattern centered in the northern part of the intrusion (Lagarde and Roddaz 1983). These concentric structures are well marked on aerial photographs and on maps (G. and H. Termier 1981); they are also visible at the outcrop scale as a more or less pronounced vertical planar structure often well expressed in dioritic domains, whereas in many late granites they are lacking or so weakly developed that their identification is uncertain.

The elliptical structure is also (parallel and) concordant with the enclosing rocks and their structure on a general scale. In detail, however, the contact is often cross-cutting, sharp and generally vertical; in the N.E. occasional sheet-like apophyses up to 100 meters thick extend horizontally into the country rocks for one or two kilometers (Bizard and Ziegler 1955). On a small scale, the margin may be irregular and is sometimes transitional with migmatitic segregations of quartzofeldspathic material in the adjacent schists. The country and in metamorphic grade towards the granite contact, from regional greenschist (chlorite, epidote, actinolite, some biotite) to coarser schist with amphiboles, andalousite, sillimanite, staurolite, garnet or to marbles with pyroxene and garnet. Lagarde and Roddy (1983) have proposed from the study of strain patterns that the intrusion results from diapiric emplacement in a domain of transcurrent shear. For the moment the origin of all the magmatic rocks of the Tichka intrusion has not received a totally satisfactory solution. Using field evidence Vogel and Walker (1975) revealed the contemporaneity and close association of acidic and basic magmas. Vogel et al. (1976) and Scott and Vogel (1980) have shown that an origin by differentiation can be ruled out on the basis of major element and RE analysis. Scott and Vogel (1980) have suggested that the granite may represent crustal melts, the diorite a mixture of low SiO_2 granite gabbro of mantle origin.

The dikes

A marked feature of the Tichka intrusion and its envelope is the presence of numerous dikes.

The dikes in the envelope

The Paleozoic of the High Atlas, which extends in an W.-E. direction from the couloir d'Argana to the river n'Fiss for about 70 km and by about 30 km in a N/S direction, is cut by numerous dikes of varying structural relationship, with compositions from basic (with olivine and pyroxenes) to acid. Most dikes appear to be vertical (the outcrop pattern may have some part in this); however, in the vicinity of the Tichka granite, especially where they are more numerous, they are also often horizontal. No systematic analysis of their orientation has been made but it seems that the great majority are more or less parallel with the general structure (folds, schistosity) with a second group at right angles, whereas only a small number have no particular relationship with the structure of the enclosing rocks.

In the vicinity of the granite, the thickness of the dikes is variable, in general a few meters or decimeters. For the moment insufficient data on intersections prevent the establishment of a time sequence of a thrustful emplacement with any confidence. Nevertheless, the basic dikes apparently always cut the andesitic ones. Many observations demonstrate that dike intrusion occurred after the main structural development (many dikes cut across folds, schistosity and deformed conglomerates Fig. 2). In the marbles, near the granite, the dikes do not maintain their post-deformation character. They are deformed, folded and stretched into boudins, but it is still possible to see that they cut across older folds (Fig. 2) When two sets of dikes intersect, even the youngest presents evidence of post-intrusion deformation, occasionnally very severe (Fig. 2) Near the granite most dikes present post-magmatic mineralogical re-equilibrations. It is not easy to separate modifications related to normal post-

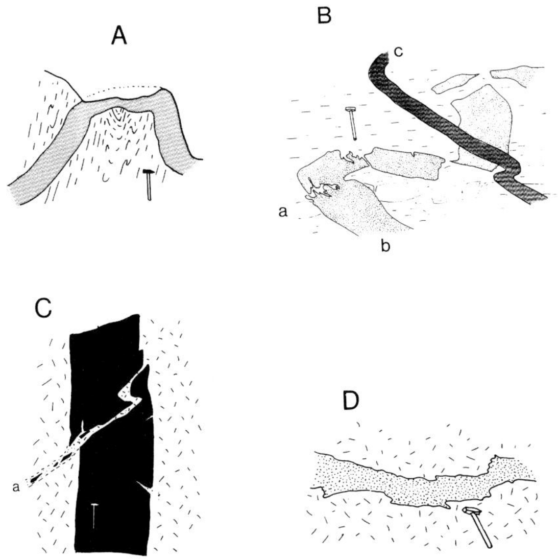

Fig. 2 A. Basic dike cutting folded metasediments near the margin of the Tichka granite. Assif n'Ait Tament, N of Tisgane
B. Two generations of folded basic dikes (b intersecting c) in a large calcareous xenolith with streched pyroxenic horizon (a), near the southern contact of the Tichka granite N of Kma, Ida Oukays.
C. In the Tichka granite, an old dike (probably generation H) is cut by a granitic vein (a) which has carried small fragments of the dike. Amendrech.
D. Dike of the oldest generation (H), with non planar geometry without chilled margins cutting granodiorite. Asif n'Fiss, SE of Azmou n'Guenfis.
In each drawing the scale is given by the ham.

magmatic and hydrothermal activities from those associated with thermal metamorphism. In some cases, however, wide development of biotites partly at the expense of magmatic amphibole in the dikes is demonstrably due to thermal and metasomatic activity associated with the main intrusion. Elsewhere, the dikes are deformed in the marbles or in the schists and it seems logical to relate their metamorphic, post-magmatic paragenesis to contact metamorphism which develops in harmony with the intensity of deformation. In conclusion for the dikes within the granite envelope, it is clear that they have intruded rocks that were already deformed, at temperatures high enough to provoke some mineralogical and structural transformation.

The dikes in the granite

In the Tichka granite numerous dikes vary in structure and in composition, from granitic through andesitic to doleritic. They are generally vertical, intersect sedimentary inclusions, granitic and dioritic foliation, but some are also parallel with the latter. The dikes in the granite, in general terms, may be grouped into three main classes, on the basis of cross-cutting relationships, lithology and structure. Their thickness varies from more than 10 meters to a few decimeters. Good exposures with many intersections allow the establishment of time sequence of intrusion. The oldest generation (H generation) is composed of a variable group of dominantly light-coloured micro-crystalline rocks, followed by light grey, finely crystallised dikes including porphyries (K generation), then by darker ones (L generation). Thin sections confirm this classification but reveal differences and transition not suspected during field work. The tendency of evolution from microcrystalline rocks of granitic to dioritic composition, to basic, vitric rocks with olivine and pyroxene through porphyritic and non-porphyritic andesites is confirmed. These classes are not homogeneous; intermediate types are present, and it is probable that the evolution from acid to basic was not linear, but included several recurrences. Field evidence shows that all dikes were emplaced in the late stages of plutonic crystallization while the main dioritic to granitic body was evolving from a crystal mush to a solid rock. The simultaneous presence of magmas of different composition is illustrated by the common occurence of amoebic contacts between dikes and granitic or granodioritic rocks (Vogel and Walker 1976). The older set of dikes (H) is often cut by aplitic, pegmatitic and granitic veins (see also Vogel and Walker 1976). In one case, a microdioritic, vertical dike with sharp contacts against the granite was observed to be cut by a late granitic vein which has carried away fragments of the dike into the enclosing granite. At the time of dike intrusion, the granite has enough rigidity to fracture but still contained nearby enough fluid material, which intruded the granite, cut the dike and removed some dike chips.

In the granite pluton as a whole the last stage of thermal and rheologic evolution is illustrated by dike behaviour in the granitoid rocks. The youngest dike generation is nearly always vertical with sharp parallel contacts and well developed chilled margins; they can easily be followed for several km sometimes as single dikes, or in swarms of 8 to 10 members. In the region of the Tichka Plateau they trend E/W, cutting the internal structure of the granite and its envelope. Within the granite, they have not been deformed after their intrusion, in opposition to the situation in the marbles of the envelope. The K dikes display a fairly regular disposition, either transecting the granite structure or parallel to it; their margins are either straight or irregular; the grain size decreases slightly towards the contacts. The oldest generation commonly (H) presents non planar geometry, subparallel to the internal structure of the granite, but it does not exhibit the mineral alignment due to post-intrusion events that has been described by Pitcher and Berger (1972) in the Donegal granite. We do not assign the wavy aspect of early Tichka dikes and the non-parallelism of their margins to folding or movements in the granite after their emplacement; we believe that their geometry and crystallization are related to the conditions of intrusion in a medium where temperatures of the dikes and of intruded material were similar. The cooling body was not rigid enough to be brittle. We believe that their geometry and crystallization history are related to the low contrast in rheology and temperature between the intruding material and host rocks.

Conclusion

The synplutonic dikes which cut the Tichka pluton and its sedimentary envelope can be used as geometric and temporal references for its late kinematic and thermal evolution. When the first dikes were emplaced, the granite and granodioritic material had not totally crystallized; the dike contacts are often irregular and the dikes are cut by aplitic, pegmatitic, granitic veins. With time, the granite acquired more rigidity and the dike geometry became more planar. In the last stage of its evolution, the pluton was still mobile with respect to its envelope, and ascended without internal deformation. After dike emplacement, ductile deformation is confined to the marbles, with intense folding and boudinage near the contact. Late deformation is concentrated within the marbles, due to their rheological properties under such a thermal regime. Where marbles lack, the deformation takes the form of fault zones cutting across the Paleozoic formations. These brittle structures are not easily separated from those related to the development of normal faults of Triassic age or from those accompanying the formation of the Atlas Range during the Cenozoic. During its last phase of intrusion, the Tichka granite moved upwards as a rigid piston with little internal deformation, while ductile structures formed within its envelope. No regional dextral ENE transcurrent shear component (Lagarde and Roddaz 1983) has been detected in the granite after the emplacement of the first dikes.

Acknowledgements

This reasearch could not have been undertaken without the active support of Dr. M. Bensaïd, director of the Division de la Géologie, and M. Dahmani, chef du Service de la Carte géologique.
I would like to thank St Ayrton who has greatly improved the english version of the manuscript.

Bibliographie

LAGARDE, J.-L. & RODDAZ, B. (1983) : le massif plutonique du Tichka (Haut Atlas occidental, Maroc) : un diapir syntectonique. - Bull. Soc. géol. France (7), 25, 3, 389-395.

SCOTT, G.B. & VOGEL, T.A. (1980) : The origin of the acidic and basic rocks of the Tichka massif, Morocco, based on rare earth elements. - Contrib. Mineral. Petrol., 75, 89-95

TERMIER, H. & TERMIER G. (1971) : Le massif granito-dioritique du Tichka (Haut Atlas occidental, Maroc). Volume 1 : Les régions et les terrains. - Notes Mém. Serv. Géol. Maroc 216, 240 p., 11 pl. & carte.

VOGEL, T.A. & WALKER, B.M. (1975) : The Tichka massif, Morocco : an example of contemporaneous acidic and basic plutonism. - Lithos 8, 29-38.

VOGEL, T.A., WILLIAMS; E.R., PRESTON, J.K. & WALKER, B.M. (1976) : Origin of the late Paleozoic plutonic massifs in Morocco. - Bull. geol. Soc. America 87, 1753-1762.

SYNTHESIS OF LATE PALEOZOIC AND TRIASSIC REDBED SEDIMENTATION IN MOROCCO

John C. Lorenz
Sandia National Laboratories, Division 6253,
Albuquerque, NM 87185 USA

Abstract

The Late Carboniferous and Permian strata of Morocco were deposited in late-orogenic, probably fault-bounded basins that formed during the last stages of the Variscan (Hercynian) orogeny. Remnants of a dozen of these basins are exposed in Morocco. They contain reddish-brown alluvial-fan and fluvial conglomerates and sandstones, as well as significant volumes of mudstone of probable floodplain origin. Several also contain intercalated acidic lava flows. Syndepositional and postdepositional tectonism caused structural deformation and erosion of these strata. The next stage of preserved deposition in Morocco began with continental rifting 30-40 million years later in Late Triassic time, and took place in isolated, fault-bounded basins that are aligned along the inherited Variscan structural trends. In southwestern Morocco, these basins were filled with brick-red alluvial-fan and fluvial conglomerates, sandstones, and mudstones; in the northern and coastal-southwestern parts of the country, marine-influenced thick halites and red evaporitic mudstones are the prevalent sedimentary deposits. As localized subsidence of the basins decreased, more regional subsidence began, causing transgression of the Tethys from the east, and of the newly formed Atlantic from the west. Widespread evaporitic mudstones were deposited across Morocco in vast nonmarine mudflats inland, and marine-influenced, muddy sabkhas nearer to the transgressive shoreline. Extensive basalt flows were extruded during deposition of these facies. As subsidence and transgression continued, open-marine Jurassic carbonates were deposited conformably over the mudstones along the High and Middle Atlas trends.

Introduction

The "Permo-Triassic" sedimentary strata of Morocco form a distinctive suite of lava flows and red conglomerates, sandstones, and evaporitic mudstones that are widespread across the country. Prior to the advent of definitive fossil and radiometric dating techniques, it was difficult to separate some of the late Paleozoic strata from the similar Mesozoic deposits. Since the age of the basal Triassic strata was unknown, these deposits were perceived as possibly being as old as the dated and somewhat similar Permian sedimentary rocks, and thus the field term "Permo-Triassic" was commonly applied to suites of both late Paleozoic and early Mesozoic redbeds. The tectonic settings of the two sedimentary assemblages are still being defined, but the Late Carboniferous and Early Permian strata are related to the complex Variscan (Hercynian) orogeny in Europe and northwest Africa, whereas the Late Triassic to Early Jurassic rocks are the sedimentary record of crustal extension and rifting associated with the separation of the continents of Africa and North America. Van Houten (1976b) noted that mild compression and vertical tectonism produced late-stage Variscan basins within the orogen, but that later, Triassic rifting created basins that flanked the Meseta (the inner, cratonized zone of the orogen).

Latest Paleozoic Structures and Deposits

Structural Inheritance

The Late Devonian through Late Carboniferous Variscan orogeny in Morocco was part of the larger-scale orogeny of western Europe, the tectonic interactions and causes for which are still subject to debate (see Hutton and Sanderson, 1984.) In Morocco, the Variscan orogenic province is separated from the relatively undeformed Saharan platform to the south by the South Atlas fault zone. In northernmost Morocco, the orogen is obscured by Tertiary sediments and by the allochthonous terrane of the Rif Mountains. Between these two areas, Michard and Piqué (1980) delineated five Variscan structural provinces, each with a northeast-trending structural fabric.

This structural grain is strongly reflected in both the late Paleozoic and Mesozoic deposits and structures discussed here. Cailleux et al. (1983) state that the late Paleozoic basins of the central Moroccan Meseta are localized along zones of weakness, commonly strike-slip faults in the Hercynian basement. During the late stages of the Variscan orogeny in both Europe and Morocco, intermontane successor basins developed on the variously deformed basement (Jowett and Jarvis, 1984). The late Carboniferous to Permian deposits which filled these basins in Morocco are only partially preserved, and some of these basins may have been destroyed completely by the 30 to 40 m.y. episode of post-orogenic/pre-Late Triassic stability and erosion in the region. Van Houten (1976a,b) plotted 15 of these basins in northwest Africa (Figure 1), suggesting that their previous extent was not significantly more than about double their present-day outcrop dimensions.

The Variscan basins in Morocco are late-orogenic and not post-orogenic basins (Van Houten, 1976a,b). Syndepositional tectonism is recorded by the different degrees of structural deformation: the older, Late Carboniferous deposits are more deformed that the closely related younger, early Permian strata. Michard and

Piqué (1980, p. 319) report that "The earlier...redbeds (Westphalian C-D) are gently folded. The later ones (Autunian) are only faulted and tilted."

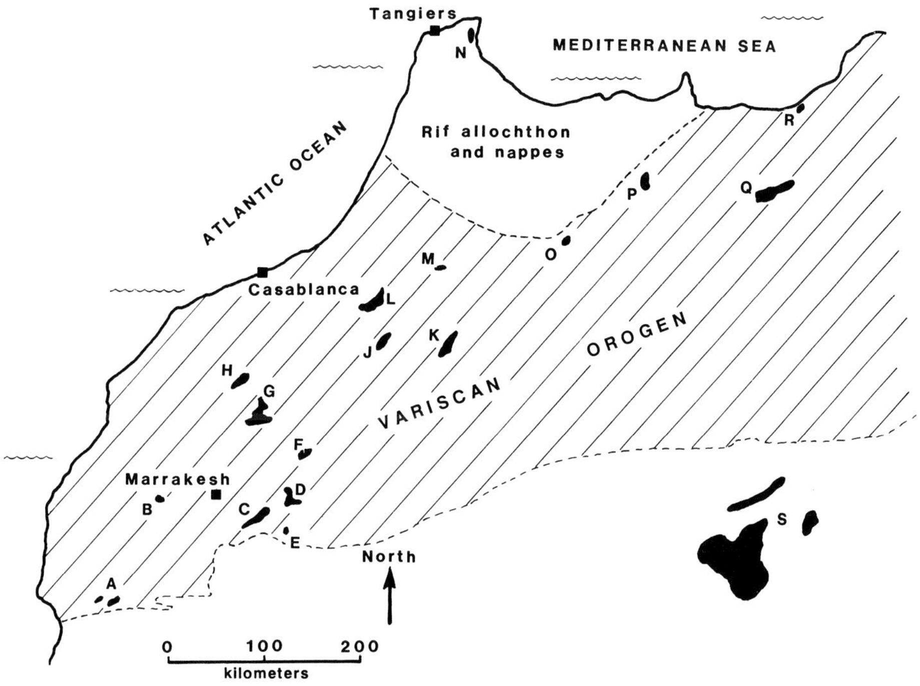

Figure 1. Generalized distribution of known late-Variscan successor basins in northwest Africa (after Van Houten, 1976b, Figure 1): A-Ida ou Zal, B-Chichaoua, C-Ait Ziffa, D-Almezi, E-Telouet, F-Hadadja, G-Nzala Arrarcha, H-Mechra ben Abbou, J-Chougrane, K-Khenifra, L-Sidi Kassem, M-Bou Achouch, N-Rif, O-Tazzeka, P-Masgout, Q-Jerada, R-Traras, S-Colomb-Bechar.

Late Carboniferous to Early Permian Deposition

The deposits that fill the Variscan basins of Morocco consist primarily of nonmarine detrital redbeds with scattered igneous and coaly intercalations. They have been dated as Late Carboniferous to Early Permian in age (Van Houten, 1976a). Nine of the 15 known basins contain thick (up to 2000 meters) clastic deposits of conglomerates, sandstones, and mudstones. The general coarseness of the deposits and the abundance of nearby faults of similar ages indicate that the basins were probably fault-bounded during sedimentation. The rest of the areas were much shallower, and probably represent deposition on the more stable interbasin areas (Van Houten, 1976a).

Khenifra basin

Strata within the Khenifra basin (Figure 2) are typical of the deposits in the larger basins, and have been relatively well studied; they will therefore be discussed in detail as an example of these basins. Termier (1936) and Cossey (1974) inferred that the strata in this basin are close to 1000 meters thick, whereas Cailleux et al. (1983) suggest that the strata are 1700 to 1900 meters thick. The rocks consist of reddish-brown conglomerates, sandstones, siltstones, and mudstones. They have been dated as Autunian, based on the presence of the plant fossil Walchia (Termier, 1936) and on palynology (Couzminer and Manspeizer, 1977). However, a fossil plant assemblage collected from the top of the section contains Lecachia c.f. pinaformis, suggestive of an age at or near the Carboniferous-Permian boundary, and the rest of the section may therefore be older (identification and interpretation by Dianne Edwards, in Cossey, 1974).

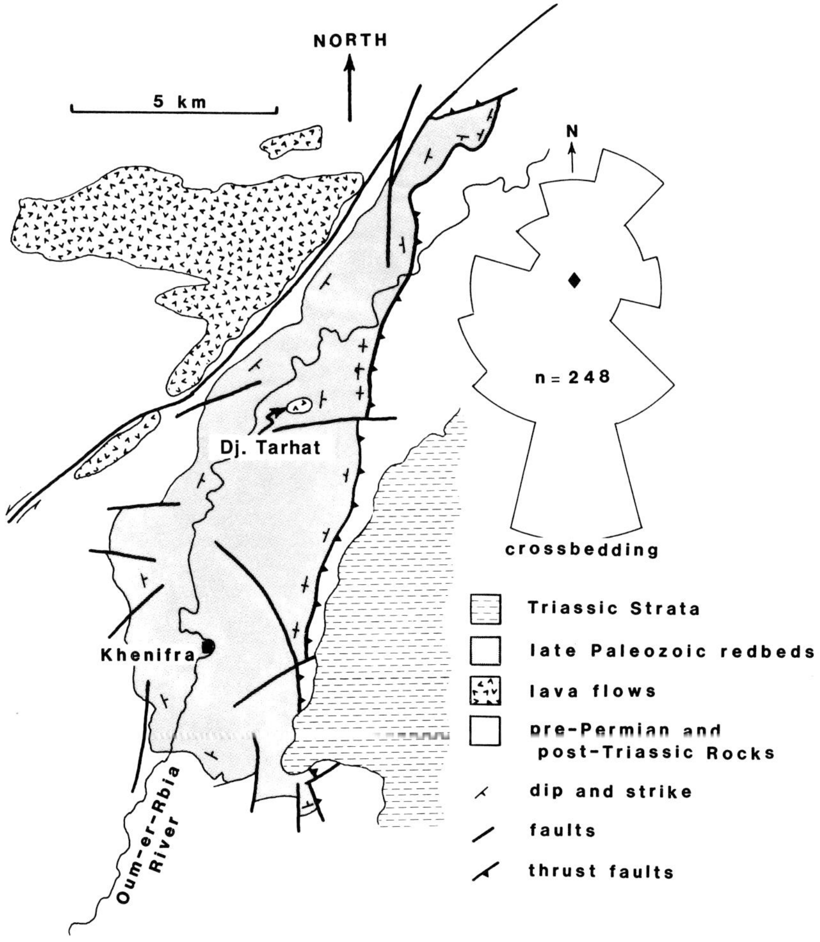

Figure 2. Geologic map of the Khenifra basin, central Morocco: crossbedding from Lorenz (1974); basin structure from Cossey (1974), Cailleux et al. (1983) and Lorenz (1974); bedrock structure from Allary et al. (1972) and Cailleux et al. (1983).

Evidence for synsedimentary volcanic activity may be preserved at the top of the section: rhyolitic flows that were formerly assigned to the Pliocene have been reinterpreted to be of Permian age based on their "tectonization," which is inferred to be more similar to the Permian lavas of Morocco than to the younger lavas of the country (Gonord et al., 1980; Cailleux et al., 1983). No radiometric ages have been published for these flows, however, and the rhyolites that cap Djebel Tarhat do not display the same easterly structural dip as the underlying conglomerates. Moreover, remnants of this rhyolite on the eastern flank of the prominence do not seem to be interbedded with the sedimentary succession, and the reassignation of these flows from the Pliocene to the Permian may be unwarranted.

Conglomerates of the Khenifra basin consist primarily of well-rounded quartz, quartzite, and metamorphic cobbles averaging 10-20 centimeters but ranging up to 30 centimeters in diameter. They commonly occur in a poorly sorted, pebbly to sandy matrix (Lorenz, 1974). The conglomerate beds are massive, 1 to 5 meters thick, and comprise about a quarter of the overall section. This proportion increases significantly at the northern end of the basin (Lorenz, 1974), and at the base of the section (Termier, 1936). The interbedded red sandstones are well rounded, and consist predominantly of metamorphic rock fragments.

Many of the conglomerates are matrix supported, very poorly sorted, and internally structureless. The 1- to 5-meter thick lenses contain few bedding planes, and can be traced laterally for a few hundreds of meters. Except at the northern end of the basin where they become amalgamated, these conglomerates are isolated units within the interbedded sandstones and mudstones. Locally, the basal contacts can be seen to be in erosional contact with underlying sandstones or mudstones, and in the latter case, large gravity-loading features can be seen. Cossey (1974) measured 14 distinct conglomerate beds within a 1050-meter-thick section in the eastern half of the basin. These conglomerates represent debris flows and braided stream deposits.

The poorly exposed, interbedded mudstones are also reddish brown and poorly sorted. Several local 10-30 centimeter-thick yellowish calcareous horizons are present in the mudstones (Termier, 1936), as well as zones of scattered 10-centimeter calcareous nodules (Lorenz, 1974). Where exposed, bedding is horizontal, even, and extensive. These deposits are composed of fluvial floodplain sediments. The calcareous horizons and nodules represent soil horizons formed during times of subaerial exposure and nondeposition.

Several thin (up to 6 centimeters thick) lignite beds were reported by Termier (1936). These occur near the top of the section and are composed of allochthonous plant fragments, without underlying rooted zones. The lignites are found within repetitive sequences of deposition. Four of these cycles were exposed in an excavation near Khenifra. Each cycle consists of a basal half-meter-thick conglomeratic sandstone, a thin lignite, a grey sandstone (0.2-4.5 m thick), and a capping calcareous claystone (0.3 to 6.0 m thick). The cycles are from one to seven meters thick.

Anomalous deposits of white, well-sorted sandstones occur at several stratigraphic intervals within the basin. The sandstones contain large and small crossbedding, and some planar-horizontal bedding. S. Cossey and G. Kelling (personal communication, 1973) described these deposits as containing one-to-two-meter-thick cyclic sequences with a basal scour, overlain by conglomerates that fine upward to crossbedded sandstones and finally to ripple-bedded sandstones. The units contain well-preserved plant fossils and some possible bivalves. The white sandstones grade laterally into the normal, barren, conglomeratic redbeds (Cossey, 1974). The fining-upward channel fills in this white-sandstone facies suggest fluvial deposition. The presence of well-preserved plant fossils, possible bivalves, and interference ripples suggest local lacustrine conditions.

The depositional environment of the rocks of the Khenifra basin was that of an intermontane fluvial complex. Conglomerate bed thickness and clast size generally decrease north to south, but crossbeds (Figure 2) indicate bimodal transport along the axis of the basin. Cossey (1974) mapped northerly crossbedding in the extreme northern end of the basin, and inferred a possible watershed just to the south. However, the predominant paleocurrent direction is generally to the south. Detritus may have been shed from upfaulted highlands along the length of the basin, with longitudinal transport down the center of the basin similar to the drainage of present-day linear basins of the Basin and Range province of the United States. The thinness of most of the conglomeratic units suggests that they were not well-developed alluvial-fan deposits, but rather were distal-fan debris flows (derived from well-developed basin-margin fans that were not preserved) and braided fluvial systems of longitudinal drainage along the basin axis. The 20 km long reverse fault that bounds the eastern edge of the basin (the Assaka fault: Cossey, 1974) may have been a syndepositional border fault. Thick, angular, border-fault conglomerates are not present adjacent to this fault, however, probably because the basin-margin alluvial-fan facies was continually reworked and destroyed during syndepositional thrusting.

Termier (1936) suggested that the distinctive tourmaline-bearing quartz clasts in the conglomerate were derived from the granitic terrane 25 to 30 kilometers to the northwest. This magnitude of transport distance would account for the well-rounded nature of most of the clasts. Termier also reported that despite the abundance of limestones in the nearby Paleozoic strata, there is an anomalous lack of carbonate clasts in the formation, and he suggested that debris was derived almost solely from the granitic source area.

The white, well-sorted sandstone facies may represent smaller parts of the established basin-axis fluvial system. These might have been periodically dammed by outflows of the conglomeratic debris-flow facies, creating the low-energy lacustrine conditions suggested by cyclic deposition of lignites and calcareous claystones, and by the unabraded nature of plant fossils found in the white sandstone facies.

Other basins

Most of the other late Variscan intermontane basins of Morocco show styles of deposition similar to the strata on the Khenifra basin. They differ principally in age, thickness, and the location of the lava flows (if present) within the stratigraphic column. All are coarse-grained, nonmarine, reddish-brown redbeds, and several are associated with interbedded acidic lava flows. Some, like the Bou Achouch deposits, are very small remnants preserved by infaulting. Carpentier (1930) reported a preserved thickness of only "a few dozen" meters of strata in this basin, noting that conglomerates occurred at different locations in the section at different localities. Locally, good evidence for lacustrine sedimentation exists in these basins, as in the Sidi Kassem area, where Termier (1936) reported several thin limestones that contain a fossil ostracod and gastropod fauna.The reddish-brown to red conglomerates with interbedded sandstones and mudstones of the Chougrane basin are also similar to those in the Khenifra basin. A lava flow is found near the base of the section in the Chougrane basin, and some of the clasts in the overlying conglomerates consist of volcanics from this flow (Termier, 1936). These basins represent the thinner, interbasin areas of deposition, where the sequence of preserved strata is generally less than 100 meters thick. In contrast, the Mechra ben Abbou area contains about 4000 meters of conglomeratic deposits, with several andesitic lava flows (Gigout, 1951).

The Late Carboniferous through Early Permian strata of the Colomb-Bechar basin offer an interesting contrast to the rest of the basins of this age in north-western Africa. Whereas the latter are situated within the orogen, the Colomb-Bechar basin is a foreland basin lying on the Saharan platform to the south of Morocco. The broad basin was probably formed as thrust sheets from the orogen (Michard et al. 1982) were pushed southward onto the platform, their weight causing broad subsidence of the lithosphere. The basin was much larger and underwent significantly less postdepositional deformation than the late-orogenic basins to the north. About 2000 meters of continuous section are present here, beginning with Late Carboniferous marine to paralic strata, grading up into nonmarine conglomerates of Late Carboniferous and Early Permian age. Coals occur in the lower conglomeratic intervals, whereas redbeds and thin beds of gypsum are found in the upper zones (Fabre, 1970; Van Houten, 1976a).

The significant difference between the Colomb-Bechar basin and the other basins of similar age in Morocco was recognized early. Termier (1936) noted that, similar to the European setting, there are "internal" limnic basins located within the orogenic province, and paralic "external" basins located along its margins.

Syn- and Post-Depositional Structural Modifications

The older strata in this suite of late Paleozoic deposits display more structural deformation than the younger ones (both from basin to basin and often within the same basin), indicating contemporaneous and ongoing structural changes associated with the last phases of the Variscan orogeny. The last of these pre-Late Triassic structural events initiated the uplift and erosion which denuded the Variscan terrane of Morocco, leaving the relatively featureless surface onto which Late Triassic sediments were deposited 30 to 40 m.y. later.

Triassic Basins and Deposits

Triassic rift-basin development

With the onset of continental rifting and the breakup of Pangea, rift basins developed on and around the wide, previously tectonized Variscan orogen of Morocco. Most of these basins were located along zones of weakness created during the orogeny (Choubert and Faure-Muret, 1962; Van Houten, 1977). Major rift basins developed along the South Atlas fracture zone, a zone of crustal weakness between two major structural provinces: the Variscan orogen to the north, and the Saharan platform to the south (Laville and Petit, 1984). Smaller rift basins developed on the flanks of the cratonized Meseta within the orogen (Van Houten, 1976a).

The major Triassic rift basins were filled with red, locally derived, sandy to conglomeratic, nonmarine sediments up to 4.5 kilometers thick. The depositional setting was one of alluvial-fan and fluvial environments, usually with longitudinal drainage along the axis of the basins. Other basins in the northern and eastern part of the country were filled with thick mudstones and evaporites in paralic environments, and contain very little coarse detrital sediment.

Recent sedimentological studies (Laville, 1981; Laville and Petit, 1984; Petit and Beauchamp, 1986) report that synsedimentary faults can be observed in the Triassic strata, and that most of the faults were dip-slip normal faults. They are connected by a subsidiary set of synsedimentary strike-slip faults. From such evidence, Laville and Petit (1984) reconstructed an extensional Triassic regime in which early extension was directed north-northwest, but where late extension was directed northwest, causing connections between the normal faults and reactivating ancient subparallel strike-slip faults.

As the basin-forming tectonic activity decreased and the rift basins were filled, more regional subsidence occurred, and evaporitic mudstones were deposited in the basins as well as over most of the low-relief surface of Morocco. During this stage of deposition, and possibly as a result of the same regional extension that caused regional subsidence, widespread basaltic lava flows were extruded, becoming interbedded with the red mudstones. Locally, marine carbonate deposits of Liassic age conformably overlie these red mudstones and basalts.

Previous work

The Triassic strata were described in the early 1900s by geologists such as Termier (1936), and Roch (1939). The general consensus of the early interpretations was expressed by Roch (1930, page 183), who stated that there were probably two facies. One was inferred to be entirely of nonmarine (desert and playa) origin, and the other facies was interpreted to be a lagoonal-marine deposit. Roch (1939) envisioned an arm of the Mediterranean [Tethys] extending as a great gulf down the trend of the Atlas Mountains as far as Marrakesh, another marine environment extending east from the Atlantic a short distance up the other end of the Atlas trend, and definitively nonmarine depositional settings between the two.

The next generation of geologists applied the more rigorous models of sedimentary depositional environments becoming available. These geologists included Busson (1974), Lucas (1962), and Salvan (1968a,b, 1974) who studied the evaporitic mudstones, and who proposed marine environments of deposition for most of these strata based on clay mineralogy and the presence of newly discovered, thick subsurface halites. Concurrently, Tixeront (1973), Mattis (1977), Brown (1980), and Lorenz (1976), working primarily with the detrital, coarse-grained graben fills, reconstructed nonmarine alluvial-fan and fluvial environments for the sedimentary strata of the local basins in central and southwestern Morocco. Most recently, Beauchamp and his colleagues have studied these strata, making new sedimentological interpretations, and finding important evidence of syndepositional tectonism in the sedimentary strata.

Stratigraphy

Most of the vertebrate fossils found in these strata indicate Carnian and younger Triassic ages (e.g., Dutuit, 1966), and palynological analysis (Couzminer and Manspeizer, 1976; Taugourdeau-Lantz 1978) confirms this. Radiometric dating of the basalts is hampered by a high degree of alteration, but generally these rocks yield Late Triassic to Early Jurassic ages (178-200 my: Krueger 1973; Manspeizer et al., 1978; Van Houten, 1977). Few age-specific invertebrate fossils have been reported, and these strata remain imprecisely dated.

Different authors have interpreted the fossils and the radiometric age data derived from the intercalated basalts in different fashions. One school of thought (Manspeizer et al., 1978; Salvan, 1984) holds that the late Middle Triassic (Ladinian) is represented in the basalt-mudstone sequences of northeastern Morocco and the Middle Atlas, whereas the lavas are close to the age of the Triassic-Jurassic boundary in the High Atlas (with the underlying clastic deposits possibly as old as Ladinian), and that they are even younger (Sinemurian) on the Meseta or in southern Morocco. Thus, up to four different episodes of basaltic volcanism are postulated to have existed, an interpretation that may be supported by the apparently different chemical compositions of the lavas between these provinces (Manspeizer et al., 1978, Figure 6).

The alternative view is presented by Van Houten (1977), in which the badly altered nature of the basalts accounts for the wide range of radiometric dates, and the poorly defined range of the fossils accounts for other imprecise dating. In this view, detrital graben filling began in early Late Triassic (Carnian) time, but the upper parts of the more widespread evaporitic mudstone and basalt facies are of early Liassic age, and are essentially isochronous across Morocco. With the exception of the basal High Atlas deposits (which are possibly as old as Early Triassic) Van Houten found no evidence for Triassic deposition earlier than the early Late Triassic in Morocco.

Regional Tectonic Setting of Morocco during Triassic Time

Prior to the breakup of Pangea in the early Mesozoic, Morocco fit against North America between Nova Scotia and Newfoundland. The nearest open seaway was the Tethys, located 1000-1500 km to the east. The area of present northwest Africa was 0 to 5 degrees above the equator, in a zone of year-round warm, dry climates (Robinson, 1973, Figure 10).

In Early Triassic time, Morocco was an area of nondeposition and erosion with an easterly paleoslope (Van Houten and Brown, 1975, 1977) (Figure 3). Paralic sediments were deposited on the margins of the Tethys to the east. Areas of the deposition spread eastward with the transgression of the Tethys during Middle Triassic time but did not reach Morocco. In late Late Triassic time, a global regression occurred, but despite this, regional subsidence allowed encroachment of paralic environments into easternmost Morocco. The Late Triassic marginal-marine evaporite and evaporitic mudstone facies in southwestern Morocco were associated with transgression of waters eastward from the newly opening Atlantic, which had been flooded from the Tethys through the Gibraltar fracture zone (Van Houten and Brown, 1977).

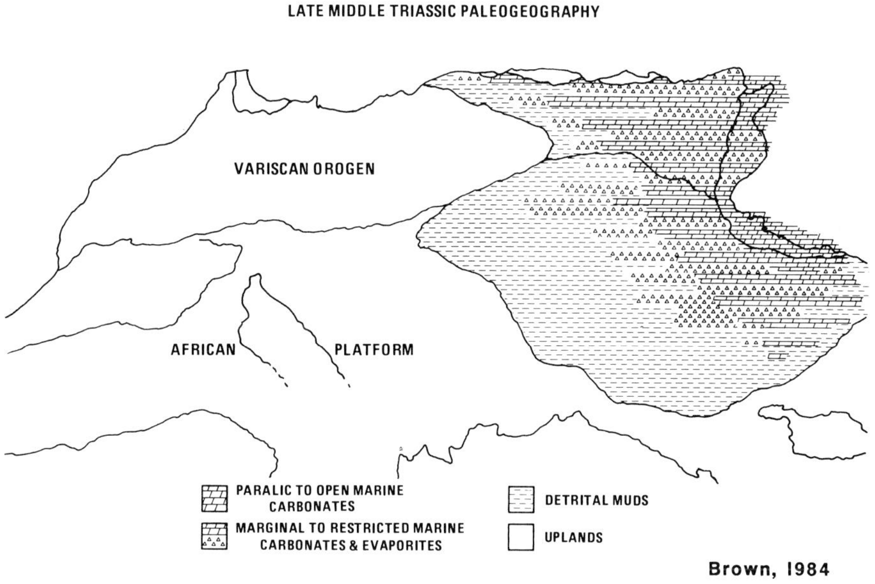

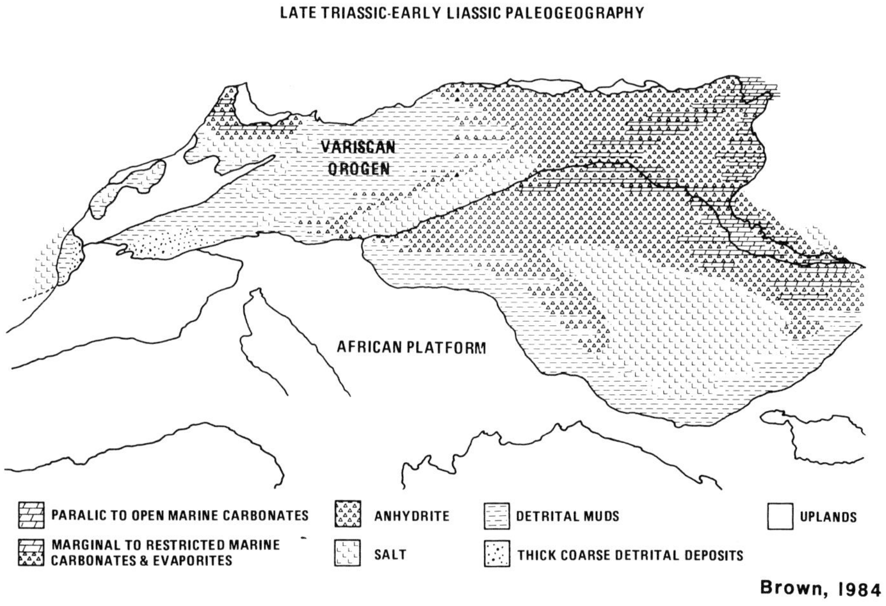

Figure 3. Mesozoic paleogeography and lithofacies of northwest Africa: (R. H. Brown, personal communication, 1983)

Sedimentology of the Detrital Rift Basins

Fifteen known basins of Late Triassic-Early Jurassic age occur in Morocco (Figure 4). Several of these may have been contiguous during the Triassic, and most have an east-northeast or northeasterly trend, although the limits of many are still poorly known. However, only about half of these basins are known to contain significant thicknesses of conglomeratic/sandy sedimentary strata, and only four were filled predominantly with coarse clastics. The other basins are filled primarily with halite and thickenings of the widespread evaporitic mudstone facies, although even these commonly contain thin, basal layers of coarse clastics in the unexposed subsurface. Many of the basins are known to contain internal basement faults that were active during deposition, accounting for abrupt lateral variation in the thickness and distribution of facies within the basins.

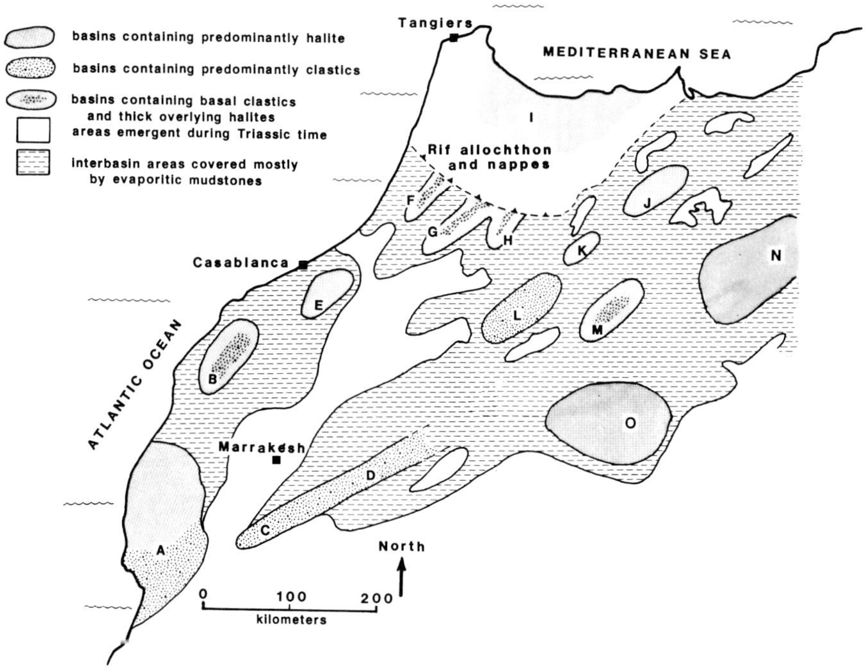

Figure 4. Generalized locations of rift basins and extent of Triassic deposition in Morocco (after Salvan, 1974, 1984; and Van Houten, 1977). A-Essaouira, B-Doukkala, C-N'Fis, D-Central, E-Berrechid, F-Rharb, G-Khémisset, H-Bou Fekrane, I-Pre-Rif depression, J-Guercif, K-Mousa ou Salah, L-Kerrouchen, M-Tamdafelt/Molouya, N-Hauts Plateaux, O-Ziz-Guir. (Moroccan Meseta is roughly coincident with the large area of nondeposition in central Morocco.)

Kerrouchen basin

The Kerrouchen basin is structurally and stratigraphically simple, and is relatively well exposed. Triassic deposits of this area overlie a highly metamorphosed (phyllitic) basement. A large granitic body adjoins the southeastern edge of the basin, directly underlying the evaporitic mudstone facies that extends beyond the basin boundaries. The faulted granite-phyllite contact trends east-northeast, forming the southeastern margin of the basin.

Prior to active faulting, a thin, erratically distributed deposit (consisting of sandy carbonates and locally derived conglomerates and sandstones) accumulated over the low-relief, eroded Variscan topography. With the onset of faulting, about 250 meters of conglomerate and arkosic sandstone were eroded from the granite and deposited in alluvial-fan and braided-river environments near the border faults of the southeastern margin of the basin (Lorenz, 1976).

The coarse-grained deposits on the southeastern side of the basin are of two types. Alluvial-fan conglomerates in the Djebel Tanourdi region consist of reddish-brown, mud-rich, matrix-supported, poorly sorted cobble and pebble conglomerates, with granitic clasts up to 30 cm and smaller metamorphic and sedimentary clasts. These deposits are debris flows and mud flows. However, the greater part of the conglomerate and coarse sandstone exposed along this edge of the basin is light-purple, somewhat better sorted, and generally mud free. These deposits, in the vicinity of the Ichoua River west of Itzer, display bedding that is laterally extensive across outcrops, but with local channel lenses. Each one- to two-meter-thick bed has a basal scour surface and may fine upward, but otherwise they are internally massive or less commonly, crossbedded. These are braided-stream deposits on distal alluvial fans. Both the stream and debris-flow facies contain irregular carbonate-rich caliche paleosols.

A sandy fluvial system prograded out from the opposite (northwestern), apparently unfaulted, margin of the basin. Petrographic character shows that this is a basal Triassic facies (Lorenz, 1974) rather than an extension of the late Paleozoic Khenifra system as suggested by Termier and Termier (1948). Maximum thickness of these deposits is 80 m. They contain several examples of small syndepositional faults, offsetting both basement and Triassic rocks by up to 4 meters. These faults (exposed in the gorge of the Srou River) are associated with local clasts derived from the contemporaneous deposits of sandstone during faulting.

The associated sandstones on this side of the basin are fine to medium grained and well sorted. Fining-upward sandstones in channel lenses 2-3 m deep and 10 m wide are common, locally interbedded with units of red to brown, finely-laminated, unburrowed, graded mudstone beds of probable lacustrine origin. Lenses and basal ripup clasts suggest fluvial deposition. Where these lenses are interbedded with lacustrine mudstones, they may represent deltaic distributary channels (Lorenz, 1976).

Paleodrainage in these two systems on opposite edges of the basin was largely parallel to the basin axis, indicating that the marginal highlands channeled drainage between them. However, opposed crossbedding in the two systems suggests that they were active at slightly different times, while the absence of granitic detritus and the high percentage of metamorphic rock fragments in the western fluvial deposits suggest that there was little overlap of the two systems. Subaerial exposure is indicated by the development of red/grey mottled caliches in some of the conglomerates, and by the presence of several generations of large desiccation polygons in the mudstones sandstones.

With the waning of tectonic activity, the highlands adjacent to the basin were worn down and provided increasingly finer material to the basin-fill deposits. A 75 m thick sandy fluvial redbed sequence marks distal facies of the fans and the transition from alluvial fan conglomerates to mudstones near the faulted basin margin. Wide shallow channel lenses (30 cm deep by 4 m wide) occur in this transitional unit. They are filled with horizontal-parallel bedded sandstones, and display basal groove and load casts (Lorenz, 1974), suggestive of ephemeral sheet flooding. Ripple-marked tops and invertebrate tracks and trails on the upper surfaces of these sandstones, as well as laterally-extensive associated beds of mudcracked, burrowed, silty mudstone, indicate that the sheet floods flowed into an ephemeral playa-lake setting. Irregular carbonate lenses and nodules, and common scattered well-rounded sand grains within the mudstones (Lorenz, 1974) suggest arid-climate caliche paleosols and local eolian deposition.

Farther up on the edge of the granite, this transitional facies is absent, and the conglomerate-mudstone contact is abrupt, possibly indicating an erosional hiatus associated with upfaulting of that section of granite. Where the mudstones directly overlie the beveled granitic surface south of the basin, a one- to two-meter-thick Triassic weathering profile of in-place granitic "grus" is commonly present at the contact (Amade, 1962; Lorenz, 1974), indicating exposure and weathering in place.

The red mudstone facies is thicker within the Kerrouchen basin than nearby, but it is part of the widespread evaporitic mudstone facies that covers much of Morocco. Within the basin the mudstones are about 275 meters thick and the basalts are about 225 meters thick (Lorenz, 1976). (As many as seven basalt flows are distinguishable locally, but pillows are exceptionally rare.) Bedding in the mudstones is laterally extensive (hundreds of meters minimum observed in outcrop), and on the order of half to several meters thick. Weathering has obscured sedimentary structures, but greenish-gray reduction mottling can be seen locally. Some silty and dolomitic layers are present, forming erosion-resistant ledges. At least one halite bed, perhaps as thick as five meters, has been exposed by small-scale underground mining in the center of the basin (Lorenz, 1976), and Termier (1936) reported a salt bed within the strata of Djebel Kerrouchen. Evidence of halite is common in mudstones along the axis of the basin, whereas gypsiferous mudstones are common at or beyond the margins of the basin (Lorenz, 1974), indicating a facies partition of the evaporites created by differential subsidence. Lorenz (1976) interpreted these strata as having formed in a nonmarine, evaporitic, playa-mudflat environment.

Essaouria basin

The Essaouria basin is located along the South Atlas fault zone, a structural discontinuity that follows the High Atlas trend, extending westward offshore to the continental shelf and eastward into Tunisia, and which existed prior to the creation of the Triassic rift system (Brown, 1980, Laville and Petit, 1984). Triassic tectonic stresses were oriented oblique to this structural trend, but its existence controlled the orientation of the basin. Because of the internal longitudinal block-faulted structure within the basin, and because of the existence of the high-standing Massif Ancien to the east and the proximity to the proto-Atlantic to the west, the stratigraphic succession and the facies patterns within the Essaouira basin are more complex than the basin margin-to-basin center, coarse-to-fine type of graben fill found in the Kerrouchen basin.

Beginning in early Late Triassic time, major fracturing occurred along the South Atlas fault zone in association with continental rifting. Fracturing created a wide zone of subsidence containing internal block faults that underwent syndepositional vertical displacement and rotation. The Paleozoic Massif Ancien, located at the east end of this irregularly subsiding basin (Figure 5), remained as a high-standing upland (Salvan, 1984) from which coarse detritus was eroded and shed westward into the Essaouria basin.

Over 4500 m of sediment was deposited in this basin in nonmarine environments that ranged from alluvial fans to lakes to meandering streams. Syndepositional tectonism produced interformational unconformities and rapid variation in unit thicknesses across faults (Brown, 1980). Erosion in the highlands exposed successive horizons in the metamorphic and igneous source area, finally unroofing the granitic batholiths of the area. This sequence can be recognized as an inverted mineralogical succession in the basal conglomerates in the adjacent basin fill (Jones, 1975). As the highlands were worn down, the lower relief produced a finer-grained sediment, and the depositional environments became those of lower-energy lacustrine and deltaic settings, in which reptiles and other amphibians lived (Dutuit, 1966).

A mudflat sequence, containing the ubiquitous basalt flows, overlies the lacustrine deposits and marks the last contributions of the severely eroded highlands. These mudstones and the overlying succession of evaporites and Jurassic carbonates record a transgression into the subsiding basin from the west by waters of the proto-Atlantic.

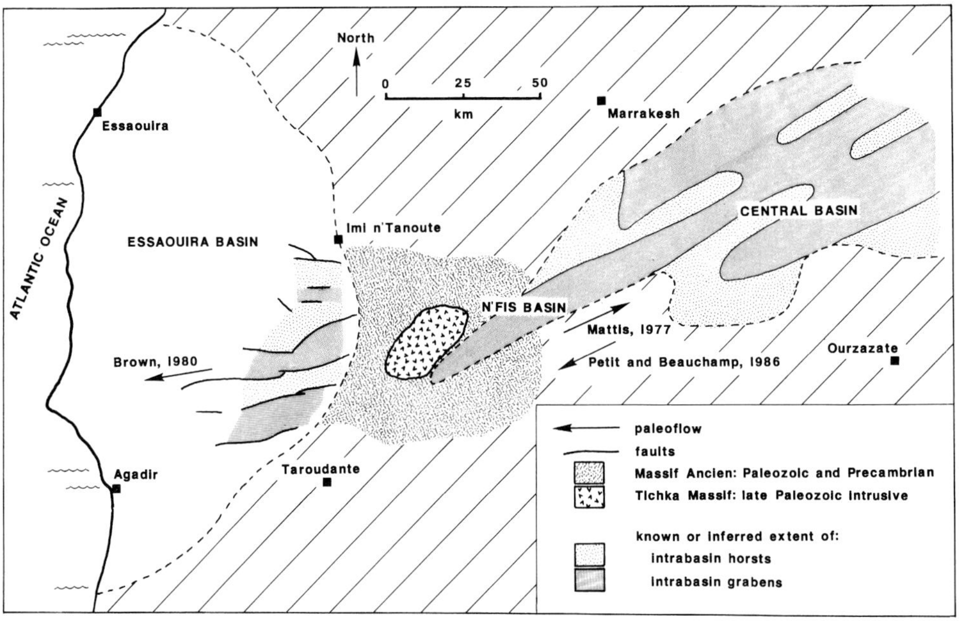

Figure 5. Triassic of the High Atlas region, Essaouira basin, Central-N'Fis basin complex, and Massif Ancien. (Data from Brown, 1980; Mattis, 1977; and Petit and Beauchamp, 1986.)

Central-N'Fis basin complex

The thick (up to 1000 m) Triassic deposits of the central High Atlas Mountains are located in a series of parallel sub-basins. The Central-N'Fis basin complex is situated along the same structural zone as the Essaouria basin (Figure 5), but it is located on the opposite (east) side of the Massif Ancien. The basin is faulted parallel to its axis, as is the Essaouria basin, but the resulting relative movement has been reconstructed in a classical horst and graben configuration (Mattis, 1977), rather than faulting with rotated blocks as in the Essaouira basin. Mattis (1977) identified different graben-fill and horst-cover facies within the basin. The former, about 1000 meters thick, begins with a conglomeratic sequence, grades up into a siltstone unit, thence to a thick sandstone, and finally to an upper siltstone/mudstone unit which contains basalt flows. Over the horsts, the sequence consists of a thin basal conglomerate, grading quickly into the siltstone/mudstone/basalt series, and the entire succession is only about 250 meters thick.

An andesitic lava flow that occurs at the base of the central High Atlas sequence has yielded radiometric age dates of 203±10 m.y. and 262±12 m.y. (Van Houten, 1977). Van Houten accepted the former date as valid as the associated strata resemble better-dated nearby Triassic deposits, and because Triassic lavas of similar composition are known from northeastern Morocco and Algeria. On this basis, Van Houten suggested that Triassic deposition started significantly earlier in the central High Atlas trend than elsewhere in Morocco. Tane et al. (1983) however, believe that the older date may be more accurate, and correlate these strata with the Permian deposits containing similar lava flows in Morocco.

Mattis (1977), Van Houten (1977), and Petit and Beauchamp (1986) have interpreted the basal graben-filling coarse-grained formations as alluvial fan-fluvial deposits. Crossbedding (Mattis, 1977, Figure 1) indicates northeasterly transport away from a source area in the Massif Ancien, along the basin axis. This paleodrainage pattern is the mirror image of that in the Essaouria basin. The easterly paleoslope trend is supported by the west to east trends of diminishing conglomerate clast size and conglomerate bed thickness (Mattis, 1977) although clast imbrication suggests west to southwesterly paleoflow (Beauchamp, personal communication, 1987). Jones (1975) plotted easterly crossbedding in the coarse conglomerates of the Tizi-n-Test region at the western extent of this basin.

Along the central basin trend, the basal conglomerates thin from 325 to 20 meters west to east, and locally fine upward to sandy meandering stream deposits (Mattis, 1977). They are overlain by a siltstone and mudstone unit that is about 250 meters thick. This unit has been interpreted as the deposits of a river-floodplain complex (Mattis, 1977) on the basis of local crossbedded channel fills and bioturbated muddy siltstone. Petit and Beauchamp (1986), however, interpreted this unit as representing a tidal-flat environment on the basis of symmetrical wave ripples and casts of inferred marine bivalves.

The overlying 200 to 400 meter-thick Okaimaden Sandstone is also interpreted differently by different authors. Fining-upward, lenticular, crossbedded sandstones are inferred to be of fluvial origin by both Mattis (1977) and Petit and Beauchamp (1986), but Petit and Beauchamp reported that the lower two-thirds of this unit contains sandstones with low-angle parallel laminations and fragments of marine fossils, suggestive of a beach environment. Similarly, the interbedded, bioturbated mudstones were assigned to either floodplain or lagoonal settings by the different authors. Petit and Beauchamp (1986, Figure 2) measured and plotted crossbedding that indicates southwesterly paleoflow. They interpreted this to reflect strong long-shore currents in an environment that was increasingly marine-influenced upsection, and inferred that the sediment source was to the southeast.

The marine interpretation is based primarily on size and sorting parameters of the sandstones that are presumed to be environmentally diagnostic, and on the interpretation of several undescribed invertebrate fossils, although Petit and Beauchamp briefly note sedimentary features such as herringbone crossbedding in support of their interpretations. Although southwesterly paleoflow was also reported by Proust (1973), this interpretation does not account for the significantly different crossbedding vectors measured earlier by Mattis (1977) or the unroofing sequence of the Massif Ancien documented by Jones (1975). This significant discrepancy in basic data has yet to be acknowledged or addressed. Petit and Beauchamp (1986) and Armand et al. (1980) suggest that the Massif Ancien was subsident during the Triassic, and that paleocurrents were continuous across it between the Central-N'Fis and Essaouria basins. Their reconstruction envisions these two basins as having formed at the southern margin of a primarily marine setting that covered most of Morocco to the north, with longshore currents sweeping along this border toward the southwest.

The final deposits of the central High Atlas area are 175-200 meters thick, and consist of evaporitic, red siltstones and mudstones with intercalated basalt flows. Petit and Beauchamp (1986) suggest that these were deposited in a lagoonal environment, whereas Mattis (1977) inferred that they were deposited in fluvial floodplain and playa environments. Peretsman (1985) concluded that the low bromine concentrations and widely variable sulfur and strontium isotope ratios in the evaporites of the Atlas region indicate that they were deposited in "continental sabkhas and playa lakes," supporting Mattis's interpretation.

Relatively pure halite beds up to 10 meters thick are found in the mudstones, most commonly above the basalt flows (Roch, 1939) and at the northern edges of the basin (Mattis, 1977). The interbedded basalt flows consist of widespread flow units 5 to 30 meters thick with oxidized vesicular flow tops indicative of subaerial extrusion. They are not pillowed in the graben areas, but local pillowed zones can be found in the thinner horst-cover facies and at the northern margin of the basin (Mattis, 1977).

Other Detrital Rift Basins

Several other Triassic basins in Morocco contain basal sequences of coarse-grained detrital sediment, but none of them contain the exceptional thickness of coarse sediment that accumulated along the High Atlas trend. In northern Morocco, the southeastern ends of the Rharb, Khēmisset, and Bou Fekrane basins extend beyond the cover of the thrust sheets and sedimentary apron of the Rif Allochthon. Burgess (1975) reported 100 meters of Triassic border-fault conglomerate in the subsurface of the Khēmisset basin, noting that it interfingered with finer-grained rocks and ultimately with evaporites away from the border fault. The central mudstone/evaporite sequence spread with time, lapping over the adjacent conglomerates and border highlands. Similar patterns are present in the Doukkala, Tamdafelt, and Guercif basins (Salvan, 1968a,b, 1974).

Most of these coarse basal units are less than 100 meters thick. This highlights the effects of the pre-existing zones of weakness at the edge of Variscan orogen (the Essaouria and Central N'Fis basins) as compared to the basins that formed within the orogen. Subsidence rates and magnitudes in the former area were maximized early in their history, when high relief was produced and coarse sediments were deposited rapidly. To the north, topographic relief was less because subsidence was slower and more even. The basins there were filled more slowly, and thus with predominantly fine-grained detrital sediments and evaporites.

Salt-filled Rift Basins

Thick sequences (exceeding 2200 meters in the Khēmisset basin; Amade, 1965) of mudstone and relatively pure halite occur in many of the northern and eastern Triassic basins. These strata are associated with a minimum volume of conglomerate, evidence for low relief in what are probably fault-bounded basins. If one accepts Van Houten's (1977) interpretation that the basalt flows are essentially isochronous across Morocco, the positioning of the basalts high in the sedimentary section of the coarse-grained graben fills, as compared to their intermediate to low stratigraphic position within the thick evaporite-mudstone basin fills, suggests that the rates of subsidence were different for the two categories of basins. Thus, whereas rapid subsidence and coarse sedimentation in those basins situated on the South Atlas fault zone produced thick deposits early in the history of the basins, slower, more uniform subsidence occurred in the basins that are located on the less tectonized basement to the north, and only finer-grained strata accumulated there prior to basalt extrusion. Therefore the northern basalts need not be older as suggested by Salvan (1968b) and Manspeizer et al. (1978), but rather may only appear to be lower in the stratigraphic section because of the differences in rates of Triassic accumulation.

Thick beds (tens to hundreds of meters) of relatively pure halite occur within the deep central zones of many basins. In several of these basins, commercial deposits of the potassium-bearing, late-stage evaporites such as sylvite and carnallite are interbedded within the halite sequences (Amade, 1965; Salvan, 1984). Some of the thicker beds of halite have been remobilized, and salt diapirs from this horizon occur in both northeastern and southwestern Morocco (Salvan, 1974). These important basins are primarily known from subsurface data, however. Although Salvan (1984) mentioned undulating horizontal bedding and "polygonal patterns" in the salt beds exposed in mine workings in the Berrechid basin, detailed sedimentological descriptions of the evaporitic basin fills have not been published. Nonetheless, the thick halite strata have been interpreted as either deep-marine deposits of starved basins, or as shallow, ephemeral, salina deposits.

Salvan (1974, 1984) and Busson (1974) used the evidence of Lucas's clay mineralogy and the presence of thick halites, as well as an apparent landward brine fractionation pattern (described below), to suggest that the halite and potash facies of the northern and eastern basins are deep-water, marine deposits. The alternative paralic to nonmarine interpretation of these strata was most recently presented by Van Houten (1977, p. 85), who stated that within the basins, the lateral "succession from sulfate to thick salt points to precipitation in structurally restricted saline lagoons fed by marginal marine incursions [where] extreme evaporation [produced] deposits of potash salts." The polygonal patterns reported by Salvan (1984) in bedded halites may be analogous to well-studied polygons in Triassic halites in England. Tucker (1981) interpreted these as indicative of shallow-water to commonly subaerial environments, which would support Van Houten's model.

Data on the bromine content and the strontium and sulfur isotope ratios of the Moroccan evaporites have not yet been published, but Peretsman (1985) reports that these data from the evaporites of the northern coastal basins suggest significant nonmarine contributions to the brines from which the evaporites precipitated (despite earlier interpretations of a more marine nature; Peretsman and Holser, 1984). Although these deposits do indicate a marine influence, they were not precipitated in open-marine environments: significant mixing of marine-derived brines and waters from nonmarine sources occurred at the sites of evaporite deposition, probably in environments of "marginal marine lagoons or coastal sabkhas." Peretsman's data also suggest, however, that the marine influence increased upsection in these areas, consistent with a Late Triassic-Early Jurassic westward encroachment of paralic environments toward Morocco. Van Houten's paralic lagoon and salina model is most consistent with these data.

Widespread, Evaporitic Mudstone and Basalt Facies

Evaporitic mudstones are found in the salt basins, in the upper sections of the detrital basins, and as a thin but widespread facies in most areas between the basins. Their color is almost exclusively a striking brick-red, although thin beds of variegated to light green/grey rocks occur locally. The mudstones are broadly bedded as seen from a distance, but are usually too badly weathered to preserve sedimentary structures on a more detailed scale. Intercalations of sand and silt are common, especially near the base. Evidence for the more soluble halite is locally present as effluorescent crusts on outcrops.

Many early workers (Coubert and Faure-Muret, 1962) considered the strata to be of marginal-marine to nonmarine origin. Salvan (1968a,b, 1974, 1983), and Busson (1974), extending their deep-marine interpretations from the graben-filling halite sequences, proposed a more widespread shallow-marine environment, although they suggest that narrow zones of paralic and nonmarine environments were present near the Atlas Mountains. Van Houten (1977) and his students (Lorenz, 1976; Mattis, 1977; Brown, 1980) used nonmarine playa interpretations in their work with the evaporitic mudstone sequences associated with the conglomeratic graben fills. The following review presents the evidence used by different workers to support their interpretations of depositional environment for these rocks.

1. Evidence from the mudstones

The work of Lucas (1962) on the clay mineralogy of the widespread red mudstones of Morocco is the primary work in this field. Lucas noted that whereas illite is present throughout the border facies and some sepiolite occurs below the basalt, attapulgite is the dominant clay mineral above the basalts. Sepiolite never occurs above and attapulgite never below the basalts. From these data, Lucas concluded that sepiolite and attapulgite are neoformed (authigenic) clays, that the conditions under which they formed were similar, but that with the outpouring of basalt, aluminum ions were made available in sufficient quantity to form attapulgite rather than sepiolite. More importantly, Lucas (p. 151) inferred that because of the absence of montmorillonite, carbonates, and authigenic silica, the most common modern environment of deposition for these two clays (that of alkaline lakes) is in fact not probable for the Triassic deposits. Instead, he proposed that they formed in a restricted epicontinental sea, "similar to a lake" in its abundance of ions and alkaline chemistry.

The neoformation of many of the component clays in these strata is an attractive hypothesis since most of the possible source terranes for detrital clays were eventually covered by the red mudstones. However, little definitive sedimentological evidence for environmental interpretations has been extracted from the mudstones. Mattis (1977; 1984, personal communication) interpreted thin beds of gypsum and dolomite within the mudstones of the High Atlas as playa deposits because of their limited lateral extent (less than 100 meters), their interfingering with fluvial deposits, and the presence of ostracods. Brown (1980) cited the boron content of clay minerals and the presence of fossil estheriids as evidence for brackish water. Termier and Termier (1948) reported unique occurrences of molds of halite hopper crystals ("le profil de troncs de pyramides en escalier": page 402), scattered ferruginous oolites, and fossil fragments of the conifer <u>Voltzia</u>, all from the red, evaporitic mudstone facies above the basalt flows in the northern Middle Atlas Mountain region. From this evidence, Termier and Termier inferred a nonmarine depositional environment for these strata; one of saline playas ("chotts") and ephemeral freshwater lakes. Further reports of these fossils (in Lucas, 1962, pp 101-102) identified ferns, also indicating a nonmarine environment. In easternmost Morocco and northern Algeria, however, the red evaporitic mudstone facies contains authigenic bipyramidal quartz crystals (Lucas, 1952), suggestive of groundwater reflux in a coastal sabkha environment.

The pattern which emerges from this poorly defined assemblage of evidence is not unlike Roch's early (1930) model: there are facies changes from paralic marine to nonmarine across North Africa. However, no evidence exists for truly shallow-marine conditions, except possibly near the eastern border with Algeria, and there remain questions of how far and how often the Late Triassic sea temporarily flooded northern Morocco in order to provide sources for the thick halites, and of how deep the marine waters were, when and where present.

2. Evidence from the basalts

The basaltic lava flows which are ubiquitous within the mudstone/evaporite facies vary from 0 to 500 m thick, but are generally on the order of 150-200 m thick (Salvan, 1974). The total basalt formation thickness is commonly uniform over distances of several tens of kilometers, although locally, individual flows can be seen to thin from 14 to 8 m across a distance of 300 m (Mossu, 1954). The basaltic sequence consists of one or more flows (up to 14 reported; Van Houten, 1977) with or without interbedded mudstones and carbonates.

Several excellent examples of pillow basalts with radial structures and vitreous rinds have been reported. Pillow basalts occur from the High Atlas (Tane et al., 1983, Mattis, 1977) to the Meseta (Cogney et al., 1971, 1974, Cogney and Faugères, 1975), but the few published measurements of lateral extent (Mattis, 1977, Cogney et al., 1971) suggest that the pillowed zones are only tens to a few hundreds of meters in extent. Several authors have in fact noted that the Triassic basalts in many areas are principally nonpillowed (western High Atlas,

Brown, 1980; central High Atlas, Tane et al., 1983, Mattis, 1977; western Middle Atlas, Lorenz, 1974). Moreover, very few pillows have been reported from the basal flow of any sequence. These observations suggest that the initial basaltic outpourings flowed across dry, subaerial surfaces, but that the resulting disruption of drainage produced pockets of water, and subsequent flows were pillowed in these areas. This ponded-drainage model was suggested by Mattis (1974, 1977) to explain localized occurrences of thin (1-2.5 m) carbonate beds above basalt flows in the High Atlas (Figure 6). Similar localized (less than 500 meters lateral extent) carbonate beds were also reported by Lorenz (1974, 1976) in the Middle Atlas. They lack fossils except for lacustrine stromatolites and gymnosperm pollen (Mattis, 1974; Taugourdeau-Lantz, 1978), and pinch out entirely or grade into red mudstones laterally. One outcrop contains a thin lignite between the two carbonate beds above a basalt (Lorenz, 1974), and silicified fossil wood is common in many of these deposits (Taugourdeau-Lantz, 1978), suggestive of nonmarine environments.

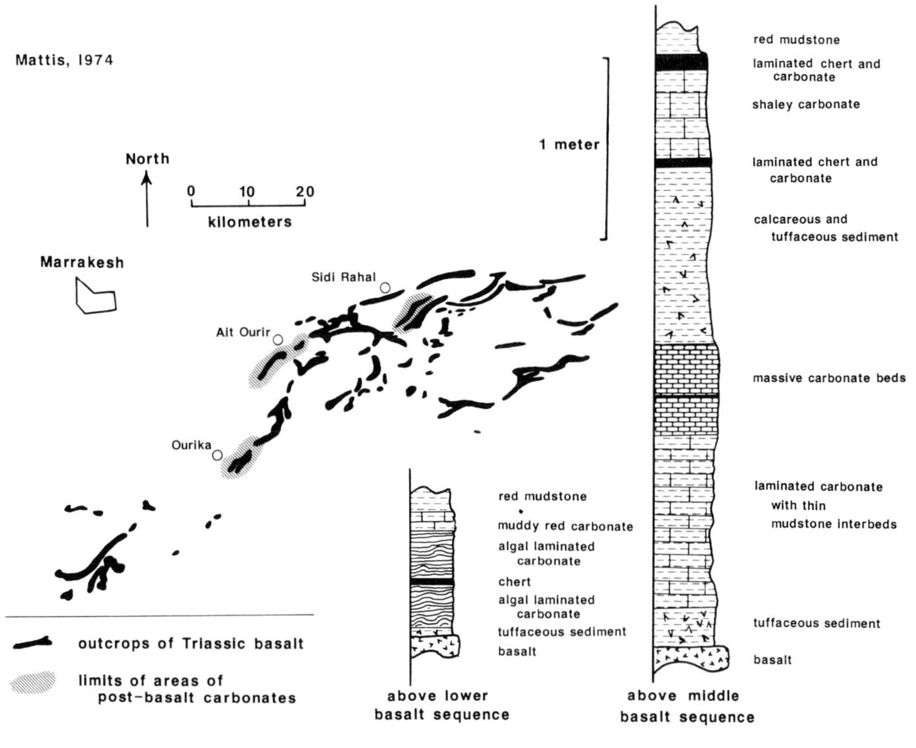

Figure 6. Map showing carbonates localized along the basin strike, and closely associated with basalt flows in the High Atlas area. This suggests a genetic relationship of ponded drainage between the basalts and the carbonates (A. F. Mattis, personal communication, 1984. See also Mattis, 1974.)

Carbonates associated with the basalts in northeasternmost Morocco may be of a different nature, containing invertebrate fossils (Taugourdeau-Lantz, 1978) and being of greater lateral extent. Cogney and Faugères (1975) reported one carbonate series northeast of the Meseta which displays stromatolites, birds' eye structures, plant debris and carbonate breccias among other features, and which they interpreted as an intertidal/supratidal facies bordering on an epicontinental sea. Pillows are present in the associated basalts, but the pillowed zones are not extensive. In easternmost Morocco and northern Algeria, the flows are disrupted into discontinuous, large-scale basalt-mudstone rock mixtures termed "ophites" (Lucas, 1952). Lucas suggested that the high-density basalts flowed out across, and sank into, plastic "lagoonal" muds, resulting in this rock fabric. Evaporitic, low-density, plastic mud is a common facies on sabkhas ("ploughed ground" or "puffy ground"), and ophites may be evidence for such unconsolidated muds in Triassic time.

The available evidence suggests that most of Morocco was subaerially exposed at the times of lava extrusion. The exceptions are in the far northeastern areas of the country, where there may have been paralic-marine influences. Data on basalt pillows from this northeastern area have not been reported, but the extensive, fossiliferous carbonates and the "ophite" lithofacies suggest a marine influence (although not open-marine conditions). In the Atlas regions, localized or absent pillows and localized unfossiliferous carbonates suggest nonmarine, subaerial deposition of the mudstone facies.

3. Evidence of the evaporites

Salvan (1974) mapped the mudstone/evaporite facies of Morocco on the basis of type and percentage of evaporite mineral. He identified several facies based on mineralogy, and inferred that the facies reflect a lateral gradation from open-marine carbonates through sulfate minerals, to an evaporitic halite facies, in the transition from the Triassic limestones facies of the Rif mountains, south through pre-Rif sulfates, into the halites of the Khemisset and Bou Fekrane basins. This pattern superficially fits the theoretical landward-evaporation brine concentration models of evaporite distribution, although the Rif mountains are allochthonous (Bosellini and Hsü, 1973; Michard, 1976). In general, Salvan's facies grade from "intermediate" to "marine" both east and west from the Massif Ancien and Meseta, areas that remained emergent throughout Triassic time.

Salvan (1974) and Busson (1974) interpreted the widespread mudstone/ evaporite lithofacies as the deposits of shallow ("more or less deep") marine environments. Salvan (1984) suggests that the overlying Liassic carbonate facies in fact represents only changes in the chemistry of an existing shallow sea, rather than a transgression. Van Houten, however, envisioned the widespread Moroccan Triassic mudstone/evaporite facies as the deposits of a "vast expanse of salt flats," commonly desiccated but susceptible to episodic marine flooding from the Tethys to the east.

Conclusions

In the absence of more rigorously definitive data, such as assemblages of sedimentary structures, environmentally diagnostic fossil evidence, or well-defined local facies transitions, it is difficult to assess the environmental models that have been presented for the evaporitic mudstones of Morocco. However, Van Houten's paralic salina and mudflat model more readily explains the known data, especially the geochemical measurements on the evaporites. At the end of Triassic time, Morocco was a vast, low-relief depositional plain, consisting of local, episodically flooded salinas in which halite precipitated, with broad, usually subaerial, clastic/evaporitic mudflats between these basins. At the time of basalt extrusion, the margins of the Tethys were located only as far west as northern Algeria and easternmost Morocco, although temporary incursions of marine brine intermittently invaded much of northern Morocco.

The global sea level curves of Hallam (1984), Vail et al. (1977), and Haq et al. (1987) show an overall eustatic sea level rise in the early Late Triassic, but Morocco was generally emergent over and above this rise, possibly due to its position on the rifted cratonic margin. In later Triassic time, regional subsidence allowed marginal marine incursions into Morocco despite a gradual eustatic sea-level fall. It was not until post-rifting regional subsidence and local foundering of the Atlas troughs, combined with eustatic sea-level rise in Jurassic time, that both the Tethys and the Atlantic were capable of flooding the High and Middle Atlas regions and creating open-marine conditions there. In Morocco, Triassic eustatic sea level changes were overshadowed by local tectonism, and therefore had no direct influence on sedimentation.

Summary

The Late Carboniferous-Permian and Triassic-Early Jurassic deposits of Morocco record a complex array of nonmarine and marine-influenced sedimentation. Despite their common characteristics of red color and commonly conglomeratic nature, they are the responses to two different structural and paleogeographic settings. During Late Carboniferous to Permian time, local late-orogenic basins on the Variscan orogen were filled with alluvial-fan, lacustrine, and fluvial sediments. Continued tectonism destroyed the basins, exposing many of these strata to erosion, and an erosional hiatus followed.

Late (and possibly Middle) Triassic time marked the renewal of sedimentation in Morocco. Rift basins formed at this time in response to crustal extension during continental rifting. The basins were filled with either coarse conglomerates and sandstones, or mudstones and evaporites, depending on whether they were located on rapidly subsiding fractured crust at the junction of two structural provinces, or on the less tectonized crust farther north. Proximity to marginal-marine environments adjacent to the Tethys also affected deposition. As local graben-forming tectonic activity ceased along the Atlas trends, coarse-grained detrital sedimentation gave way to fine-grained evaporitic playa mudstones. However, paralic evaporitic mudstone facies continued to be deposited in the northern and eastern grabens, and as regional subsidence began, this facies expanded beyond the borders of the rift basins, spreading to cover much of Morocco. Widespread basaltic lavas were extruded at this time. With continued subsidence and eustatic sea-level rise in Jurassic time, the evaporitic mudstone facies gave way to marine carbonates along the Atlas trends.

Acknowledgments

I would like to express deep appreciation to R. Brown, S. Cossey, and A. Mattis for generously providing me with important unpublished data from their studies of the Middle and High Atlas regions. These three, as well as J. Beauchamp and F. Van Houten, have provided reprints and numerous photographs of the Moroccan formations, vastly improving my knowledge of these strata. F. Van Houten provided a valuable review of this manuscript. Many thanks also to L. Bloomberg, L. Clay, and M. A. Padilla, who astutely interpreted my longhand, and were able to make sense out of and type this manuscript.

References

Allary, A., Andrieux, J., Lavenu, A., and Ribeyrolles, M., 1972, Présence de décrochements dans la Meseta sud-oriental du Maroc central: C.R. Acad. Sci. Paris, v. 274, pp. 653-656.

Amado, E., 1962, Les gisements de plomb de Zeïda et de Bou-Mia: Notes et Mémoirs Service Géologique de Maroc, no. 181, pp. 175-184.

Amade, E., 1965, Le gisement de potasse triasique de Khémisset: Maroc, Mines et Géologie, Rabat, no. 23, pp. 23-48.

Armand, C., Beauchamp, J., Ganeval, J., And Petit, J. P., 1980, Quelques aspects de l'environment tectono-sedimentaire des molasses permo-triasiques du Haut Atlas: l'exemple du bassin du N'Fis: Mines, Géologie et Energie, no. 48, Rabat, pp. 69-75.

Beauchamp, J., 1983, Le permien et le trias marocains, <u>in</u> Colloque sur le permien et le trias du Maroc: Bulletin de la Faculté des Sciences, Université Cadi Ayyad, Section Sciences de la Terre, numéro special 1, pp. 1-21.

Bosellini, A., and Hsü, K. J., 1973, Mediterranean plate tectonics and Triassic paleogeography: Nature, v. 244, pp. 144-146.

Brown, R. H., 1980, Triassic rocks of the Argana Valley, southern Morocco, and their regional structural implications: Am. Ass'n Petrol. Geol. Bull., v. 64, pp. 988-1003.

Burgess, C. L., 1975, Geology of the Sais basin: unpublished Master's thesis, Univ. of South Carolina, 36 pp.

Busson, G, 1974, Le Trias évaporitique d'Afrique du Nord et d'Europe occidentale: données sur la paléogéographie et les conditions de depot: Bull. Soc. Géol. France, t. 16, no. 6, pp. 653-665.

Cailleaux, Y., Gonord, H., Le Guern, M., and Sauvage, M., 1983, Taphrogénèse et magmatisme Permian dans le Maroc central, in Colloque sur le Permian et le Trias du Maroc: Bulletin de la Faculté des Sciences, Université Cadi Ayyad, Marrakech, Section Sciences de la Terre, numéro special 1, pp. 24-39.

Carpentier, A., 1930, La flore permienne du Bou Achouch: Notes et Mémoires de la Service Géol. de Maroc, t. 8, pp. 3-43.

Choubert, G., and Faure-Muret, A., 1962, Evolution du domaine atlasique marocain depuis les temps paleozoiques, in Livre à la memoire du prof. Paul Fallot: Soc. Géol. France Mem., v. 1, pp. 447-514.

Cogney, G., Termier, H., and Termier, G., 1971, Sur la presence de "pillow-lavas" dans le basalt du Permo-Trias au Maroc Central: C. R. Acad. Sci. Paris, v. 273, pp. 446-499.

Cogney, G., Normand, M., Termier, H., and Termier, G., 1974, Observations sur le basalt du bassin triasique de Rommani-Maaziz (Maroc occidental): Notes Service Géologique Maroc, v. 36., no. 264, pp. 153-174.

Cogney, G., and Faugères, J. C., 1975, Précisions sur la mise en place des epanchements basaltiques des formations triasiques de la bordure septentrionale du Maroc central: Bull. Soc. Géol. France, v. 17, pp. 721-733.

Cossey, S. P. J., 1974, The geology of the eastern Khenifra basin, Morocco: unpublished B.Sc. thesis, University College of Swansea, 73 pp.

Couzminer, H. L., and Manspeizer, W., 1976, Triassic pollen date Moroccan High Atlas and the incipient rifting of Pangea as middle Carnian: Science, v. 191, pp. 943-945.

Couzminer, H. L., and Manspeizer, W., 1977, Autunian and Carnian palynoflorules: Contribution to the chronology and tectonic history of the Moroccan pre-Atlantic borderland, in Swain, F. M. (ed), Stratigraphic Micropaleontology of Atlantic Basin and Borderlands, Elsevier, Amsterdam, pp. 185-201.

DuTuit, J. M., 1966, Apport des déscovertes de vertébrés à la stratigraphie du Trias continental du couloir d'Argana (Haut-Atlas occidental, Maroc): Maroc Service Géol. Notes, v. 26, no. 188, pp. 29-31.

Fabre, J., 1970, Le Paléozique terminal à faciès grès rouge au Sahara central et occidental: Internat'l. Stratig. and Géol. Carbonifère Cong., 6th, Comptes Rendus, v. 2, pp. 737-744.

Gigout, M., 1951, Etudes géologiques sur la Meseta Marocaine occidentale: Notes et Mémoires no. 86, Division des Mines et de la Géologie, Service Géologique du Maroc.

Gonord, H., LeGuern, M., Turiot, D., and Rebours, H., 1980, Mise en évidence d'un volcanisme rhyolitique stéphano-permien sur la bordure nord du massif hercynien central du Maroc; extension et importance du volcanisme tardi-hercynien: C. R. Acad. Sci. Paris, t. 291, pp. 51-54.

Hallam, A., 1984, Pre-Quarternary sea-level changes: Ann. Rev. Earth Planet. Sci., v. 12, pp. 205-243.

Haq, B. U., Hardenbol, J, and Vail, P. R., 1987, Chronology of fluctuating sea levels since the Triassic: Science, v. 235, pp. 1156-1167.

Hutton, D. W. H., and Sanderson, D. J., 1984, Preface, in Hutton, D. W. H., and Sanderson, D. J. (eds), Variscan tectonics of the North Atlantic region: Blackwell Scientific Publications, Oxford, pp. vii-ix.

Jones, D. F., 1975, Stratigraphy, environments of deposition, petrology, age, and provenance of the basal redbeds of the Argana Valley, western High Atlas Mountains, Morocco: unpublished Masters thesis, New Mexico Inst. Mining and Technology, 148 pp.

Jowett E. C., and Jarvis, G. T., 1984, Formation of foreland rifts: Sedimentary Geology, v.40, pp. 51-72.

Krueger, H. W., 1973, K-Ar dating of Moroccan dolerites [abs.]: Geol. Soc. Am., Abstracts with Programs, v. 5, p. 700.

Laville, E., 1981, Rôle des décrochements dans le mécanisme de formation des bassins d'effondrement du Haut Atlas marocain au cours des temps triasique et liasique: Bull. Soc. Géol. France, t. 23, no. 3, pp. 303-312.

Laville, E., and Petit, J.-P., 1984, Role of synsedimentary strike-slip faults in the formation of Moroccan Triassic basins: Geology, v. 12, pp. 424-427.

Lorenz, J. C., 1974, Triassic sediments and basin structure of the Kerrouchen basin, Morocco: unpublished M.Sc. thesis, Univ. of S. Carolina, 45 pp.

Lorenz, J. C., 1976, Triassic sediments and basin structure of the Kerrouchen basin, central Morocco: J. Sediment. Petrol., v. 46, pp. 897-905.

Lucas, G., 1952, Bordure nord des Hautes Plaines dans l'Algérie occidentale: 19th Int'l Geol. Cong., Algeria, Mon. Régionales, ser. 1, Algérie, no. 21, 139 pp.

Lucas, J., 1962, La transformation des minéraux argileaux dans la sédimentation: études sur les argiles du Trias: Mémoirs du Service de la Carte Géologique d'Alsace et Lorraine, Strasbourg, no. 23, 202 p.

Manspeizer, W., Puffer, J. H., and Couzminer, H. L., 1978, Separation of Morocco and eastern North America: a Triassic Liassic stratigraphic record: Geol. Soc. Am., Bull., v. 89, pp. 901-920.

Manspeizer, W., 1981, Early Mesozoic basins of the central Atlantic passive margins, in Bally, A. W., et al. (eds.), Geology of Passive Continental Margins: Am. Ass'n. Petrol. Geol., Education Course Note Series no. 19, pp. 4-1 to 4-60.

Mattis, A. F., 1974, Upper Triassic lacustrine carbonates, central High Atlas Mountains, Morocco [abs.]: Geol. Soc. Am., Abstracts with Programs, v. 6, p. 53.

Mattis, A. F., 1977, Nonmarine Triassic sedimentation, Central High Atlas Mountains, Morocco: J. Sediment. Petrol., v. 47, pp. 107-119.

Michard, A., 1976, Eléments de géologie Marocaine: Notes et Mémoirs Service Géologique de Maroc, v. 252, 408 pp.

Michard, A., and Piqué, A., 1980, The Variscan belt in Morocco: Structure and developmental model, in Wones, D. R. (ed.), The Caledonides in the USA: IGCP Project 27, Virginia Polytechnic Institute and State University, Memoir no. 2, pp. 317-322.

Michard, A., Yazidi, A., Benziane, F., Hollard, H., and Willfert, S., 1982, Foreland thrusts and olistostromes on the pre-Sahara margin of the Variscan orogen, Morocco: Geology, v. 120, pp. 253-256.

Mossu, R., 1954, Les basaltes permo-triasiques à indices de cuivre d'Agourir et de Sidi Tair: Notes et Mémoirs de Service Géologique de Maroc, t. 12, no. 122, pp. 109-121.

Peretsman, G., 1985, Geochemical and petrographic analysis of early Mesozoic evaporites from Morocco: Implications for the evolution of the North Atlantic rift: unpublished M.Sc. thesis, Univ. Oregon, 87p.

Peretsman, G. S., and Holser, W. T., 1984, Geochemistry of evaporites from the North Atlantic rift--progress report [abstract], in Reports and Abstracts, The West-African Connection, Evolution of the Central Atlantic Ocean and its Continental Margins: SGA Penrose Conference and CNRS-NSF Colloquium, Giens, France, p. 28.

Petit, J. P., and Beauchamp, J., 1986, Synsedimentary faulting and paleocurrent patterns in the Triassic sandstones of the High Atlas (Morocco): Sedimentology, v. 33, pp. 817-829.

Proust, F., 1973, Etude stratigraphique, pétrographique, et structurale du bloc oriental du Massif Ancien du Haut Atlas (Maroc): Notes Service Géol. Maroc, v. 34, no. 254, pp. 15-53.

Robinson, P. L., 1973, Palaeoclimatology and continental drift, in Tarling, D. H., and Runcorn, S. K. (eds.), Implications of Continental Drift to the Earth Sciences: Academic Press, London, pp. 451-476.

Roch, E., 1930, Etude géologique dans la region méridionale du Maroc oriental: Service des Mines et de la Carte géologique du Maroc: Notes et Mémoirs, no. 9.

Roch, E., 1939, Description géologique des montagnes à l'Est de Marrakech: Mém. Serv. Mines et Cartes Géol., Maroc, no. 51, 438 pp.

Salvan, H. M., 1968a, L'évolution du problem des évaporites et ses conséquences sur l'interprétation des gisements marocains: Mines et Géologie, no. 27, Rabat, pp. 5-30.

Salvan, H. M., 1968b, Les niveaux salifères marocains, leurs charactéristiques et leurs problèmes: Geology of Saline deposits, UNESCO Earth Sci. Series, no. 7, pp. 147-159.

Salvan, H. M., 1974, Les séries salifères du Trias marocain; caractères generaux et possibilités d'interprétation: Bull. Soc. Géol. France, v.7, pp. 724-731.

Salvan, H. M., 1983, Les évaporites triasiques du maroc, caractères généraux-répartition--interprétation, in Colloque sur le permien et le trias du Maroc: Bulletin de la Faculté des Sciences, Université Cadi Ayyad, Marrakech, Section des Sciences de la Terre, Numéro Special 1, pp. 73-84.

Salvan, H., 1984, Les formations évaporitiques du Trias marocain. Problèmes stratigraphiques, paléogéographiques, et paléoclimatologiques. Quelques réflexions: Revue de Géologie Dynamique et de Géographie Physique, v. 25, fasc. 3, pp. 187-203.

Tane, J. L., Pachtere, P., Ferrandini, M., Bertrand, H., and Girod, M., 1983, Le volcanisme permien et triasique de l'Atlas de Marrakech dans son contexte sédimentaire, in Colloque sur le permien et le trias du Maroc: Bulletin de la Faculté des Sciences, Université Cadi Ayyad, Section des Sciences de la Terre, numéro spécial 1, pp. 101-112.

Taugourdeau-Lantz, J., 1978, Pollens des niveaux sédimentaires associés aux basaltes du Trias sur la bordure septentrionale du Maroc central; précisions stratigraphiques: Service Géol. Maroc, Notes, t. 40, no. 275, pp. 135-146.

Termier, H., 1936, Etude géologique sur le Maroc central et le Moyen Atlas septentrional: Notes et Mémoirs du Service des Mines et de la Carte Géologique, Rabat, Maroc, no. 33, 4 volumes.

Termier, H., and Termier, G., 1948, Observations nouvelles sur le Permo-Trias et la base du Lias dans le Maroc Central et le Moyen Atlas septentrional: Soc. Geol. France, Bull., 5th series, v. 8, pp. 395-405.

Tixeront, M., 1973, Lithostratigraphie et minéralization cuprifères stratiformes syngenetiques et familières des formations detritiques permo-triasiques du couloir d'Argana (Haut-Atlas occidental, Maroc): Maroc Service Géol. Notes, v. 33, no. 249, pp. 147-177.

Tucker, M., 1981, Giant polygons in the Triassic salt of Cheshire, England: a thermal contraction model for their origin: J. Sed. Petrol., v. 51, pp. 779-786.

Vail, P. R., Mitchum, R. M., Jr., and Thompson, S., III, 1977, Seismic stratigraphy and global changes of sea level, Part 4: Global cycles of relative changes of sea level: Ass'n Petrol. Geol. Memoir 26, pp. 83-97.

Van Houten, F. B., 1976a, Late Variscan nonmarine deposits, northwestern Africa: Implications for pre-drift North Atlantic reconstructions: Am. J. Sci., v. 276, pp. 671-693.

Van Houten, F. B., 1976b, Late Variscan nonmarine basin deposits, northwest Africa: Record of Hercynotype orogeny, in Falke, H. (ed.), The Continental Permian in central, west, and southern Europe, D. Reidel Publishing Co., pp. 215-224.

Van Houten, F. B., 1977, Triassic-Liassic deposits of Morocco and eastern North America: comparison: Am. Ass'n Petrol. Geol. Bull., v. 61, pp. 79-99.

Van Houten, F. B., and Brown, R. H., 1975, Triassic redbeds and evaporites, northwest Africa [abs]: Geol. Soc. Am., Abstracts with Programs, v.7,p. 1306.

Van Houten, F. B., and Brown, R. H., 1977, Latest Paleozoic-early Mesozoic paleogeography, northwestern Africa: J. Geology, v. 85, pp. 143-156.

JURASSIC CARBONATE FACIES OF THE CENTRAL AND EASTERN HIGH ATLAS RIFT, MOROCCO

John E. Warme
Department of Geology, and Exploration Geosciences Institute, Colorado School of Mines, Golden, Colorado 80401 U.S.A.

ABSTRACT

The present Central and Eastern High Atlas mountains of southern Morocco represent a short-lived Mesozoic rift, 500 x 100 km, associated with an ancient fracture zone between the Saharan Craton to the south and the smaller microplate mesetas to the north. A continental rift opened in the Upper Triassic, accumulating redbeds, evaporites and basalts. Marine carbonate deposits were initiated abruptly in the Lower Liassic, and continued until the Middle Dogger, when the basin ceased to subside and became continental again.

In addition to regional subsidence and eustatic sealevel changes, individual fault blocks of the rift contain stratigraphic sequences that mirror the tectonic history of that block. A representative tectonic-stratigraphic geotraverse, transverse across the middle of the range, crosses a southern shore and shelf, southern deep, central platform, northern deep and northern shore and shelf.

Blocks representing paleogeographic shelves exhibit stratigraphic sections of shallow-marine limestones and reefs. Adjacent deeps contain thicker sections of allodapic limestones, other gravity deposits, and carbonate olistostromes. A central platform possesses a Liassic upward-shallowing sequence capped by gigantic coralgal buildups.

The sequences on all the fault blocks are punctuated by Toarcian marls, representing a eustatic sealevel rise followed by upward-shoaling Aalenian and Bajocian sequences. Post-Jurassic uplift and tectonic inversion created the superb exposures of the modern Central and Eastern High Atlas.

INTRODUCTION

The Atlas mountain ranges in the Kingdom of Morocco are outstanding regions for the study of carbonate rocks. Almost all carbonate facies are represented within the ancient Jurassic seaways that are now exposed in the Moroccan Atlas ranges (Fig. 1).

Specifically, the carbonates of the Central and Eastern High Atlas mountains (CEHA, for brevity) represent facies that developed in a short-lived Jurassic marine trough whose framework was a continental rift, the Central and Eastern High Atlas Rift (CEHAR). The carbonates exhibit unparalleled examples of carbonate facies relationships. Many of the carbonate units are not dolomitized, and possess well-preserved sedimentary textures, structures, and fossils. Superb exposures are provided by the dry climate, lack of vegetation, and high topographic relief. Stream gorges provide access to thick and continuous stratigraphic sections.

The High Atlas of Morocco are part of the Alpine mountain system, but lie on the periphery of that system and are not severely tectonized. The Moroccan segments provide a model for the more severely deformed regions in the main Alpine system to the north and east.

The purpose of this paper is to describe and interpret some of the carbonate facies in the CEHA, which were originally deposited in a basin created by the CEHAR. The carbonates are put in perspective by means of a representative geotraverse across the High Atlas, showing the response of carbonate facies to their structural and eustatic controls. As a preface to describing the carbonates in the CEHAR, its relationship to the general geology and geologic history of Morocco is described.

GEOLOGIC FRAMEWORK FOR JURASSIC CARBONATES

The major geologic elements of Morocco (Fig. 1) include three main stable areas called platforms, plateaus, blocks, or, in plate-tectonic terms, plates or microplates: the Saharan Shield or Craton, the Moroccan Meseta, and the Oran Meseta. They are separated by two major Atlas orogenic belts: the High Atlas, and the Middle Atlas. The Rif is an overthrust belt on the north. Jacobshagen (1986) reviewed the structural history and interrelationships of these elements.

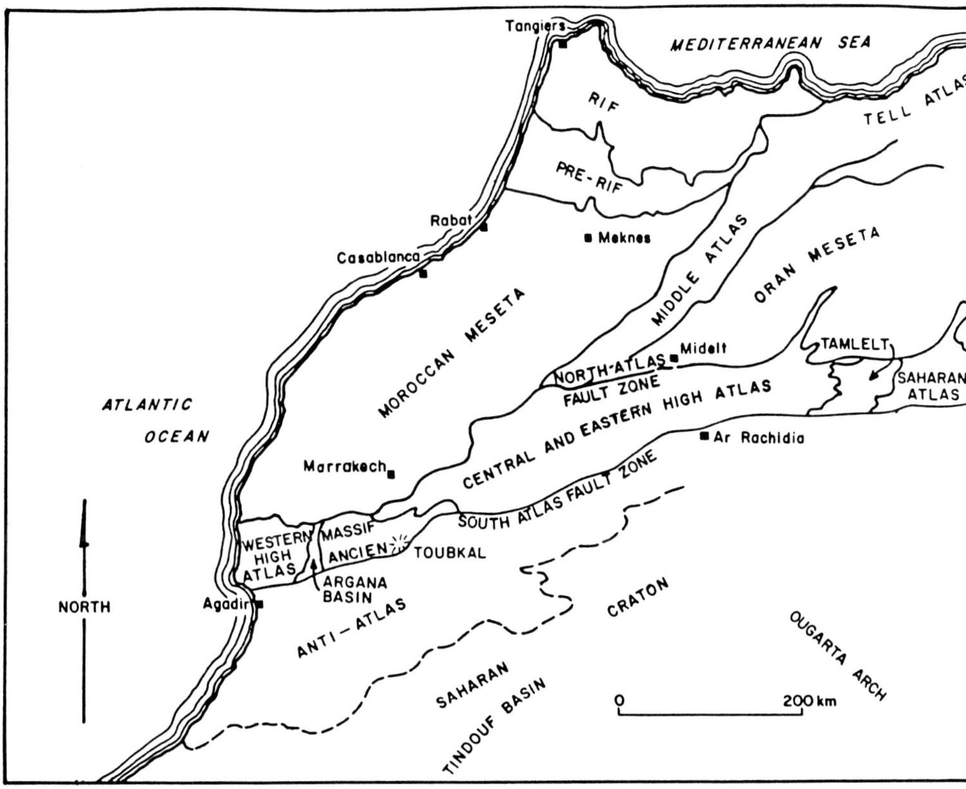

Figure 1. Index map of Morocco showing cities and major geologic provinces.

Ancient Fracture Zones

The Middle and High Atlas orogenic belts are comprised of discrete major long-lived fault- or fracture-zones that have two major trends. One is approximately NE/SW, parallel to the Atlantic, and the other is approximately ENE/WSW or E/W, parallel to the Mediterranean (Rod, 1962; Kanes et al., 1973). These major zones commonly have divergent salients or offsets within them. The fracture zones were apparently reactivated within or along the margins of the sites of Caledonian and/or Variscan deformation. During late Paleozoic and early Mesozoic continental breakup these major fault zones were reactivated by rifting into a horst-and-graben mosaic, evident from the distributions and thicknesses of both continental and marine sediments deposited in and between the down-faulted elements. In Morocco the major fracture zones also represent the boundaries of tectonic plates or microplates.

As emphasized by many students of North African geology (e.g., Choubert and Marcais, 1956; Van Houten and Brown, 1977), these zones are permanent and important. The tectonic style within and bordering them, from one orogenic episode to the next, is either inherited from, or deeply influenced by, the previous episodes.

Plate-Tectonic History

With the onset of Jurassic marine deposition, the distribution and character of the Atlas carbonates reflect the basic plate-tectonic framework of the country. The relationships shown by Van Houten and Brown (1977) and Manspeizer et al. (1978) show a Triassic-Jurassic westward migration of the earliest presence of carbonates across North Africa. This sequential development is consistent with the concept of a thermal bulge and deep erosion in the late Paleozoic and early Mesozoic along the Atlantic margin of northwest Africa, which collapsed and subsided gradually in the Upper Triassic and Lower Jurassic, allowing marine flooding progressively westward from the Tethys Sea, through the Straits of Gibraltar area, and southwest along the developing western Moroccan margin.

The sea preferentially flooded the major early Mesozoic rifting fracture zones in Morocco. The zone between the Moroccan and Oran mesetas became the northeast-trending trough that was later uplifted to become the Middle Atlas. The zone between the Oran Meseta and the Saharan Craton was an almost east-west rift basin that later became the CEHA. The Oran Meseta appears also to have rotated clockwise, asymmetrically widening and deepening the basins at the northern end of the Middle Atlas rift and the western end of the CEHAR (Fig. 1).

These rifted troughs are characterized internally by a mosaic of fault-bounded blocks, similar to the Red Sea or Gulf of Suez rifts today (Heybroek, 1965; Lowell et al., 1975). Carbonate development on and between these blocks provides a key for unravelling their relative vertical motions throughout the Jurassic, and for understanding the relationship between tectonic and eustatic controls on the carbonate facies development.

Major Geologic Elements

Saharan Platform

The southern part of Morocco (Fig. 1) is occupied by the Saharan

Craton, a vast, stable region of broad, large-scale arches (e.g., Ougarta) and basins (e.g., Tindouf). Surficial rocks are mainly Precambrian, Paleozoic and Cretaceous. The craton was a rigid structural block bordering the long-lived High Atlas fracture zone immediately to the north. This fracture zone opened in the Triassic and Jurassic, to contain the Western Atlas (or "HaHa" or "Essaouira") basin and the CEHAR. During post-Jurassic tectonics the Saharan Craton was the stable block against which the High Atlas basins were compressed to create the present High Atlas ranges that span the country from west to east (Fig. 1).

Mesetas

The Moroccan Meseta and Oran Meseta microplates are stable, eroded, Paleozoic terranes, separated from the Saharan Platform by the High Atlas and from each other by the Middle Atlas. During the Jurassic the mesetas were wholly or partially covered by epicontinental seas, and have thin or incomplete sedimentary sections in marked contrast to the thick sections in the Atlas troughs. Their northern margins are obscured by the Rif and Tell Atlas nappes that border the Mediterranean from Morocco to Tunisia (Fig. 1). No comprehensive summaries exist on the stratigraphy of the mesetas, although they contain widespread marine Jurassic as shown on the geologic maps of Morocco.

Rif

The Rif and Pre-Rif thrust belt includes Jurassic carbonates, largely or wholly allochthonous from the north. They are different in character from the Jurassic of the Middle and High Atlas south of the Rif (Faure-Muret and Choubert, 1971; Du Dresnay, 1979), and are not discussed herein.

Middle Atlas

The Middle Atlas mountains represent the uplifted Jurassic marine basin between the mesetas. Its northern margin is now covered by the nappes of the Rif or otherwise poorly exposed. The trough collected Lower, Middle and Upper Jurassic carbonates. Colo (1961) summarized the stratigraphy of the Middle Atlas, and other references treat local areas within the range.

Atlas Fracture Zone and the High Atlas Ranges

The topographic High Atlas ranges represent the current status of an ancient fracture zone that was reactivated in the late Paleozoic Variscan orogeny, and probably one or more times before (Van Houten, 1977). The fracture zone extends from the Atlantic Coast north of Agadir eastward to beyond the Algerian border (Fig. 1), and is now about 60 to 100 km wide (before palinspastic restoration). From west to east it contains the Western High Atlas mountains, the Massif Ancien, and the CEHA (Fig. 1). The Western High Atlas, and the CEHA, are topographic ranges that coincide with two discrete Triassic-Jurassic basins, separated by the Massif (Figs. 1, 2).

Massif Ancien--The Massif Ancien is a late Paleozoic (240-320 Ma) shallow intrusion (Vogel et al., 1976), comprised of teardrop-shaped granitic pods, with abundant basic intrusions and a well-defined aureole of late Paleozoic metamorphosed sedimentary rocks. It contains the highest elevation in Morocco (Jebel Toubkal, 4165 m; Fig. 1), and during Tertiary uplift carried with it surrounding deposits of mainly non-marine Mesozoic rocks which were originally shed from it, mainly east and west into the High Atlas basins.

High Atlas Mesozoic Basins--Figure 2 diagrammatically shows the ages, general facies and stratigraphic relationships between the two Moroccan High Atlas basins on either side of the Massif Ancien. The latest Triassic and earliest Jurassic is comprised of continental redbeds, evaporites and basalts (see Lorenz, this volume).

The western part of the High Atlas is now topographically relatively low, and comprised of plateaus of Jurassic and Cretaceous rocks interrupted by sharp folds and reverse faults. This Western Atlas Basin opened to the fledgling Atlantic on the west and/or northwest. The carbonates of the Western Atlas were studied by Ambroggi (1963), and Huessner (1985) who reviewed past work and related the High Atlas to their plate-tectonic framework. The eastern edge, bordering the Massif Ancien, is known as the Argana Basin (Figs. 1, 2). It has been erosionally stripped down to lowermost Liassic and Triassic redbeds, with some exposures of metamorphosed Paleozoics of the Massif (Ambroggi, 1963; Reyre, 1965; Brown, 1980).

Redbeds bordering the massif thin westward but reach the present Atlantic coast, interrupted by three marine pulses represented by thin

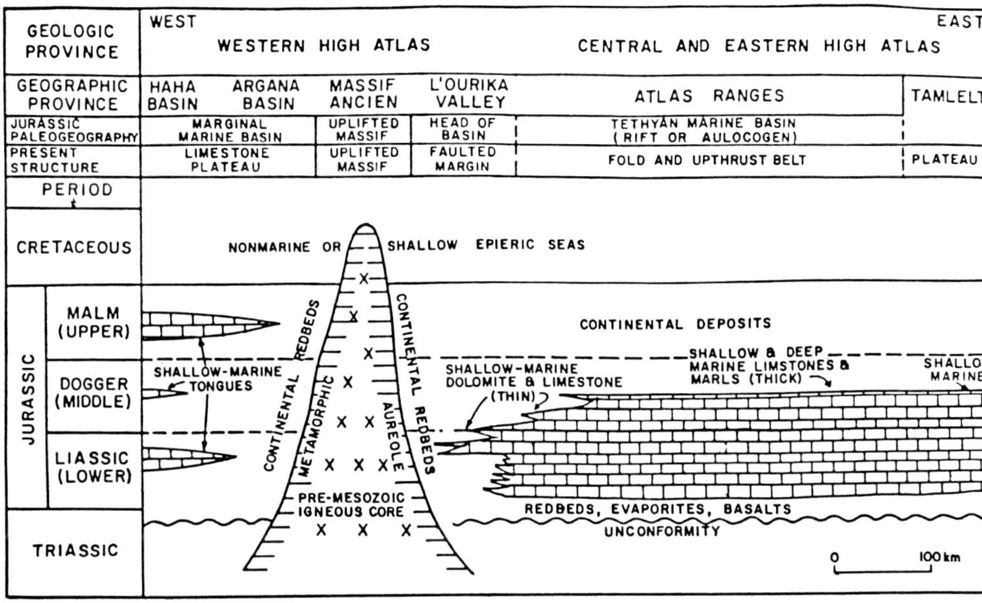

Figure 2. Diagrammatic cross-section along the High Atlas fracture zone of Morocco, showing the relationship of Jurassic carbonates to geologic and geographic provinces, paleogeography and present structure. Carbonate strata of the HaHa Basin and the Central and Eastern High Atlas rift were not physically connected along the fracture zone.

limestones and related lithologies. The Lias and Malm pulses are apparently the best developed, but all three are of shallow- and marginal-marine facies only (Fig. 2).

East of the Massif Ancien the present CEHA mountains are comprised of parallel to sub-parallel, genetically similar sharp anticlines or upthrusted ridges, separated by broad plateaus, stretching for 500+ km to the east. Today this area exhibits high and rugged topographic relief, and represents the basin created by the Mesozoic CEHAR. It accumulated late Triassic and early Liassic continental sediments and volcanics, and Lower and Middle Jurassic marine limestones and marls (Fig. 2).

Carbonate deposition began in the CEHAR by the Sinemurian (perhaps Hettangian), and ended by the Middle Dogger. The upper half of the Jurassic is represented by a poorly-dated continental sequence, now largely stripped away by uplift and erosion.

Thus, within the High Atlas fracture zone of Morocco two basins contain Jurassic marine strata, mainly limestones (Fig. 2). The Western Atlas basin exhibits three thin sequences separated by redbeds, one each in the Lower, Middle and Upper Jurassic. The CEHAR contains a thick and continuous sequence of limestones and marls that span most of the Lias and about the first half of the Dogger, with no evidence of Malm marine environments. The western basin is related to the embryonic Atlantic, and the CEHAR to the evolving Tethys.

MESOZOIC CONTINENTAL TO MARINE TRANSITION

Salvan (1974), Van Houten (1977), Manspeizer et al. (1978), and Lorenz (this volume), among others, summarized the early Mesozoic non-marine facies in Morocco. The Triassic and earliest Jurassic record is one of deposition in a mosaic of rifted troughs, containing characteristic volcanics, redbeds, and evaporites that signaled oncoming continental breakup. For two or three prior geologic periods, encompassing the Upper Paleozoic and most of the lower part of the Triassic, the Tethys Sea was forced away from a linear bulge that later collapsed to become the locus of the newly forming Atlantic. On the west, marine conditions were initiated with deposition of a Toarcian limestone, bounded by redbeds, but well within the Jurassic. In contrast, the Middle Atlas and CEHAR were apparently abruptly flooded in Sinemurian (Hettangian?) time, followed by continuous marine deposition (Fig. 2).

It is likely that rifting, operating across the Triassic-Jurassic boundary, continued until fault-trough floors, or even the whole CEHAR and Middle Atlas basins, were below sea level such as the Dead Sea today, providing a scenario for synchronous and abrupt marine flooding in the Lower Liassic as the rift broke through to the Tethys Sea.

In the CEHAR the earliest carbonates overlie olivine and quartz tholeiites, giving radiometric dates near the Triassic-Jurassic boundary (Manspeizer et al., 1978), and in turn resting on or interbedded with late Triassic and/or the early Jurassic nonmarine siliciclastics and evaporites.

THE CENTRAL AND EASTERN HIGH ATLAS RIFT (CEHAR)

The present CEHA mountains occupy the location of a Triassic and Jurassic rift basin approximately 500 km long and 60 to 100 km wide. Although this is a large and remote area, access is possible by road, and several papers have been published from field studies in the region.

Past Work

The most comprehensive review of Middle and High Atlas Jurassic stratigraphy is Du Dresnay (1979); no summary is readily available in English.

The CEHAR area is covered by six 1:200,000 scale geologic maps that are exceedingly useful for chronostratigraphy of the rock units, largely using zonation by ammonites. Facies distributions also are generally indicated on the maps, but not in detail and, for the older maps, uninterpreted in light of modern facies concepts. From west to east, the maps are Demnat and Telouet (1941), Ouaouizarht and Dades (1942), Midelt (1939), Ksar es Souk and Boudenib (1956), Anoual and BouAnane (1976), and Haut-Atlas Oriental (1963). Explanatory notes for some of the maps have not been published, a multitude of local names represent correlative formations, and not until recently were any formation names formally proposed. Several newer maps cover local areas of special interest, and significantly modify earlier correlations and interpretations. Map lists are available from the Moroccan Service Geologique in Rabat, who also publish the valuable "Notes et Memoirs" series. Annotated bibliographies of Moroccan geology are also available from the Service Geologique.

Boundaries of the CEHAR

The northern and southern boundary fault zones (Figs. 1, 3, 4) of the present topographic CEHA generally delimit the north and south margins of the Jurassic CEHAR, as evidenced from stratigraphic thinning and facies changes toward the CEHA margins (Fig. 4). To the north, tidal flat, shelf and shelf-edge reef environments faced the CEHAR from the southern edges of the Moroccan and Oran mesetas. There were only minor areas of subaerial erosion and dispersal of siliciclastic sediment from the north.

On the southern margin, shelf carbonates are interbedded with red sandstones and argillites shed from the Saharan Platform, together with rare, thin, non-red, fine-grained, siliciclastic marine sandstones. In places, a relatively wide tract of shelf facies is present, ending northward with an abrupt shelf-slope break (see Fig. 4).

To the east the carbonate-filled trough shallowed into the general area of the Tamlelt (Fig. 1); it is not certain if the trough terminated with a shoreline on the east, but it shallowed in that direction. The structural rift crossed the present Moroccan-Algerian border and contained Jurassic and younger basins along the trend of the present Saharan and Tunisian Atlas ranges (Fig. 1).

To the west, the trough terminated against the Massif Ancien, which sits astride the fracture zone containing the CEHAR.

Topographic Expression of the Central and Eastern High Atlas

The modern CEHA mountains are topographically a series of parallel to en echelon fault- or fold-controlled sharp ridges and intervening broad synclinal valleys. They extend parallel or subparallel to the margins of the mountain belt (Fig. 3). The ridges are approximately evenly-spaced, providing exposures of the Liassic carbonates. Eroded slopes of the intervening synclinal plateaus expose the upper portions of the Lias and the Dogger.

Geologic History

Table 1 is an outline of the history of the CEHAR. Marine deposition in the trough lasted about 35 million years, from approximately the early Sinemurian into the Bajocian.

Paleozoic Basement

Basement rocks are early to late Paleozoic mildly metamorposed sandstones and mudstones, in places exhibiting marine fossils and trace fossils. The basement was uplifted and eroded in the late Paleozoic and early Mesozoic, apparently with very little relief remaining over broad areas.

Mesozoic Dry Rift

Unconformably overlying the Paleozoic basement are two unmetamorphosed continental redbed sequences. The first is late orogenic Carboniferous and Permian, and the second is early ("dry") rift late Triassic and earliest Jurassic. The sequences are separated by a major unconformity (Fig. 2). The early sequence is deformed in

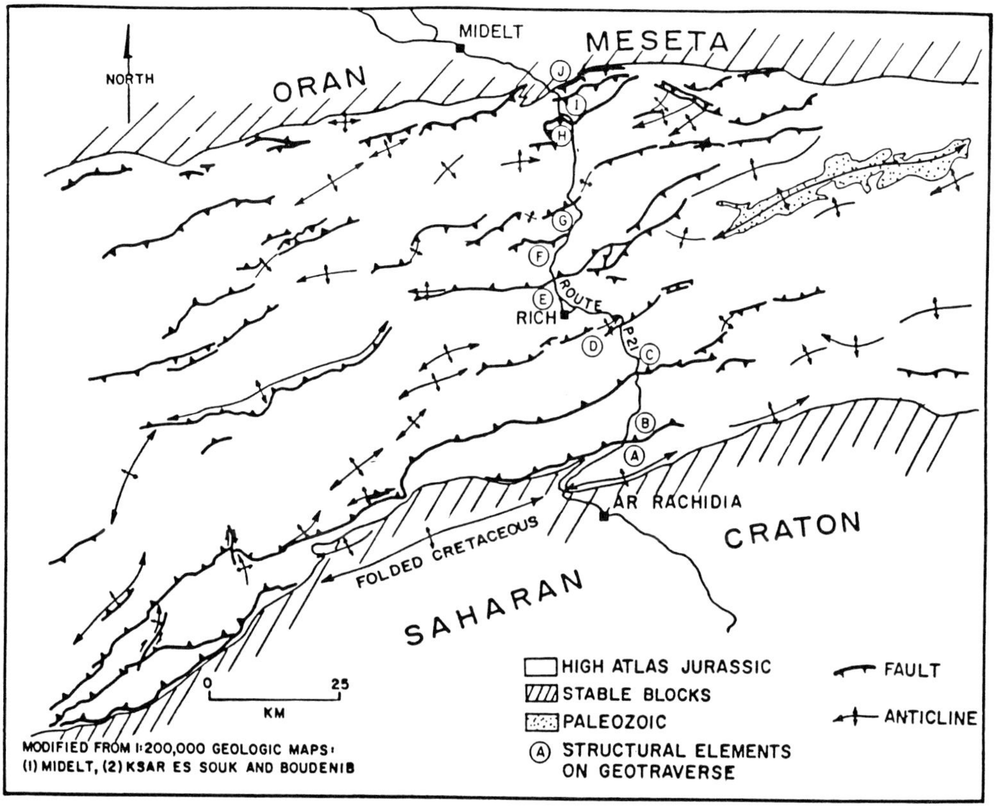

Figure 3. Portion of Central and Eastern High Atlas showing mapped faults and folds. Those that are lettered are shown on the geotraverse cross-section of Figure 4B. Note that most faults verge toward the margin closest to them, a consequence of structural inversion.

places, deeply eroded, and contains acidic extrusives. The later sequence contains extensive basalt flows (Lorenz, this volume). Salvan (1974) and Van Houten (1977) indicated that the CEHAR constituted a major continental basin for the later sequence.

The best dry rift CEHAR exposures exist on the west end of the basin, against the Massif Ancien, where Mattis (1977) measured up to 1000 m of uplifted redbeds and related lithologies. Fluvial sandstones with only rare conglomerates suggest that paleorelief in the massif was not as great as today. Thin Upper Triassic carbonates are present in this sequence, interpreted by Mattis (1974) as lacustrine deposits impounded behind basalt flows.

An unique erosional window next to an upthrust in the southwestern part of the Midelt map (1939), near the village of

Toumliline, provides a view of the continental sequence near the center of the CEHAR. It reveals several hundred meters of interbedded redbeds, basalts, and evaporites, including halite deposits of local economic importance. One volcanic neck is also exposed there. Elsewhere within the core of the CEHAR the dry rift sequence is present as disturbed redbeds and basalts, smeared along with Paleozoic basement against the resistant Jurassic carbonates along many of the steep folds and upthrusts shown on Figure 3.

TABLE 1

GEOLOGIC HISTORY OF CENTRAL AND EASTERN HIGH ATLAS

AGE	GEOLOGIC EVENTS
Late Cenozoic	Uplift and erosion -- High Atlas mountains.
Cenozoic	"Alpine" deformation -- compression; faults rejuvenated and reversed along en echelon folds and upthrusts.
Late Mesozoic (Malm -- Cretaceous)	Mild deformation, uplift, erosion; Cretaceous shallow marine deposits.
Middle to late Dogger -- Malm	Continental deposits; fluvial, floodplain.
Middle Dogger (Late Bajocian -- Bathonian	Trough filled -- end of carbonate deposition; "wet" rift filled.
Early Dogger (Aalenian -- Bajocian)	Subsidence greater than sedimentation, then equal to sedimentation, followed by less than sedimentation.
Late Lias (Toarcian)	Abrupt eustatic deepening; marls ubiquitous.
Middle Lias (Sinemurian -- Pliensbachian)	Subsidence generally greater than sedimentation -- development of entire facies mosaic controlled by fault blocks, with deep- and shallow-marine deposits.
Early Lias (Hettangian(?) -- Sinemurian	Rifting continues; abrupt initiation of "wet" rift; stratified water column.
Late Triassic -- Early Liassic	"Dry" rift -- fault-block mosaic with redbeds, evaporites, and basalts.
Permo-Triassic	Erosion and peneplanation.
Permo-Carboniferous	Variscan Orogeny -- mild metamorphism of sandstones and mudstones; late orogenic fault-troughs.

Jurassic Marine Carbonates

Following is a generalized sketch of the carbonate facies distributions throughout the Jurassic CEHAR. Because of the size of the trough, not all of it has been studied, but this review, in addition to previous papers (e.g., Du Dresnay, 1979), provides an introduction to the specific geotraverse presented as Figure 4 below.

The earliest marine ("wet rift") rocks overlying the dry rift sequence are Hettangian or Sinemurian and difficult to date because of the lack of diagnostic ammonites (Table 1). The dry rift/wet rift transition is abrupt, showing no obvious interfingering (Table 1); marine flooding of the trough was apparently exceedingly rapid. Because of the tensional tectonic regime within the fracture zone, as mentioned above, the floor of the trough may have been partly or wholly below sea level during redbed deposition, such as the Dead Sea is today; rapid marine flooding may represent a breakthrough from the Tethys Sea to the east, or encroachment over the Oran Meseta along the northern margin of the rift.

At the western terminus of the basin the Liassic sequence was analyzed in a series of papers by Burgess and Lee (1978), Lee and Burgess (1978), Burgess (1979) and more recently by Septfontaine (1986). The westernmost 100 to 150 km of the trough shallowed toward the Massif Ancien, and contained Jurassic shallow marine dolomitic limestone that thinned and interfingered with redbeds of the Massif. Variations in thickness of the shallow-water facies on the western end of the basin indicates that the trough was rifted into several horst and graben elements, with slightly thicker and deeper marine facies in the grabens and thinner and shallower marine and shoreline facies over the horsts. Correlation of the shallow marine carbonates was attempted on the basis of lithologic similarities by Burgess and Lee (1978), and corrected and refined by Septfontaine (1986) who zoned the column on the basis of larger foraminifera.

Away from the margins of the basin, the earliest marine carbonates are everywhere laminated to thin-bedded limestones. On the margins of the rift these laminites are commonly light gray to yellow, cracked, tepeed, and show abundant supratidal features. However, across the central parts of the CEHAR (as well as the Middle Atlas basin; Du Dresnay, 1979), they are dark gray to black. About 40 m of these laminated, black, unfossiliferous carbonates comprise the initial carbonate deposition over very wide areas, likely representing an hypersaline and/or stratified-anoxic water mass.

Above the laminated carbonates, but within the Sinemurian (Table

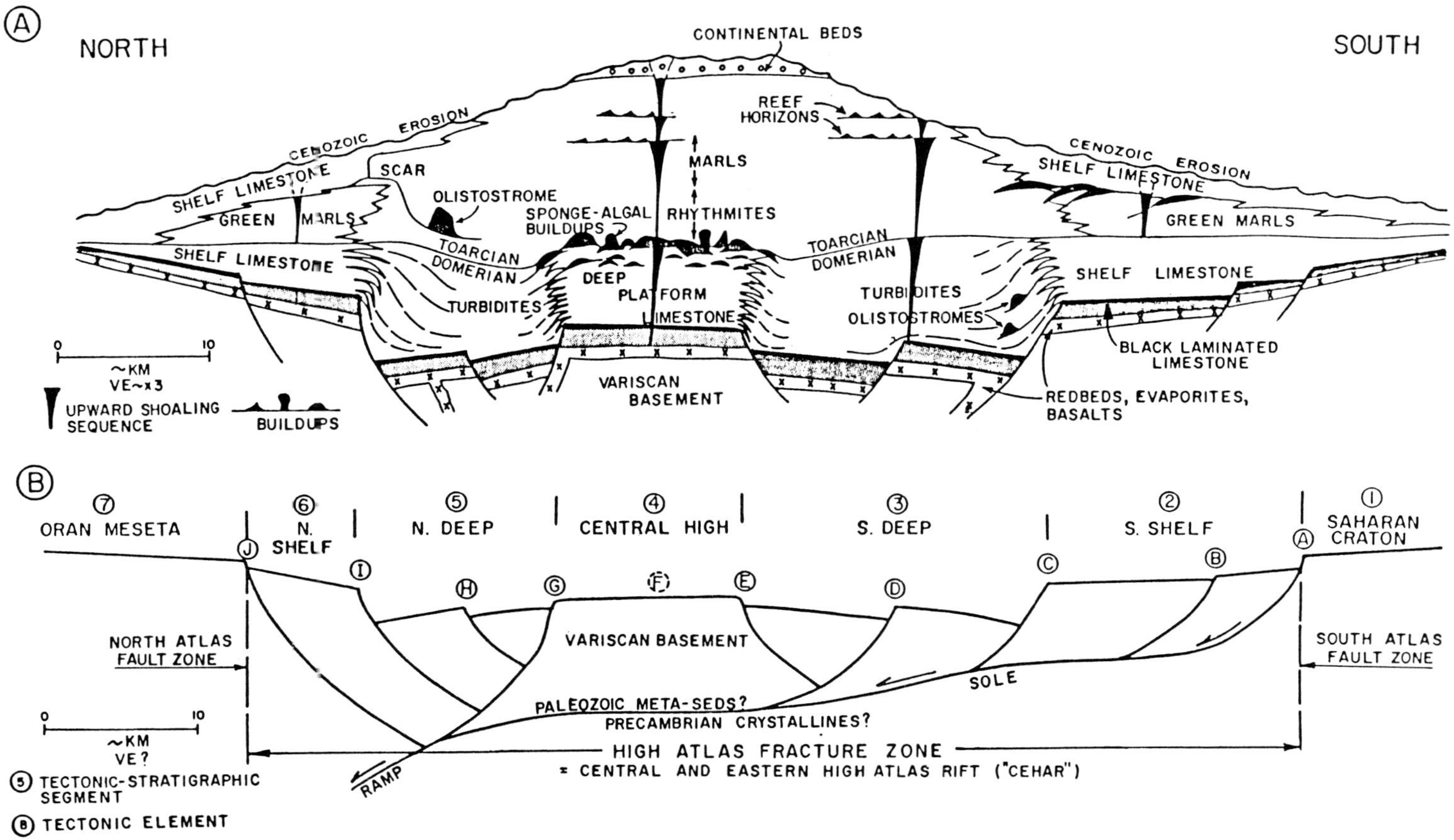

Figure 4. Diagrammatic cross-section along the tectonic-stratigraphic geotraverse described in text. Section B shows the simplest possible configuration of basement faults and their possible relationship to a master sole fault. Section A is drawn on a Domerian-Toarcian datum, showing the distribution of major carbonate facies above each paleographic segment of Section B. Entire basin is now shortened and inverted by reverse movement along faults.

1), three general facies developed within the CEHAR. Du Dresnay (1979) diagrammatically showed these relationships, synthesizing his own and much previous work. He described: (1) a bordering facies, generally equivalent to shoreline and shelf, (2) a central facies, generally deeper-water, and (3) an intermediate or "mixed" facies, containing both shallow- and deep-water paleoenvironmental indicators.

Shoreline, shelf and shelf-edge environments are easily recognized in the bordering zones. However, the mixed and central facies are not everywhere simply slope and deepwater equivalents, respectively (see Evans and Kendall, 1977). In some regions, such as the geotraverse of Figure 4, the mixed facies is clearly comprised of turbidites and other gravity deposits resedimented mainly from the shelf edge. However, during the Sinemurian and Pliensbachian stages, the central facies, over large areas of the trough, contains shallow-water indicators such as algal stromatolites, algal-coated grains, and rich accumulations of fossil brachiopods, bivalves, gastropods, and crinoids. In addition, sponges are an important component, creating significant sponge-algal buildups. This unique assemblage, commonly flanked to the north and south by turbidite "mixed facies" sequences, suggests the presence of one or more central basin highs, separated from the rift margins by deeper troughs (see Fig. 4). Accepting this, a problem still remains to explain why the flourishing scleractinian coral buildups present on the northern and southern margins of the CEHAR trough (Du Dresnay, 1977) are not present in coeval shallow-water strata on the central basin high(s). This paleoecological problem is unresolved.

Throughout most of the CEHAR, the stratigraphic sequence abruptly changes character near the beginning of the Toarcian. The shelf areas are dominated by green-weathering dark gray or black marls, the turbidite-bearing sections become very fine-grained, and the axial zone shifts from large-scale sponge-algal buildups and related facies to alternating medium-bedded limestones and marls (rhythmites) of deepwater character (Table 1). The entire rift zone appears to deepen, probably from a eustatic sealevel rise independent of individual fault-block tectonics. Not until the Aalenian and Bajocian do the shorelines prograde seaward, into the rift zone proper, to displace the fine-grained marls from the shelf areas. The axial part of the basin does not develop obviously shallow-water limestones until the Bajocian (Table 1).

In the Bajocian abundant mid-basin coralgal buildups developed. Cenozoic erosion reveals reef masses that are oval to rhomboid in plan, resembling exhumed atolls. The buildup horizons may be stacked

(Fig. 4), separated by hundreds of meters of fine-grained marls, suggesting that the trough was generally only a few tens of meters to a few hundreds of meters deep at most times; subsidence barely outpaced deposition. The immense volume of marls deposited in the Dogger indicate that a very muddy seaway existed over millions of years in the late stages of marine deposition in the CEHAR.

The youngest ammonite found in the CEHAR is Bajocian. Thus, by the Late Bajocian or earliest Bathonian deposition outpaced subsidence, and the trough was sheeted over with continental deposits of red and green marls, cross-bedded sandstones, and sedimentary structures such as mudcracks, raindrop imprints, and dinosaur tracks, indicating nonmarine conditions (Table 1).

It is likely that the significant eustatic rises of the Cretaceous (Haq et al., 1987) inundated most or all of the area of the present CEHA with shallow epeiric seas. Herbig (1986) showed that little relief existed in the central High Atlas region in the Paleogene, and it was partly flooded by shallow seas. However, there is no evidence for deep-marine sedimentation in this region since the Middle Jurassic. The Alpine Orogeny created the modern CEHA, with uplift stripping away much of the Jurassic and younger deposits and tectonic activity continuing today, as evidenced by the present topographic relief and frequent earthquakes (Table 1).

SOUTH TO NORTH CEHAR GEOTRAVERSE

Introduction

The map of Figure 3 covers the central part of the CEHA, and shows the location of the geotraverse of Figure 4, which parallels Moroccan Route P21 from the military outpost of Ar Rachidia (formerly Ksar es Souk) on the south to the mining town of Midelt on the north (Figs. 1, 3). This is one of many possible transects, but was chosen because these two towns are connected by Moroccan Route P21, an all-season paved highway, and the carbonate sequences along the traverse have been more thoroughly studied than most of the areas immediately east or west. The traverse is situated at approximately the junction between the Midelt (1939) and Ksar es Souk and Boudenib (1956) geologic maps. Evans and Kendall (1977) discussed the facies along the southern part of this traverse, interpreting basin geometries and facies relationships, and providing a framework for the analysis shown in Figure 4. Brede and Heinitz (1986) focused on the structural relationships and tectonic history along the same traverse.

Structural Framework

Because the genesis of the CEHAR incorporates a component of wrenching rather than pure rifting, the major structural elements of the basement--faults and fault blocks--are en echelon and strike at an angle with the boundaries of the rift. The relationships shown on the map of Figure 3 imply left-lateral movement, as also suggested by Huessner (1985). Wrench faulting, even with a tensional component, produces uplifted blocks as well as downdropped ones (e.g., Cristie-Blick and Biddle, 1985). Along the length of the rift, different transverse traverses will each intersect an unique configuration of fault blocks. Thus, the geotraverse described here is unique; however, it generally represents the kinds of tectonic-stratigraphic and facies relationships exhibited elsewhere within the CEHAR.

For convenience, the geotraverse is divided into 7 segments (Figure 4B). Each segment contains an overall similar paleogeographic depositional environment. Each is bounded by a mapped fault or sharp fold, assumed to be related to a basement fault, and each is depicted internally as comprised of only 1 or 2 structural blocks. The number of segments, blocks and faults diagrammed on Figure 4B is the lowest number that explains the carbonate facies documented along the geotraverse. Each fault is lettered on Figure 3, for comparison with Figure 4B.

The major faults bounding the blocks appear as high-angle reverse faults on Figure 3. They are interpreted as originally normal faults of the Jurassic CEHAR, inverted during post-Jurassic compression (Table 1). The inter-block relationships shown on Figure 4B are highly diagrammatic, but are substantiated by the distribution of overlying carbonate facies, as described below.

The cross-section of Figure 4B has not been rigorously balanced. Subsurface relationships, showing the faults connecting to a master fault, are conjectural. The concept stems from papers by Wernicke and Burchfiel (1982), McClay and Ellis (1987) and discussions with B. C. Burchfiel and W. H. Kanes (1974, pers. comm.). The concept for Figure 4 assumes (1) a stable Saharan Craton to the south, (2) a rigid Oran Meseta to the north, that pulled away to the north, and (3) a low-grade metamorphosed Paleozoic basement in the High Atlas Fracture Zone. It is likely that a listric fan dominated the structural pattern as the Oran Meseta moved away, and that the surface of detachment along the master fault coincides approximately with lithologic discontinuity between phyllitic to schistose Paleozoic metasediments and Precambrian crystalline rocks.

Tectonic-Stratigraphic Segments

Figure 4 shows the following tectonic-stratigraphic segments:

(1) The Saharan Craton, now covered with post-Jurassic rocks.

(2) The southern shelf, composed of two rotated blocks covered with Jurassic shallow-water, sabkha, and minor continental facies. The boundary between segments 1 and 2 is usually covered by alluvial fans from the southern edge of the Atlas. The South Atlas Fault Zone (structural element "A" on Figures 3 and 4) is assumed to separate the two segments, and may possess low-angle faults rather than the vertical faults shown. Segment 2 is depicted as two blocks because of the presence of a fault (element B) within the shelf facies (Figs. 3, 5), assumed to be a reactivated Jurassic fault. The best exposures of the shelf sequence are along the Oued Ziz gorge (Figs. 5, 6).

(3) The southern deep, containing two half-grabens filled with allodapic limestones and other mass-flow gravity deposits. The sequence is bounded on the south by the Tizi n' Firest fault (element C), a major reverse-fault zone that exposes the shelf/deepwater transition. Another uplift in the middle of segment 3, Jebel Bou Hamid (element D), exposes thinner, distal equivalents of the deepwater carbonate package; thus, Figure 4 shows a half-graben tilted toward the shelf to the south, containing submarine fans or aprons prograding northward. Segment 3 is divided into two structural blocks to account for a paleofault, reactivated to become the modern faulted ridge of Jebel Bou Hamid. Excellent exposures of the resedimented carbonates occur where the Tizi n' Firest fault zone (element C) crosses the Oued Ziz (Fig. 7).

(4) The central high, comprised of one or more blocks that contain a sequence of carbonates that are very dissimilar to the deepwater facies to the south and north. The sequence is exposed in three upthrusted ridges (elements E, F, G) along Route P21. If all these evenly-spaced faults are

Figure 5. View south down the Oued Ziz valley; Domerian cyclic limestones overlain by Toarcian marls (in distance) overlain by Aalenian limestones.

Figure 6. Oued Ziz valley showing shingled coralgal reefs (8 to 10 m thick) and reef talus prograding northward, overlain by shelf limestones.

Figure 7. View westward along Tizi n' Firest fault zone (left), separating shelf carbonates from 1000+ m sequence of carbonate turbidites and other mass-flow deposits.

Figure 8. View west along uplift "E" (Figs. 3, 4). Beds dip steeply to north (right). Massive beds on left are sponge-algal mounds overlain with irregular contact by Toarcian rhythmites.

reactivated Jurassic normal faults, as is probable, then the central high was broken into two or more blocks rather than a single horst as shown for simplicity in Figure 4. The best exposures of the central high facies are in stream gorges cut through these upthrusted ridges (Fig. 8).

(5) The northern deep, containing two rotated blocks covered with resedimented limestones, marls, and a massive slump-block derived from the northern shelf (described below). It is bounded by elements G and I, with element H internal to the segment and evidenced by a mapped fault.

(6) The northern shelf, exhibiting open-marine shallow-water limestones and marls, bounded by elements I and J.

(7) The Oran Meseta, representing the northern shore of the rifted trough (Fig. 4). Jurassic strata are either buried or eroded away, except for a stream gorge north of Midelt exhibiting a thick sequence of intertidal and sabkha carbonates (Fig. 9).

Carbonate Facies Mosaic

Within each of the segments of Figure 4B is an unique stratigraphic sequence, diagrammed in Figure 4A. Deposition of each sequence is controlled by (1) subsidence of the rift as a whole, with carbonate sequences totally 6 km or more in some segments, (2) differential movement of the fault block(s) comprising each segment, and (3) relative sealevel changes, herein assumed to be eustatic. The rift-wide Toarcian deepening, marked everywhere across the traverse of Figure 4A by bathymetrically deeper and finer-grained deposits, provides a natural division for discussion between the generally shallower pre-Toarcian Liassic sequences below and the generally shallower post-Toarcian Dogger sequences above. Such Toarcian deepening has been noted worldwide (e.g. Haq et al., 1987; Hallam, 1987).

Figure 4A is drawn on a Domerian-Toarcian datum (approximately horizontal) and shows the basement blocks more deeply subsided than in Figure 4B. Erosion is actually deeper and much more irregular than diagrammed.

Liassic Sequences

Southern Shore and Shelf (Segments 1 and 2)--On this geotraverse, the most continuous exposures of shelf facies are in the gorge of the Oued

Ziz north of Ar Rachidia (segment 2) Late Pliensbachian (= Domerian) cyclic limestones are abruptly overlain by Toarcian green marls (Fig. 5), which generally shallow upward to be replaced by small coral buildups, carbonate lagoonal sediments, oolite shoals and other shallow-water limestones representing a shelf prograding from the south (Fig. 6). This area of shelf carbonates is currently under study by P. D. Crevello (Marathon Oil Company, Littleton, Colorado, U.S.A.).

The Domerian limestones are 100+ m of stacked upward-shallowing sequences each 1 to 2 m thick. They are rich in corals, megalodont and lithiotid bivalves, and capped by fenestral fabric and thin reddish marls suggesting subaerial exposure between cycles.

In the gorge of the Oued Todra, 100 km west, tepee structures are interbedded with other shallow-water, intertidal and supratidal dolomites and/or dolomotized limestones several hundred meters thick (Burri et al., 1973), mapped as Sinemurian through Pliensbachian. On the geotraverse, such facies are not exposed, but may be present in the buried early Lias or developed more shoreward on the Saharan Craton.

The Domerian limestones are separated from the Toarcian green-weathering marls by a condensed bed crowded with mussel-like bivalves and coated with a black ferruginous and/or manganiferous deposit a few millimeters thick. The overlying marls are 100 to 200 m thick, rich in bivalves, gastropods, ammonites and brachiopods, and exhibit several thin-bedded, persistent nodular limestones per meter.

Near the northern edge of segment 2, the marls are overlain by a 200+ m sequence of prograding coral reefs (Fig. 6), oolite shoals, and shallow-water limestones, culminating in shoreline and continental beds. The prograding reefs and oolites of Figure 6 were close to the paleo-shelf edge, at the juncture of segments 2 and 3.

Southern Shelf Edge and Deep (Segment 3)--The Tizi n' Firest fault zone is now a high-angle reverse fault, placing resedimented slope and apron carbonates over shelf sediments (Fig. 7). It was originally a normal fault separating tectonic-stratigraphic segments 2 and 3. Segment 3 accumulated 1000+ m of channelized and non-channelized turbidites, debris flows and other gravity deposits (Fig. 7; Evans and Kendall, 1977; Halliwell, 1985). The bounding fault was active early in the Sinemurian, with relatively coarse-grained gravity deposits generally becoming finer-grained through the Pliensbachian, and very marly in the Toarcian (Fig. 4A). The coarser fragments are comprised of corals and other shelf-derived lithoclasts, as well as ripped-up slope and apron debris. Halliwell (1985) documented that the entire sequence becomes thinner and finer-grained at the next exposure to the north at Jebel Bou Hamid, 10 km from the Tizi n' Firest fault zone, suggesting segment 3 was one or more half-grabens dipping to the south.

Central Basin High (Segment 4)--The pre-Toarcian sequence of segment 4, between the turbidite troughs of segments 3 and 5, is well-exposed in stream cuts through three upthrusted ridges (Fig. 8). The northern and southern ridges are regarded as edges of the central high. The middle ridge may be a reactivated fault within the central high, but not indicated on Figure 4. The sequence varies from 300 to 450 m thick. The basal 100 to 200 m is dominantly laminated unfossiliferous black limestone, increasingly interbedded upward with gray, fine-grained, vaguely mottled, bioturbated limestone. The central part of the section is increasingly bioturbated, thicker bedded, and fossiliferous. Chert-bearing beds appear to represent silicified sponges. Crinoid plates, brachiopods and thin-shelled pelagic bivalves are common.

The upper third of the section contains increasingly larger sponge-algal buildups, which cap the section exhibiting tens of meters of relief (Fig. 8). The flanks of the mounds contain cascaded beds of brachiopods, bivalves, crinoid plates and other fossil debris. Corals are very rare and mainly solitary. Algal-coated grains and fossil fragments are common.

The buildups are mainly dense carbonate mudstones. Favorable exposures exhibit a variety of sponges and a kind of algal stromatolite (Fig. 9) that are the only identifiable framework components. Sponge spicules are present in thin-sections. These sponge-algal buildups are probably amongst the largest and best preserved anywhere (Warme, 1986).

The contact between the mounds and the overlying rhythmites is sharp but irregular (Fig. 4A, 8); the tops and flanks of some of the uppermost mounds are littered with numerous large ammonites that lived on and around the buildups before they were smothered by the succeeding fine-grained rhythmites of the Toarcian.

Northern Shelf Edge and Deep (Segment 5)--Along this geotraverse the northern margin of the CEHAR is structurally complicated, and also confused by the presence of an apparent giant olistostrome derived from the northern shelf edge (Figs. 4A, 10). The stratigraphic sequence is dominated by dark gray marls at least 1650 m thick. Letsch (1985) described and interpreted the sequences in segments 5 to 7 of Figure 4, summarized in Figure 10.

The deepwater marls of segment 5 are generally much finer-grained than those in segment 3 with only minor beds of coarser-grained turbidites and debris flows. However, over a large area the sequence is punctuated by a mass of light-gray recrystallized limestone and dolomite containing abundant shallow-water indicators such as corals and oolitic grainstones. A discrete flat plane separates the base of the mass from the underlying Aalenian or Bajocian marls, being regarded as a plane of detachment along which the overlying mass slid. The western end and upper surface of the mass is enveloped in channelized and non-channelized debris beds that were presumably shed off the slump scar left by the block (Fig. 11); they represent the coarsest deposits in the entire sequence. The displaced block was previously mapped as a mid-basin carbonate reef at least 22 km long and 2 to 3 km wide (Ksar es Souk and Boudenib map, 1956). Although undated, the mass probably represents a detached portion of the post-Toarcian prograding shelf margin, as diagrammed in Figures 4A and 10.

Northern Shelf and Shore (Segments 6 and 7)--A tectonically-complicated sequence of shallow-water limestones, at least a few hundred meters thick and generally dated as lower to middle Lias, is exposed just south of the North Atlas fault zone (Fig. 4). The medium-gray limestones contain crinoids, echinoids, brachiopods and mollusks, with no evidence of intertidal or supratidal environments. The shoreline was to the north of the present North Atlas fault zone.

One small fault-slice outcrop of green-weathering shale, mapped as Toarcian, is lithogically similar to its southern counterpart on segment 2, described above. No post-Toarcian shelf environments escaped erosion, except for the displaced gravity slide of segment 5.

An erosional window, on the Oran Meseta about 20 km north of the North Atlas fault zone, exposes 100+ m of intertidal and supratidal cyclic dolomites, with abundant algal laminations, fenestral fabrics, and tepee structures (Fig. 12; Letsch, 1985). Intermediate environments, between these shoreline features and the deeper shelf

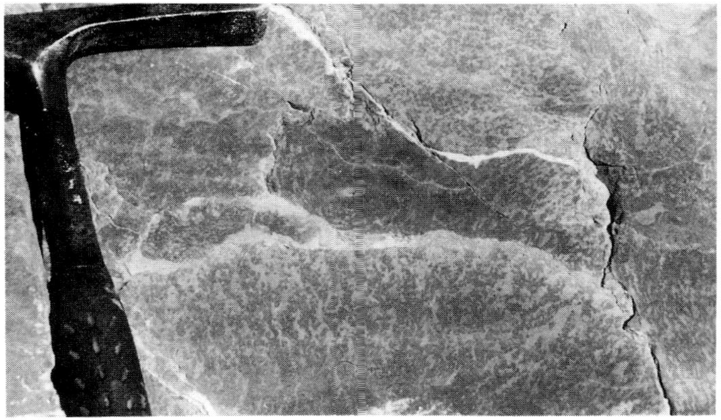

Figure 9. Heads of stromatolitic algae interbedded with sponges to form buildups as shown on left side of Figure 8.

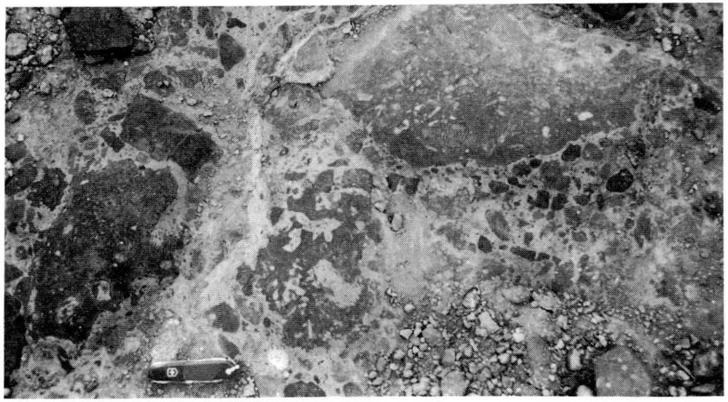

Figure 11. Disorganized debris bed of unit 4, Figure 10, containing coral heads (large clasts), bored and burrowed shallow-water limestones, and other lithoclasts.

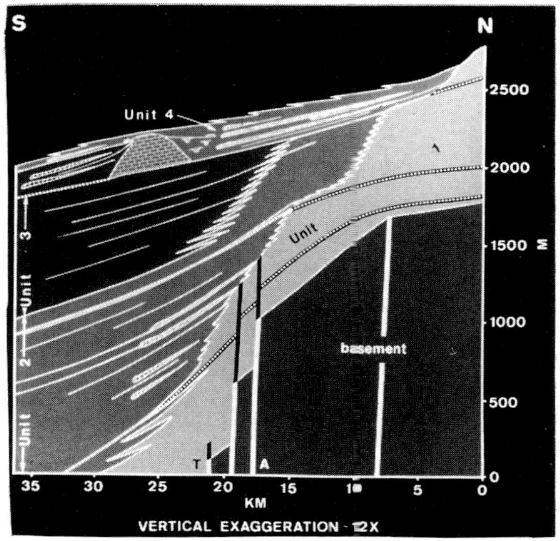

Figure 10. Diagram of evolution of northern shelf-slope-basin transition, showing synsedimentary shelf-margin faulting, shallow- to deep-water facies transitions, Toarcian sealevel rise (unit 3), and displaced shelf-edge olistostrome covered with shelf-edge debris (unit 4; see Figure 11). From Letsch (1985).

exposed on the northern edge of the Atlas, are eroded away or buried on the Oran Meseta between these two areas.

Dogger Sequences

Along the southern and northern margins of the CEHA the upper parts of the marine Dogger are largely eroded away. The transition from marine to continental is preserved in only a few scattered areas. However, toward the center of the Atlas ranges increasingly larger areas contain Aalenian and lower Bajocian shelf carbonates, especially as flat-lying capping beds over the Toarcian shales of segment 2 (for example, Figure 6).

In the center of the CEHAR, on tectonic-stratigraphic segments 3 to 5, the Dogger is represented by thick marls continuing upward from the fine-grained Toarcian rhythmites (Fig. 4), now preserved in the broad synclines between upthrusts. The fine-grained sections have been measured to be 1200+ m thick (Stanley, 1981; see Figs. 13, 16), and probably range to 2000+ m. These sections contain volumetrically minor but spectacularly exposed coralgal reef horizons (Figs. 13, 14).

Coralgal Reefs--Stanley (1981) documented one generally upward-shallowing sequence within segment 3 (Fig. 16). It culminates in a 100+ m thick reef horizon that crowns the edges of an eroded syncline (Fig. 14) due east of the town of Rich (Fig. 3). The center of the syncline preserves a higher reef horizon, separated by several hundred meters of marls from the lower one (Ksar es Souk and Boudenib map, 1956). These relationships are diagrammed on Figure 4A, and indicate that a shallow, muddy, marine gulf represented the final 10 to 15 million years of the CEHAR, producing immense volumes of fine-grained carbonate debris (Fig. 15). The stratigraphically restricted coralgal horizons are comprised of buildups that are very similar to modern Caribbean scleractinian reefs (Figs. 14, 15), exhibiting diversity of about 40 species (Beauvais, 1970), a rich community of mollusks, brachiopods, echinoids, encrusters and skeletal borers (Warme, 1977), and a lateral zonation that implies subtle control by basement faults (Warme et al., 1975). Buildup colonizers were a pioneer community of one or a few coral species that were mud-tolerant, succeeded by a more diverse coral community (Fig. 15).

Figure 12. Laminated, cracked and deformed intertidal to supratidal dolomites exposed on the Oran Meseta north of Midelt.

Figure 13. Aalenian and Bajocian strata typical of the central part of the CEHAR. Coralgal buildup is approximately 50 m thick, surrounded by fine-grained limestones and marls. Location west of Rich.

Figure 14. Upward-shallowing sequence described by Stanley (1981) east of Rich. Sequence is capped by coralgal reef horizon about 100 m thick. See Figures 15 and 16.

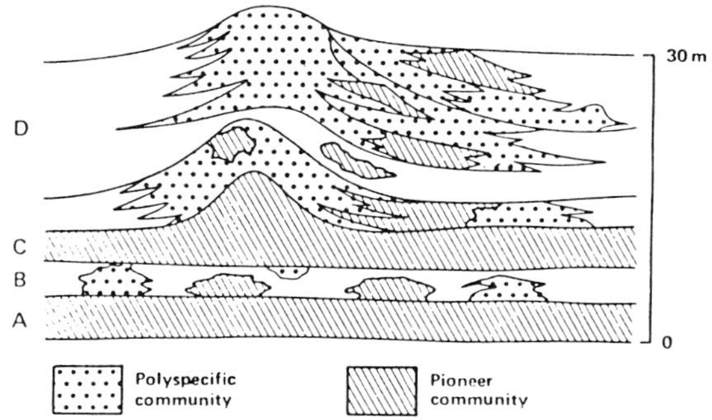

Figure 15. Diagram of typical Bajocian reef geometry and components (Warme et al., 1975).

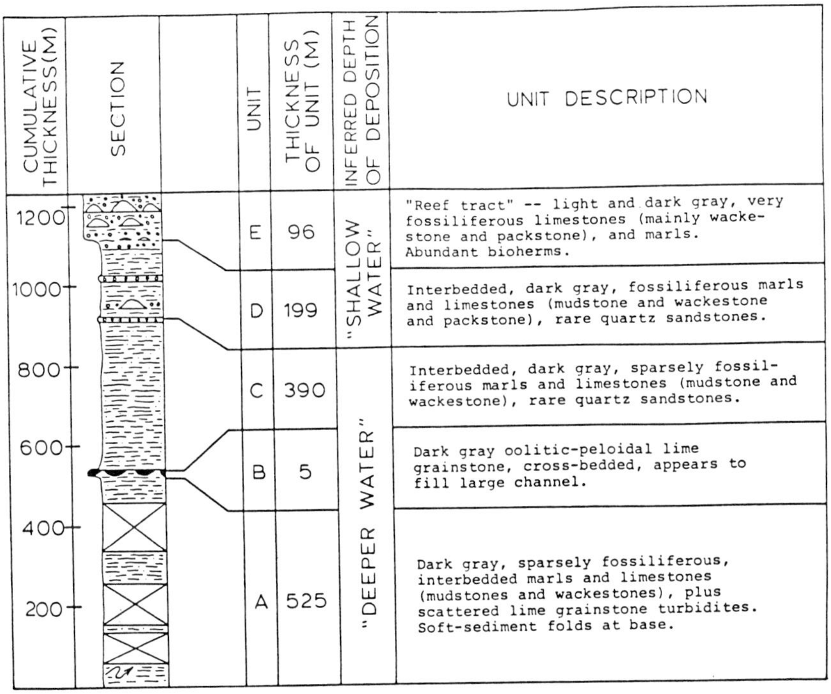

Figure 16. Upward shallowing stratigraphic sequence representative of Aalenian-Toarcian strata in the center of the CEHAR (Stanley, 1981). Grainstone of unit B may correspond to the Aalenian-Toarcian boundary. Top of unit D and most of unit E shown in Figure 14.

<u>Capping Continental Facies</u>--In the axes of several synclines along the CEHA, time-transgressive siliciclastic sandstones and mudstones contain mudcracks, raindrop imprints, dinosaur tracks and bones (Jenny et al., 1981), red coloration and other continental indicators, signaling the end of the marine rift (Table 1).

SUMMARY

The present Central and Eastern High Atlas Mountains of Morocco (Fig. 1) represent the boundaries of a long lived fracture zone that contains a sequence of Variscan mildly metamorphosed sediments, overlain by a rift sequence of late Triassic and earliest Jurassic continental deposits followed by Liassic and Dogger marine carbonates (Table 1).

Carbonate facies were controlled primarily by subsiding and rotating basement fault blocks (Fig. 4B), creating an evolving mosaic of shallow- and deep-water facies across the rift (Fig. 4A). Differential movement produced unique sequences on each segment, modified by eustatic drowning most notable in the Toarcian and Aalenian. Subsidence ceased in the Middle Dogger (Fig. 2), and post-Jurassic orogenesis inverted the original normal faults into upthrusts that created the modern topographic ranges of the High Atlas (Fig. 3).

On a selected geotraverse from south to north, as few as eight basement blocks (Fig. 4), all now evidenced by inverted bounding faults (Fig. 3), explain the carbonate facies encountered along the traverse. One or two blocks comprise each of five tectonic-stratigraphic segments, plus two segments on the bounding stable blocks to north and south (Fig. 4). Each segment contains a distinct suite of carbonate environments, representing northern and southern shelves and deeps bracketing a central-basin platform.

ACKNOWLEDGEMENTS

I acknowledge the countless Moroccan, French and other geologists who have contributed to the understanding of the geology of Morocco. I am indebted to Ian Evans, Chris Kendall, James Lee Wilson, Tony Ekdale, Rick Stanley, Michael Brady and Martin Schuepbach, among others, who discussed High Atlas geology in the field. However, I take full responsibility for the interpretations herein. Initial support for the work was by U.S.A. National Science Foundation Grant Number GF 32510, to the University of South Carolina, William H. Kanes, project director. Subsequent work by Warme, Halliwell and Letsch was by U.S.A. NSF Grant Number EAR 8214122 to the Colorado School of Mines. Sincere thanks to the Directorates of the Moroccan Service Geologique and National Petroleum Organization (ONAREP) for their friendly cooperation and permission to work in the High Atlas. Thanks also to R. Du Dresnay, who willingly shared his immense knowledge of Moroccan geology. Prof. Dr. V. Jacobshagen kindly invited inclusion of the contribution in this volume. This is Colorado School of Mines Exploration Geosciences Institute Publication Number 7.

REFERENCES

Ambroggi, R., 1963, Etude Geologique du versant meridional du Haut Atlas Occidental et de la Plaine du Sous: Service Geol. Maroc, Notes et Mem., no. 206, p. 37-116.

Beauvais, L., 1970, Etude de quelques polypiers Bajociens du Maroc Oriental: Service Geol. Maroc, Notes et Mem., v. 30, no. 225, p. 39-50.

Brede, R. and Heinitz, W., 1986, Zum Bau des Hohen Atlas zwischen Errachidia und Midelt (Marokko): Berliner geowiss. Abh., (A), v. 66, p. 433-454.

Brown, R. H., 1980, Triassic rocks of Argana Valley, southern Morocco and their regional structural implications: AAPG Bull., v. 64, no. 7, p. 988-1003.

Burgess, C. J., 1979, The development of a Lower Jurassic carbonate tidal flat, Central High Atlas, Morocco. 2: diagenetic history: Journal of Sed. Pet., v. 49, no. 2, p. 413-428.

Burgess, C. J., and Lee, C. W., 1978, The development of a Lower Jurassic carbonate tidal flat, Central High Atlas, Morocco. 1: sedimentary history: Journal of Sed. Pet., v. 48, no. 3, p. 777-794.

Burri, P., du Dresnay, R., and Wagner, C. W., 1973, Tepee structures and associated diagenetic features in intertidal carbonate sands (Lower Jurassic, Morocco): Sed. Geol., v. 9, p. 221-228.

Choubert, G., and Marcais, J., 1956, Apercu structural, in Geologie du Maroc: Service Geol. Maroc, Notes et Mem., no. 100.

Colo, G., 1961, Contribution a l'etude du Jurassique du Moyen-Atlas Septentrional: Service Geol. Maroc, Notes et Mem., no. 139, 226 p.

Cristi-Blick, N., and Biddle, K. T., 1985, Deformation and basin formation along strike-slip faults: SEPM Spec. Pub. no. 37, p. 1-34.

Dresnay, R. du, 1977, Le milieu récifal fossile du Jurassique inférieur (Lias) dans le domaine des chaînes Atlasiques du Maroc: Mem. Bur. Rech. Geol. Min., no. 89, p. 296-312.

Dresnay, R. du, 1979, Sediments Jurassiques du domaine des chaînes Atlasiques du Maroc: Symposium "Sédimentation Jurassique W. Européen", ASF Spec. Pub., no. 1, p. 345-365.

Evans, I., and Kendall, C. G. St. C., 1977, An interpretation of the depositional setting of some deep-water Jurassic carbonates of the Central High Atlas Mountains, Morocco; in Cook, H. E., and Enos, P., eds., Deep-Water Carbonate Environments: SEPM Spec. Pub. no. 25, p. 249-261.

Faure-Muret, A., and Choubert, G., 1971, Le Maroc; Domaine rifain et atlasique, in Tectonique de l'Afrique: UNESCO, Earth Science Series, no. 6, p. 17-46.

Hallam, A., 1987, Radiations and extinctions in relation to environmental change in the marine Lower Jurassic of northwest Europe: Paleobiology, v. 13, p. 152-168.

Halliwell, B. A., 1985, Deep-water carbonate deposits of the southern margin of the Jurassic central High Atlas trough, Morocco: Master of Science thesis, Colorado School of Mines, 226 p.

Haq, B. U., Hardenbol, J., and Vail, P. R., 1987, Chronology of fluctuating sea levels since the Triassic: Science, v. 235, p. 1156-1167.

Herbig, H-G., 1986, Lithostratigraphisch-fazielle Untersuchungen im marinen Alttertiär sudlich des zentralen Hohen Atlas (Marokko); Berliner geowiss. Abh., (A), v. 66, p. 343-380.

Heybroek, P., 1965, The Red Sea Miocene evaporite basin, in Salt Basins Around Africa: Inst. of Petroleum, London, p. 17-40.

Huessner, H., 1985, Jurassic carbonates of the Western High Atlas (Morocco); Microfacies analysis and plate tectonic framework: Facies, v. 12, p. 141-218.

Jacobshagen, V., 1986, Zur Bau-Geschichte des zentralen Hohen Atlas (Marokko); Berliner geowiss. Abh., (A), v. 66, p. 455-470.

Jenny, J., Le Marrec, A., and Monbaron, M., 1981, Le couches rouges du Jurassique moyen du Haut Atlas central (Maroc): correlations lithostratigraphiques, éléments de datations et cadre tectono-sédimentaire: Bull. Soc. geol. France, (7), v. 23, p. 627-639.

Kanes, W. H., Saadi, M., and Ehrlich, R., 1973, Moroccan crustal response to continental drift: Science, v. 180, p. 950-953.

Lorenz, J. C. (this volume), Synthesis of late Paleozoic and Triassic redbed sedimentation in Morocco, 31 p.

Lee, C. W., and Burgess, C. J., 1978, Sedimentation and tectonic controls in the Early Jurassic Central High Atlas Trough, Morocco: Geol. Soc. America Bull., v. 89, p. 1199-1204.

Letsch, D. K., 1985, Early Jurassic depositional history of the northern margin of the central High Atlas, Morocco: Master of Science Thesis, Colorado School of Mines, 171 p.

Lowell, J. D., Genik, G. J., Nelson, T. H., and Tucker, P. M., 1975, Petroleum and plate tectonics of the Southern Red Sea, in Fischer, A. G., and Judson, S., eds., Petroleum and Global Tectonics: Princeton University Press, p. 129-153.

Manspeizer, W., Puffer, J. H., and Cousminer, H. L., 1978, Separation of Morocco and eastern North America: A Triassic-Liassic statigraphic record: Geol. Soc. America Bull., v. 89, p. 901-920.

Mattis, A. F., 1974, Upper Triassic lacustrine carbonates, central High Atlas mountains, Morocco (abs.): Geol. Soc. Am., Abstracts with Programs, v. 6, p. 53.

Mattis, A. F., 1977, Nonmarine Triassic sedimentation, Central High Atlas Mountains, Morocco: Journal of Sed. Pet., v. 47, no. 1, p. 107-119.

McClay, K. R., and Ellis, P. G., 1987, Geometries of extensional fault systems developed in model experiments: Geology, v. 15, p. 341-344.

Reyre, D. (ed.), 1965, Le Bassin du sud-ouest Marocain, in Sedimentary Basins of the African Coasts: Association of African Geological Surveys.

Rod, E., 1962, Fault pattern, northwest corner of Saharan shield: AAPG Bull., v. 46, p. 529-534.

Salvan, H. M., 1974, Les séries salifères triassiques du Maroc, comparaison avec les séries homologues d'Algérie et de Tunisie, nouvelles possibilities d'interpretation: Service Geol. Maroc, Notes et Mem., v. 35, no. 255, p. 8-25.

Septfontaine, M., 1986, Depositional environments and associated foraminifera (lituolids) in the Middle Liassic carbonate platform of Morocco: Revue de Micropaleontologie, v. 28, p. 265-289.

Stanley, R. G., 1981, Middle Jurassic shoaling of the central High Atlas sea near Rich, Morocco: Journal of Sed. Pet., v. 51, no. 3, p. 895-907.

Van Houten, F. B., 1977, Triassic-Liassic deposits of Morocco and Eastern North America: comparison: AAPG Bull., v. 61, p. 79-99.

Van Houten, F. B., and Brown, R. H., 1977, Latest Paleozoic-early Mesozoic paleogeography, northwestern Africa: Journal of Geol., v. 85, p. 143-156.

Vogel, T. A., Williams, R. E., Preston, J. K., and Walker, B. M., 1976, Origin of the Late Paleozoic plutonic massifs in Morocco: Geol. Soc. America Bull., v. 87, p. 1753-1762.

Warme, J. E., 1977, Carbonate borers - their role in reef ecology and preservation: AAPG Studies in Geology, no. 4, p. 261-279.

Warme, J. E., 1986, Giant Liassic sponge-algal buildups, High Atlas, Morocco (abs.): Proc., 12th Int. Sed. Cong., Canberra.

Warme, J. E., Stanley, R. G., Wilson, J. L., and Ekdale, A. A., 1975, Middle Jurassic reef tract, Central High Atlas, Morocco: Proc., 9th Internat. Sed. Congress, Nice.

Wernike, B., and Burchfiel, B. C., 1982, Modes of extensional tectonics: Jour. Structural Geology, v. 4, p. 105-115.

JURASSIC SEDIMENTS OF THE CENTRAL HIGH ATLAS OF MOROCCO : DEPOSITION, BURIAL AND EROSIAN HISTORY.

Brechbühler Y.A., Bernasconi R. and Schaer J.-P.
Institut de Géologie, Université de Neuchâtel
11, rue E.-Argand, CH-2007 Neuchâtel

Abstract

During the beginning of the Jurassic, the Central and Eastern High Atlas of Morocco was a sea way open to the East and connected with the Tethys but not with the newly formed Atlantic. Limestones and marls were deposited in a narrow trough (40 km) where up to 9 km of sediments have been locally mesured. There, subsidence was very active (average : 100-200 m/m y). On the bordering platform of that narrow basin, the sediments are much thinner (0,8 km) and composed essentially of pure limestone (subsidence : env. 20 m/m y). In the trough sedimentation starts with thick intertidal limestones, but with time becomes more influenced by slumps, mudflows, turbidites and local erosion. At the end of basin filling (Mid-Jurassic) shallow terrigenous sediments and red sandstones indicate the transition to continental deposits. As erosion has been active, ever since the Upper Jurassic, it is often difficult in many regions to estimate the original sedimentary thickness. Illite crystallinity, IC, has proved to be a reliable method for tracing the burial evolution of these carbonate rocks, from diagenesis to the beginning of metamorphism. Our results suggest that the lower part of the sedimentary pile in the synclines have reached 300° C with a normal gradient. In the anticlines, where the sedimentary pile was 4-5 km thick, an abnormal gradient of 60° C /km was probably related to the circulation of hot fluids.

Introduction

The Atlas mountains extend from Atlantic Morocco to Tunisia over more than 2000 km but are rarely more than 100 km wide. The Eastern part of the range, the High Atlas of Morocco is usually divided into two parts : to the West, the Western High Atlas represents a fragment of the passive continental margin of the Atlantic Ocean; to the East, following a region of Paleozoic and Precambrian formations, SE of Marrakech, the range is mostly composed of a folded Mesozoic cover. Our study will be restricted to an analysis of the sedimentary evolution of a part of the range, S of Midelt (Fig. 1), where the Central High Atlas merges into the so-called Eastern High Atlas which extends to the Algerian border.

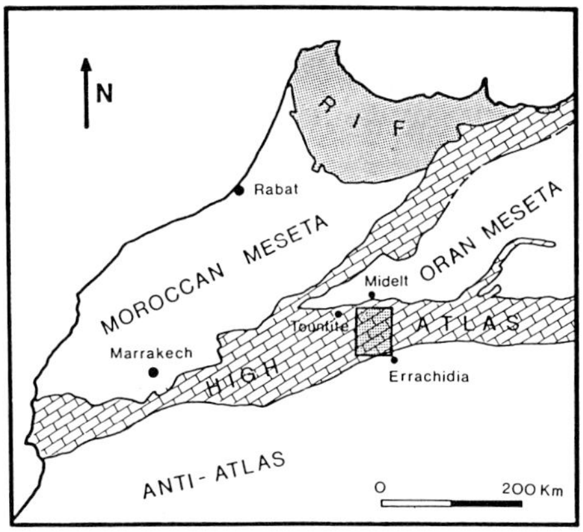

Fig. 1. The High Atlas of Morocco and the studied area.

In the Central and Eastern High Atlas, at the beginning of the Jurassic, the sea transgressed Triassic continental formations capped by basalts. A long and narrow, subsiding trough formed. In its central part, the accumulation of sediments was considerable; on its southern and northern edges carbonate reef shore facies developed with decreasing thickness away from the basin. The future range coincides rather closely with the active, sinking trough and, especially to the south, the beginning of the platform (Dubar 1962, Choubert et Faure-Muret 1962, Du Dresnay 1972 and 1979). In the western extremity of the trough, sediments of shallow subtidal environment with carbonate silts, sands and evaporites have only a moderate thickness (Burgess and Lee 1978, Jenny et al. 1981).

The Jurassic sediments of the Central and Eastern High Atlas trough consist mainly of limestones and marls. The lowest part of the formation records shallow water environments passing upward into moderately deep water facies. Progressive shoaling led to subsequent the development of coral reefs, and finally, when preserved, to redish continental deposits. For convenience of presentation we have divided the total sequence into six formations initially described by Studer (1980b) on the basis of lithologic, faunal and environmental considerations.

The sediments

Starting from the bottom, the Lower Liassic formation ("dolomies et calcaires massifs d'Idikel") includes laminated micrites of supratidal to subtidal environment followed by thick limestones with dolomite containing oolites, bivalves, brachiopoda and sponge reefs indicating a moderate increase in bathymetry. The age of this formation is considered to be Sinemurian.

The second formation ("calcaires lités de l'Aberdouz") consists of a succession of dark micritic undulating beds 10 to 20 cm thick, separated by very thin (some millimetres) of marly to shaly layers. Nodular cherts are common. The faunal content is rather similar to that of the previous formation with a variety of benthic organisms : the presence of ammonites and planctonic foraminifera suggest more open marine conditions. The age is Sinemurian.

The next formation is normally composed of a succession of limestones alternating with marly horizons ("calcaires et marnes de l'Ouchbis"). Near the lower part of the section, the marly intervals are thin (a few centimetres) but they increase gradually in number and thickness. These sequences have a turbiditic origin (Du Dresnay 1972 and 1979, Evans and Kendall 1977, Studer 1980). In the basin they are mainly distal, whereas more proximal facies were observed along the northern margin of the trough with flow direction toward the S. The alternating micrites, biomicrites and marls represent mainly basin plain deposits winnowed by currents from the carbonate platforms where the degree of production and destruction was high, in a environment marked by common well preserved reefs. The most spectacular and best studied example is the Bou-Dahar reef where coral and large bivalves were the most active organisms (Dubar 1948, Agard et Du Dresnay 1965). There and along the margin of the shelf, large slabs of reef limestone have slid into the basin where they appear as exotic blocs in lithologies of deeper facies. Carixian in the lower part and the entire Domerian have been recognised in this formation which marks a significant deepening of the basin.

The fourth formation ("Grès calcaires et marnes gréseuses de Tagoudite") is a very distinctive unit characterized by thin-bedded sandstones and fine-grained siltstones introduced into the basin by density currents and then deposited from dilute suspensions. Sand grains of detrital origin are composed of quartz and feldspar; their relative concentration in the clay matrix is greater in the northwestern regions where the formation is thicker (200 meters). Near the southeastern edge of the basin the detrital character is less prominent with less quartz and feldspar, more calcite, and a reduction in thickness to less than 70 m and sometimes only 35 m. No conglomerate has been reported. The arrival of this terrigenous material often accompanied by fragments of coal could come from the Moroccan Meseta and be partly reworked Triassic sandstone and silt driven into the basin during a period of low sea level. We have no evidence to relate this detrital sequence to an increase in tectonic activity in the vicinity of the basin. The age of the formation is Lower and Middle Toarcian.

From the beginning of the early Jurassic marine transgression up to the Tagoudite formation, the paleogeography remained rather simple : between two stable platforms, the subsiding basin is undifferentiated; despite the presence of some lateral facies variations essentially linked with reef development, the formation can be traced without too much difficulty. The depocenters are in the northwestern part of the study area. Sediment thickness decreases towards the East and the South. Indications of changes in the paleogeography of the basin first appear in the Tagoudite formation, the thickness of which, normally 100 to 150 meters in the northeastern sector, can locally decrease to less than 50 m or even 10 m. From that time on, the sedimentation was much more diversified. The presence of local ridges induced discordances, with local conglomerates; slumps and gravity deposits became widespread and filled the lower parts of the basin.

The Tagoudite formation is followed by a very thick sequence of argillaceous marls and limestones named the "Agoudim formation". Its lower part is mainly marly with only a few dark gray limestone beds up to 50 cm thick. The upper part of the formation contains more massive limestones which form one or several resistant cliffs. In the lower part of the formation the fossil content is mainly pelagic, whereas biodetrital limestones with oolites, brachiopoda, echinoderms, corals become more abundant up section. Patch reefs and major reefs occur along and around submarine ridges near the edges and also in the middle of the basin (Fig. 2 ; Stanley 1981).

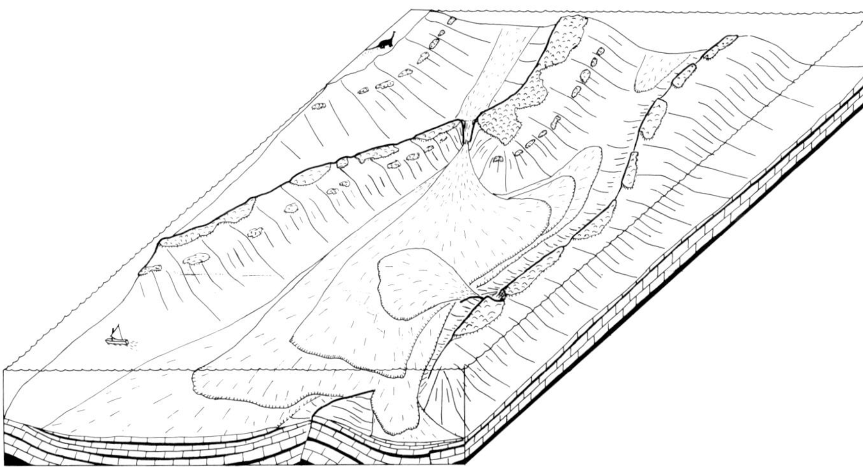

Fig. 2. Cartoon illustrating the paleogeography of the northern margin of the trough during Aalenian-Bajocian time. Coral reefs spread along the edge of the platform and on ridges. A large fan of turbidites disperses sediments produced around the reef into the basin.

Precise bathymetry as expressed by sedimentary structures is lacking. However, the increase in biological activity, the replacement of pelagic faunas by benthic assemblages and by coral reefs support a model of a shallowing basin during the beginning of the Middle Jurassic. In the lower part of the section the sediments are formed by deposition of fine terrigenous particles mixed with hemipelagic fallout and limestone mud from the active reefs on the basin edges. The degree of resedimentation by slumps, turbidites and mud flows is thought to be large. Along the northern edge of the basin,

Brechbühler (1984) described a large turbidity fan which has dispersed sediments over more than 30 km in a direction parallel with the general basin axis (Fig. 2). Large blocks of coral reef limestone up to 10 m^3 have been transported by mud flows no more than 2 m thick. These local turbidite accumulations can be more than 200 m thick in the deep basin and they consist of more than 50 % of shallow water sediments. Along some ridges, clasts of lower Sinemurian and Triassic basalts reveal the importance of erosion (Studer 1980b) whereas the thickness of the associated conglomerates (up to 2000 m) shows that the subsidence was still active as erosion continued. Bernasconi (1983) has shown the large and spectacular facies variations in a small pull-apart basin where conglomerates are common, especially along its marginal ridges (Fig. 3).

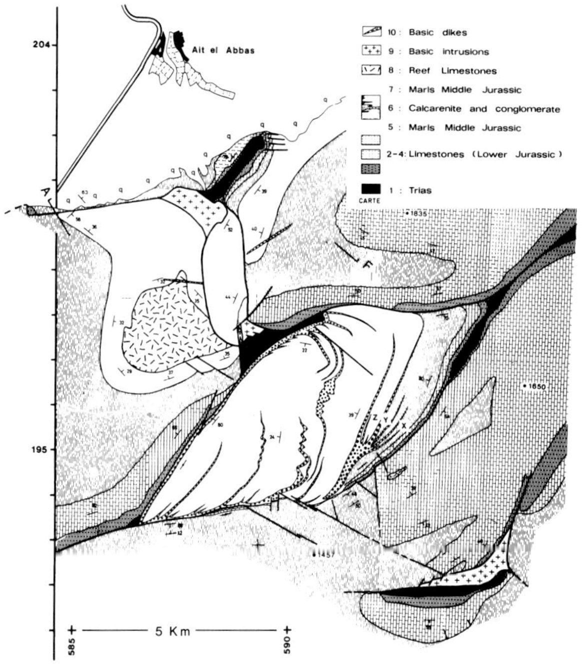

Fig. 3. Geological map of a small pull-apart basin northeast of Rich with development of conglomerate along its margins.

From that position, the relative abundance of the clasts as well as their grain size decreased towards the center of this small basin, where the bulk of the sediment has a marly composition. The origin of the large amount of terrigenous minerals, clays, quartz and feldspar, in the Agoudim formation is not clear. They could have been transported from the nearby continent (from the NW or the S) across the platform, which seems unlikely in view of a nearly constant coral barrier. A large part of this fine detrital material could have been carried by currents along the axis of the basin and deposited in the most rapidly subsiding part. The Agoudim formation extends from the Upper Toarcian to the base of the Bathonian.

The Anemzi formation, which is only preserved to the west of the study area, represents the last episode of basin filling. At the base, near the Bajocian-Bathonian boundary, some limestones are conspicuous in shallow water assemblages (Du Dresnay 1979); up section the sediments become more terrigenous with cross-bedding, local discordances, ripple-marks and mud-cracks. The continental character becomes dominant as the color changes with many recurrences, from gray-green to reddish. Further to the West, the development of these often discordant continental facies has been well studied by Jenny et al. (1981) and Monbaron (1981 and 1982). Following Du Dresnay, it is now accepted that they are of Middle Jurassic age.

Variations in thickness and sedimentation rates

It has long been recognised that the Jurassic accumulation of sediment was moderate on both platforms bordering the very thick Central High Atlas trough. The survey of different sections published by Studer (1980b), Bernasconi (1983), Brechbühler (1984) and personal communications from Heitzmann are integrated in Fig. 4, which shows variations in thickness of the different formations of the lower part of the Jurassic up to the Tagoudite formation. It is apparent that the variations in thickness do not follow a systematic and simple pattern with respect to basin geometry. Even at the beginning of the sedimentation, when the formations had only slight variations of facies, the most active depocenters were changing with time.

It is considered that large changes of sea level did not take place during the beginning of the Jurassic (Haq et al. 1987). In the Cen-

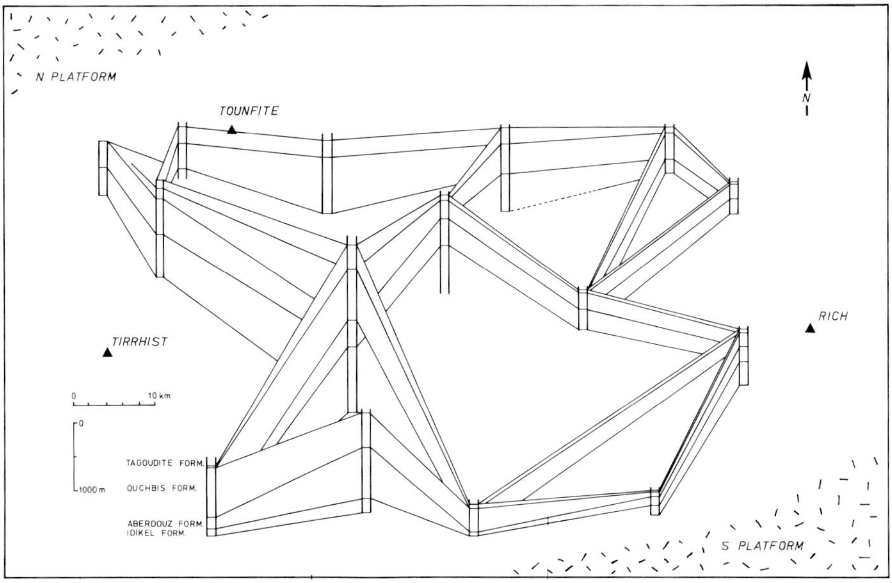

Fig. 4. Fence diagram showing the variation in thickness of the different formations of the Lower Jurassic in the Central Atlas trough south of Midelt.

tral High Atlas, the bathymetry was not constant, but the variations were moderate compared with the thicknesses of the different formations. We therefore conclude that these thicknesses can be used to make a first attempt to follow the history of subsidence of the basin and its margins (Fig. 5).

Thickness of sediments in the studied area

In the Central High Atlas some kilometres West of the studied area Studer (1980) measured along a continuous profile more than 8 km of Lower and Middle Jurassic sediments. Around Rich, erosion has been active since the Tertiary (and even earlier) until recent times, and it is therefore often difficult to estimate the original thickness of the Jurassic sediments as only 2 to 3 km of rock have been preserved. We have tried to use the evolution of illite crystallinity in these carbonate and marly rocks to estimate the diagenetic evolution, which is itself a function of the burial and thermal regime.

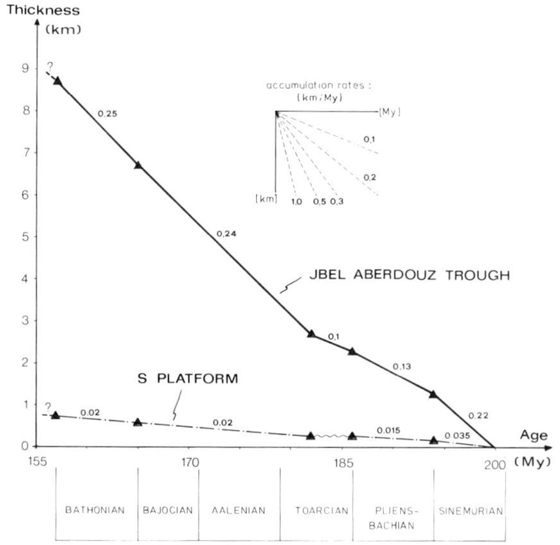

Fig. 5. Evolution of sedimentation rate in the trough and on the platform in the Central High Atlas. Time scale after Haq et al.(1987)

Illite crystallinity, method and previous studies

Illite "crystallinity" (IC) (Kübler 1964) is the sharpness of the (001) illite peak measured at half height on x-ray diffractograms. This peak represents the reflection of the 10 Å-spaced crystallographic planes of illites. A sharp peak (low IC value) indicates that all the micaceous material has the theoretical lattice space of 10° A, which is found in epimetamorphic rocks. Conversely a broad peak (high value) means that the illite is not well ordered and that the lattice layers are not equidistant; this is typical of weathered illites.

(Fig. 6) shows the general pattern of IC evolution as observed in boreholes. The IC values decrease with depth (Dunoyer 1970, Kübler 1980). The beginning of classical metamorphism (anchi/epizone limit) is marked by average IC values of 2.5. It is accepted that they reflect the effect of temperatures on the order of 300° C (Kübler 1966). Schaer and Persoz (1976) made a first attempt to use this method to assess the metamorphic evolution of limestones and marls in the High Atlas. Accepting a thin sedimentary cover (3 km), having identified epizone IC values near small plutons, a thin anchimetamorphic zone and only a few values of the diagenetic zone, they

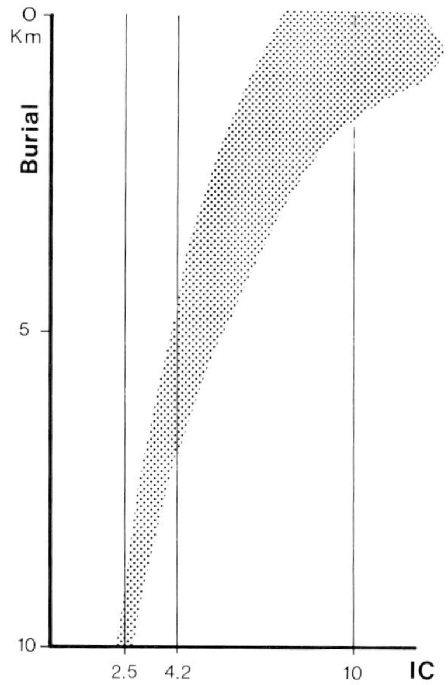

Fig. 6. Model of evolution of illite crystallinity versus depth (Kübler, 1966)

attributed a strong thermal effect to the intrusive material in the Central High Atlas. To the West of that study, where the erosion is less pronounced, Studer (1980b) was able to follow the evolution of IC along a continuous stratigraphic profile (8 km) of Jurassic sediments. He showed that IC evolves regularly from the youngest to the oldest sediments on a way which is only compatible with a normal thermal gradient of about 30 c/km (Fig. 7).

The IC has been examined systematically in a large sector of the High Atlas S.E. of Midelt, and its evolution has been studied in a traverse through the basin using a marly layer of early Toarcian age. This marker horizon is well dated and extends throughout the basin without marked changes in composition. Thus lithological influence on IC measurements can be kept to a minimum (Burst 1959). The result is presented on Fig. 8; two major domains can be recognized and they correspond to the main paleogeographic divisions. On the southern shelf high IC values are characteristic of early diagenesis, whereas low IC values occur in the basin and show a symmetrical decrease towards both margins. Several lithological profiles (A to F in Fig. 9) were also sampled in the basin and on the southern platform.

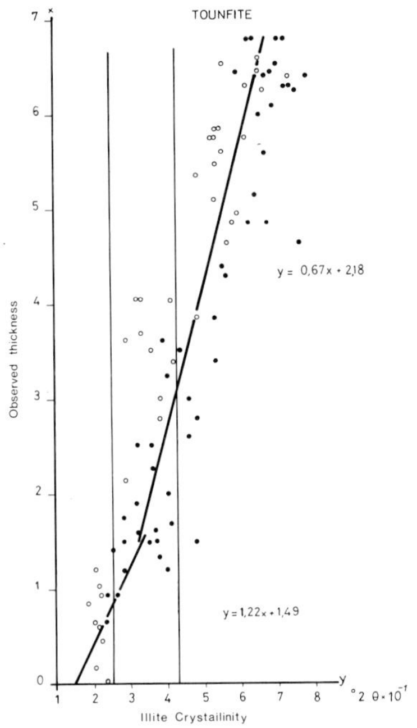

Fig. 7. Illite crystallinity versus depth in the Tounfite area (after Studer 1980a)

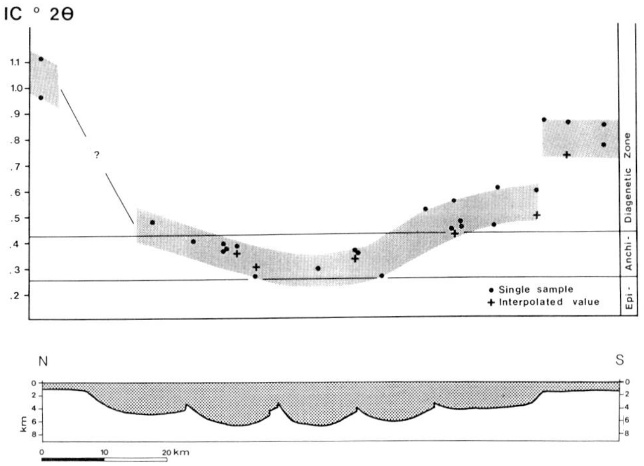

Fig. 8. Illite crystallinity in the marls of the lower part of Tagoudite formation through the High Atlas; for localisation of sampling position, see Fig. 9. At the bottom, proposed reconstruction of the post-Tagoudite burial.

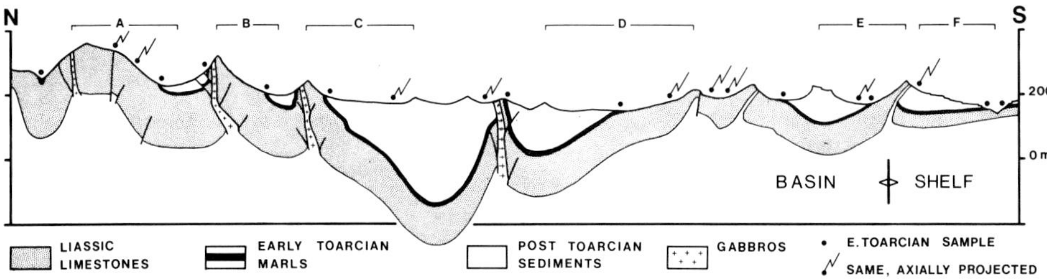

Fig. 9a. Profile through the High Atlas (vertical exaggeration) with localisation of the sampling area (A-F).

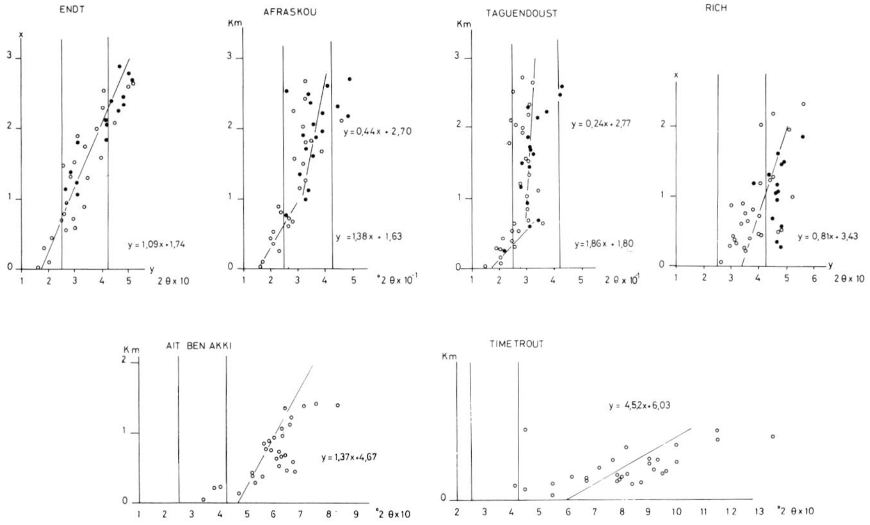

Fig. 9b. Position of illite crystallinity in different sectors versus thickness of exposed sections.

The local and structural conditions restricted sampling possibilities because limestones of Lower and Middle Lias are only present in the anticlines, whereas the marly sequence (Upper Lias and Dogger) is generally restricted to the synclines.

Three main statements can be made :

a) In each local profile the IC values decrease regularly from top to bottom.

b) In different profiles rocks of the same age have different IC.

c) Correlation of IC with increased depth (illite gradient) has been established statistically (Brechbühler 1984); it varies from profile to profile and sometimes within a profile. Illite gradients were calculated and drawn in each profile or in different parts of a profile.

Interpretation

As the observed thickness in a single profile never exceeds 3 km, it is impossible to follow in one profile the entire evolution, from the least affected rocks to those of the epimetamorphic zone. As the illite evolution depends on local factors such as the sedimentary history, the geothermal gradient, the lithology, etc. (Kübler 1964), neither is it possible to use correlations established in an other basin. The total evolution of IC in the Central High Atlas around Rich has been obtained by using four local profiles (F, E, D and C) representing investigations made mainly in synclinal positions. The southern platform profile (F) is used to describe near surface conditions. There the geological evidence allows an erosion estimate of 500 $\pm$ 200 m.

Deeper conditions of burial are reprensented by profile E where IC values are lower. They come from sediments deposited just outside the platform, along the slope to the basin. Part D and C of Fig. 9 come from successive synclines, they represented sediments from the fast subsiding part of the basin. C records the lowest IC values; they are very similar to those obtained by Studer (1980) in the lowest part of the Tounfite profile (Fig. 7). Our results suggest that the low part of section C in Fig. 9, which was sampled in the Toarcian, had been covered by 6 km of sediment of which 4 have been eroded. In that part of the High Atlas, where the Lias limestones make up about 1 km, the total subsidence was of the order of 7 km. In the lowest part of sections B and C, and in section

A of Fig. 9, the illite gradients are much higher than in the syncline. They are the expression of IC evolution in anticlines. Fig. 11 tries to present a composite profile in that situation. The epizone is reached under a cover of 4 or 5 km, giving a paleogeothermal gradient of 50 to 60° per km. These high values were probably induced by fluid circulations along the faulted anticlines. Temperatures around 300° have been obtained from fluid inclusion analysis in quartz crystals collected in calcite veins. The thermal influence of the small plutons or of the dikes seems to be restricted in space (Brechbühler 1984).

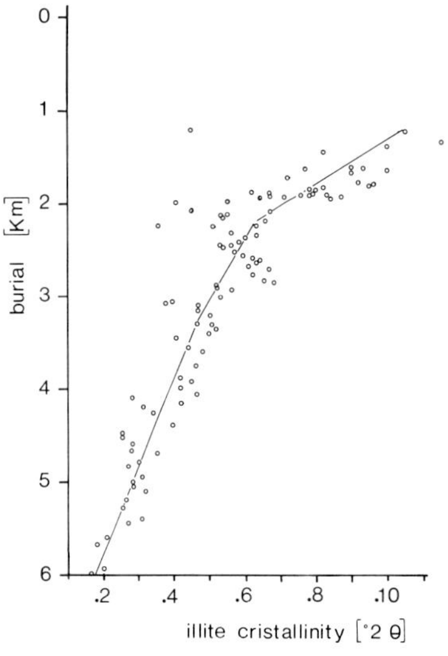

Fig. 10

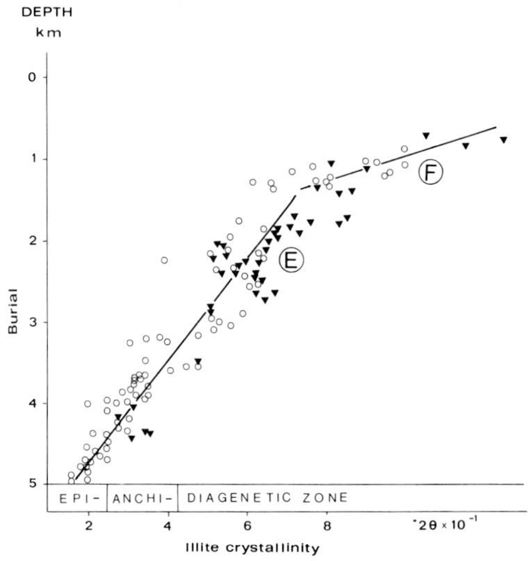

Fig. 11

Fig. 10. Composite diagram of illite crystallinity in syncline position.

Fig. 11. Composite diagram of illite crystallinity in anticline position

Conclusions

The sedimentation in the High Atlas trough S of Midelt was very active during the Jurassic. Subsidence was highest in the middle of the trough and along its northern margin. Up to almost the end of the Lower Jurassic, the paleogeography was rather uniform. From the Toarcian, ridges began to be active. They were the first manifestation of anticline development. According to our study and interpretation of illite crystallinity, the metamorphic evolution of the clay minerals in the limestones and marls was activated by important hydrothermal systems, which produced abnormal geothermal gradients. In the synclines, the geothermal gradients were normal. The use of illite crystallinity to decipher the different thermal regimes and to appreciate the degree of erosion turned out to be an appropriate method. It is important, however, to bear in mind that for a good approach, it is necessary to study a large number of samples.

Acknowledgements

The authors thank B. Kübler for his helpful suggestions throughout this research. Support for this study was provided by the Fonds National Suisse de la Recherche Scientifique (requête 2 450.0.79) and by Drs. M. Bensaïd and M. Dahmani of the Geological Survey of Morocco.

We are very grateful to S.N. Ayrton for critical review and improvement of the English.

References

AGARD, J. & DU DRESNAY, R. (1965) : La région minéralisée du J. Bou-Dahar, près de Beni Tajjite (Haut Atlas oriental) : Etude géologique et métallogénique.- Colloque sur des gisements stratiformes de plomb, zinc et manganèse du Maroc (2-14 Mai 1962).- Notes Mém.Serv.géol. Maroc 181, 135-152.

BERNASCONI, R. (1983) : Géologie du Haut Atlas de Rich (Maroc). Thèse, Sciences, Neuchâtel, 107 p.

BRECHBÜHLER, Y.-A. (1984) : Etude structurale et géologique du Haut Atlas calcaire entre le Jebel Ayachi et Rich (Maroc). Thèse, Sciences, Neuchâtel, 128 p.

BURGESS, C.J. & LEE, C.W. (1978) : The development of a lower jurassic carbonate tidal flat, Central High Atlas, Morocco, 1 : sedimentary history.- J. sediment. Petrol. 48, 777-794.

BURST, J.-F. (1959) : Postdiagenetic clay mineral environmental relationship in the Golf Coast Eocene.- Clays and Clay Miner. (6th Nat Conf. 1959), 328-341.

CHOUBERT, G. & FAURE-MURET, A. (1960-1962) : Evolution du domaine atlasique marocain depuis les temps paléozoïques. Livre mémoire P. Fallot.- Mém.Soc.géol. France, h.sér. 1, 447-527.

DUBAR, G. (1948) : Etudes paléontologiques sur le Lias du Maroc: Faune domérienne du Jebel Bou-Dahar près de Beni Tajjite. Etude suivie de celle de quelques mollusques d'autres gisements marocains.- Notes Mém.Serv.géol. Maroc 68, 250 p.

DUBAR, G. (1960-1962) : Notes sur la paléogéographie du Lias marocain. (Domaine atlasique).- Livre Mém. P. Fallot, 529-544.

DU DRESNAY, R. (1972) : Sédiments carbonatés intercotidaux et supracotidaux du Lias inférieur dans le domaine atlasique marocain. - C.R.Acad.Sci. (Paris), 275 D., 341-344.

DU DRESNAY, R. (1979) : Sédiments jurassiques du domaine des chaînes atlasiques du Maroc.- Symposium "Sédimentation jurassique W-Européen". A.S.F. Publ. spec. 1.

DUNOYER DE SEGONZAC, G. (1970) : The transformations of clay minerals during diagenesis and low grade metamorphism : a review.- Sedimentology, 15, 281-346.

EVANS, I., KENDALL, C.G.St.C., BUTLER, J.C. (1977) : Genesis of shallow water rhytms, Central High Atlas Mountains, Morocco.- J. Sed. Petrology, 47, 120-128.

HAQ, B.U., HARDENBOL, J., VAIL, P.R. (1987) : Chronology of fluctuating sea levels since the Triassic.- Science, 235, 1156-1166.

JENNY, J., LE MARREC, A. & MONBARON, M. (1981) : Les couches rouges du Jurassique moyen du Haut Atlas central (Maroc) : corrélations lithostratigraphiques, éléments de datations et cadre tectonosédimentaire.- Bull.Soc.géol. France (7), 23, 627-639.

KÜBLER, B. (1964) : Les argiles indicateurs de métamorphisme.- Rev. Inst.franç.Pétrole, 19, 1093-1112.

KÜBLER, B. (1966) : La cristallinité de l'illite et les zones tout à fait supérieures du métamorphisme.- Colloque sur les Etages Tectoniques, Neuchâtel, La Baconnière, 105-122.

KÜBLER, B. (1980), les premiers stades de la diagenèse organique et de la diagenèse minérale : une tentative d'équivalence. 2ème partie : Zonéographie par les transformations minéralogiques, comparaison avec la réflectance de la vitrinite, les extraits organiques et les gaz absorbés.- Bull.Ver.schweiz. Petroleum- Geol. u.Ing. 46, 110, 1-22.

MONBARON, M. (1981) : Sédimentation, tectonique synsédimentaire et magmatisme basique : l'évolution paléogéographique et structurale de l'Atlas de Beni Mellal (Maroc) au cours du Mésozoïque : ses incidences sur la tectonique tertiaire.- Eclogae geol. Helv. 74/3, 625-638.

MONBARON, M. (1982) : Précisions sur la chronologie de la tectogenèse atlasique : exemple du domaine atlasique mésogéen du Maroc.- C.R. Acad.Sci. (Paris) 294, II, 883-886.

MONBARON, M. (1982) : Un relief anté-Bathonien enfoui sur la ride du Jbel La'bbadine (Haut Atlas central, Maroc : conséquences pour la chronologie de l'orogenèse atlasique.- Bull.Ver. schweiz.Petroleum-Geol. u.Ing. 48, 114, 9-25.

SCHAER, J.P. et PERSOZ, F. (1976) : Aspects structuraux et pétrographiques du Haut Atlas calcaire de Midelt (Maroc).- Bull.Soc. geol. France, 5, 1239-1250.

STANLEY, R.G., (1981) : Middle Jurassic shoaling of the Central High Atlas sea near Rich, Morocco.- J. Sed. Petrology, 51, 895-907.

STUDER, M. (1980a) : Métamorphisme d'enfouissement dans le Haut Atlas central (Maroc). Essai sur l'évaluation de l'épaisseur des couvertures sédimentaires.- C.R.Acad.Sci. (Paris) 291 D, 457-460.

STUDER, M., (1980b) : Tectonique et pétrographie des roches sédimentaires, éruptives et métamorphiques de la région de Tounfite-Tirrhist (Haut Atlas Central, Maroc).- Thèse, Sciences, Neuchâtel, 95 p., et vol. illustrations.

ASPECTS OF WESTERN HIGH ATLAS TECTONICS

NIKOLAUS FROITZHEIM*, JOHANNES STETS, and PAUL WURSTER
Geologisches Institut der Universität Bonn
Nußallee 8, D 5300 Bonn 1
*) New address: Geologisches Institut, ETH-Zentrum, CH-8092 Zürich

Abstract

An overview of the structural provinces of Morocco is given, showing the position of the Western High Atlas in its geotectonic context. Recent results are presented concerning the structural development of the Hercynian (Variscan) ancient massif as well as of the Atlassic (Alpine) mountain range of the Western High Atlas.

The paleogeographic setting of the Cambrian, Ordovician and Lower Carboniferous series in the ancient massif shows close relations to the Paleozoic of the Moroccan Meseta. Strong volcanic activity in parts of the massif during the Early Paleozoic points to rifting processes affecting the foreland of the African craton. This stage of the tectonic evolution ended in the Silurian to Devonian. There is little evidence for orogenic processes during the Caledonian era. During the Hercynian (Variscan) orogeny, folding, subsequent thrust faulting, and intrusion of granites were most intense in the central part of the Variscan belt which later formed a relatively stable axis.

Locally, the contact between Triassic and Upper (?) Jurassic strata is developed as an angular unconformity. Lacking of Triassic series in parts of the ancient massif may be a consequence of pre-Upper (?) Jurassic erosion following Early to Middle Jurassic deformation. This hiatus marks the end of the Triassic Atlas Rift stage and the beginning of the Upper Jurassic to Cretaceous Atlas Gulf stage. Local unconformities have also been observed between Senonian and Tertiary series, giving first indications of the uplift and inversion process in the High Atlas mobile belt. The uplift of the present Atlassic mountain range was accommodated mainly by steeply dipping, anti-listric faults. Vertical relative movements clearly predominate over horizontal ones. Different kinematic models for the Atlassic uplift are shortly discussed.

1. Introduction

For travellers as well as for geologists, the Western High Atlas is one of the most fascinating landscapes in Morocco. Vast, desert mountain chains contrast with valleys and oases of deep green. Tectonic deformation has created an impressive architecture whose lines are often accentuated by a surprisingly colourful sequence of rocks. - The following paper is meant to give some recent results and to show the present state of the "Bonner Atlasforschungen", a research program pursued by the "Geologisches Institut der Universität Bonn". Our activities in the High Atlas started in 1975 with investigations on the Cretaceous of the Moroccan coastal basins. A precise stratigraphy was established for the Cretaceous continental, lagoonal and shelf sediments onshore, which thus could be compared with the deep sea sediments drilled during the IPOD/DSDP project in the eastern Atlantic Ocean off the NW African continental margin. Additionally, this extensive stratigraphic work enabled us to reconstruct the Cretaceous paleogeography of the Western High Atlas.

First results of the onshore activities gave rise to the idea that the present High Atlas mountain range originated from a Triassic rift structure ("Atlas Rift") branching off the Atlantic rift as a failed rift arm directed towards the east. This rift was converted into a deltoid embayement during the Upper Jurassic and Cretaceous, called "Atlas Gulf". Fig.2 (foreground) shows the Atlas Rift/Atlas Gulf crossing the Atlantic passive margin in the area of the Agadir-Essaouira coastal basin. Investigations on the Cretaceous Atlas Gulf during 1976 to 1978 showed the interdependence between facies distribution of the sedimentation area and a distinct rhomboidal block fault pattern in the subsurface. These basement faults are very well exposed in the strongly uplifted central part of the Western High Atlas (fig.2, second block). According to our results, the Atlas Gulf underwent inversion processes which began in the late Cretaceous, increased during the Tertiary, and resulted in the present mountain range. Subsequent marginal basins formed along the northern and southern flank of the rising mountain range and were filled with molasse-like clastic sediments coming down from the High Atlas.

In the ancient massif, erosion has uncovered the pre-Triassic basement. Here, the previous history back to the infra-Cambrian and Precambrian can be studied in detail. In 1980, another research program started concerning, among other topics, the Paleozoic history of the ancient massif. By means of geological mapping and tectonic work in limited, representative areas, the Paleozoic structural evolution as well as the relationships between Variscan and "Atlassic" (Alpine) deformation were studied. Both for the Variscan and the Alpine history, the close vicinity to the border of the African platform played an important role. Starting in the Cretaceous of the near-shore Atlantic coastal basins and gradually going down to the Paleozoic and even earliest Paleozoic, we found this vicinity to be a fundamental determinant in Western High Atlas tectonics.

The authors are indebted to Dr. M.BENSAID and M.DAHMANI at the Ministry of Energy and Mines in Rabat, Morocco, for facilitating the field work during the whole time span in every respect. The work done in 1975 to 1986 was largely supported by the "Deutsche Forschungsgemeinschaft" (DFG) in Bonn under the grants Wu 32/15-21. We thank all the students and geologists involved in the "Bonner Atlasforschungen", especially BERNHARD BECKER, FRANZ BINOT, Dr. GEORG DRESEN and PETER LATH who have contributed to the results presented in the following paper.

2. Structural outline of the Atlas system: Provinces, relationships and problems

In Morocco, six main structural provinces can be defined, characterized by different geotectonic mobilization and consolidation processes during the Phanerozoic earth history (see fig.1). Five of them have played a role in the development of the Western High Atlas. Different structural levels ("Stockwerke"), different tectonic stiles, different orientations of structural features, and different timing of the tectonic events have been involved in their evolution.

The **NW African craton** is one of the Precambrian nuclei of the African continent. It is mostly covered by latest Precambrian and Paleozoic platform sediments and volcanics. The Tindouf basin represents a large syneclise filled with epicontinental Paleozoic sediments. The Anti-Atlas at the northern border of the craton is an anteclise, characterized by several domal uplifts of the crystalline basement, so-called "boutonnières" (buttonholes). In the Graara, Siroua, and Tafraoute "boutonnières", the ancient rock series are well exposed revealing an excellent record of the late Precambrian history of North Africa. Towards the NW, the craton is clearly bounded by the Mauretanides, a Paleozoic fold belt pointing to trans-Atlantic communications. The northernmost edges of the craton are cut by the "accident sud-atlasique" (South Atlas fault), a very complicated and heterogeneous structural feature. Thus, the Siroua block and the Precambrian of the Ouzellarh promontory, situated N of the South Atlas fault, represent portions of the NW African craton which became incorporated into "Atlassic" (Alpine) deformation and uplift of the Western High Atlas. Here is the only location where we can study the margin of the old African platform in detail. Towards the west and north, the nearest Precambrian outcrops of comparable quality are situated far away in America (Avalon), and in Europe (Brittany, Bohemia, Scandinavia). Therefore the continuation of the Precambrian socle and the nature of the deep subsurface beneath the adjacent northern structural provinces of Morocco are fairly unknown.

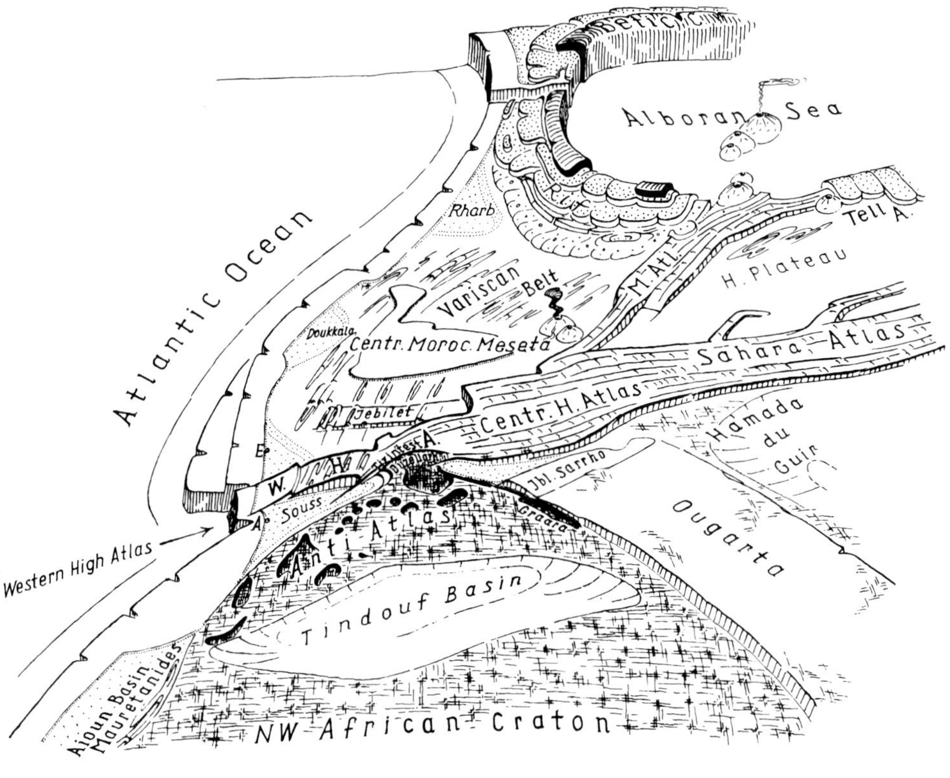

Fig.1: The main structural provinces of Morocco: NW African craton, Ougarta block, High Atlas (W.H.A.: Western High Atlas), Central Moroccan Meseta, NW African continental margin, Northern Atlas (Rif, Tell). A: Agadir, E: Essaouira.

The **Ougarta block** in the SE is another very important feature of the African platform. Its geotectonic position is still a matter of speculation. Recently, geochemical data were interpreted in terms of an active continental margin at a Panafrican suture zone. Perhaps one should also envisage another model: The Ougarta block could well be explained as a SE-NW oriented rift system branching off the transcontinental Panafrican arch. In the NW prolongation of the Ougarta block *sensu stricto*, huge lava flows and sheets of the Late Precambrian are wide-spread and well exposed in the Jbel Sarrho and in the volcanoclastic sequences of the Ouzellarh promontory. This wide-spread volcanic activity is in line with the assumption of late Precambrian continental rifting. The inverted and uplifted aulacogen remained a consolidated stable block. It was covered by a thin blanket of Paleozoic sediments. Phanerozoic syneclises, like the Tindouf, the Hamada du Guir, and the extended Sahara basin, are bordering the Ougarta block.- It remains an open question whether this supposed Ougarta aulacogen ends somewhere beneath the Central High Atlas or can be traced farther towards the NW, beneath the Central Moroccan Meseta. Extensive magmatic mobilization during the Mesozoic rifting process and the abnormally high subsidence of the Central High Atlas during the Jurassic might show the influence of a deep-seated prolongation of the Ougarta block. Some NW-SE-striking faults near Demnate point in the same direction.

The **Central Moroccan Meseta** represents a denudated segment of a "Hercynian" or Variscan fold belt. The NE-SW-striking fold belt is accentuated by late-orogenic granite intrusions. These granite plutons are concentrated in the main deformation belts, the central anticlinoria. The Cambrian and Ordovician sequences are thick marine sediments with few indications of volcanic activity. The basins have been preserved almost in their original dimensions as is indicated by paleo-rim positions at the present Atlantic coast near El Jadida as well as in front of the Central High Atlas and the Middle Atlas. The distribution of the orogenic facies during the Carboniferous is wide-spread, representing a rather shallow level of the orogenic processes. As the denudation level is also a relatively shallow one, there is no evidence for an extreme thickening of the crust as it is usual at active margins or within collision fronts. The Permian peneplain of the Meseta is covered by extremely thin Mesozoic and Tertiary platform sediments including the famous Eocene phosphate deposits. This flat-lying cover shows that the crust has remained in a post-orogenic, epirogenic, at least undisturbed structure and composition since Carboniferous-Permian times.

The ancient massif of the Western High Atlas is a part of the Variscan belt which has been uplifted by "Atlassic" (Alpine) tectonics. The Variscan structures are definitely cut off at the important Tizi-n'Test fault. Nevertheless, a narrow connection with the Mauretanides in the SW may exist beneath the Souss basin, and perhaps there is some continuation to the NE beneath the Algerian High Plateaus and, strongly remobilized, in the Paleozoic nappes of the Rif and Tell Atlas. Beyond the enigmatic oceanized pit of the Alboran Sea, the Spanish Meseta on the Iberian microplate represents another segment of the Variscan belt. Both the Moroccan and the Spanish Meseta are elements of the peri-Atlantic realm, passively rotated and involved in deformation processes during the opening-closure-openings of the Atlantic Ocean.

The **High Atlas system**, as has been stated above, is a convincing example of an inverted and uplifted rift complex. Mesozoic extension processes produced the graben structures of the Western High Atlas, the Central High Atlas, the Sahara Atlas and the bifurcation of the Middle Atlas. The extensional faults probably used pre-existing zones of weakness: late Precambrian and/or Variscan lines. These relationships can be studied in the excellent outcrops of the Western High Atlas where Precambrian, Variscan, and

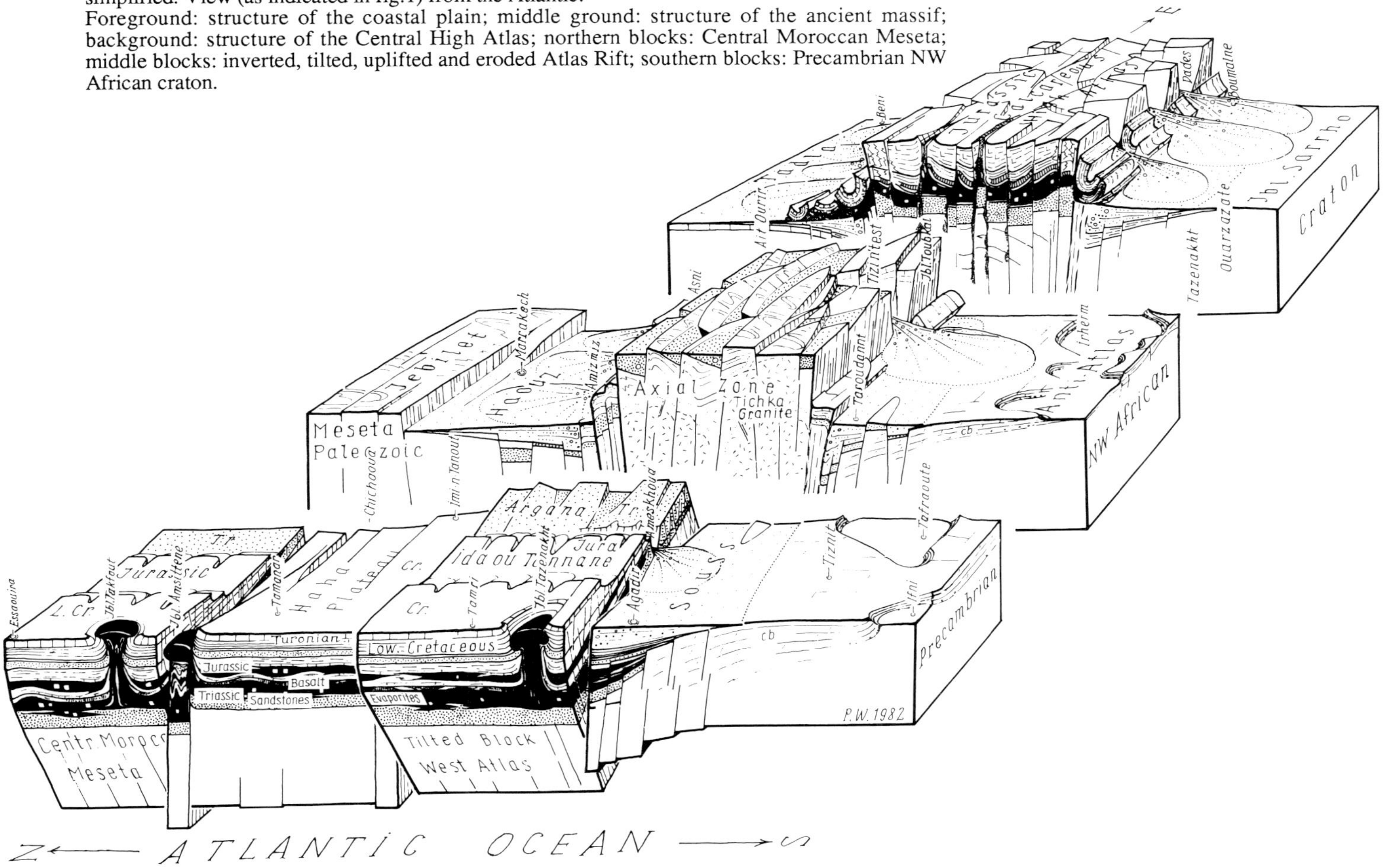

Fig.2: Block diagram of the Western High Atlas, strongly generalized, vertically exaggerated and simplified. View (as indicated in fig.1) from the Atlantic:
Foreground: structure of the coastal plain; middle ground: structure of the ancient massif; background: structure of the Central High Atlas; northern blocks: Central Moroccan Meseta; middle blocks: inverted, tilted, uplifted and eroded Atlas Rift; southern blocks: Precambrian NW African craton.

Meso-Cenozoic fault and fold patterns overlie each other, representing a unique example of superimposed tectonics.

The block diagram (fig.2) illustrates the geological situation of the Western High Atlas. The view is from the Atlantic Ocean towards the E. The foreground shows the tectonic situation in the coastal plain between Essaouira (N) and Agadir (S). The tilted block of the westernmost Atlas mountains is pierced by salt diapirs and bordered by marginal basins (Souss, Haouz). The Mesozoic sediments which have filled the former rift consist of Triassic sandstones, red beds, evaporites, and basaltic lavas, Jurassic mudstones, carbonates, and evaporites, Cretaceous sandstones, siltstones, and carbonates.- The middle ground shows the uplifted and denudated pre-Triassic peneplain of the ancient massif, consisting of the axial zone and the narrow northern and southern Subatlas zones. The Variscan structures trend N-S to NE-SW while strike of the main "Atlassic" faults is about E-W.- In the background, the Mesozoic sediments have been preserved on top of the mountain range. The main normal faults bordering the rift were transformed into steeply dipping thrusts by the inversion process. Sheets of the Mesozoic cover were sheared off and, by the mechanism of gravitational gliding, moved towards the lowlands of the Meseta (N) and the Anti-Atlas (S). This block diagram shows the impressive simplicity of the edifice, but provokes a fundamental and yet open question: How can we explain the block-like inversion of the failed rift into a mountain range? Which were the driving forces creating a mountain belt up to 100 km wide, 4000 m high and 1000 km long?

We consider the **NW African continental margin** as another autonomous structural province of Morocco. As indicated by the results of deep-sea drilling and by many geological and geophysical data, rifting between the North American and North African plates started in early Mesozoic time. The sedimentological and tectonic development of the passive margin during the Jurassic and Cretaceous is recorded in coastal basins and shelf wedges. It was controlled by the structural pattern of the NW African craton, the Mauretanides, the High Atlas system, the Meseta, and the Rharb basin south of the Rif Atlas.

The **Rif and Tell Atlas belt** is an excellent example of alpine nappe tectonics and therefore fundamentally different from the rest of Morocco. Mesozoic facies distributions, sequences of internal and external nappes, the Tertiary timing of orogenic phases, and the direction of tectonic transports indicate that this thrust belt is a part of the Alpine system - whose roots, unfortunately, lie in the Mediterranean sea.

3. The Western High Atlas in Paleozoic time

3.1. Adrar-n'Dren, the ancient massif of the Western High Atlas

Within the complex mountain range of the Western High Atlas, series of Paleozoic age are best exposed in the ancient massif (massif ancien: ROCH 1950), called Adrar-n'Dren by the native people. This Paleozoic-Precambrian massif which cumulates in the Jbel Toubkal region is bounded to the west and to the east by a Triassic sedimentary cover successively overlain by younger Mesozoic strata. Towards the north and south, mainly Cretaceous sediments cover the Paleozoic socle in the northern and southern Subatlas zones bending downward beneath the Tertiary and Quaternary filling of the Souss and Haouz basins (fig.3).

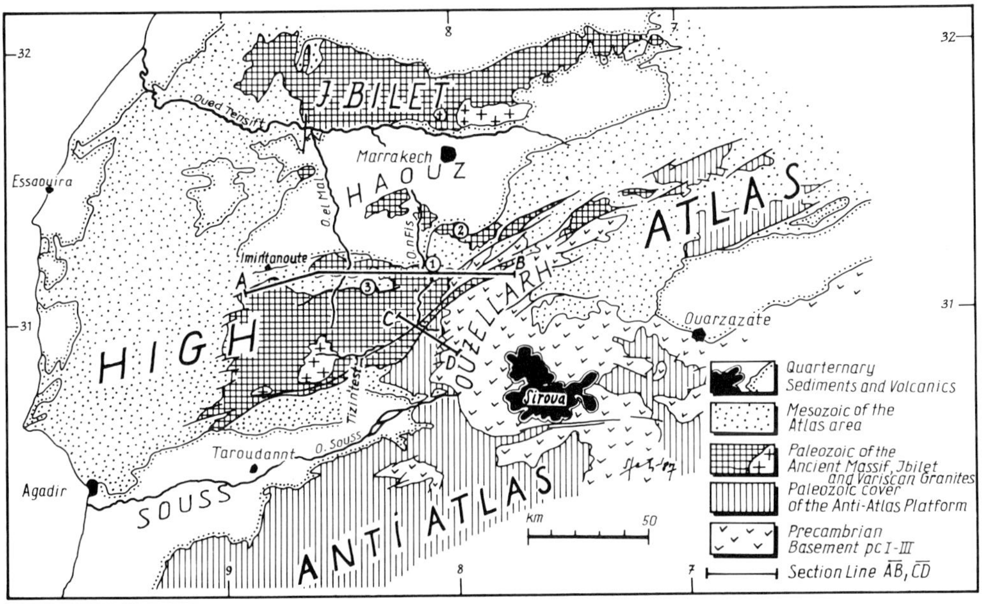

Fig.3: Geology of the Western High Atlas, Morocco; localities cited in the text: 1 Tizi Ouzla region, 2 Tahnaout (Rhiraïa valley), 3 Al Medinat (Erdouz region).

The age of the Paleozoic rock sequence in the Adrar-n'Dren ranges from the infra-Cambrian Adoudou Formation (HOLLARD 1985) to the Lower Carboniferous. Here, the Paleozoic shows strong changes in thickness varying from a few hundred to more than 7 000 and up to 10 000 m. Little is known on the basin development in this branch of the Variscan orogenic belt. Analysis is complicated because of strong tectonic deformation in parts of the massif. In the eastern part ("bloc oriental"), no appreciable deformation of Variscan age can be observed. Infra-Cambrian rocks unconformably overlap on a Precambrian basement complex which belongs to the Ouzellarh promontory ("promontoire d'Ouzellarh"). In the western part ("bloc occidental") which wholly consists of Paleozoic strata, folding and thrusting processes occurred, accompanied with low grade metamorphism and intrusion of granitic and dioritic rocks. These two blocks of the ancient massif are separated by the rather complex Tizi-n'Test fault system (AMBROGGI & NELTNER 1952, PETIT 1976, PROUST et al. 1977) recently discussed by BINOT et al. (1986). The Paleozoic of the western block can be followed from the ancient massif northward to the Moroccan Meseta which is built up of the same rock sequence, the Variscan structures striking in a northnortheasterly direction (MICHARD 1976). The eastern block, however, shows a close relationship to the Anti-Atlas, i.e. to the consolidated African platform (CHOUBERT 1963), as far as the Paleozoic history is regarded.

Results of geological mapping and field data compiled by FROITZHEIM (1986) in the northern Subatlas zone, together with data from MORET (1931), ROCH (1950), SCHAER (1964 a, b), PROUST (1962), TERMIER & TERMIER (1971), and DRESEN (1985), enable us to draw a rough picture of the paleogeographic and structural setting of the ancient massif - some aspects of the Paleozoic history of the Adrar-n'Dren.

3.2.Paleozoic basin development

Infra-Cambrian. At the southwest border of the Ouzellarh promontory (fig.3), a variegated sequence of sedimentary rocks about 1400 m thick which belongs to the infra-Cambrian Adoudou Formation is exposed in the famous Agoundis valley section (HOLLARD 1985). It stretches farther south towards the Ouneihn valley. From here, correlation is possible with equivalent series of the Anti-Atlas (BOUDDA & CHOUBERT 1972). In the Agoundis valley, this sedimentary pile unconformably overlies a volcanic series including lavas, tuffs, conglomerates, and siltstones of up to 2000 m thickness. This latter series belongs to the Precambrian (pc III) and forms the base of the Paleozoic in this region. According to our field data (BINOT et al. 1986), the infra-Cambrian series comprises the "série de base", "calcaire inférieur" and "série lie-de-vin" formations of the French authors (HOLLARD in DESTOMBES et al. 1985).

By SDZUY (1978) and his colleagues (SCHMITT 1979 a, b), the infra-Cambrian/Lower Cambrian boundary was established at the top of the "série lie-de-vin" formation in the Ouneihn valley. In the lower part of the section, i.e. in the "calcaire inférieur" formation, pyroclastic rocks at various levels announce volcanic activity in the nearby Jbel Wankrim region NE of the Agoundis river. Relics of this infra-Cambrian series can be followed towards the NE along the Tkent-Tacheddirt furrow (see below) which is part of the Tizi-n'Test fault zone (BINOT et al. 1986).

Farther to the west and northwest, the whole Adoudounian rock sequence disappears and is replaced by conglomerates which probably belong to the Lower Cambrian lying on top of the pc III volcanic series. In the field, the infra-Cambrian formation nowadays is situated in a graben structure formed by Variscan and Alpine deformation. It was called "graben of Tkent-Tacheddirt" by DRESEN (1984). Paleogeographic reconstruction gave rise to the idea that this structure originated already from infra-Cambrian time. The eastern margin slightly bent down towards the northwest, representing the border of an ancient Ouzellarh massif. The western margin was formed by steeply dipping syngenetic normal faults with high amounts of vertical displacement. The resultant asymmetric half-graben geometry of the Tkent-Tacheddirt structure is shown in fig.5 (line C-D).

Lower Cambrian. Considerable portions of the Adrar-n'Dren are built up of Lower Cambrian (Georgian) rocks. Two different provinces can be distinguished:

- A volcanoclastic sequence consists of alternating lavas, tuffs, conglomerates, sand-, and siltstones. On the top, archaeocyathid-bearing limestones, dolomitic limestones, and calcareous slates occur in places. The thickness of this calcareous series ranges from 20 up to 400 m in the Jbel Aborji region (see fig.6), while the volcanoclastics locally reach more than 1000 m.

A siliciclastic facies, on the other hand, consists of dark grey and black mudstones with sandy layers and laminae in the lower part and of medium-grained sandstones and quartzites alternating with mudstone sequences in the upper part.

Fig.5 (line A-B) is a hypothetic section running E-W from today's Jbel Toubkal region to the Aït Mohand ou Moussa block south of Imi-n'Tanout. Along this section line, the occurrence of the volcanoclastic facies is restricted to the region between Azegour and the Oued n'Fis, i.e. the proximal foreland of the Ouzellarh promontory. Strong variations in thickness are caused by syngenetic block faulting which affected the western margin of the promontory and gave rise to the volcanic activity. Farther towards the east, near the immediate border of the Ouzellarh promontory, the basement is exposed. It is built up of the pc III volcanic series except the graben of Tkent-Tacheddirt where relics of the infra-Cambrian sequence occur. The pc III volcanics are also exposed in the gorges of Oued n'Fis, NW of Wirgane. Here, they are overlain by Lower Cambrian conglomerates, sandstones, dolomitic limestones, and volcanic rocks.

Towards the west, the basement dips farther down and cannot be observed anywhere in the western Adrar-n'Dren. Rocks of infra-Cambrian age are also not found. In the western block of the ancient massif and farther to the north, there is no evidence for infra-Cambrian rocks in a facies and thickness comparable to that of the Agoundis valley section or of the Anti-Atlas.

According to DE KONING (1957) and SCHAER (1964 a), the Azegour volcanic province stretches towards the SW along the northern border of the old African platform. West of the Assif al Mal (fig.5), the volcanoclastic facies is replaced by the siliciclastic facies. As the base of this latter series is not exposed in the western block of the ancient massif, its thickness remains uncertain. Nevertheless, we agree with SCHAER (1964 a) who supposed a thickness of up to 5000 m for the Lower Cambrian in this part of the High Atlas.

Middle Cambrian. Middle Cambrian (Acadian) bluegreen slates are the most characteristic rocks in the Paleozoic of the ancient massif. They are also found in the Anti-Atlas ("Schistes à Paradoxides") and in the Moroccan Meseta (see MICHARD 1976). In the Azegour region, the clastic development of the Acadian is replaced by a facies abnormally rich in volcanic rocks (Upper volcanoclastic series, FROITZHEIM 1986). It consists of dacitic, andesitic, and basaltic lavas (PERMINGEAT 1957), and pyroclastics alternating with siltstones and conglomerates.

The paleogeographic configuration is shown in fig.4. The remarkable concentration of volcanic rocks, the occurrence of coarse-grained clastic sediments, and strong variations in thickness seem to be caused by syngenetic normal faulting as was shown for the Georgian. We tend to interpret the Azegour region in terms of a fault-controlled horst-like high and a centre of volcanic activity which in some way continued the Lower Cambrian tradition. This "Azegour-Tichka-High" is flanked by two troughs with high thickness of monotonous bluegreen slates. The eastern trough is represented in the Jbel Tighardine region and in the graben of Tkent-Tacheddirt where Acadian bluegreen slates are still preserved (DRESEN 1984). The other

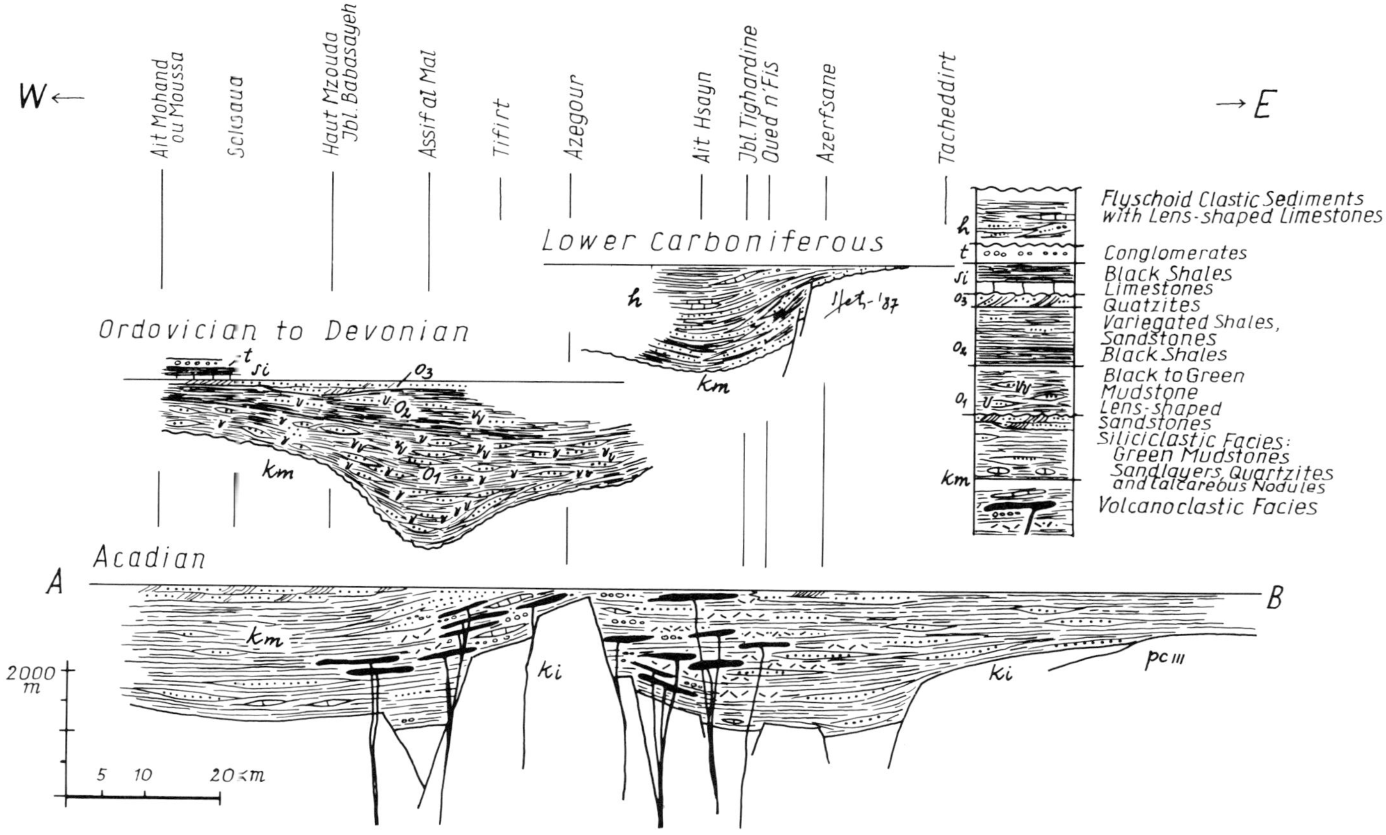

Fig.4: Diagrammatic section across the basin west of the Ouzellarh promontory during Middle Cambrian (Acadian), Ordovician to Devonian, and Lower Carboniferous; abbreviations: km Middle Cambrian (Acadian), o_1-o_3 Ordovician, si Silurian, t Devonian, h Lower Carboniferous.

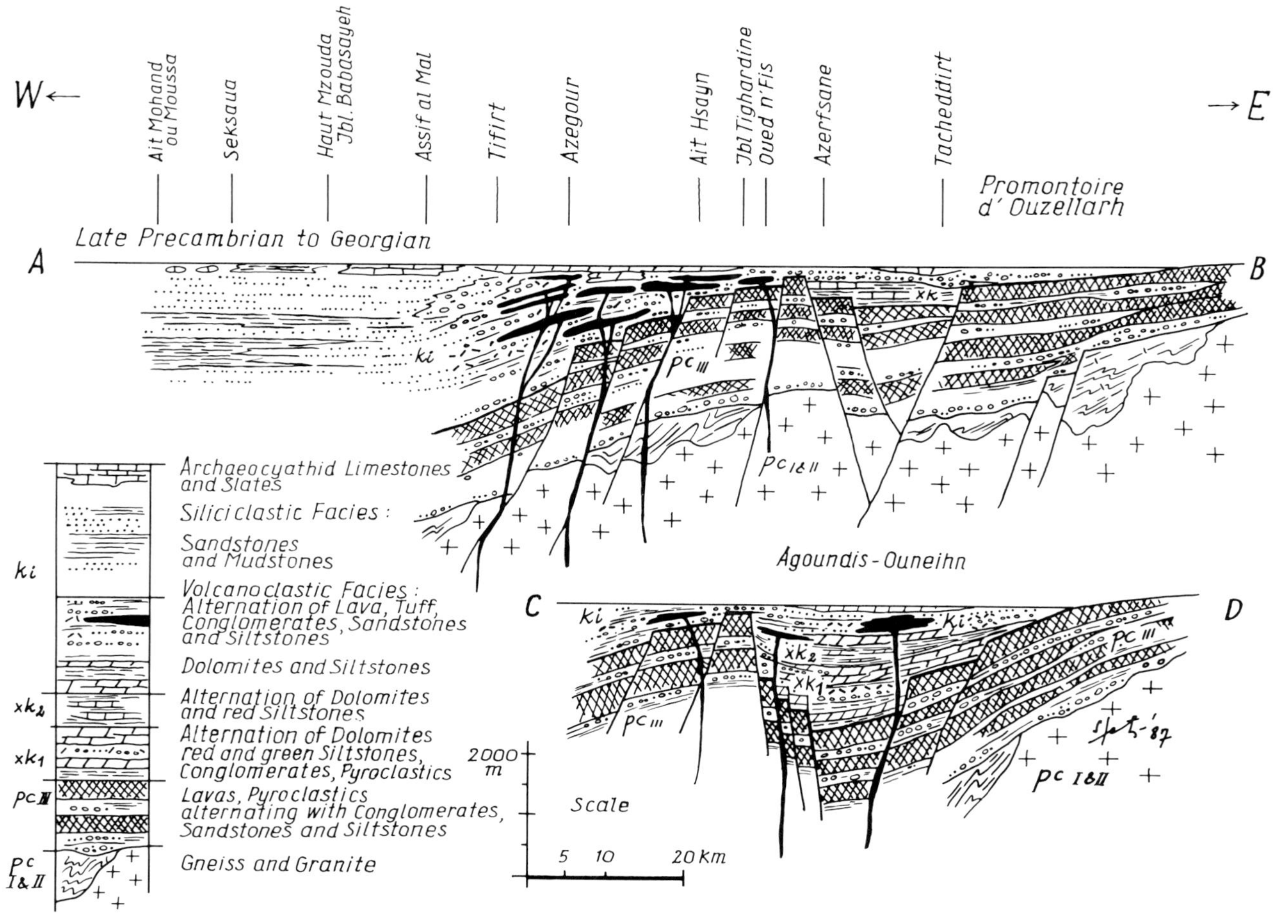

Fig.5: Diagrammatic cross-section of the western border of the Ouzellarh promontory and adjacent basin from the Late Precambrian to Lower Cambrian (Georgian); abbreviations: pc I - pc III Precambrian; xk, xk_1, xk_2 infra-Cambrian, ki Lower Cambrian.

trough is situated in the west where the bluegreen slates are followed by a sequence of sandstones up to 300 m thick, cropping out in the valleys of Aït Lahsen and Seksaua.

Ordovician to Devonian. The next series which can be recognized on top of the Middle Cambrian (Acadian) belongs to the Ordovician. Until now, Upper Cambrian rocks have not been found in the ancient massif nor elsewhere in Morocco (DESTOMBES et al. 1985). Ordovician series conformably overly the Acadian. The Ordovician is exposed near Azegour and towards the west, in the area of Assif al Mal, Mzouda and Seksaua (fig.4). Here, an Ordovician trough can be reconstructed coinciding with the Acadian one.

The Ordovician sequence starts with micaceous slates and fine-grained lense-shaped sandstone layers passing into black-and-white striated slates which, according to DE KONING (1957), are called "schistes demi-deuil". Trace fossils are abundant in this sequence. The following series consists of black, partly sandy slates. Equivalent sediments in the Ida ou Zal region belong to the Llandeilo (DE KONING 1957). The roof of the Ordovician is formed by up to 40 m thick quartzites of Caradocian age.

In the Seksaua region (fig.4), a thin sequence of black limestones alternating with black shales follows above the Ordovician sequence. This series probably belongs to the Ludlowian. On top of it, olive-coloured slates follow. They may be Silurian, too, but the exact age is unknown. The youngest formation in this western region S of Imi-n'Tanout consists of red, green, and black conglomerates and coarse-grained sandstones. The pebbles are of quartz and quartzite. These clastic rocks are assigned to the Gedinnian (SCHAER 1981).

The paleogeographic evolution in the Silurian and Lower Devonian can hardly be reconstructed as this sequence is lacking all over the western block of Adrar-n'Dren except the Seksaua region. Nevertheless, there is much evidence that the basin evolution which started in the Early Cambrian and persisted during the Middle Cambrian and Ordovician came to an end at the beginning of the Silurian.
As all the series described above conformably follow on top of each other, there are no indications for orogenic processes during the Caledonian era. The stratigraphic gaps probably were caused by "epirogenic" movements.

Lower Carboniferous. Lower Carboniferous sediments are found only in the northwestern and northern surroundings of the Ouzellarh promontory. In the Jbel Tighardine and Aït Hsayn area (fig.4), more than 1500 m of black slates with sandy layers build up a distal flyschoid sequence. In the middle part of the sedimentary pile, a series of massive sandstones and intercalations of crinoid limestones are observed which yielded the Lower Carboniferous fauna described by NELTNER (1938). The Lower Carboniferous in the Aït Hsayn area conformably overlies Acadian sediments. Series of Ordovician, Silurian, and Devonian age are unknown in this area.

Similar series of Lower Carboniferous age are found at the Ouzla pass (Tizi Ouzla) NW of Wirgane, in the Rhiraïa valley section, in the Souktana massif, and in the Guemassa area south and southwest of

Marrakech. According to results of DOCHY (1980 in: DRESEN 1984), the Lower Carboniferous east of the Ourika valley at the northern border of the Ouzellarh promontory shows a thickness of only 40 m. Towards the Rhiraïa valley, the thickness suddenly increases to more than 2000 m. Thus, the deposition of a thick sedimentary pile in a relatively limited area during the Lower Carboniferous can again be explained by syngenetic normal faulting at the northern and northwestern border of the promontory (see fig.4). Nevertheless, as the sedimentology and stratigraphy of the Carboniferous in the High Atlas have not yet been investigated in detail, other possible models for the tectonic environment of this "flyschoid" basin should not be ignored.

3.3. Structural setting and evolution

The structural setting within the ancient massif is best illustrated by another E-W running section across the northern Subatlas zone (fig.6; fig.3: line A-B). Going down the slope of the Ouzellarh promontory, the graben of Tkent-Tacheddirt is crossed at Adrar el Hajj. This ancient structure has been preserved by Atlassic (Alpine) reverse faults. They follow the former normal faults and were reactivated as well in Variscan as in Alpine time (PROUST 1962, DRESEN 1984). Farther downslope, west of the basin of Asni-Wirgane, a synclinorial structure is reached between Oued n'Fis and Assif Anougal (fig.6). It is filled with Acadian and Lower Carboniferous rocks and coincides with the centre of the Acadian respectively Lower Carboniferous troughs mentioned above. Thrust faults (most of them of Variscan age) show a fan-like arrangement centered in the cores of neighbouring anticlinorial structures.

In the upper part of fig.6, the structural setting of the Azegour area is illustrated by detailed cross-sections based on geological mapping by FROITZHEIM (1986). These sections show several anticlines, especially that of Jbel Aborji, and the post-kinematic Azegour granite. Two further synclinorial structures follow farther W, separated by a heavily disturbed anticline west of Assif al Mal. Again, thrust faults show a fan-like orientation: they are mostly directed from the cores of the anticlines towards their flanks.

Strongest deformation is observed in the centre of the Variscan belt, i.e. in the Azegour region. Low grade metamorphism is stated here decreasing towards the east and west. The age of fold and thrust tectonics in the western block of the ancient massif can be confined to the Namurian and Westphalian. Radiometric age determinations of the synkinematic Tichka granite (318±10 and 305±10 Ma; TERMIER & TERMIER 1971) and of the post-kinematic Azegour granite (269±11 Ma; CHARLOT et al. 1967) show good coincidence with the geological field data.

The rather complex fold pattern corresponds with the previous paleogeographic setting. The situation immediately at the border of the Ouzellarh promontory is representing that of the western Anti-Atlas, too. Here, extension of continental crust by syngenetic block faulting led to a mobile belt in the Lower Paleozoic. This block pattern controlled the later deformation processes during the Variscan orogeny. Progressive shortening of the crust produced fold tectonics at first, followed by thrust faults later on. Intrusion of granitic melt was caused by further crustal thickening. Mobilization processes probably affected Precambrian, partly Lower Cambrian rocks, thus causing a great variability of rock types (TERMIER & TERMIER 1971) as well as dyke swarms and vein type ore deposits (PERMINGEAT 1957).

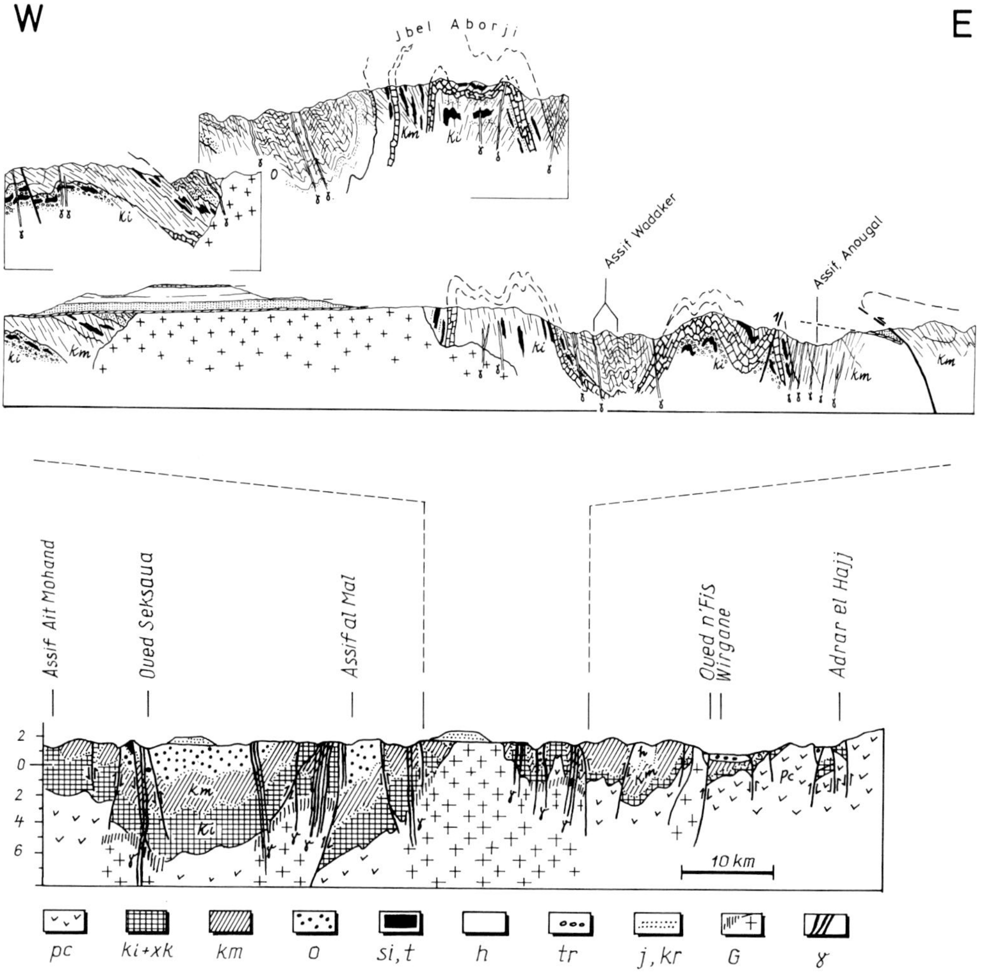

Fig.6: Section along line A-B (fig.3) across the northern Subatlas zone; abbreviations: pc Precambrian, ki + xk infra- and Lower Cambrian, km Middle Cambrian, o Ordovician, si,t Silurian, Devonian, h Lower Carboniferous, tr Triassic, j,kr Jurassic, Cretaceous, G granite, γ dyke swarm.

The section presented above crosses the "couloir de déformation" (MICHARD 1976), i.e. the belt of intense Variscan deformation in Morocco. It is flanked by areas of lower deformation, that is the "môle cotier" to the west and the African craton, here represented by the Ouzellarh promontory ("bloc oriental"), to the east. This *couloir* stretches in a northnortheasterly direction towards the Moroccan Meseta. Towards the south, it closely joins the northern margin of the African platform, which underlines the important role of the Tizi-n'Test/South Atlas fault zone.

There is no evidence that an active plate margin existed during the Paleozoic in the Adrar-n'Dren. The presumed ensialic setting of the Paleozoic belt mentioned above could have formed by thinning and rifting of continental crust. Possibly, subduction and collision processes farther to the west affected the weakened crust during the Variscan orogeny. The full history has not yet been revealed in detail, but, after all, the events and aspects reviewed here seem to be closely related to the evolution of the Paleozoic proto-Atlantic Ocean.

4. Recent observations concerning the Mesozoic and Cenozoic development

On first sight, Alpine or "Atlassic" tectonics of the Western High Atlas, especially of the ancient massif, seem to be rather clear: Large basement blocks of the Variscan edifice, partly carrying remainders of the Mesozoic "couverture", have been uplifted by various amounts, the relative movements being accommodated by steeply dipping reverse faults. When did that happen? Sometime in the Tertiary... On a closer investigation, many complications appear: "Atlassic" deformation does not originate from a single "phase" but from a long process, the nature of the reverse faults is dubious, the structure of the deeper subsurface is almost completely unknown, and the mechanism of the "Atlassic" uplift is far from being understood. The following results, derived from geological field work in the northern Subatlas zone SW of Marrakech, provide some new informations which may improve our knowledge of the "Atlassic" development.

4.1. Jurassic tectonics and paleogeography

At the northern border of the ancient massif between Imi-n'Tanout and the Oued n'Fis valley (fig.3), dolomites, marls and gypsum of (probably) Jurassic age unconformably overlie folded and partly metamorphic Paleozoic rocks. Triassic sediments are lacking, while farther to the W (Couloir d'Argana) and to the E (Asni-Wirgane basin, Talat-n'Yakoub graben) thick clastic red beds and basalts of Triassic age have been preserved underlying the Jurassic. This led several authors to the assumption that, in Triassic times, the western half of the ancient massif had been an area of non-deposition or erosion, part of a SW-NE trending isthmus separating the basins of Agadir-Essaouira and of the Central High Atlas (see MICHARD 1976, fig.83). A similar situation was supposed for the Jurassic ("terre des Almohades": CHOUBERT & FAURE-MURET 1962). STETS & WURSTER (1981), however, assumed one coherent basin both for the Triassic (Atlas Rift) and for the Jurassic/Cretaceous (Atlas Gulf).

Reconstruction of the Triassic as well as the Jurassic paleogeography and tectonic evolution is difficult, because little is known until now about the stratigraphic correlation of the fossil-poor, mostly lagoonal Jurassic sediments in this part of the High Atlas: FERRANDINI & PETIT (1981) inferred a Lias age for sandy dolomites and marls in the region between Oued n'Fis and Oued Ourika. DRESCH & ROCH (1934), like other French authors, used the term "eocrétacé" (= uppermost Jurassic?) for similar rocks in the Anougal valley (S' of Amezmiz). The Carte géologique du Maroc 1 : 500 000, sheet Marrakech

(CHOUBERT 1955/56), shows no Jurassic at all E of the Assif al Mal valley. In fact, the thickness of the Jurassic strata decreases rapidly in this region: from Imi-n'Teghli (Seksaua) in the W to Ighermane (Assif al Mal) in the E the decrease is from 580 m (HUDEC 1984) to 76 m over a horizontal distance of about 25 km. Nevertheless, some horizons of the Jurassic can be traced, by means of mapping and correlation of stratigraphic sections, from the Assif al Mal valley towards the E as far as Auzer / Oued n'Fis and (probably) Tahnaout in the Rhiraïa valley. Only small "islands" remained free from Jurassic sediments, e.g. a part of the Azegour granite mass. By connecting these Jurassic sections with the well-known and fossil-bearing sections of the westernmost High Atlas (Haha region), we come to the assumption of a Malm age for the Jurassic rocks in the northern and western ancient massif, the lower parts possibly being Middle Jurassic (Bathonian / Callovian: see Carte géologique du Maroc 1 : 100 000, sheet Imi-n'Tanout). This is supported by the fact that in all sections the Jurassic strata are overlain by the same series of Lower Cretaceous red clastic sediments without any unconformity separating these two members.

Fig.7: Jurassic marls unconformably overlying tilted Triassic siltstones. The crest of the hill is built up of upper (?) Jurassic limestone. W of Tahnaout / Oued Rhiraïa (see fig.3). W is to the left.

As has been stated by DRESEN (1984) and WILDBERGER (1984), the contact between Triassic and Jurassic strata near Tahnaout/Rhiraïa (see fig.3: loc.2) is an angular unconformity of about 20° (fig.7). LATH (1985) found unconformity angles of 40 to 50° in the Tizi Ouzla region NW' of Wirgane/Oued n'Fis (fig.3: loc.1, fig.8). Following our assumptions on the age of the Jurassic rocks, we consider the tectonic deformation which led to these unconformities to be of Lower to Middle Jurassic age. This fits very well with observations from the Central High Atlas (e.g. STUDER & DU DRESNAY 1980).

The Tizi Ouzla region is of special interest for reconstruction of Lower/Middle Jurassic deformation. Restoration of the pre-Malm situation (LATH 1985) yields a dip angle of 40 to 50° ESE and a strike of 20-30° for the Triassic siltstones, but only within a very restricted area. Farther to the E, the contact between Triassic and Jurassic becomes conformable (see fig.8). Obviously, the deformation of the Triassic strata was related to an ENE trending structure, either fault or monocline over a basement fault. The same fault zone was reactivated later on, but with opposite sense of movement, in the Tertiary Atlas orogeny to form the Kik fault, a steeply dipping fault at the western border of the Plateau du Kik. This is just one example of fault reactivation in Atlas tectonics, a feature typical for this region since the earliest Cambrian (see for example PROUST 1962, BINOT et al. 1986).

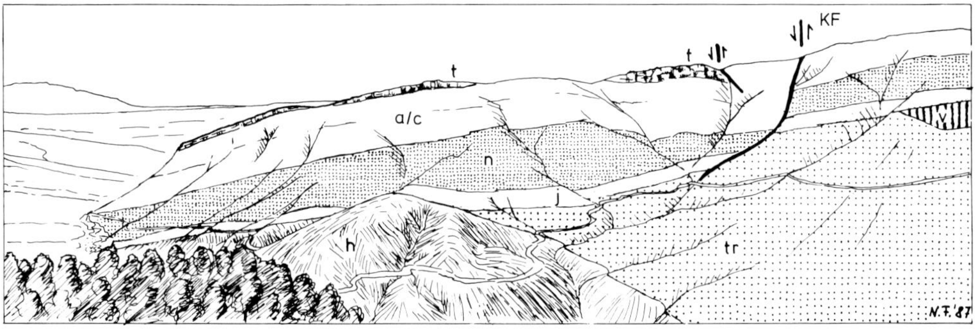

Fig.8: The southern slopes of Plateau du Kik in the Tizi Ouzla region (see fig.3) NW of Wirgane/Oued n'Fis, seen from S (Jbel Tikida). Angular unconformity of Jurassic marls and gypsum (j) over Lower Carboniferous slates (h), Triassic siltstones (tr) and Triassic-Jurassic boundary basalt (v). n: Lower Cretaceous clastic red beds; a/c: Aptian, Albian, Cenomanian; t: Turonian limestones and dolomites; KF: Kik fault. Background, left: Haouz plain and Jbel Aborji near Amezmiz. The road in the middle ground leads from Auzer/Oued n'Fis (left) to Tizi Ouzla (right).

We can summarize our observations concerning the Jurassic history of the western High Atlas as follows:

- Lacking of Triassic strata in parts of the ancient massif is not necessarily due to non-deposition, but may as well be a consequence of Early to Middle Jurassic erosion.
- Jurassic tectonics were, at least partly, controlled by the same fault pattern as the Tertiary / Quaternary uplift of the High Atlas.
- The "terre des Almohades" did not remain free of Jurassic sediments, but was covered by a thin, reduced series of Late Dogger (?) and Malm age.

The Lower to Middle Jurassic events separate two stages in the tectonic evolution of the Western High Atlas: The Triassic "Atlas Rift" stage with fault-controlled clastic sedimentation and volcanism, and the Upper Jurassic/Cretaceous epicontinental "Atlas Gulf" stage.

4.2. Upper Cretaceous deformation and the Al Medinat fault

Alpine or "Atlassic" tectonic deformation and uplift of the Western High Atlas started already in Late Cretaceous times. As a result of Tertiary-Quaternary uplift and erosion, however, relics of the Upper Cretaceous structures can be studied only in a few locations like the one shown in fig.9, NW of the little village Al Medinat in the Erdouz region.

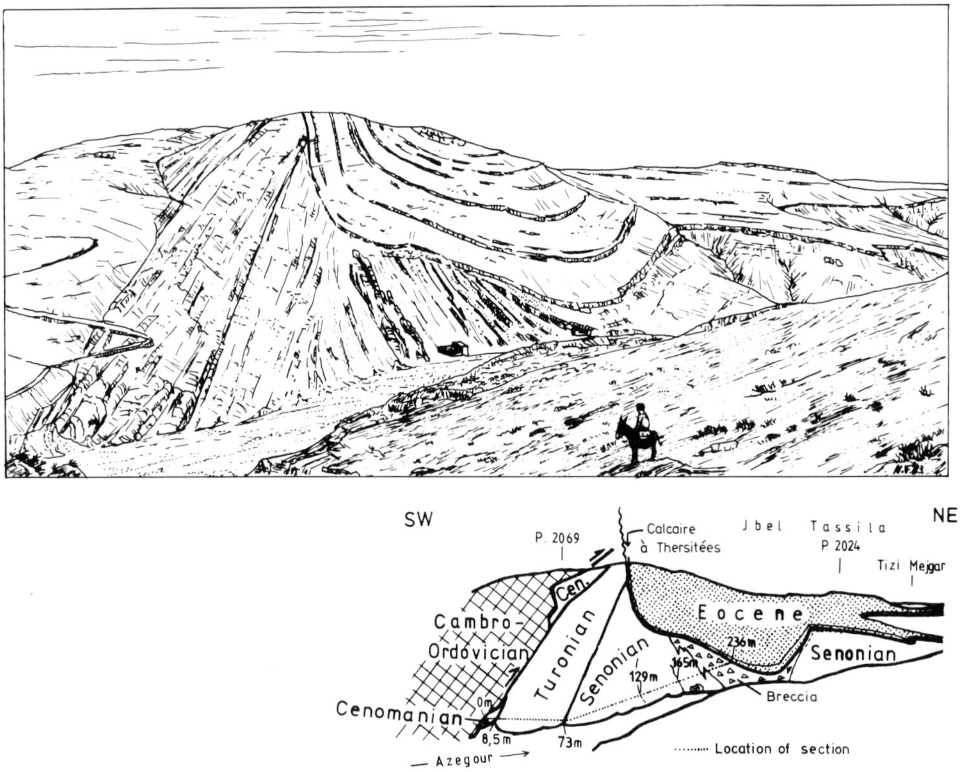

Fig.9: Jbel Tassila near Al Medinat, seen from SE. Background, right: Plateau de Medinat; left: Paleozoic rocks of the axial zone. Location: see fig.3.

The important reverse fault zone separating the axial zone from the northern Subatlas zone was named "accident de Medinat" by MORET (1931). It consists of one, two or locally more, curved, upward-convex ("anti-listric") reverse faults which have accommodated uplift of the axial zone relative to the northern Subatlas zone by 2500 to 3000 m. Some segments of the *accident* strike WNW-ESE (~105°) and others WSW-ENE (~65°), but the resultant overall trend is W-E. In fig.9 one branch of the *accident* is seen to the left, separating Cambro-Ordovician slates of the axial zone from Cretaceous carbonate rocks of the Northern Subatlas zone. Typically for Atlas tectonics, deformation has been localized to this narrow fault zone since the late Cretaceous, while the neighbouring blocks have suffered only little internal deformation. Late Cretaceous movements are documented by the angular unconformity of Tertiary (probably Paleocene) "calcaire à Thersitées" over Turonian and Senonian carbonate rocks. Fig.10 is a columnar section of the Cretaceous strata below the unconformity plane. Reconstruction of the late Cretaceous situation in fig.11/I is based on sedimentological evidence gained from this section and on geometrical restoration. - The Turonian dolomites and limestones ("barre calcaire turonienne"), of shallow marine facies, are relatively thick in this section which is situated near the supposed trough axis of the Cretaceous Atlas Gulf in the axial zone.

Like the Turonian, the lower Senonian strata show marine facies, but with an overall regressive trend and increasing influx of terrigenous clastic material (pelites, fine-grained conglomeratic layers). The top of the Senonian sequence is built up of breccias with sandy and clayey intercalations. Components of the breccias are Cenomanian, Turonian and Senonian limestones, dolomites and siltstones, which were transported over a very short distance under terrestrial conditions. Towards the N, these breccias decrease in thickness very rapidly and disappear only some 1 or 2 km away from the Al Medinat fault zone. In the Turonian as well as in the Senonian carbonates, we found cavities filled with coarse-grained calcareous sandstone identical with the "calcaire à Thersitées" both in lithology and biofacies. Obviously, deposition of the above-mentioned breccias was contemporaneous with the formation of karst cavities, which were "flooded" later on during the Tertiary transgression (see fig.11).

The transgressive "calcaire à Thersitées" is a littoral sediment rich in fossil debris and coarse-grained, well-rounded clastic components. There has been some confusion about the exact age of these limestones, mostly due to the fact that "calcaires à Thersitées" occur both at the base and at the top of the phosphate-bearing Eocene marls (see MORET 1931). According to ARAMBOURG's monographic study of the fossil vertebrates in Moroccan phosphate deposits (1952), we tend to assume a Paleocene age for the "lower", transgressive "calcaire à Thersitées". The phosphatic marls, of open marine facies, overlying these limestones (see fig.9) are of Paleocene (?) and Eocene age. Probably beginning with the Middle Lutetian, fault activity of the "accident de Medinat" went on and the Paleogene sediments were folded themselves.

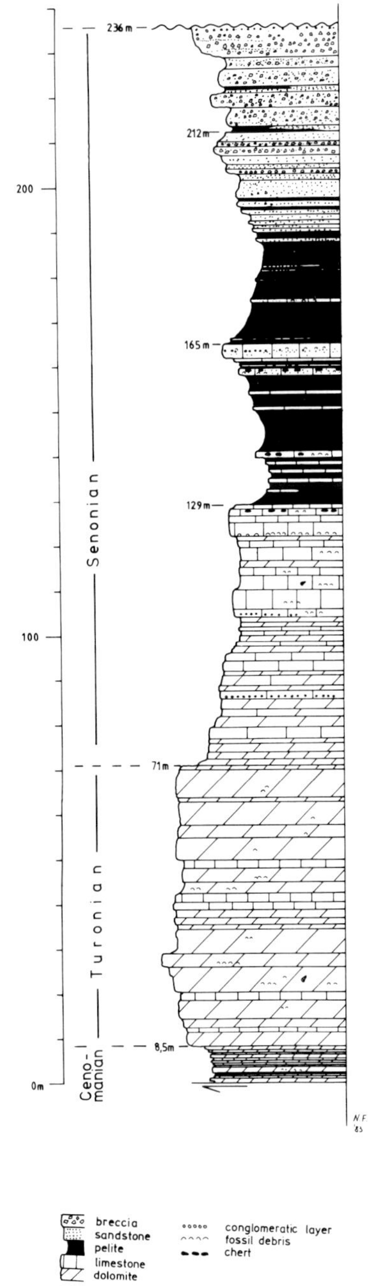

Fig.10: Sedimentological section of the Cenomanian -Turonian - Senonian set of strata. Location of section: see fig. 9.

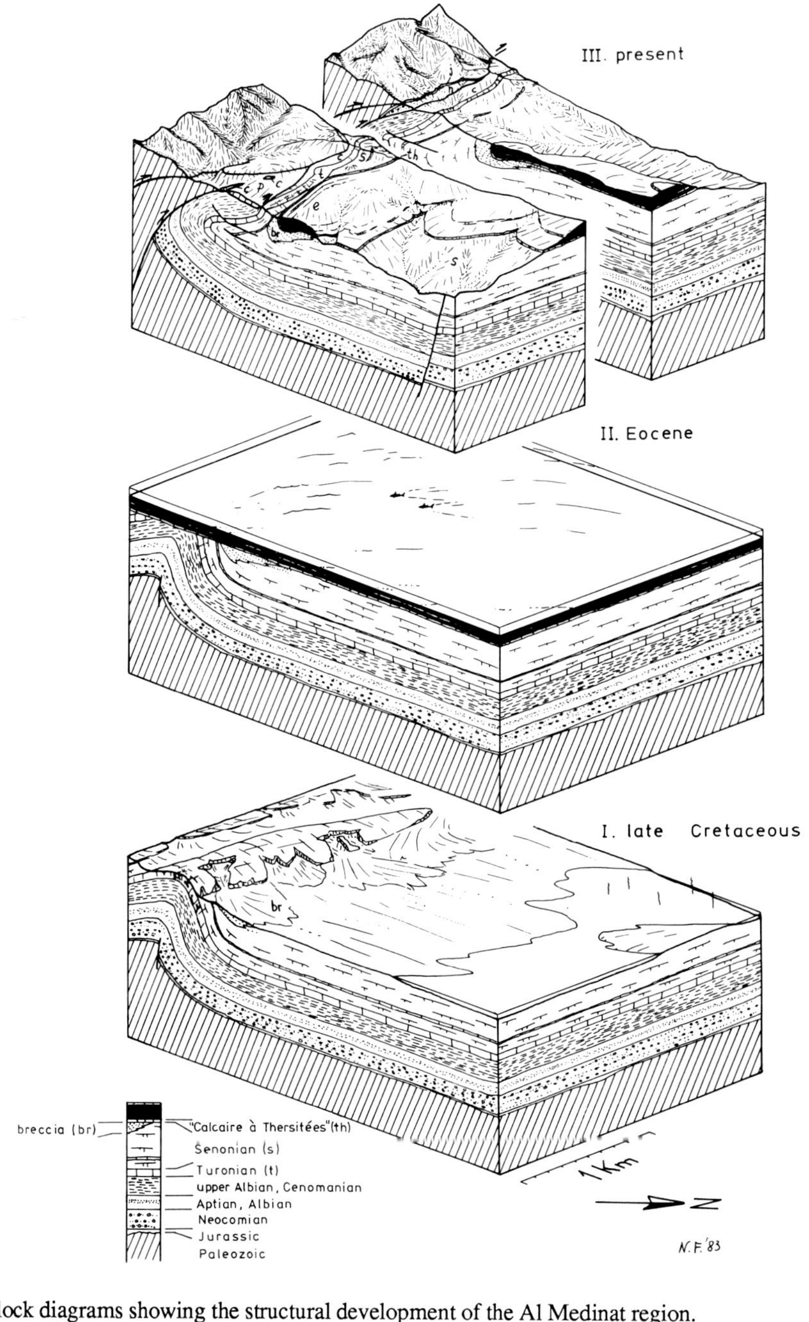

Fig.11: Block diagrams showing the structural development of the Al Medinat region.
I: Late Senonian: Formation of a monocline, development of karst features, erosion of Turonian and Senonian carbonate rocks, debris cone sedimentation.
II: Eocene: Transgression and marine sedimentation.
III: Present situation: Paleozoic rocks of the axial zone have been thrusted onto the southernmost part of the Plateau de Medinat by curved reverse faults.

Unconformable contacts between Senonian and Tertiary strata have been reported also from the Chichaoua region W of Marrakech (see ROCH 1930) and from Tahnaout / Rhiraïa (WILDBERGER 1984, DRESEN 1984). In other regions such as the Amezmiz fault zone at the northern border of the northern Subatlas zone, there is no evidence for Cretaceous deformation.

4.3. How do the High Atlas' faults continue to depth?

As stated above, the tectonic evolution of the Western High Atlas since Triassic times has been dominated by block movements, most deformation being localized to the same narrow fault zones over long periods of time. Some of these fault zones were inherited from the Variscan faulting (see SCHAER 1967, FROITZHEIM 1986), while the Tizi-n'Test fault zone is even older and can be traced back to the Late Precambrian (BINOT et al. 1986). The rhomboid fault pattern characteristic of the present High Atlas is probably of Triassic age: Many of today's steeply dipping reverse faults have functioned as normal faults in the Triassic Atlas rift system, as can be seen, for example, in DU DRESNAY & PROUST's Ourika valley section (MICHARD 1976, fig.123).

The main problem of block uplift tectonics is valid for the High Atlas as well as for the Rocky Mountains foreland (see for example STEARNS 1978), the Pyrenees and several other mountain ranges: Do the reverse faults shallow or steepen at depth? Do they correspond to a shallow detachment horizon or do they die out, the displacements being taken over by ductile strain, or, as a third alternative, do they cut through the whole of the crust or even deeper? We cannot answer any of these questions definitively, but make some remarks in favour of one or the other theory, which are based on field observations in the Western High Atlas. What is true for this region is not necessarily true for other regions of block uplift.

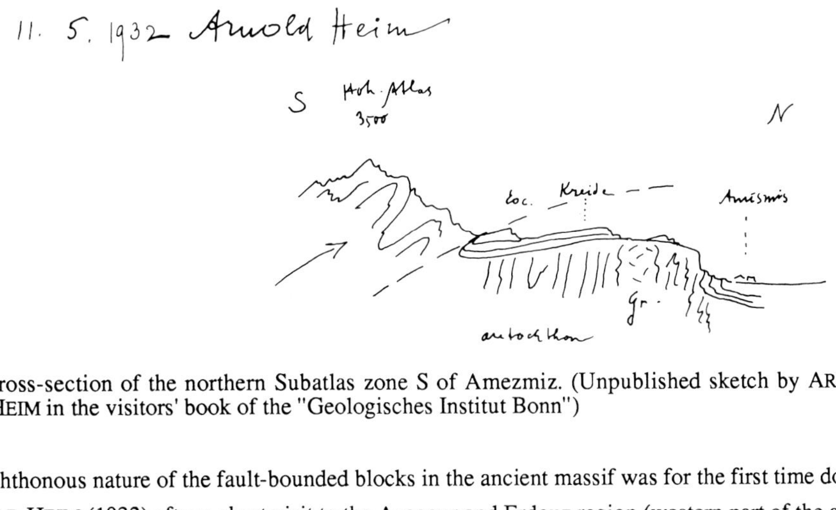

Fig.12: Cross-section of the northern Subatlas zone S of Amezmiz. (Unpublished sketch by ARNOLD HEIM in the visitors' book of the "Geologisches Institut Bonn")

The autochthonous nature of the fault-bounded blocks in the ancient massif was for the first time doubted by ARNOLD HEIM (1932) after a short visit to the Azegour and Erdouz region (western part of the ancient massif). Fig.12 is an unpublished sketch of the Al Medinat section by HEIM. He considered the "accident de Medinat" and the Anougal fault to be parts of a shallow thrust plane at the base of the Erdouz "nappe".

is no evidence for "young" strike-slip movements for any of the *accidents* in the ancient massif except the Tizi-n'Test fault zone, and even there the amount of lateral displacement seems to be small (BINOT et al. 1986).

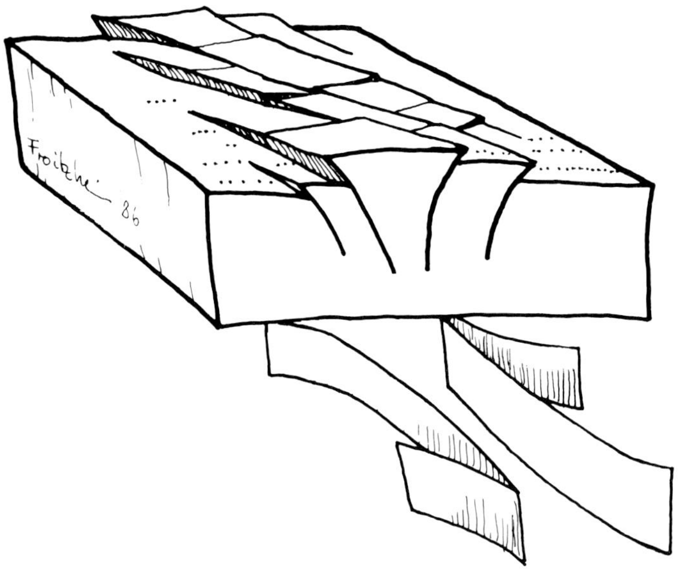

Fig.15: The Western High Atlas as a hypothetical transpression zone, seen from NW, with the Tizi-n'Test fault zone diagonally cutting the mountain chain (Idealized sketch after an experiment with wet clay deformed under transpression-like conditions).

The "accident de Medinat", for example, cuts Variscan folds and faults of the Paleozoic basement without causing any lateral offset of these structures, which excludes important post-Variscan strike-slip components of movement for this fault. Evaluation of slip vectors by analysis of slickensides etc. for some of the *accidents* in the rhomboid fault pattern of the ancient massif is necessary before transpression in the Western High Atlas can be either proved or rejected.

References

AMBROGGI, R. & NELTNER, L. (1952): Haut Atlas et Antiatlas occidentaux. Fascicule 2: Déscription d'itinéraire.- 19ᵉ Congr. géol. intern. Alger 1952, livret-guide, Maroc, **13**: 83 pp; Alger.

ARAMBOURG, C. (1952): Les vertébrés fossiles des gisements de phosphates (Maroc-Algérie-Tunisie). - Not. Mém. Serv. géol Maroc, **92**: 396 pp; Paris.

BALLY, A. W. (1984): Tectogenèse et sismique réflexion. - Bull. Soc. géol. France, (7), **26**: 279-285; Paris.

BELOW, R. (1976): Geologische Aufnahmen am Südrand des Hohen Atlas bei Taroudannt etc. - Diplomarbeit Bonn (unpubl.): 89 pp; Bonn.

BINOT, F., DRESEN, G., STETS, J. & WURSTER, P. (1986): Die Tizi-n'Test-Verwerfungszone im Hohen Atlas (Marokko). - Geol. Rdsch., **75**: 647-664; Stuttgart.

BOUDDA, A. & CHOUBERT, G. (1972): Sur la limite inférieure du Cambrien au Maroc.- C. R. Acad. Sci.Paris, **275**: 5-8; Paris.

CHARLOT, R., TISSERANT, D., VIDAL P. & F. (1967): Rapport technique et quelques résultats. -C. R. Activ., Serv. carte géol. Maroc, **1967** (unpubl.): 126-137; Rabat.

CHOUBERT, G. (1955-56): Carte géologique du Maroc, feuille Marrakech 1: 500 000. - Serv. géol. Maroc; Rabat.

-- (1963): Histoire géologique du Précambrien de l'Anti-Atlas.- Not. Mém Serv. géol. Maroc, **162**: 352 pp; Rabat.

-- & FAURE-MURET, A. (1962): Evolution du domaine atlasique marocain dépuis les temps Paléozoiques. - In: DURAND-DELGA, M. (ed.): Livre à la mémoire P. FALLOT, Mém. h.-s. Soc. géol. France, **1**: 447-527; Paris.

COWARD, M. P. (1983): Thrust tectonics, thin skinned or thick skinned, and the continuation of thrusts to deep in the crust. - J. Struct. Geol., **5** (2): 113-123.

DESTOMBES, J., HOLLARD, H., WILLEFERT, S. (1985): Lower paleozoic rocks of Morocco. - In: HOLLAND, C.H. (ed.): Lower Paleozoic of north-western and west-central Africa: 91-336; (WILEY & Sons) Chichester - New York - Brisbane - Toronto - Singapore.

DRESCH, J. & ROCH, E. (1934): Esquisse tectonique de la région d'Amismiz. - Bull. Soc. géol. France, (5), **4**: 165-178; Paris.

DRESEN, G. (1984): Bruchtektonik, Faltung und Schollenbau im Westlichen Hohen Atlas bei Marrakech/Marokko. - Diss. Univ. Bonn: 171 pp; Bonn.

-- (1985): Bruchtektonik und Schollenbau im Hohen Atlas südlich von Marrakech (Marokko). - Geol. Rdsch., **74**: 95-108; Stuttgart.

FERRANDINI, J. & LE MARREC, A. (1982): La couverture jurassique à paléogène du Haut Atlas de Marrakech est allochthone dans la «zone des cuvettes» d'Aït Ourir (Maroc). - C. R. Acad. Sci., **295**: 813-816; Paris.

-- & PETIT, J. P. (1982): Sur la tectonique récente de la bordure Nord du Haut Atlas de Marrakech (Maroc). - Bull. Fac. Sci. Marrakech, **1** (1): 87-96; Marrakech.

FROITZHEIM, N. (1984): Oberkretazische Vertikaltektonik im Hohen Atlas SW' von Marrakech/ Marokko - Rekonstruktion eines Bewegungsablaufs im Frühstadium der Atlas-Orogenese. - N. Jb. Geol. Paläont. Mh., **1984** (8): 463-471; Stuttgart.

-- (1986): Untersuchungen zur Tektogenese in der Nördlichen Subatlas-Zone, SW' von Marrakech (Hoher Atlas, Marokko). - Diss. Univ. Bonn: 141 pp; Bonn.

HEIM, A. (1932): Tektonische Beobachtungen im Hohen Atlas. - Eclogae geol. Helvetiae, **25**:262-264; Zürich.

HOLLARD, H. (1985): Pre-Cambrian - Cambrian boundary and Lower Cambrian.- In: HOLLAND, C.H. (ed.): Lower Paleozoic of north-western and west-central Africa: 101-157; (WILEY & Sons) Chichester - New York - Brisbane - Toronto - Singapore.

HUDEC, B. (1984): Geologische Aufnahmen am Nordrand des Hohen Atlas bei Imi N'Tanoute (Marokko) etc. - Diplomarbeit Bonn (unpubl.): 165 pp; Bonn.

JENNY, J. (1983): Les décrochements de l'Atlas de Demnat (Haut Atlas central, Maroc): Prolongation orientale de la zone de décrochement du Tizi-n-Test et clef de la compréhension de la tectonique atlasique. - Eclogae geol. Helvetiae, **76**: 243-251; Basel.

KONING, G. DE (1957): Géologie des Ida ou Zal (Maroc). Stratigraphie, pétrographie et tectonique de la partie SW du bloc occidental du massif ancien du Haut Atlas. - Leidse geologische Mededelingen, **23**: 209 pp; Leiden.

LATH, P. (1985): Zur Geologie des westlichen Kik-Plateaus südlich Marrakech (Hoher Atlas, Marokko). - Diplomarbeit TU München (unpubl.): 110 pp; München.

LAVILLE, E., LESAGE, J. L., & SEGURET, M.(1977): Géométrie cinématique (dynamique) de la tectonique atlasique sur le versant sud du Haut-Atlas marocain. - Bull. Soc. géol. France, (7), **19**: 527-539; Paris.

MATTAUER, M., TAPPONIER, P., & PROUST, F. (1977): Sur les mécanismes de formation des chaînes intracontinentales. L'exemple des chaînes atlasiques du Maroc. - Bull. Soc. géol. France, (7), **19**: 521-526; Paris.

MICHARD, A. (1976): Eléments de géologie marocaine. - Not. Mém. Serv. géol. Maroc, **252**: 408 pp; Rabat.

MORET, L. (1931): Recherches géologiques dans l'Atlas de Marrakech. - Not. Mém. Serv. Mines Carte géol. Maroc, **18**: 262 pp; Rabat.

NELTNER, L. (1938): Etudes géologiques dans le Sud marocain (Haut-Atlas et Anti-Atlas). - Not. Mém. Serv. Mines Carte géol. Maroc, **42**: 298 pp; Rabat.

PERMINGEAT, F. (1957): Le gisement de molybdène, tungstène et cuivre d'Azegour (Haut Atlas). - Not. Mém. Serv. géol. Maroc, **141**: 284 pp; Rabat.

PETIT, J.-P. (1976): La zone de décrochements du Tizi-n'Test (Maroc) et son fonctionnement depuis Carbonifère.- Thèse 3e cycle: 99 pp; Montpellier.

PROUST, F. (1962): Tectonique de socle par failles inverses, en liaison avec d'anciennes failles normales, dans le Haut-Atlas (Maroc). - C. R. somm. Soc. géol. France, **1962** (1): 9-11; Paris.

-- , PETIT, J.-P., & TAPPONIER, P. (1977): L'accident du Tizi-n'Test et le rôle des décrochments dans la tectonique du Haut Atlas occidental (Maroc). - Bull. Soc. géol. France, (7), **19**: 541-551; Paris.

ROCH, E. (1930): Etudes géologiques dans la région méridionale du Maroc occidental. - Not. Mém. Serv. Mines Carte géol. Maroc, **9**: 542 pp; Macon.

-- (1950): Histoire stratigraphique du Maroc. - Not. Mém. Serv. géol. Maroc, **80**: 435 pp; Toulouse.

SANDERSON , D.J. & MARCHINI, W.R.D. (1984): Transpression. - J. Struct. Geol., **6** (5):449-458.

SCHAER, J.P. (1964a): Volcanisme cambrien dans le massif ancien du Haut Atlas occidental. -C. R. Acad. Sci. Paris, **258**: 2114-2117; Paris.

-- (1964b): Aspècts de la tectonique dans le bloc occidental du massif ancien du Haut Atlas. - C. R. Acad. Sci. Paris, **258**: 2353-2356; Paris.

-- (1967): Interférence entre les structures du socle et celles de la couverture dans le Haut-Atlas Marocain. - In: «Etages tectoniques» (Colloque Neuchâtel 1966): 297-305; (La Baconnière) Neuchâtel.

-- (1981): Carte géologique du Maroc, feuille Imi n'Tanout 1: 100 000.- Serv. géol. Maroc; Rabat.

SCHMITT, M. (1979a): The section of Tiout (Precambrian/Cambrian boundary beds, Anti-Atlas, Morocco): Stromatolithes and their biostratigraphy.- Arb. Paläont. Inst. Würzburg, **2**: 188 pp; Würzburg.

-- (1979b): New stromatolithes from the late Precambrian of the Anti-Atlas and from the lower Cambrian of the High Atlas, Morocco.- Senckenbergiana lethaea, **60**: 39-49; Frankfurt a.M.

SDZUY, K. (1978): The Precambrian-Cambrian boundary beds in Morocco (preliminary report).- Geol. Mag., **115**: 83-94, Cambridge.

STEARNS, D. W. (1978): Faulting and forced folding in the Rocky Mountains foreland. - In: MATTHEWS, V. (ed): Laramide folding associated with basement block faulting in the Western United States. - Geol. Soc. America Mem., **151**: 1-37; Boulder.

STETS, J. & WURSTER, P. (1981): Zur Strukturgeschichte des Hohen Atlas in Marokko. - Geol. Rdsch., **70**: 801-841; Stuttgart.

-- (1982): Atlas and Atlantic - structural relations. - In: RAD, U. VON et al. (eds.): Geology of the Northwest African continental margin: 69-85; (SPRINGER) Heidelberg-Berlin-New York.

STUDER, M. & DU DRESNAY, R. (1980): Déformations synsédimentaires en compression pendant le Lias supérieur et le Dogger, au Tizi n'Irhil (Haut Atlas central de Midelt, Maroc). - Bull. Soc. géol. France, (7), **22**: 391-397, Paris.

TERMIER, H., & TERMIER, G. (1971): Le massif granito-dioritique du Tichka (Haut Atlas occidental, Maroc). Les régions et les terrains. - Not. Mém. Serv. géol. Maroc, **216**: 240 pp; Rabat.

WILDBERGER, J. (1984): Geologische Aufnahmen am Nordrand des Hohen Atlas bei Tahnaout (Marokko) etc. - Diplomarbeit Bonn (unpubl.): 114 pp; Bonn.

STRUCTURE AND POST-PALAEOZOIC EVOLUTION OF THE CENTRAL HIGH ATLAS

Volker Jacobshagen, Rainald Brede, Manfred Hauptmann,
Werner Heinitz and Renate Zylka
Institut für Geologie, Freie Universität Berlin,
Altensteinstr. 34 A, D-1000 Berlin 33

ABSTRACT

In the central High Atlas, two stockworks can be distinguished: A basement that was part of a Variscan belt and a cover composed of Mesozoic and Palaeogene sediments accompanied with magmatites. The deformation of the cover was dominated by a fractural pattern within the basement, which was repeatedly reactivated from the Triassic onward, while the continental stress-field gradually changed its orientation. Contrary to the western High Atlas, where the deformation was predominantly caused by vertical movements, compressional structures are more frequent in the central High Atlas. Here, the amount of shortening increases continuously to the east from 7% to 20%. In its northeastern part, compressional deformation is, moreover, indicated by axial plane foliation.

The post-Palaeozoic evolution of the central High Atlas started with the Triassic break-up of Pangaea and the origin of continental basins. Early Jurassic subsidence produced the Atlas Gulf, which was a Tethys embayment. Its formation was accompanied by block-faulting and by mafic volcanism, both culminating during the Upper Dogger. The Malm-Eocene interval was a period of calm and retarded subsidence. Contemporaneously, mafic or intermediate alkaline magmas were intruded. The Cenozoic inversion of the rift graben started with a compressional phase during the Upper Oligocene-Lower Miocene interval. Rapid uprisal in Miocene and again in Pliocene times caused gravity sliding on both flanks of the western part of the central High Atlas.

The period of rifting and subsidence (Triassic-Eocene) was synchronous with the opening of the Western Tethys and of the Atlantic, whereby the peaks of Cenozoic inversion coincide with the main tectogenic events of the Betides and the Maghrebides. Thus, inversion was probably induced by collisional processes in the Western Mediterranean region.

1. Introduction

The High Atlas, extending to about 800 km in length and 40-100 km in width, separates southern Morocco from the northern provinces of the country. The western High Atlas between the Atlantic Ocean and the Tichka defile displays the highest summit (J. Toubkal 4165 m). The central part with similar elevations (Irrhil Mgoun 4077 m, J. Ayachi 3751) is bordered to the E by the Ziz valley and is followed by the eastern High Atlas that ends at the elevated plateau of Tamlelt near the Algerian boundary. Farther east, the Sahara Atlas prolongates the High Atlas to Tunesia (for an overview see JACOBSHAGEN, this vol., fig. 1).

From a geotectonic point of view, the High Atlas is an intracratonic belt which does not show the typical features of an orogen such as: ophiolites, flyschs, nappes regional metamorphism and granitoid intrusions. Two stockworks are to be distinguished: The basement being the lower of the two apparent stockworks consists of Precambrian and Palaeozoic rocks. In the western High Atlas, it is identical with the central Anti-Atlas, and in the central and eastern segments parts of a Variscan orogen are exposed (MICHARD et al. 1982). The basement is covered by weakly deformed sediments and magmatites of Mesozoic to Lower Tertiary age. Cretaceous and Palaeogene rocks are mostly confined to the northern and southern rims of the High Atlas. Among the tectonic structures of the cover, wide-spanned folds, monoclines, steep faults and low-angle strike-slip faults prevail, whereas narrow or inclined folds, upthrusts and overthrusts seem to play a minor role. Nappes formed by gravity sliding on the flanks of the High Atlas (LAVILLE et al. 1977, LAVILLE 1980, FERRANDINI and LE MAREC 1982) are additional features.

The elevation of the basement changes remarkably along the mountain belt. In its western part, it is completely buried in the Atlantic region, but rises quickly to the E, forming the "massif ancien", which is connected with the basement of the Anti-Atlas. East of the Tichka defile, the basement dips to the E where it is overlain by thick Jurassic carbonate strata ("Calcareous Atlas") and only exposed in a few domes at the western end of the central High Atlas. Farther to the east in the eastern High Atlas the updomed basement appears again in places, indicating an elevation. Finally it crops out in the plateau of Tamlelt.

It is generally agreed that the deformation of the cover of the High Atlas was induced by block faulting of the basement. DUBOURDIEU (1962) and MATTAUER et. al. (1972) have paid special attention to NE and ENE striking wrench-faults of late Variscan origin. NNE and WNW striking fractures are also common (JENNY 1983, 1984; ZYLKA and JACOBSHAGEN 1986). The same fractural pattern is also developed in the central Anti-Atlas (HEINITZ 1984). There, it had originated at the end of the Panafrican orogeny during late Proterozoic times. This fault pattern plays a key role in the latest interpretations of the geotectonic development of the High Atlas. MATTAUER et al. (1977) proposed the theory that it has been repeatedly reactivated since the beginning of the Mesozoic, while the sense of the movements varied with changing orientation of strees fields in the northwestern part of the African plate from the Triassic up to present. These changes reflected the relative movements of the African and the European plates initiating extensional processes in connection with the opening of the Atlantic and the Tethys and later compressional ones due to West Mediterranean collisions. Strike-slip movements along diagonal faults are believed to have caused pull-apart basins or en echelon folds (MICHARD et al. 1975, COURBOULEIX et al. 1981).

Deviating from these opinions, STETS and WURSTER (1981) interpreted the geotectonic evolution of the High Atlas merely with respect to the opening of the Atlantic. From thickness distributions of sediments and from the early Mesozoic volcanism these authors deduced that from the Triassic to the Turonian the High Atlas was on the whole a rift graben. During the Cretaceous the rift development was disrupted and replaced by inversion, which continued troughout the Cenozoic. Former downthrusts along the borders of the Atlas rift were then changed to upthrusts. In addition, the rapid uplift of the mountain belt produced block rotation and gravity sliding on both flanks. The latter process was believed to be the origin of scales and inclined folds which had been interpreted as compressional features. The model of STETS and WURSTER was based on stratigraphic and palaeogeographic studies and on tectonic analyses in the western High Atlas.

2. The central High Atlas

The central High Atlas (fig. 1) is bordered by the following geological units: In the N, there are the Upper Cenozoic basins of the Tadla and of the Haute Moulouya, with the deviating Middle Atlas in between. The southern foreland consists of the Upper Cenozoic Ouarzazate Basin in the W and, from Tinerhir to the E, of undeformed Mesozoic and Tertiary sediments of the Sahara platform (HAMADA de Meski). The High Atlas is, on the whole, upthrust or overthrust towards both the S and the N. Along the southern border, they are accentuated by a small zone of folds and scales with southerly vergencies (South Atlas Marginal Zone, ZYLKA and JACOBSHAGEN 1986).

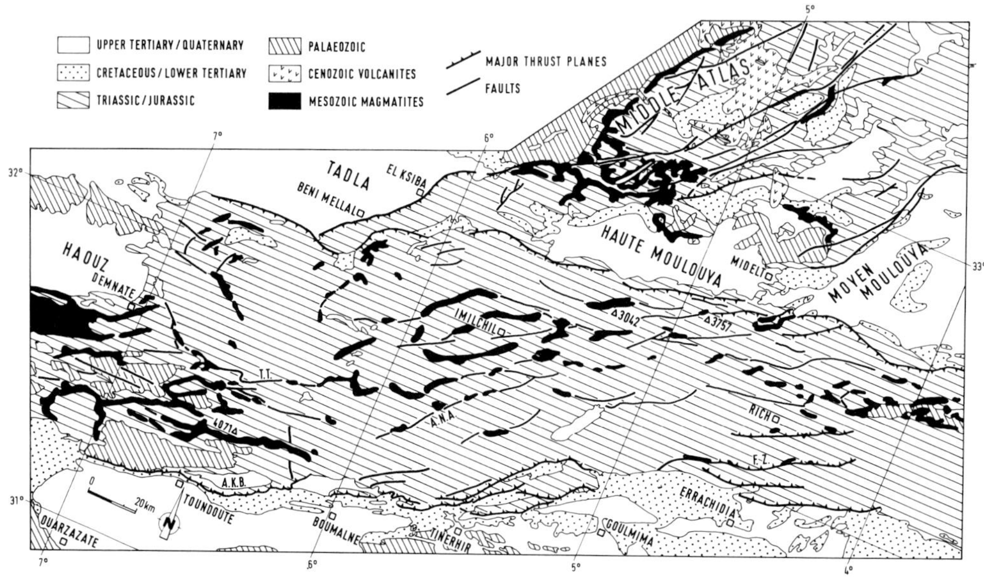

Fig. 1: Tectonic sketch-map of the central High Atlas. A.K.B. Ait Kandoula Basin, A.N.A. accident nord-atlasien, F.Z. Foum Zabel fault, T.T. Tizi n'Test fault.

Our investigations on the structural evolution of the central High Atlas started at its southern border (ZYLKA and JACOBSHAGEN 1986). Later, two traverses were studied. One crosses two thirds of the range from Boumalne/Tinerhir in the south to Imilchil in the north (SEUFERT 1986), whereas the second from Errachidia to Midelt represents a complete cross-section (see also BREDE and HEINITZ 1986). Field work is finished, but the evaluation of data is not yet complete. This paper intends, however, to present first results on a few major problems of the structural evolution in Mesozoic and Cenozoic times. In the following chapters, we shall, therefore, confine the discussion to the post-Palaeozoic cover of the central High Atlas. As for the Variscan basement, which is exposed in a few uplifts, see JENNY and LE MARREC (1980) and DESTOMBES et al. (1983).

2.1 Post-Palaeozoic stratigraphy

Our studies were focussed on the structural history of the central High Atlas. Therefore, we confine ourselves here, in reference to recent publications, to a short description of the post-Palaeozoic sequences.

Triassic

The Triassic sequence (MATTIS 1977; MANSPEIZER et al. 1978, 1982) consists of continental redbeds, up to 1000 m thick. Sandstones and siltstones predominate, with intercalations of conglomerates or pelites and, in the middle, an evaporitic horizon of dolomites, gypsum and halite exists. On the contrary to SALVAN (1984), we agree with CHOUBERT and FAURE-MURET (1962) that sedimentation of evaporites was restricted to certain parts of the Triassic depositional area of the central High Atlas. Locally Ladinian and Carnian fossils have been found. Near the top, a dolerite formation reaches a maximum thickness of 200 m. It is overlain by a thin band of red pelites with a few lacustrine limestone beds. The dolerites are of prevailingly tholeiitic character and provided radiometric ages of 212 - 180 m. y., i. e. from Rhaetian to Toarcian. But as the overlying Jurassic carbonate sequence yields, Sinemurian fossils, volcanic activity must have ceased at least by the lowermost Jurassic. The evaporites of the Triassic as well as the pelites on top of the dolerites provide excellent conditions for tectonic shearing.

According to MANSPEIZER (1982), the Triassic redbeds and dolerites originated in pull-apart basins and small rift grabens during the breakup of the North Atlantic Ocean, when the thinned crust of the Variscan Alleghanian-Mauretanide orogen was deformed by wrench tectonics and transform faulting.

Jurassic

At the beginning of the Jurassic, subsidence accelerated in the mobile belt of Morocco. A large "Atlas Gulf" extending from the opening Western Tethys in a SW direction was created by the formation of deep rift basins in the Middle Atlas and in the central and eastern High Atlas (CHOUBERT and FAURE-MURET 1962, DU DRESNAY 1971, ELMI 1984, CARIOU et al. 1985). It was bordered by continental regions to the W (Meseta) and to the S (Anti-Atlas), that were linked together by the "massif ancien". The idea of the Atlas Gulf being connected with the relatively small Atlantic Essaouira basin during the Jurassic (STETS and WURSTER 1981), cannot be approved of, since it neither complies with the facies distribution nor with faunal relationships.

The Lias and large parts of the Dogger of the central High Atlas are

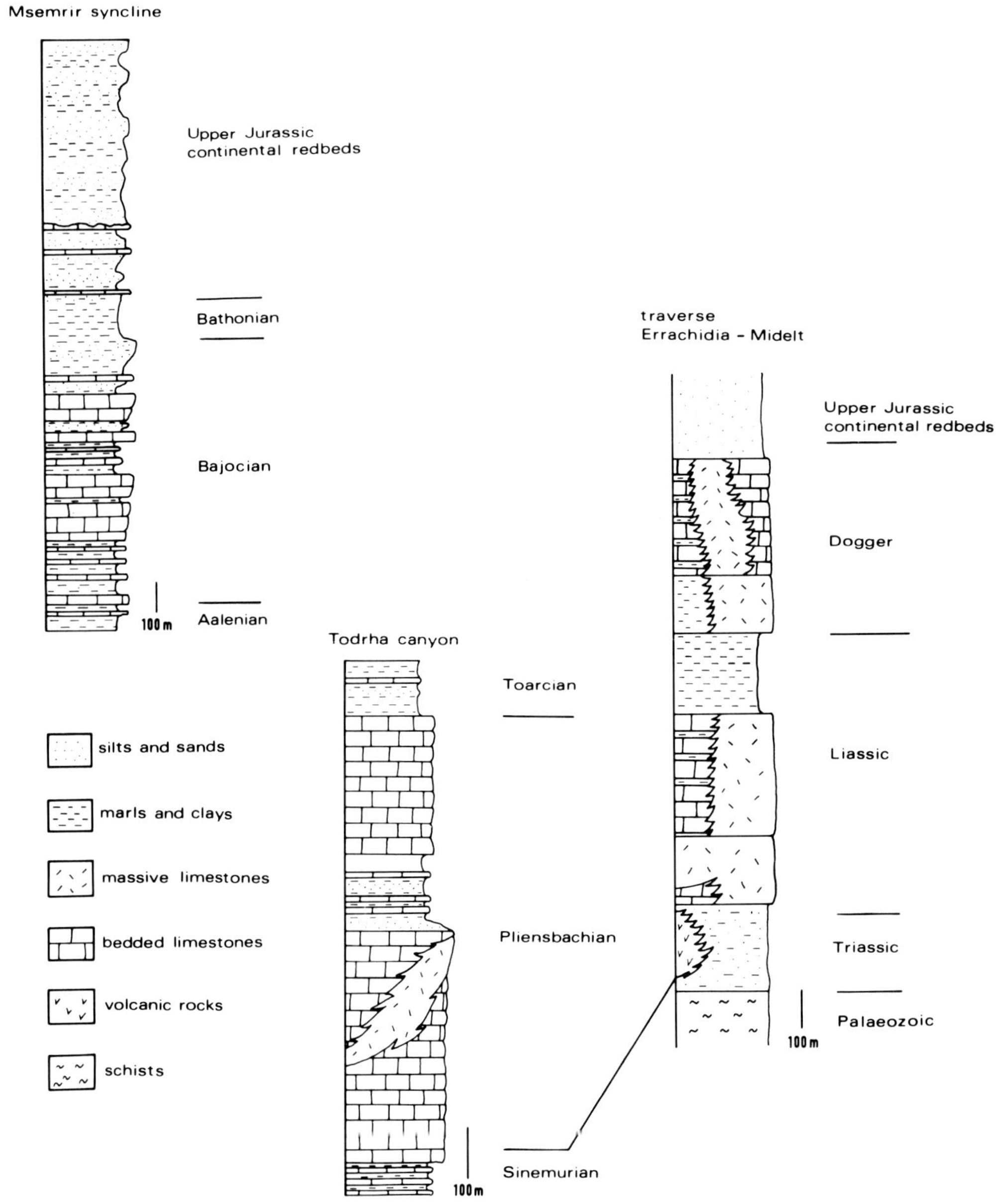

Fig. 2: Stratigraphic columns of the Jurassic of the central High Atlas.

represented by very thick carbonate deposits, with a few marly horizons intercalated (fig. 2). Marine development culminated during the Pliensbachian and the Bajocian. For an overview see DUBAR (1962),

DRESNAY (1964, 1971 a,b), EVANS and KENDALL (1977), BURGESS and LEE (1978), STANLEY (1981) and SEPTFONTAINE (1984). Special reference to the eastern part of the central High Atlas is given by WARME et al. (this volume). this whole region was situated at the southwestern end of the "Atlas Gulf". Thus, when going from SW or S to NE or N, one observes along the southern rims a lateral sequence of variegated lagoonal deposits merging into a relatively broad carbonate shelf with changing facies, which finally grades into deep-basin marls with slumping structures. The northern carbonate shelf in the Moulouya region was probably small. From the Pliensbachian onward, the basin was subdivided by several swells from which recifal debris was transported to the subbasins. From the reconstruction of Upper Pliensbachian palaeoslopes it may be derived that these swells developed in regions which suffered a strong tectonic deformation. The axis of the Jurassic trough moved northwards during the Toarcian to the Bathonian. In the Aalenian, a steep slope originated on the northern flank of the trough which is testified by sliding of large carbonate bodies. One of them, situated SW of Taklat-n-Imiter, is about 8 km long and 1,5 km in width. It consists of thick-bedded Liassic dolomites and was emplaced into Lower Dogger marls. Local facies changes were, moreover, caused by synsedimentary faults, especially wrench-faults (STUDER and DRESNAY 1980, BERNASCONI 1984, SEUFERT 1986). According to FAUGERE et al. (1986) the assymetric shape of the Jurassic trough itself indicates a pull-apart basin.

Nevertheless, as a rule, compact carbonate series of the Sinemurian-Domerian and the Bajocian-Bathonian intervals are widespread, whereby marly deposits are typical of the Toarcian-Aalenian interval. Except for local deviations, the thickness of the Jurassic rock units decrease from a maximum in the Lower and Middle Liassic to a minimum in the Middle Dogger. Towards the end of the Dogger carbonate sedimentation remained in confined areas (ZYLKA 1987). A remarkable change of depositional conditions took place during the Bathonian when lagoonal redbeds, formerly restricted to the southern rim, spread to the center of the basin (JENNY et al. 1981). In a few locations they yield relics of dinosaurs and fossil wood. It is generally agreed that these redbeds comprise the Upper Jurassic.

In many places, the Jurassic carbonate sequence contains mafic lavas and dykes (e.g. MONBARON and JUST 1980, BREDE and HEINITZ 1986), whose radiometric ages point to upper parts of the Dogger (Băthonian?). Locally these volcanics have been observed in the redbeds, too (MON-

BARON and JUST 1980).

Summarizing the Jurassic evolution, we emphasize that in the central High Atlas carbonate sedimentation came to an end in the Bathonian, whereas in the western High Atlas we find continental deposits of Lower and Middle Dogger age in addition to marine carbonate formations prevailing in the Callovian and in the Malm (ADAMS 1979, HÜSSNER 1985). These facts strengthen the opinion of CHOUBERT and FAURE-MURET (1962) that the Atlantic Essaouira basin was not connected with the Tethysian Atlas Gulf during the Jurassic.

Cretaceous

Generally, the sedimentary record of the central High Atlas ceases with the redbeds mentioned above ("Rouge terminal"). It is an open question whether they also comprise a part of the Lower Cretaceous, because of the lack of fossils (JENNY et al. 1981). In a few synclines situated in the northwestern part of the range and along its southern margin, the redbeds are, however, intercalated by yellow dolomite layers and gypsiferous marls of Aptian age ("barre aptienne"). The redbeds are covered by a marked limestone horizon of Cenomanian-Turonian age, dated by FERRANDINI et al. (1985). Along the southern margin of the central High Atlas, redbeds of probably Cretaceous age (pre-Cenomanian) directly overlie Anti-Atlas rocks (GAUTHIER 1957, ZYLKA 1987).

On top of the Turonian limestones redbed sedimentation continued at least in the Southern Marginal Zone (GAUTHIER 1957, ZYLKA and JACOBSHAGEN 1986) up to the Maastrichtian.

It should be mentioned that all conglomerates of the "Rouge terminal" as well as of the Lower and of the Upper Cretaceous redbeds only yield Anti-Atlas components showing that the central High Atlas was not yet eroded.

Lower Tertiary

The Upper Cretaceous redbeds of the Southern Marginal Zone are conformably overlain by a Palaeogene series of marine and lagoonal deposits (GAUTHIER 1957). HERBIG (1986) distinguished two fossiliferous carbonate formations of Danian and Ypresian age, separated by a pink siliciclastic formation, and an interbedded siliciclastic-carbonate formation on top, which he contributed to the Lutetian. According to HELMDACH (1986 and oral communication), the latter formation reaches up to the Lower Oligocene. The uppermost formation of the Palaeogene consists of red and violet marls with gypsum and halite. They are intercalated by fanglomerates of varying thickness, which contain the first Jurassic carbonate pebbles indicating an uplift of the High Atlas.

Post-Liassic intrusions

The Jurassic deposits of the central High Atlas were intruded by magmas different both in chemistry and age. SCHAER and PERSOZ (1976) and STUDER (1980) described gabbros, diorites, plagioclasites and syenites. K/Ar data of these rocks are scattered over a range of 175 - 135 m. y. (HAILWOOD and MITCHELL 1971). It is quite uncertain, whether all data indicate crystallization. The majority are, however, 175 - 165 m. y. old (MONBARON and JUST 1980), which would point to Middle or Upper Dogger. But also younger intrusions exist. TISSERANT et al. (1976) e. g. dated nepheline syenites and monzonites near Midelt as Eocene by concordant Rb/Sr and K/Ar ages.

Neogene

Neogene rocks are confined to a few basins within the central High Atlas, among which the Ait Kandoula Basin was investigated in detail by GÖRLER and ZUCHT (1986). New results are presented in this volume by the contributions of GÖRLER et al. and FRAISSINET et al.

2. 2. Tectonic structure

As already mentioned, the central High Atlas shows only moderate deformation except for its marginal zones in the S and in the N. For an overview of its tectonic structure, we shall first describe three sections across the Calcareous Atlas and then discuss the marginal zones.

2.21 The Calcareous Atlas

A first glimpse on the geological map of Morocco 1 : 500.000, (sheets of Ouarzazate and Rabat), shows that the central High Atlas is subdivided by major anticlinal structures, the distance between which is relatively large in the W and becomes narrower to the E. The same can be recognized from the distribution of Mesozoic intrusions on the tectonic sketch-map (fig. 1), as they are usually exposed in the core of the anticlines. Thus, it is to be expected that deformation and possibly, crustal shortening increase from W to E. as has been confirmed by our investigations.

A generalized section crossing the whole range in the N from Toundoute to the Tadla Basin (fig. 3), shows the basement dome of Skoura in its southern part and, as an outlier, the "Nappe of Toundoute", which originated by gravity sliding from that dome. To the NNE, the basement gradually dips down, being cut by several fractures. Only a few of them have also deformed the Mesozoic cover. Generally, block-faulting of the basement is reflected by shallow synclines and occasional box folds in the cover. Structures with northwestern vergency are confined to the northernmost 15 km. On the whole, tectonic shortening does not exceed 5 %, but increases suddenly to 20 % near the northern margin.

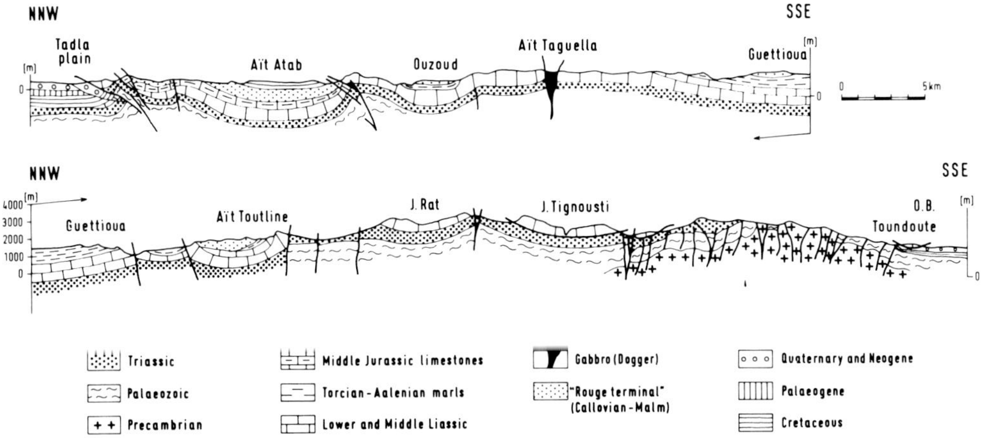

Fig. 3: Cross-section of the western segment of the central High Atlas (from JENNY in DESTOMBES et al. 1983).

Farther to the E, a combination of two sections from SEUFERT (1986) represents a traverse of two third of the central High Atlas (Fig. 4 a, b). From S to N a marginal strip of scales (which will be discussed later) is followed by a zone 25 km in width, with the Jurassic carbonate sequence forming a plateau, locally with weak bending of the beds or gentle box-folds. This zone is disrupted by a few anticlinal structures of diapiric character, which consist of Triassic redbeds and evaporites and of Middle Jurassic intrusions. The anticlines strike to the NE or even NNE and are quite far reaching. One of them - the "accident nord-atlasien" (ROCH 1939) can be followed through the whole range (SEUFERT 1986). In the median area, i. e. in the Imilchil region, moderate folds prevail many of them with northern vergencies. Among them, a strip of very narrow folds separates the Syncline of the Lakes from the Syncline of Sountat. Note that these folds were, according to SEUFERT, intruded by a gabbro dated at 155± 5 m. y. (HAILWOOD and MITCHELL 1971), i. e. Upper Dogger (ODIN 1982) or Lower Malm (HARLAND et al. 1982). This would mean that folding happened as soon as the end of the Dogger, as the Bathonian beds were already deformed.

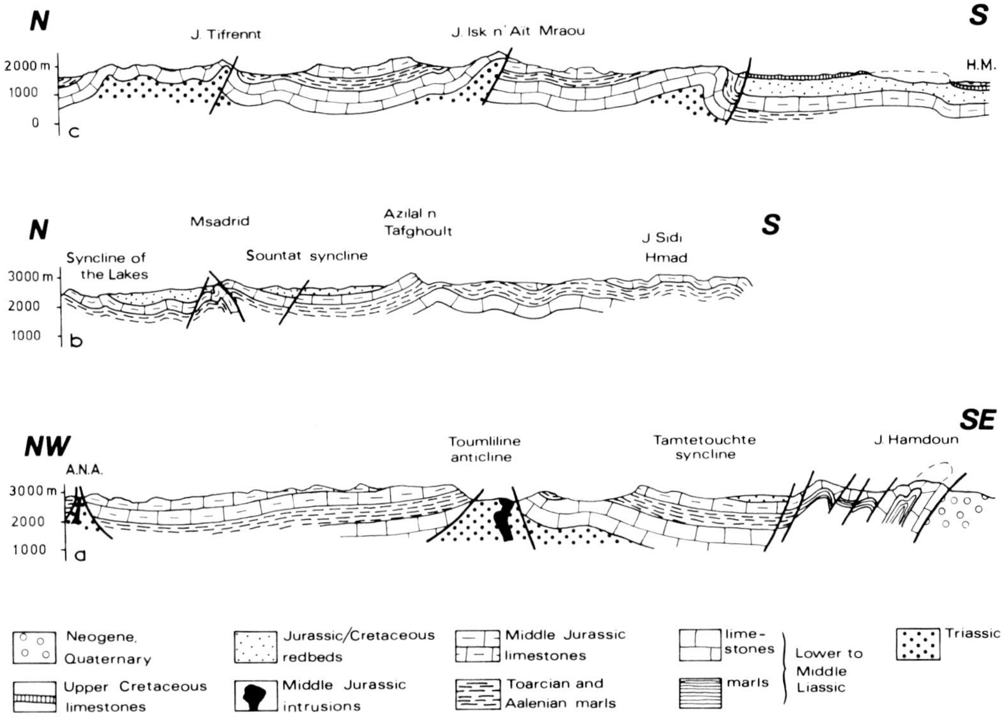

Fig. 4: Cross-sections of the middle (a,b) and the eastern (c) segments of the central High Atlas. 4 a, E of Tinerhir, is approximately prolonged to the Imilchil region by 4 b (both after SEUFERT 1986). A.N.A. accident nord-atlasien, H.M. Hammada de Meski.

In the eastern part of the central High Atlas a traverse from Errachidia to Midelt, has provided a great many new observations, concerning tectonic structures as well as stratigraphy and magmatic events (BERNASCONI 1983, BRECHBÜHLER 1984, BREDE and HEINITZ 1986, BREDE 1987). The southern part of this traverse is shown in fig. 4 c. The border between the High Atlas and the Hammada de Meski is formed by a simple monocline, which passes laterally into subordinate faults.

Within the High Atlas, deformation increases in a northerly direction, showing a poorly deformed plateau along its southern margin followed by broad synclines and narrow anticlines striking at 070°. Within the anticlines, upthrusts or overthrusts are common. They face to the SSE in the southern regions, as opposed, to the NNW in the northern part of the section. At the Foum Zabel overthrust, even Palaeozoic schists are exposed. Liassic limestones rest nearly horizontally on Upper Jurassic redbeds with an offset of at least 1 km.

The increase of deformation from S to N is, moreover, documented by an axial-plane foliation in restricted areas of the central and northern part of the traverse. The nature of this foliation was discussed by SCHAER and PERSOZ (1976), BERNASCONI (1983), BRECHBÜHLER (1984) and BREDE and HEINITZ (1986). Although mineral recrystallization parallel to the foliation could not be observed, we believe it to be a real schistosity on account of its orientation parallel to the anticlinal axes as well as of the phenomena of cleavage refraction and pencil structures. Schistosity is often developed in the surroundings of Upper Mesozoic intrusions, but also in a clear distance to them. Therefore, we do not agree with LAVILLE and HARMAND (1982) who argue that foliation was produced by the process of intrusion. Internal deformation of the sediments is, moreover, testified by elongated ammonoids. The median value of shortening of the Mesozoic cover is about 20 % for the eastern section.

2.22 The Southern Marginal Zone

The southern margin of the central High Atlas is characterized by a small zone of folds and scales with southerly vergencies. Its western part forms the northern rim of the Ouarzazate basin, whereas from Tinerhir to Errachidia it borders directly on the plateau of Hammada de Meski. From a structural point of view, three segments can be distinguished, which are briefly described in the following paragraphs.

The western segment (Toundoute - Tinerhir)

In this segment, the Calcareous Atlas is overthrust onto folds and scales which mainly consist of Cretaceous and Lower Tertiary rocks, but also include Neogene sediments of the Ouarzazate basin. Recent investigations have revealed different thrusting mechanisms. For instance in its western part, the Jurassic carbonate pile was sheared off from the Upper Triassic and seems to have slid down from the rising Calcareous Atlas onto the marginal zone ("nappe of Toundoute", LAVILLE et al. 1977, LAVILLE 1980, GÖRLER and ZUCHT 1986, GÖRLER et al., this vol.). The construction of the southern marginal zone, showing scates and narrow folds in this region, is still under discussion as to whether it too might be ascribed to gravity sliding. On account of our results farther to the east, is seems likely that gravity sliding happened only during the Neogene and has merely overprinted the structures of a foregoing compressional deformation. In this context, it may be important to point out that about 1500 m of Neogene sediments were deposited upon the Toundoute nappe in the Ait Kandoula basin (GÖRLER and ZUCHT 1986), both forming a syncline at present, which is upthrust to the south. This suggests subrecent compressional deformation, because gravity sliding could not produce major structures in a depth of more than 1000 m. Post-Neogene compression can, furthermore, be derived from the fold structures of a section W of Boumalne du Dadès (fig. 5a). In the Dadès valley, the tectonic style of the southern marginal zone changes. Along its northern rim, scales of Triassic and Jurassic rocks of the Calcareous Atlas are overthrust onto Neogene deposits, whereas in the central part wide-spanned synclines of Upper Cretaceous and Palaeogene beds predominate. Towards the S folds become narrower and, finally, Eocene limestones are overthrust onto the Ouarzazate Basin (ZYLKA and JACOBSHAGEN 1986). Farther to the E, between the Dadès and the Todrha valleys, the southern marginal zone consists of up to four scales. The boundary to the Calcareous Atlas is marked by overthrusts which laterally grade into undisturbed anticlines.

The southern border of the Southern Marginal Zone is of changing structural character. As apposed to the frequent overthrusts found in the western part, several areas show normal sedimentary contacts with structures dipping beneath the Ouarzazate basin. Further structures separating both units are minor vertical faults.

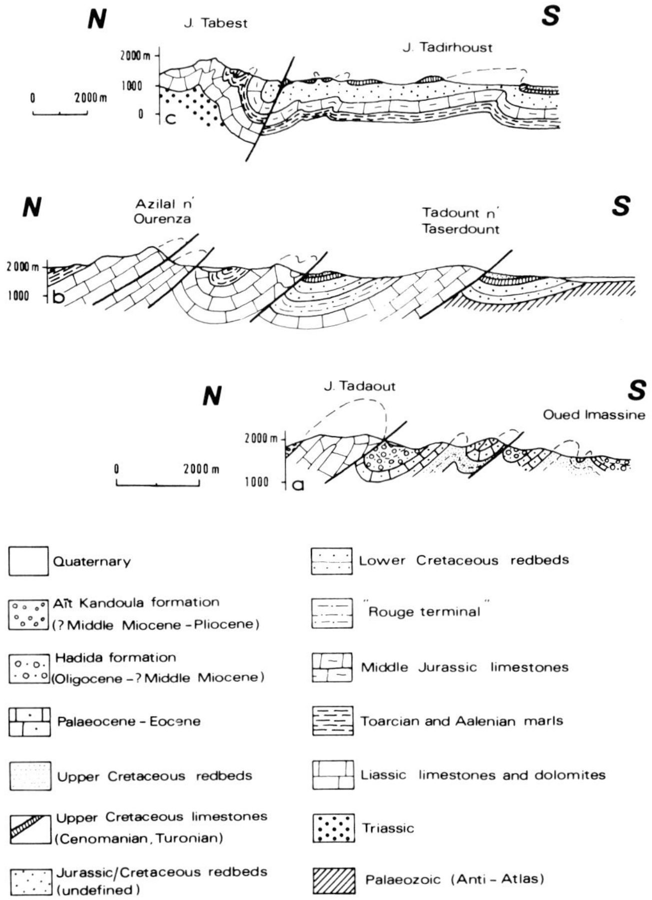

Fig. 5: Cross sections of the Southern Marginal Zone a) W of Boumalne, b) E of Tinerhir, c) W of Errachidia.

The central segment (Tinerhir-Goulmima)

The central segment of the Southern Marginal Zone belongs mainly to the Calcareous Atlas. Only a small rim of the foreland was deformed, which consists mainly of Cretaceous sediments. The distribution of Lower Tertiary deposits, which are very common in the western segment, ends a few kilometers E of Tinerhir. Contrary to the western segment , we could not find any indications of gravity sliding.

Although the foreland strip shows only occasional moderate folds (locally only one syncline), we observe in the southernmost Calcareous

Atlas four Jurassic scales which are overthrust up to 3 km to the SE (fig. 5 b). All of them develop from anticlines which alternate along the southern border of the High Atlas. The easternmost overthrust, about 20 km long, ends near the oasis of Taltfraout within a Jurassic anticline, whose axis dips to the east under flat-lying Cretaceous beds of the foreland (ZYLKA and JACOBSHAGEN 1986). This means that the present southern boundary of the High Atlas does not coincide exactly with the former boundary of the Atlas rift-graben.

The eastern segment (Goulmima-Errachidia)

North of Goulmima, the scales described grade to the E into moderate anticlines or even monoclines. The southern boundary of the High Atlas is made up of minor upthrusts, vertical faults or even monoclines, as shown in fig. 5 c. The foreland is completely undeformed and even the southern part of the High Atlas is very moderately deformed, showing only a few upthrusts, monoclines or upright anticlines within a nearly tabular region. As mentioned previously (2.21 and fig. 4), major upthrusts or overthrust with southern vergencies follow farther to the N.

The South Atlas Fault Zone

Since the publication of RUSSO and RUSSO (1934), geologists have accepted the idea that the High Atlas is bordered to the S by a large fault zone which was called "accident sud-atlasien" by French authors. The opinions on the character of this fault zone are, however, extremely divergent ranging from a zone of limited faults to a huge wrench-fault, which occasionally was even believed to be the northwestern boundary of the African plate in Mesozoic and Cenozoic times (DEWEY et al. 1973).

From our observations of the Mesozoic cover at the southern edge of the central High Atlas it is evident that there is no continuous fault zone, neither along the southern border of the calcareous Atlas nor within the Southern Marginal Zone. The study of satellite images and aerial photographs has supplied the same results. On the other hand there is good reason to assume an important suture zone within the basement, which probably originated in late Palaeozoic times and was repeatedly reactivated by different movements later on. The direction of primary displacement along that zone is still unknown.

The existence of a pre-Mesozoic suture zone can be deduced from the frequency of minor and even major faults of different character along the southern rim of the High Atlas. A second argument is the sudden increase of thicknesses in the Mesozoic cover (STETS and WURSTER 1981), for which a striking example exists N of Goulmima near the Taltfraout oasis. In this area, Cretaceous foreland redbeds are exposed resting with sedimentary contact on Palaeozoic rocks, without any Jurassic or Triassic rocks in between. But only 1 km to the N, at the southern margin of the Calcareous Atlas, about 800 m of Jurassic carbonate rocks are exposed. Taking into account the marginal overthrust, it is evident that the distance between both outcrops was only a few kilometers in Jurassic times. The different sedimentary evolutions from Tertiary to Quaternary in adjacent regions and the actual steep southern slope of the central High Atlas demonstrate the enormous uplift of the Jurassic carbonate pile to more than 1000 m above its level in the Southern Marginal zone and indicate that this bordering fault system has been active up to recent times.

2.23 The northern margin

Along the northern margin of the central High Atlas, several upthrusts and overthrusts with northern vergencies are observed (fig. 1). Gravity sliding as described by FERRANDINI and LEMARREC (1982) was apparently confined to the northern slope of the "massif ancien".

The intensity of deformation and the amount of thrusting are variable, but, in general, do not reach the same degree as in the Southern Marginal Zone. In the cross-section Errachidia-Rich-Midelt only a few upthrusts of minor importance exist (Fig. 6), whereas more to the W, in the region of Beni Mellal-El Ksiba, offset grows to about 1 km (BENZAQUEN 1963, MICHARD 1976). It is still under discussion, where to draw the boundary between the High and the Middle Atlas in that region (see DRESNAY, this volume). Detailed palaeogeographic and structural data, which may shed some light on this problem, were presented by MONBARON (1981).

2.24 Fractures, joints

As already mentioned, the fault pattern within the sedimentary cover of the central High Atlas is ruled by late Palaeozoic or even late Precambrian fractures of the basement. One of the main systems of these

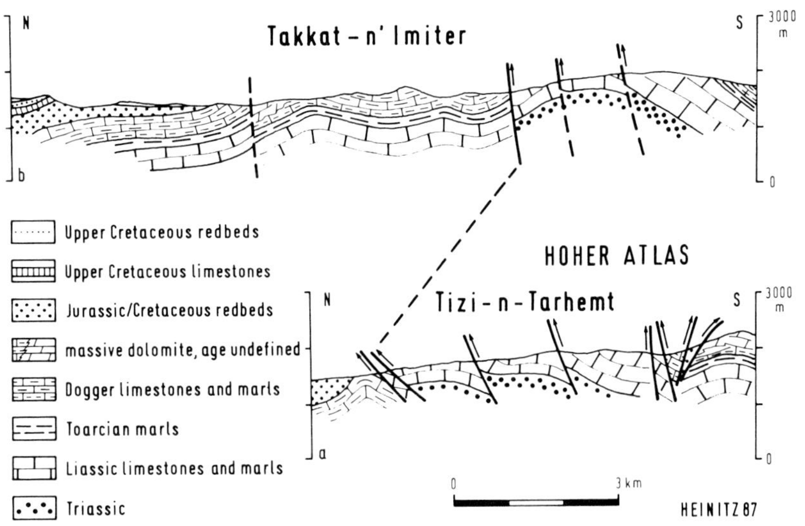

Fig. 6: Cross-sections of the northern margin of the central High Atlas, near the Midelt-Errachidia road (a) and east of it (b).

fractures shows the same orientation as the major folds, i.e. NE to ENE, further systems strike to the NNE and to the E, respectively. They separate rhomboidal blocks which can be recognized on geological maps and satellite images as well as by local facies patterns e.g. of the Jurassic, thus demonstrating repeated reactivation over a very long period.

The sense of post-Palaeozoic movements along these fractures is still under discussion, as detailed analyses are rare. In the north-western part of the central High Atlas, JENNY (1983) observed major wrench-faults with righthand movements in a WNW direction (accident de Demnate) and sinistral ones running to the ENE, among them a prolongation of the Tizi-n-Test fault (PROUST et al. 1977). Normal faults have been observed in a SW/NE direction, and vertical ones recorded striking E/W. The directions of these faults were combined with the directions of box folds and other structures to a general scheme which was related to a stress field with maximum compression in a NNE direction. These results seem, however, not to be representative for the whole region under discussion. Thus, BINOT et al. (1986) were able at least to exclude major strike-slip movements of post-Triassic age along the Tizi-n-Test fault from its type-area.

In the vicinity of the Dadès and Todhra valleys, SEUFERT (1986) observed two predominant fault systems: Sinistral wrench-faults and normal faults striking NW, and dextral wrench-faults, upthrusts and overthrusts of ENE direction. NNW and NNE striking joints and faults seem, however, to be of minor importance.

In the eastern part of the central High Atlas, we find the same orientation of faults and joints, although some of them show different directions of movements (BREDE and HEINITZ 1986). These are upthrusts and overthrusts striking ENE, dextral wrench-faults running to the NE and satellite lineations striking WNW. This pattern might point to a main compressional stress in a NNW direction.

Summarizing these results, we find that the directions of the fractural pattern are almost uniform in the central High Atlas, whereby the different types of fractures are not bound to the same directions, everywhere. This fits well to the hypothesis mentioned above that an old fractural pattern was repeatedly reactivated from the Triassic onward, while the orientation of the continental stress-fields changed with time. At present, it is, however, not possible to reconstruct the details of this development.

3. Geotectonic evolution

We shall now describe the stages of the geotectonic evolution of the central High Atlas from the Triassic to Recent. Special attention will be paid to relations between tectonic and magmatic events.

The Triassic break-up of Pangaea is reflected by large scale block-faulting, strike-slip movements and by the origin of pullapart basins (MANSPEIZER 1981). It is, moreover marked by the wide-spread effusion of basaltic lavas at the Triassic/Jurassic boundary. During the early Lias, rifting proceeded forming the "Atlas Gulf" as a southwesterly embayment of the Tethys. Sedimentation was dependent upon the reactivation of a Palaeozoic or even Panafrican fractural pattern within the basement which produced wrench-faults, block rotations and horsts and grabens on a minor scale. In the Upper Lias-Dogger interval, the intensity of block-faulting increased, as shown by rapid changes of facies and thicknesses or by local unconformities (STUDER and DRESNAY 1980, SEUFERT 1986). As previously mentioned, a strip of narrow folds developed in the Imilchil area, between two major fault zones, towards the end of the Dogger (SEUFFERT 1986). Similar observations were also recorded of the western High Atlas (FROITZHEIM et al., this volume). We agree with STETS and WURSTER (1981) and MONBARON (1982) that these folds do not reflect regional compression, but probably originated at the margins of rotating blocks, while extension continued. The culmination of Jurassic block-faulting in the Upper Dogger was underlined by a second magmatic phase, with intrusions 175-165 m.y. in age (2.1). Subsidence, however, slowed down as demonstrated by the general change from marine to lagoonal sedimentation (JENNY 1984). Locally, even emersion occured (CHEVREMONT 1977).

The following interval from Malm to Eocene was a period of moderate subsidence. On both flanks of the central High Atlas, we find undisturbed, conformable sequences from the Cenomanian/Turonian limestones up to the Eocene. This is in contrast to unconformities between Cretaceous and Lower Tertiary beds which were observed in some places near the northwestern margins of the western High Atlas (FROITZHEIM 1984, FROITZHEIM et al., this volume) and of the Middle Atlas (MARTIN 1981). We assume that these unconformities do not signalize the beginning of a general inversion of the Atlas rift system in Senonian times, but might be caused by local upheaval or block rotations. We mention here that according to HERBIG (1986) most of the central High Atlas was flooded by Palaeocene and Eocene seas.

In the central High Atlas, inversion started about the end of Oligocene. According to GÖRLER et al. (this vol.), the first breccias and conglomerates with debris of the Calcareous Atlas form the base of the Ait Kandoula formation and are probably of Lower Miocene age. Locally they seal a marked relief (DOMEYER 1985). Uplift must have been associated with compressional tectonics which culminated during the Lower Miocene, since even Oligocene sediments have been deformed by the recumbent folds of the Southern Marginal Zone. We assume that the schistosity mentioned above (2.21) originated at this time. The Lower Miocene age of the main compressional deformation may also be deduced from certain unconformities at the base of the Miocene deposits along the southern margin of the central High Atlas (fig. 7).

According to GÖRLER et al. (this vol.), two stages of uplift are reflected by the sedimentary sequence of the Mio-Pliocene Ait Kandoula Formation at the southern margin of the central High Atlas. Its lower conglomerates were deposited in front and on top of the down-

Fig. 7: Unconformity between Jurassic/Cretaceous red-beds (R) and the Neogene Ait Kandoula Formation (N), jl Lias, Tv Triassic volcanics, Ts Triassic sandstones.

sliding "nappe of Toundoute". These phenomena testify to rapid upheaval. Gravity sliding was confined to the western part of the central High Atlas. Here, high summits consist of the Hercynian basement which is, however, deeply buried beneath the adjacent Tadla and Ouarzazate basins thus demonstrating a maximum of vertical displacement. After a calm period from the Serravallian to the Pliocene, which is testified by lacustrine marls, a second stage of uplift started probably in the Upper Pliocene. It is documented by thick conglomerates which contain also mega-breccias, caused by land-slides. An extremely high relief energy is, moreover, denoted by the gravels of the oldest glacis in the Ouarzazate Basin ("Villafranchian", GAUTHIER 1957) yielding boulders of 1 m Ø. Younger gravels of the Quaternary do not show boulders of that size.

Thus, inversion of the central High Atlas lasted from Oligocene to present, with a maxima in the Lower Miocene and in the Upper Pliocene. According to MICHARD (1976), it coincides exactly with the inversion of the Anti-Atlas and also of the Middle Atlas. CHOUBERT & FAURE-MURET (1962) concluded, however, that uplift of the Middle Atlas started not earlier than in Pliocene times.

Finally, it may be observed that the central High Atlas was free of volcanic activity during Neogene and Quaternary times, which is contrary to the other parts of the Atlas system of Morocco where volcanoes with alkaline lavas poor in SiO_2 are wide-spread. These volcanics vary from rhyolites to phonolites, andesites, shoshonites and even ultramafic rocks, e.g. ankaratrites (HERNANDEZ and BELLON 1985, SCHERMERHORN and GROTHE 1986). Radiometric data of these rocks cover an interval of 13.1 - 0.7 m.y.

Conclusions

The tectonic structures and the post-Palaeozoic geotectonic history, infer that the central High Atlas is a product of two major geotectonic processes, interacting with each other at certain stages in time: These are the opening of the western Tethys and of the Atlantic on one hand and West Mediterranean collisions on the other.

The first stage from the Triassic to the Eocene is dominated by the break-up of Pangaea and the opening of the oceanic basins. Upper Triassic fracturing was associated with the effusion of doleritic lavas and the formation of limited sedimentary basins filled by redbeds. During the Lower Liassic the "Atlas Gulf" originated as a southwestern embayment of the Tethys. Jurassic sedimentation was influenced by vertical and strike-slip faulting due to a reactivation of a Panafrican or Variscan fault pattern within the basement. The rapid subsidence of the Lias slowed down during the Dogger. The passage from marine to continental sedimentation (Bathonian-Malm) coincided with a culmination of block-faulting and magmatic intrusions. Both tectonic and magmatic activities ended in the Malm. A period of moderate subsidence followed, which lasted up to the Eocene.

Inversion started in the Oligocene and was probably induced by West Mediterranean collisions. Uplift and compressional deformation of the central High Atlas during the Oligocene to Lower Miocene coincide with the main phase of deformation of the internal zones of the Rif and of the Betic Cordilleras (CHOUBERT and FAURE-MURET 1962, RONDEEL and SIMON 1974, MICHARD 1976, BOUILLIN et al. 1986). This is in harmony with the fact, that the degree of horizontal shortening in the cover of the High Atlas decreases from E to W, i.e. with increasing distance from the Rif orogen. The Pliocene uplift of the central High Atlas was synchronous with the last nappe movements in the Rif (MICHARD 1976) and probably continues even today, since uplift and compressional stress are documented throughout the Quaternary (DUTOUR & FERRANDINI 1985). Therefore, it is likely to assume that both were genetically linked. It can, however, not be excluded that uplift was in relation to the Upper Cenozoic volcanism which is widespread in North Africa. Geochemical and geochronological investigations of its rocks as well as current geophysical studies in the Atlas system of Morocco (SCHWARZ and WIGGER, this vol.) may provide a better base to discuss these problems in the near future.

Acknowledgements: The authors have been very effectively supported by the Moroccan Ministère de l'Energie et des Mines, Division de la Géologie. They feel deeply indebted especially to M. Bensaid, M. Dahmani, J. Destombes, R. du Dresnay, J.-A. Jossen, M. Khalek, A. Milhi, M. Septfontaine and S. Tougent. Technical support by R. Cousino, S. Huhn, W. Jung and D. Reich (Berlin). The English version of the manuscript was reviewed by G. Lindemann (Berlin).

REFERENCES

ADAMS, A.E. (1979): Sedimentary environments and palaeogeography of the western High Atlas, Morocco, during the Middle and Late Jurassic.- Palaeogeogr., Palaeoclimat., Palaeoec., 28: 185-196, Amsterdam.

BENZAQUEN, M. (1963): La bordure septentrional de l'Atlas de Beni-Mellal. Contribution à l'étude de la région d'Elksiba.- Notes Serv. géol. Maroc, 22: 45-80, Casablanca.

BERNASCONI, R. (1983): Géologie du Haut Atlas de Rich (Maroc).- Thèse Univ. Neuchâtel, 107 S., Neuchâtel.

BINOT, F., DRESEN, G., STETS, J. & WURSTER, P. (1986): Die Tizi-n' Test-Verwerfungszone im Hohen Atlas (Marokko).- Geol. Rdsch., 75: 647-664, Stuttgart.

BOUILLIN, J.-P., DURAND-DELGA, M. & OLIVIER, Ph. (1986): Betic-Rifian and Tyrrhenian arcs: Distinctive features, genesis and development stages.- in F.A. WEZEL (ed.): The origin of arcs. Developments in tectonics, 21: 281-304, Amsterdam (Elsevier).

BRECHBÜHLER, Y.-A. (1984): Etude structurale et géologique du Haut Atlas calcaire entre le Jebel Ayachi et Rich (Maroc).- Thèse Univ. Neuchâtel, 128 S., Neuchâtel.

BREDE, R. (1987): Strukturelle Entwicklung des zentralen Hohen Atlas nordwestlich von Errachidia (Marokko).- Diss. Freie Univ. Berlin (in preparation).

BREDE, R. & HEINITZ, W. (1986): Zum Bau des Hohen Atlas zwischen Errachidia und Midelt (Marokko).- Berliner geowiss. Abh., (A), 66: 433-454, Berlin.

BURGESS, C.J. & LEE, C.W. (1978): The development of a Lower Jurassic carbonate tidal flat, Central High Atlas, Morocco. 1. Sedimentary history.- J. Sed. Petrol., 48, 777-794, Tulsa.

CARIOU, E., CONTINI, D., DOMMERGUES, J.-L., ENAY, R., GEYSSANT, J.R., MANGOLD, Ch. & THIERRY, J. (1985): Biogéography des ammonites et évolution structurale de la Téthys au cours du Jurassique.- Bull. Soc. géol. France, (VIII), 1:679-697, Paris.

CHEVREMONT, P. (1977): Les roches éruptives basiques des boutonnières de Tassent et Tasraft et leurs indices métallifères (Haut Atlas central, Maroc).- C.R. Acad. Sci. Paris, (D), 284: 1493-1496, Paris.

CHOUBERT, G. & FAURE-MURET, A. (1962): Evolution du domaine atlasique marocain depuis les temps paléozoiques.- Livre Mém. P. Fallot, 1960, 1: 447-527, Paris (Soc. géol. France).

COURBOULEIX, S., DELPONT, G. & DESTEUCQ, CH. (1981): Un grand décrochement est-ouest au nord du Maroc à l'origine des structures plissées atlasiques. Arguments géologiques et expérimentaux.- Bull. Soc. géol. France, (VII), 23: 33-43, Paris.

DESTOMBES, J., JENNY, J. & YAZIDI, A. (1983): L'avant-pays présaharien (Anti-Atlas central) et la partie méridional de la chaîne hercynienne du Maroc (Haute Atlas central et Jebilat).- Le Maroc et l'orogène paléozoique: Symposium de Rabat 1983, Livret-guide de l'excursion B5, 1273 p., Rabat.

DEWEY, J.F., PITMAN, W.C., RYAN, W.B. & BONIN, J. (1973): Plate tectonics and the evolution of the Alpine system.- Geol. Soc. Amer. Bull., 84: 3137-3180.

DRESNAY, R. du (1964): Les discontinuités de sédimentation pendant le Jurassique, dans la partie orientale du domaine Atlasique Marocain, leurs conséquences stratigraphiques et leurs relations avec l'orogenèse Atlasique.- Colloque du Jurassique Luxembourg 1962, C.R. et Mém., 899-912, Luxembourg.

\- (1971a): Relations paléogéographiques entre sillons atlasiques (Haut et Moyen-Atlas) et plate-forme des Hauts-Plateaux (Maroc oriental) pendant l'Aalénien et le Bajocien.- Colloque du Jurassique, Luxembourg 1967, Mém. B.R.G.M. France, 75: 147-161, Paris.

\- (1971b): Extension et développement des phénomènes récifaux Jurassiques dans le domaine atlasique marocain, particulièrement au Lias moyen.- Bull. Soc. géol. France, (VII), 13: 46-56, Paris.

DUBAR, G. (1962): Notes sur la paléogéographie du Lias marocain (Domaine atlasique).- Livre mém., P. Fallot, (Mém. Soc. géol. France h, série), 1: 529-544, Paris.

DUBOURDIEU, G. (1962): Dynamique wegenérienne de l'Afrique du Nord.- Livre Mém. Professeur Paul Fallot, 1: 627-644, Paris (Soc. géol. France).

DUTOUR, A. & FERRANDINI, J. (1985): Nouvelles observations néotectoniques dans le Haut Atlas de Marrakech et le Haouz central (Maroc). Apports sur l'évolution récente d'un segment du bâti atlasique.- Rev. Géol. dyn. Géogr. phys., 26(5): 285-297, Paris.

ELMI, S. (1984): Jurassic paleogeography of North Africa and its implications.- Internat. Sympos. Jurassic Stratigr., Erlangen, September 1-8: 629-639, Kopenhagen (Geol. Survey Danmark).

EVANS, I. & KENDALL, C.G.St. (1977): An interpretation of the depositional setting of some deep-water Jurassic carbonates of the Central High Atlas Mountains, Morocco.- in H.E. COOK & P. ENOS (eds.): Deep-water carbonate environments, SEPM Spec. Publ., 25: 249-261.

FAUGERE, E., BRUN, J.-P. & DRIESSCHE, J. van den (1986): Bassins asymétriques en extension pure et en décrochement: modèles expérimentaux.- Bull. Centres Rech. Explor.-Prod., Elf-Aquitaine, 10: 13-21, Pau.

FERRANDINI, J. & LE MARREC, A. (1982): La couverture jurassique à paléogène du Haut Atlas de Marrakech est allochthone dans la "zone des cuvettes" d'Ait Ourir (Maroc).- C.R. Acad. Sci. Paris, (II), 295: 813-816, Paris.

FERRANDINI, M., PHILIP, J., BABINOT, J.-F., FERRANDINI, J. & TRONCHETTI, G. (1985): La plate-forme carbonatée du Cénomano-Turonien de la région d'Erfoud-Errachidia (Sud-Est marocain): stratigraphie et paléoenvironnements.- Bull. Soc. géol. France,(8), 1: 559-564, Paris.

FROITZHEIM, N. (1984): Oberkretazische Vertikaltektonik im Hohen Atlas SW' von Marrakech/Marokko-Rekonstruktion eines Bewegungsablaufs im Frühstadium der Atlas-Orogenese.- N. JB. Geol. Paläont. Mh., 1984: 463-471, Stuttgart.

GAUTHIER, H. (1957): Contribution à l'étude géologique des formations post-liasiques du bassin du Dadès et du Haut-Todra (Maroc méridional).- Notes Mém. Serv. géol. Maroc, 119: 212 S., Casablanca.

GÖRLER, K. & ZUCHT, M. (1986): Stratigraphie und Tektonik des kontinentalen Neogens im Süden des zentralen Hohen Atlas, Provinz Ouarzazate (Marokko).- Berl. geowiss. Abh., (A), 66: 471-494, Berlin.

HAILWOOD, E.A. & MITCHEL, J.G. (1971): Palaeomagnetic and radiometric dating results from Jurassic intrusions in South Morocco.- Geophys. J. R. astr. Soc., 24: 351-364, London.

HARLAND, W.B., COX, A.V., LLEWELLYN, P.G., PICKTON, C.A. G., SMITH, A. G. & WALTERS, R. (1982): A geological time scale.- Cambridge Earth Sci. Ser.: 131 p., Cambridge.

HELMDACH, F.F. (1986): Zur Alterseinstufung des höchsten marinen Tertiärs im Bereich der Südatlas-Randzone (Provinz Ouarzazate, Marokko).- Berl. geowiss. Abh., (A), 66: 495-502, Berlin.

HERBIG, H.-G. (1986): Lithostratigraphisch fazielle Untersuchungen im marinen Alttertiär südlich des zentralen Hohen Atlas (Südmarokko).- Berl. geowiss. Abh., (A), 66: 343-380, Berlin.

HERNANDEZ, J. & BELLON, H. (1985): Chronologie K-Ar du volcanisme miocène du Rif oriental (Maroc): implications tectoniques et magmatologiques.- Revue Géol. dyn. Géogr. phys., 26: 85-94, Paris

HÜSSNER, H. (1985): Jurassische Karbonate des westlichen Hohen Atlas (Marokko): Mikrofaziesanalyse und plattentektonischer Rahmen.- Facies, 12: 141-218, Erlangen.

JACOBSHAGEN, V. (1986): Zur Baugeschichte des zentralen Hohen Atlas (Marokko).- Berl. geowiss. Abh., (A), 66: 455-470.

JENNY, J. (1983): Les décrochements de l'Atlas de Demnat (Haut Atlas central, Maroc): prolongation orientale de la zone de décrochement du Tizi-n-Test et clef de la compréhension de la tectonique atlasique.- Eclogae geol. Helv., 76: 243-251, Basel.

- (1984): Dynamique de la phase tectonique synsédimentaire du Jurassique moyen dans le Haut Atlas central (Maroc).- Eclogae geol. Helv., 77: 143-152, Basel.

JENNY, J. & LE MARREC, A. (1980): Mise en évidence d'une nappe à la limite meridionale du domaine hercynien dans la boutonnière d'Ait Tamlil (Haut Atlas central, Maroc).- Eclogae geol. Helv., 73: 681-696, Basel.

JENNY, J., LE MARREC, A. & MONBARON, M. (1981): Les couches rouges du Jurassique moyen du Haut Atlas central (Maroc): correlations lithostratigraphiques, éléments de datation et cadre-tectono-sédimentaire.- Bull. Soc. géol. France, (VII), 23: 627-639, Paris.

LAVILLE, E. (1980): Tectonique et microtectonique d'une partie du versant sud du Haut Atlas marocain (boutonnière de Skoura, nappe de Toundoute).- Notes Serv. géol. Maroc, 41: 81-184, Rabat.

LAVILLE, E. & HARMAND, C. (1982): Evolution magmatique et tectonique du bassin intracontinental mésozoique du Haut Atlas (Maroc): un modèle de mise en place synsédimentaire de massifs "anorogeniques" liés à des décrochements.- Bull. Soc. géol. France, (VII), 24: 213-227, Paris.

LAVILLE, E., LE SAGE, J.C. & SEGURET, M. (1977): Géométrie, cinématique (dynamique) de la tectonique atlasique sur le versant sud du Haut Atlas marocain. Aperçu sur les tectoniques hercyniennes et tardi-hercyniennes.- Bull. Soc. géol. France, (7), 19: 527-539, Paris.

MANSPEIZER, W. (1982): Triassic-Liassic basins and climate of the Atlantic passive margins.- Geol. Rdsch., 71: 895-917, Stuttgart.

MANSPEIZER, W., PUFFER, J.H. & COUSMINER, H.L. (1978): Separation of Morocco and eastern North America: A Triassic-Liassic stratigraphic record.- Geol. Soc. Amer. Bull., 89: 901-920.

MARTIN, J. (1980/81): Le Moyen Atlas central. Etude géomorphologique.- Notes Mém. Serv. géol. Maroc, 258(1980): 10-91, and 258 bis: 268-416, Rabat.

MATTAUER, M., PROUST, F. & TAPPONIER, P. (1972): Major strike slip fault of Late Hercynian age in Morocco.- Nature, 237, 160-162.

MATTAUER, M., TAPPONIER, P. & PROUST, F. (1977): Sur les mécanismes de formation des chaînes intracontinentales. L'exemple des chaînes atlasiques du Maroc.- Bull. Soc. géol. France, (7), 19: 521-526, Paris.

MATTIS, A.F. (1977): Nonmarine Triassic sedimentation, Central High Atlas Mountains, Morocco.- J. sediment. Petrol., 47: 107-119, Tulsa.

MICHARD, A. (1976): Eléments de géologie marocaine.- Notes Mém. Serv.

géol., 252: 1-408, Rabat.

MICHARD, A., WESTPHAL, M., BOSSERT, A. & HAMZEH, A. (1975): Tectonique de bloc, dans le socle atlaso-mésétien du Maroc: une nouvelle interprétation des données géologiques et paléomagnétiques.- Earth planet. Sci. Lett., 24: 363-368, Amsterdam.

MICHARD, A., YAZIDI, A., BENZIANE, F., HOLLARD, H. & WILLEFERT, S. (1982): Foreland thrusts and olistromes on the pre-Saharan margin of the Variscan orogen, Morocco.- Geology, 10: 253-256, London.

MONBARON, M. (1981): Sédimentation, tectonique synsédimentaire et magmatisme basique: l'évolution paléogéographique et structurale de l'Atlas de Beni Mellal (Maroc) au cours du Mésozoique; ses incidences sur la tectonique tertiaire.- Eclogae geol. Helv., 74: 625-638, Basel.

- (1982): Précisions sur la chronologie de la tectogenèse atlasique: exemple du domaine atlasique mésogéen du Maroc.- C.R. Acad. Sci. Paris, (II), 294: 883-886, Paris.

MONBARON, M. & JUST, E. (1980): Les roches magmatiques basiques de la région de Tagalft (Haut Atlas central): relations spatiales et génétiques.- Mines, Géol. Energie, 48: 45-50, Rabat.

ODIN, G.S. (1982): The Phanerozoic time scale revisited.- Episodes, 1982(3): 3-9, Ottawa.

PROUST, F., PETIT, J.-P. & TAPPONIER, P. (1977): L'accident du Tizi n'Test et le rôle des décrochements dans la tectonique du Haut Atlas occidental (Maroc).- Bull. Soc. géol. France, (VII), 19: 541-551, Paris.

ROCH, E. (1939): Description géologique des montagnes à l'Est de Marrakech (Maroc).- Notes Mém. Serv. Mines Carte géol. Maroc, 80: 438 S., Casablanca.

RONDEEL, H.E. & SIMON, O.J. (1974): Betic cordilleras.- in A.M. SPENCER (ed.): Mesozoic-Cenozoic orogenic belts. Data for orogenic studies.- Geol. Soc. London Spec. Publ., 4: 23-35, Edinburgh.

RUSSO, P. & RUSSO, L. (1934): Le grand accident sud-atlasien.- Bull. Soc. géol. France, (5), 4: 375-384, Paris.

SALVAN, H.M. (1984): Les formations évaporitiques du Trias marocain. Problèmes stratigraphiques, paléogeographiques et paléoclimatologiques. Quelques réflexions.- Rev. Géol. dyn. Géogr. phys., 25: 187-203, Paris

SCHAER, J.P. & PERSOZ, F. (1976): Aspects structuraux et pétrographiques du Haut Atlas calcaire de Midelt (Maroc).- Bull. Soc. géol. France, (VII), 18: 1239-1250, Paris.

SCHERMERHORN, L.J.G. & GROTHE, Ch. (1986): Vulkanismus im Anti-Atlas (Marokko).- Berl. geowiss. Abh., (A), 66, 333-342, Berlin.

SEPTFONTAINE, M. (1984): Biozonation (à l'aide de foraminifères imperforés) de la plate-forme interne carbonatée liasique du Haut Atlas (Maroc).- Rev. Micropaléont., 27: 9-229, Paris.

SEUFERT, G. (1986): Zur strukturellen Entwicklung des zentralen Hohen Atlas: Die Region zwischen Boumalne du Dadès, Tinerhir und Imilchil (Südmarokko).- Diss. Freie Univ. Berlin: 143 S. (microfiches), Berlin.

STANLEY, R.G. (1981): Middle Jurassic shoaling of the Central High Atlas sea near Rich, Morocco.- J. Sed. Petrol., 51: 895-907, Tulsa.

STETS, J. & WURSTER, P. (1982): Atlas and Atlantic-structural relations.- In U. von RAD et al. (eds.): Geology of the Northwest African continental margin: 69-85, Berlin etc. (Springer).

STUDER, M. (1980): Tectonique et pétrographie des roches sédimentaires, éruptives et métamorphiques de la région de Tounfite-Tirrhist (Haut Atlas central, Maroc).- Thèse Univ. Neuchâtel: 95 S., Neuchâtel.

STUDER, M. & DRESNAY R. du (1980): Déformations synsédimentaires en compression pendant le Lias supérieur et le Dogger, au Tizi n'Irhil (Haut Atlas central de Midelt, Maroc).- Bull. Soc. géol. France, (VII), 22: 391-397, Paris.

TISSERANT, D., THUIZAT, R. & AGARD, J. (1976): Données géochronologiques sur le complexe de roches alcalines du Tamazeght (Haut Atlas de Midelt), Maroc.- Bull. B.R.G.M., 1976(3): 279-283, Paris.

ZYLKA, R. (1987): Die südliche Randzone des Hohen Atlas zwischen Tounfoute und Goulmima (Marokko).- Ph.D. Thesis Freie Univ. Berlin, in preparation.

ZYLKA, R. & JACOBSHAGEN, V. (1986): Zur strukturellen Entwicklung am Südrand des zentralen Hohen Atlas (Marokko).- Berl. geowiss. Abh., (A), 66: 415-432, Berlin.

STRUCTURAL EVOLUTION OF THE SOUTHERN AND NORTHERN CENTRAL HIGH ATLAS IN PALEOGENE AND MIO-PLIOCENE TIMES

Cyril Fraissinet*, El Mustapha Zouine**, Jean-Luc Morel*, André Poisson*, Jean Andrieux* et Anne Faure-Muret*

*Laboratoire de Géologie Historique, Structurale et Appliquée, Université de Paris-Sud, Bat. 504, 91405 ORSAY Cedex, France.

**Département des Sciences de la Terre, Ecole Normale Supérieure Rabat - Souissi, BP 773, RABAT-AGDAL, Maroc.

Abstract - The High Atlas is an intracontinental folded belt which evolved from narrow subsiding basins in the Early Mesozoic times; folds and thrusts with a double vergence formed since the Upper Cretaceous till the Pleistocene.

The southern and northern borders of the Central High Atlas show a different scheme of development. In the South, during the Cenozoic times, the effects of a whole compressive state appear as an alternation of paroxystic phases of folding and thrusting, and of thick sedimentation in narrow, elongated parallel to the chain, synformal basins; four tectonic phases, each defined by a marked unconformity, are described during the Oligocene-Pliocene time span. In the North, the main deformations are younger than the Mio-Pliocene.

For the later events, the whole submeridian shortening suffers some local changes in orientation, due to the reactivation of ancient faults.

Résumé - Le Haut Atlas est une chaîne intracontinentale qui s'est développée à l'emplacement de bassins étroits subsidents au Mésozoïque inférieur; les plis et chevauchements, doublement déversés, se sont formés depuis le Crétacé terminal jusqu'au Pléistocène.

Les bordures sud et nord du Haut Atlas central ont évolué différemment. Au Sud, le régime général compressif au Tertiaire est responsable d'une alternance de paroxysmes tectoniques avec plis et chevauchements, et de périodes de sédimentation dans des bassins étroits en synclinaux parallèles à la chaîne; quatre phases tectoniques, soulignées chacune par une discordance, ont été définies entre l'Oligocène et le Pliocène. Au Nord, les déformations majeures sont plus récentes que le Mio-Pliocène et, parallèlement, les seuls dépôts syntectoniques sont mio-pliocènes ou plus récents.

Pour les évènements tectoniques les plus jeunes, la compression d'ensemble subméridienne se traduit localement par des directions de raccourcissement qui peuvent varier, en raison de la réactivation de failles plus anciennes.

I - INTRODUCTION

The High Atlas is a typical intracontinental symmetrical mountain belt which developed during the Meso-Cenozoic times between the north-western border of the West-African Platform and two stable blocks with a Variscan basement, the Moroccan Meseta and the "High Plateaux"; these blocks are separated by the Middle Atlas belt. The High Atlas reaches 4165 m (J. Toubkal) and its mean elevation is about

3000 m; its width varies from 50 km to 120 km, in contrast to its 800 km lenght in Morocco. This narrowness is related to the early history of the belt: during the Early Mesozoic, narrow basins, bordered by high angle faults, developed on the em - placement of the High Atlas (Mattauer et al., 1977; Stets and Wurster, 1982); these basins were subsequently compressed so that the previously normal faults now appear as steep reverse faults in the basement; the infilling of the basins is folded as cover folds and is thrusted towards the stable borders, thanks to a detachment with the Upper Triassic argilites.

At its western end, the High Atlas joins the Essaouira Basin, a part of the passive margin of Morocco, which began to differentiate in the Triassic (Hinz et al., 1982); this western termination of the High Atlas is the less deformed and the lowest part of the belt. South of Marrakech, the High Atlas is called "Massif Ancien". Here the belt is very narrow because it is cut across by a system of strike-slip faults, the Tizi n'Test Fault Zone (TTFZ), active since the Upper Paleozoic (Mattauer et al., 1977; Proust et al., 1977). During the Mesozoic times, the TTFZ acted as a structural high. Later on, this zone suffered typical block tectonics.

Whereas the Central High Atlas keeps the same orientation (N060) as the TTFZ, the East High Atlas, eastwards from Midelt, is characterized by east-west structures. The High Atlas overrides the African stable platform (Ouarzazate basin) by means of flat south-verging thrusts which steepen in the basement; Laville et al. (1977) have shown that these compressive structures initiated during Upper Cretaceous times and were still active during the Mio-Pliocene.

At the northern border, the situation is more complicated. In fact, East of Beni Mellal , the High Atlas links with the folded Middle Atlas and here the frontal thrust has a N050-N060 strike; westwards, this front is orientated at N080-N090 and seems to damp down North of the Jebilet, a part of the Moroccan Meseta. Although some authors argued about Eo-Oligocene tectonic events (Benzaquen, 1963; Huvelin, 1973), it seems that the most important compressive movements occured during the Pliocene (Montbaron, 1982); the High Atlas northern border is still shortening now (Coisy and Frogneux, 1980).

In the axial zone, most of the faults, as the TTFZ, have a long history which can be traced back until the Triassic. Choubert and Faure-Muret (1960-1962) and Du Dresnay (1975) emphasized the intra-Jurassic movements. As pointed out by Laville (1985), the Mesozoic development of the High Atlas basins was greatly influenced by the strike-slip faults which cut across the southern border of the Variscan orogen (Arthaud and Matte, 1977). These faults are distributed along three main directions: E-W and NE-SW, with a dextral sense of offset during the Late Paleozoic, and WNW-ESE with a probable sinistral sense at the same time.

During the rifting which preceded the opening of the Atlantic Ocean, both fault systems guided the subsidence of ENE-WSW basins which were filled by detritic and evaporitic sediments and also some basaltic floods. Structural evidences reported by Laville (1981) confirm that extensional conditions prevailed at this time, the direction of extension being N160.

During the Early-Middle Jurassic the subsidence was still very active: carbonate platforms developed on both sides of the High Atlas while a deep furrow appeared on the northern part of the axial zone (Mattauer et al., 1977). For this deeper part, Laville (1985) argues for the coeval formation of S-shaped basins, anticlinal ridges (mainly inferred from progressive unconformities) and gabbro-dioritic intrusions (elongated in a mean N060-N070 direction). The shifts between the intrusives bodies, the basins and the ridges (Fig. 1A) are explained by sinistral movements on the previous faults (like the TTFZ or its eastward extension) and are consistent with a general NNW-SSE extension and an overall N060-N070 compression. The importance of these synsedimentary tectonic and plutonic events is well demonstrated (Laville, 1985), and they reached their climax during the Middle Jurassic (Jenny et al., 1981). But the model proposed by Laville fails to explain every structural pattern, for instance that of the High Atlas northern border: previously, in that re-

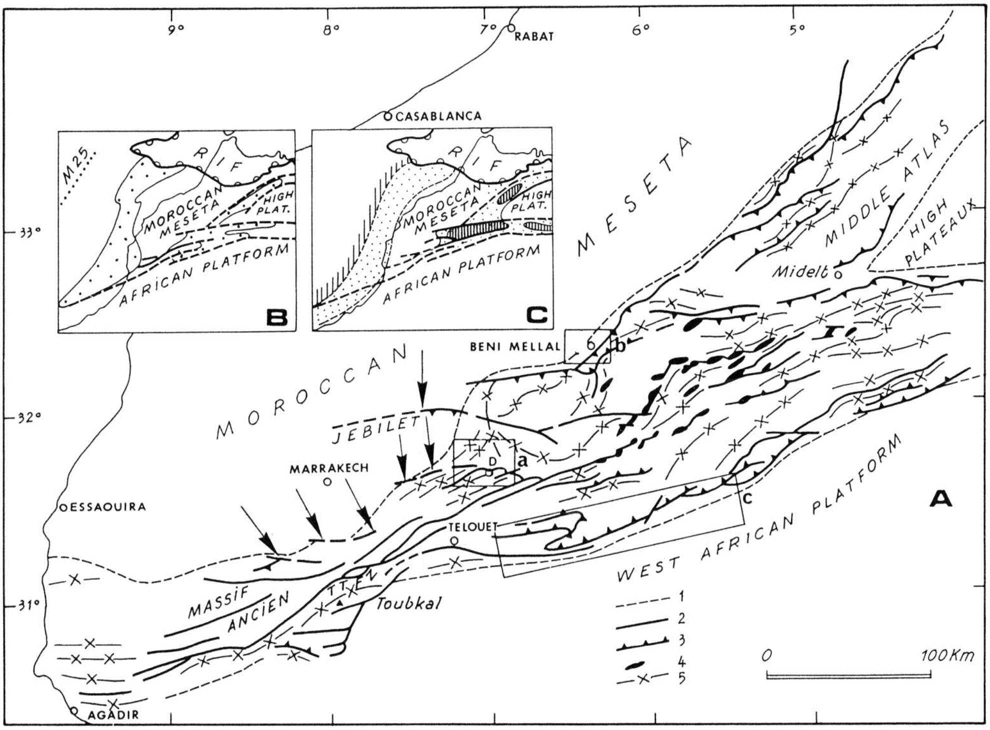

Fig. 1 - Structural setting of the High Atlas. A : Simplified tectonic map. 1:limit of the folded belt. 2:faults. 3:reverse faults and thrusts. 4:main jurassic intrusives. 5:synclines. TTFZ: Tizi n'Test Fault Zone. D:Demnate. arrows:recent shortening directions. B and C : Zones of high subsidence during the Triassic (B , dots) and the Middle Jurassic (C , deep basins hatched, carbonate platform dotted)

gion, WNW-ESE to NW-SE compressions of Jurassic age have been reported (Laville, 1978; Gratier, 1981). Possibly because of the complexity at the junction of the High Atlas and Middle Atlas, the state of stress could have been different. One of the purposes of our work is to define precisely the Neogene structures and dynamics in order to understand the Jurassic ones better.

At the time when the Atlantic Ocean floor began to spread (Upper Jurassic, Olivet et al., 1984), the mobility of the High Atlas ceased. After the Turonian, a tectonic inversion began (Stets and Wurster, 1982) and this allowed the Paleozoic basement to outcrop. Slightly later the southern thrusts started to develop in the Maastrichtian (Laville et al., 1977) and continued to advance until recently. The southern border of the High Atlas and the Subatlasic zone display numerous unconformities (Laville, 1975); here we describe the relationships between the tectonic events and the sedimentation during the Oligocene and Neogene times; then we try to characterize the latest movements in order to decipher the earlier ones.

II - THE SOUTHERN HIGH ATLAS BORDER

A - THE FORMATIONS (Fig. 8)

For structural studies in continental domains one of the main critical point is the obtention of a detailed chronology. In the High Atlasic domain an important point under discussion concerns the existence, or not, of a continuous deformation during Neogen time.

In the High Atlas of Boumalne, the existence of unconformities which separate numerous formations is in favour of distinct tectonic phases. This does not mean that no deformations could have occured during the sedimentary filling up of the basins due, for instance, to subsidence, intra-formational deformations, slumpings... It actually means that there was distinct sedimentary sequences corresponding to distinct periods of sedimentation separated by periods of more active uplift and deformation. These deformations occured preferentially along ancient faults where they were more intensively registered.

We shall describe here different Oligocene and Neogen formations corresponding to the preceeding criteria: the existence of unconformities joined to the general sedimentary characteristics. We have retained three North-South transverse sections offered by the three main rivers (from East to West: O. Dades, O. M'Goun and O. Tabia-Toundout). The chronology is based on ancient (Gauthier, 1957) and recent biostratigraphical studies (Görler and Zucht, 1986).

1 - THE DADES GEOTRAVERSE

a - The Aït el Arbi formation

This formation was named by Choubert (1946) and studied later by Gauthier (1957). It conformably overlies the Eocene limestones and represents the first continental deposits. It includes clays, sandstones and conglomerates.Pebbles are mainly coming from the Anti-Atlas basement and not from the calcareous High Atlas. Along the northern limb of J. Imlin anticlinorium (Fig. 3A), the Aït el Arbi formation is about 600 m thick.

Gauthier (1957) has attributed an Upper Eocene age to the lowermost part of this formation, on the basis of regional considerations and pelecypod faunas. Recently these levels were more precisely dated of the Upper Eocene-Oligocene limit (Mohr and Fechner, 1986; Görler and Zucht, 1986; Helmdach, 1986; Herbig,1986).

b - The Aït Ouglif formation (Fig. 3A and B)

This new formation lies on top of Liassic limestones (Fig.4). These limestones are interpreted as tectonic slices thrusted onto the Aït el Arbi formation (Fig. 3A). On the section B (Fig. 3), the Aït Ouglif formation unconformably overlies the Lower Cretaceous and the Cenomano-Turonian beds.

The Aït Ouglif formation presents many sedimentary similarities with the Aït el Arbi formation: clays, clayed sandstones and pinkish conglomerates (more or less coarse). But the presence in the Aït Ouglif conglomerates of limestone pebbles from the High Atlas Liassic limestones is striking. The change in the nature of the detritic input is interpreted as indicative of a tectonic event, yet impos-

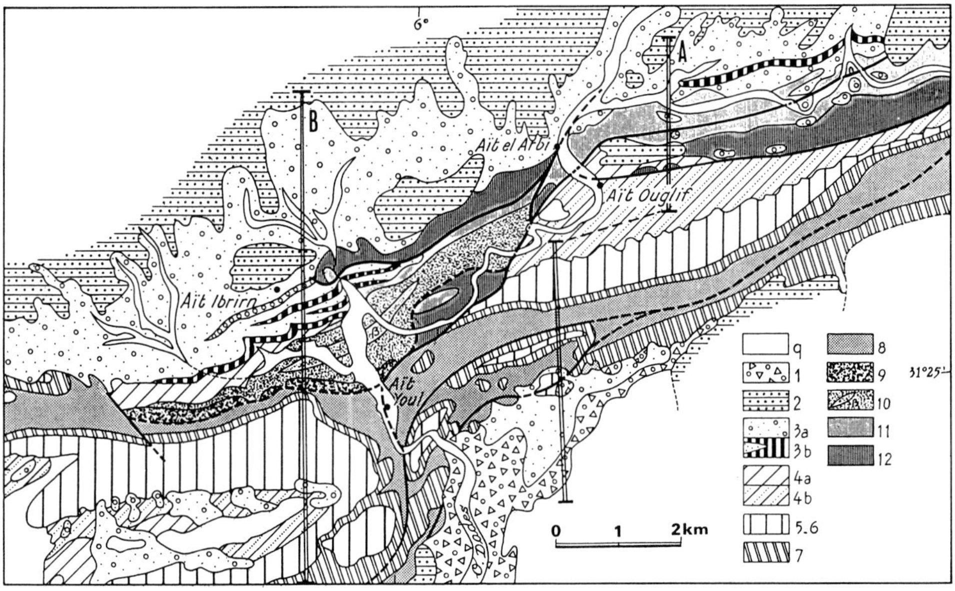

Fig. 2 - Geological map of the Dades geotraverse area. q: Quaternary. 1: Boumalne formation (conglomerates, Villafranchian?). 2: Ait Seddrat fm (Pliocene ?). 3a: Tamassint fm. (Mio-Pliocene). 3b: Aït Ibrirn fm (Lower Miocene). 4a: Aït Ouglif fm (Upper Oligocene). 4b: Aït el Arbi fm (Oligocene ?). 5&6: Thanetian , Ypresian and Lutetian limestones. 7: Dano-Montian limestones. 8: Upper Cretaceous formation. 9: Cenomano-Turonian limestones. 10: Lower Cretaceous red formations. 11: Upper Lias. 12: Middle Lias.

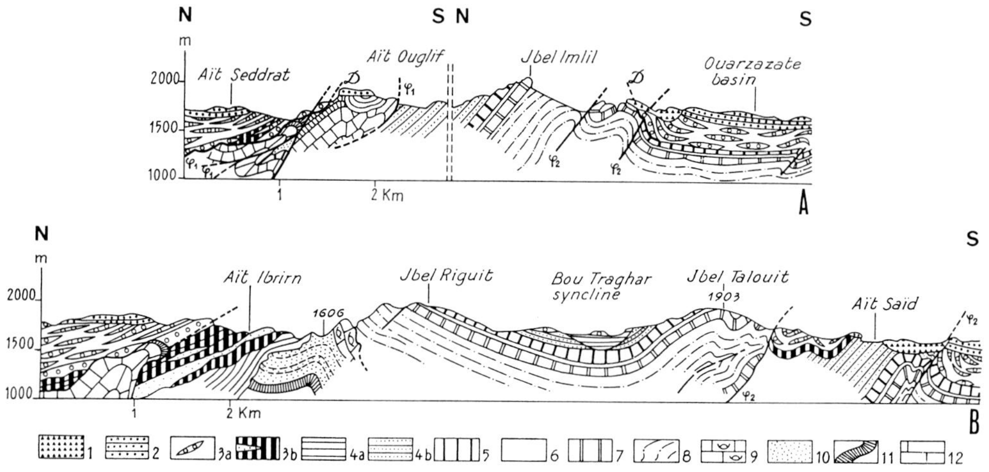

Fig. 3 - Dades geotraverse. Representative general profiles. (same caption as fig.2)

Fig. 4 - Panorama behind the Aït Ouglif douar showing the structural relationship between the Aït el Arbi formation, the Liassic limestones, the Aït Ouglif formation and the unconformable upper conglomeratic formation.

sible to define precisely, but taking place during the Middle Oligocene.

Another interpretation (Jacobshagen, 1986) suggested that the Liassic limestones could be olistolithes within the Aït el Arbi formation.

c - The Aït Ibrirn formation

This formation was named by Gauthier (1957). It includes lacustrine limestones (with traces of roots at the base), conglomerates, marls and gypsiferous lenses from place to place. The pebbles exhibit various affinities (Mesozoic and basement). The Aït Ibrirn formation unconformably overlies various previously deformed and more or less eroded reliefs. The basal unconformity is well marked here but it becomes weaker westwards.

On the basis of palaeogeographical and climatic arguments, Gauthier (1957) attributed an Oligocene age to this formation. Its gasteropod faunas were recently revised; they present many similarities with the lowermost Aït Kandoula faunas (Görler and Zucht, 1986). The unconformity between it and the overlying Tamassint formation is better exposed farther West in the basin.

d - The Tamassint formation

This formation is equivalent to the "Miocene formation" (Gauthier,1957). It is composed of red conglomerates and sandstones. Coarse lenticular conglomerates at the base contain pebbles of various origin: High Atlas Liassic limestones, basement rocks and often Eocene limestones. According to Görler and Zucht (1986), we consider that the deposits are of alluvial fans and braided river type.

This formation can be correlated with the Aït Kandoula formation of Middle-Upper Miocene age (Görler and Zucht, 1986). It seems to conformably overlie the Aït Ibrirn formation, but it clearly seals the reverse fault between the Liassic limestones and the Aït Ibrirn deposits near Sidi Bou Ahmed (Fig.2 and Fig.3B). In the South, it unconformably again overlies folded and sometimes faulted Cretaceous-Eocene beds. Along the Northern boundary of the Ouarzazate basin a lateral equivalent of this formation is flexured and folded. These structures can be observed, for instance, east of Boumalne (Fig. 2 and 3A).

In the Aït Seddrat basin, the Tamassint formation becomes vertical along the Dades fault.

e - The Aït Seddrat formation

This formation corresponds to the coarse conglomerates that unconformably overly the Tamassint formation (Fig. 5) previously deformed. We would retain a Pliocene age for this formation as a working hypothesis.

f - The Boumalne conglomerates

These conglomerates are well exposed in the Dades Valley. They appear to have been deposited in an ancient valley which had been excavated during a period of active uplift of the northern margin of the Ouarzazate basin.

2 - THE M'GOUN GEOTRAVERSE

The M'Goun geotraverse is located in the western part of the Boumalne area. The cross sections (Fig.8) show two main tectonic units separated by more or less important faults (reverse faults or thrusts).

Due to the lack of precise stratigraphic data (excepted for the Hadida formation) the correlations with the Boumalne and Toundout areas remain approximative. We have used similarities of facies and the existence of unconformities.

a - The Aznag section (Fig.6) is localized along the normal limb of the J. Tazilna anticlinorium. In the lower part of the section, the red formation conformably overlies the Eocene deposits. It is composed of conglomerates (more or less clayey), of increasing size towards the West. Due to this conformity and facies similarities, this formation could be correlated with the Aït el Arbi formation. The unconformity between it and the overlying formation (a lateral equivalent of the Tamassint formation) is in good agreement with this hypothesis. We have to notice the lack of the Aït Ouglif and Aït Ibrirn formations.

b - The Aït Saïd section is located along the northern side of the Aït Saïd anticlinorium.

- The Hadida formation was first described in this region (Gauthier, 1957; Görler and Zucht, 1986). It conformably overlies the Eocene limestones and is composed of continental red deposits. It may be correlated with the Aït el Arbi formation (Oligocene) (Görler and Zucht, 1986).

- Southern, the Ifar formation corresponds to sandstones and conglomerates containing "spathic limestones" (Liassic?) pebbles (Gauthier, 1957). This formation could be correlated either with the Hadida-Aït el Arbi formations or with the Aït Ouglif formation.

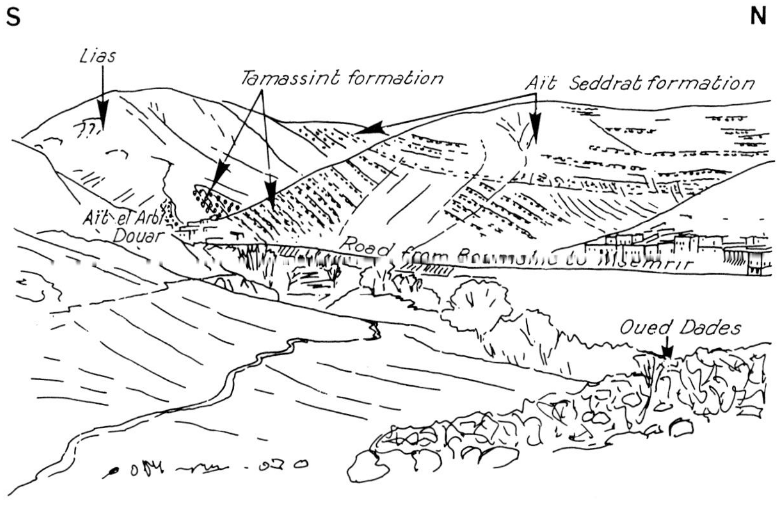

Fig. 5 - Panorama of the Dades valley near Aït el Arbi showing the unconformable Aït Seddrat formation above the Tamassint formation.

- The conglomerates of lacustrine limestones.

They directly overlie the preceding formations with a cartographic unconformity.This formation may be followed from the M'Goun valley towards the East in the direction of the Dades valley. It has been mapped as "lacustrine limestones" (1/200.000 J. Sahro-Dades). Actually it corresponds to a thick bed of conglomerate or breciated lacustrine limestones. The reworking does not seem very important in the western part of the basin (badly sorted breccias with angular blocks), while it is noticeable in the eastern end (rounded pebbles). This formation is considered as an equivalent to the Aït Ibrirn formation. The genesis of this brecciated and conglomeratic formation is probably the result of a tectonic instability prior to the deposition of the overlying Aït Saïd conglomerates.

We noticed also the lack of an equivalent to the Aït Ouglif formation.

- The Aït Saïd and upper conglomerates formations.

The Aït Saïd formation is composed of red sandstones and conglomerates. The detritals came from the High Atlas. We consider that this formation may be correlated with the Tamassint (in agreement with Gauthier, 1957).

The Upper conglomerates unconformably overlie the Aït Saïd formation. They are not highly deformed and seem to cover a preexisting topography. They occupy a position equivalent to the Aït Seddrat conglomerates.

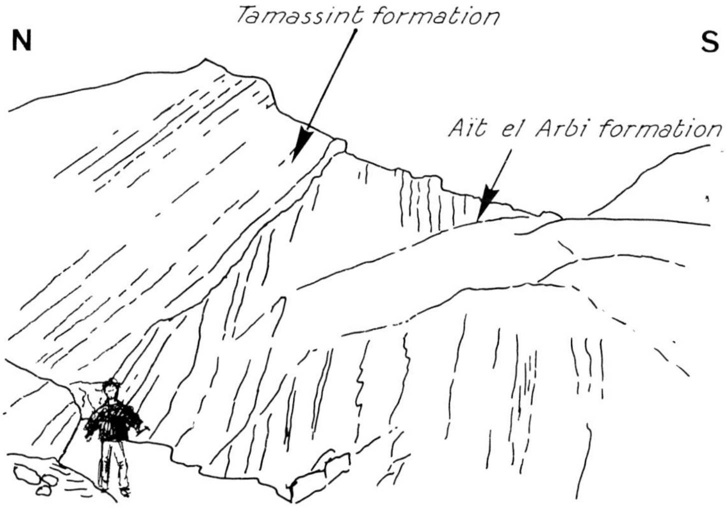

Fig. 6 - Panorama of the Aznag unconformity between the Aït el Arbi and the Tamassint formations.

3 - THE AÏT KANDOULA-TOUNDOUT-AMEKCHOUD GEOTRAVERSE

Four different sections are distinguished along this geotraverse.

a - The Aït Kandoula basin has been studied in detail and its litho-and biostratigraphy is one of the best known (Görler and Zucht, 1986). The Aït Kandoula formation may be used as a guide for Neogene stratigraphic correlations along the High Atlas southern border.

The Aït Kandoula formation, as defined by Görler and Zucht (1986), can be subdivided into four members (if we separate the unconformable highest conglome-

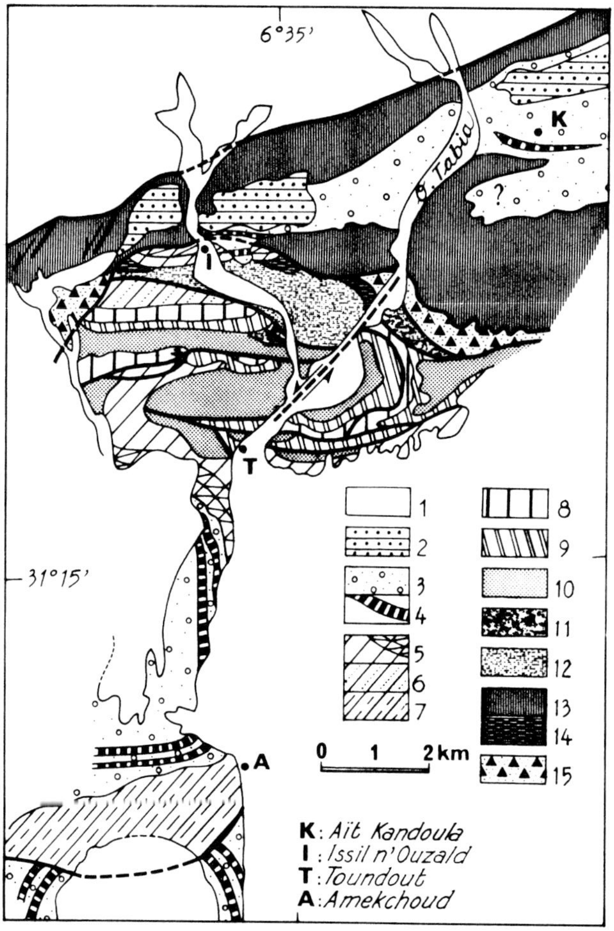

Fig. 7 - Schematic geological map of Toundout-Amekchoud area (after Laville, 1975 and a new field survey). 1: Quaternary. 2: Aït Seddrat formation. 3: Aït Kantoula, conglomeratic fm. 4: Aït Kantoula, lacustrine fm. 5: Issil n'Ouzald fm. 6: Aït el Arbi fm. 7: Amekchoud lower fm. 8: Ypresian and Lutetian limestones. 9: Dano-Montian and Thanetian limestones. 10: Upper Cretaceous red fm. 11: Cenomano-Turonian. 12: Lower Cretaceous. 13: "autochthonous" Liassic limestones. 14: Liassic limestones of the Toundout structure. 15: Trias.

rates probably Pliocene in age). From base to top:

. lower conglomerate member,

. gypsum, light coloured marls and lacustrine limestones,

. pinkish to red clays,

. upper conglomerates.

A Miocene age has been determined on the basis of vertebrate faunas, ostracods, Charophytae and gastropods extracted from the second member of the formation. Comparing the series of the Aït Kandoula basin with the formations described in the Boumalne area, we propose the following correlations:

. The highest (Pliocene) conglomerates of the Aït Kandoula basin probably are the equivalent of the Aït Seddrat formation.

. The upper conglomerate member and the red clays of the Aït Kandoula formation probably correspond to the Tamassint formation.

. The lacustrine (gypsum, marls and limestones) member was correlated with the Aït Ibrirn formation (Görler and Zucht, 1986) owing to similarities in the Characeae associations. Nevertheless, in the Boumalne area, the Tamassint formation unconformably overlies the tectonized Aït Ibrirn deposits. Such an unconformity does not appear in the Aït Kandoula basin.

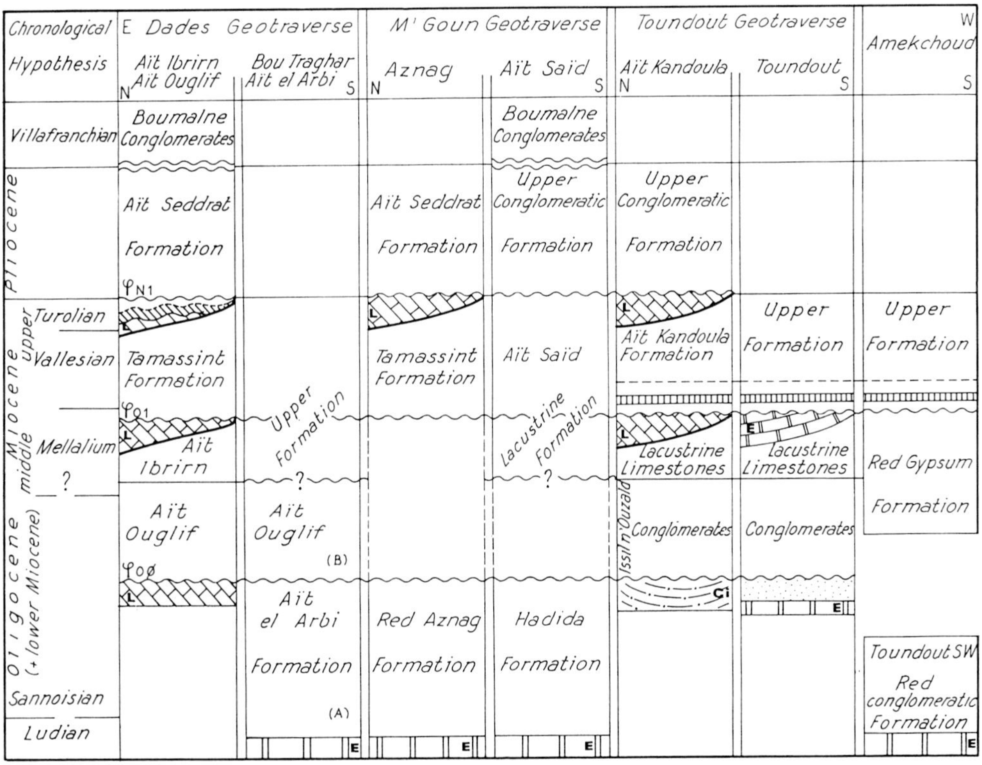

Fig. 8 - Correlation table showing the chronological evolution of tectonic events

The section across (Fig. 8) the western end of the Aït Kandoula basin, near Issil n'Ouzald, is not fully representative of the central part of the basin (described in Görler and Zucht, 1986). The main interest of this section is that it clearly shows the major unconformities that may be correlated with different compressional tectonic events. From base to top, the following formations have been recognized:

. The Toundout formation crops out in the North of Toundout, conformably overlying the Eocene limestones. Its sandstones, pelites and conglomeratic lenses may be correlated with the Aït el Arbi and Hadida formations (Fig. 8).

. The Issil n'Ouzald formation is composed of two members (Fig. 7) :

- a basal conglomerate unconformably overlying detritic and volcano-sedimentary formations of the Lower Cretaceous. This formation is highly deformed and according to Laville (1975), we correlate it with the Aït Ouglif formation.

- a lacustrine limestone overlies the conglomerate in an apparent conformity. Görler and Zucht (1986) obtained a Middle to Upper Miocene age in the Aït Kandoula formation in this part of the basin. This would suggest that the Aït Ibrirn formation, which is older than the Aït Kandoula formation, would have been affected by the $\varphi 0_1$ tectonic event.

b - The Toundout section. In the West of Toundout a thick conglomeratic formation is conformably overlain by lacustrine limestones. Following Laville (1975),we propose a correlation with the Issil n'Ouzald formation; this suggests that the corresponding basin was opening towards the South.

c - The Amekchoud section. The Amekchoud anticline is the most important folded structure of the Ouarzazate basin. Görler and Zucht (1986) have studied the region and have distinguished the following succession (Fig. 7) :

. a lower formation composed of thick conglomerates surrounded by gypsiferous marls, limestones and brown sandstones. The primitive fold was disrupted by diapyrism (gypsiferous). This formation can be correlated with the Issil n'Ouzald formation (Fig. 8).

. an upper formation composed of lacustrine limestones containing Ostracods and Characeae of Miocene age. Görler and Zucht correlate this formation with the Aït Kandoula formation.

B - SUMMARY OF THE TECTONIC PHASES

The main subatlasic structural features are large N070 synclines separated by narrow anticlines. These anticlines are generally southerly overturned;their inverted limb is stretched by reverse or thrust-faults and thus is often missing.

The N070 main folds and faults are deformed from place to place by sinistral N030 wrench faults. These tectonic features resulted from four main tectonic phases which occured between Oligocene and Villafranchian times (Fig. 8 and 11).

φ Oo : Middle Oligocene (?)

This phase can be characterized by the unconformity which is separating the Aīt el Arbi and Aīt Ouglif formations. The arrival of High Atlas derived detritals is one of the main sedimentary criterium to differentiate the Ait Ouglif deposits from the previous ones.

φ O1 : Post-Upper Oligocene to Middle Miocene (?)

This phase probably corresponds to the main initiation of the thrusts and folds in the sub-atlasic zone (south of the Toundout thrust). The age of this phase is not yet well enough constrained. It seems younger in the western part.

φ N1 : Post-Turolian

This event occured before the deposition of the Upper conglomerates (Aīt Seddrat formation)..., for the essential it can be characterized by the recurrence of the main steep faults and a remove of the Toundout thrust.

φ N2 : Ante-Villafranchian (?)

After a tectonic relaxation during a part of the Pliocene the erosional processes were reactivated by the High Atlas uplift. One consequence was the creation of the Boumalne valley which was excavated and later filled up by the Boumalne conglomerates.

During the Quaternary, the High Atlas uplift was in progress while the Ouarzazate basin subsided slowly (Schmidt, 1986, and our observations).

III - THE NEOTECTONIC COMPRESSIONAL DIRECTIONS ON THE BORDERS OF THE HIGH ATLAS - MICROTECTONIC RESULTS -

A - INVESTIGATION PROCEDURE

In order to know what could have been the directions of the stresses during the early history of the belt, we have tried to characterize the younger tectonic events registered in the most recent deposits. In a future work we will try to study the deformations in the axial part of the High Atlas, and at that time we hope we are able to separate these younger events from the other (intra-Jurassic) ones.

The recent deposits on both borders of the High Atlas are mainly conglomerates. In some places these conglomerates are made up with unperfectly rounded limestone pebbles in an argilaceous sandy matrix. The faces of the pebbles represent discontinuities in the material, relative to applied stresses: numerous slicken - sides develop and dissolution processes push quartz grains into the limestone, so the sense of the slip is easily determined. After measuring the orientation of the pebble faces and of the slickensides, we used an inversion method (Carey, 1976) which allowed us to find the orientations and the relative values of the principal deviatoric stresses: these are presented by the ratio $R = \frac{\sigma'_2 - \sigma'_1}{\sigma'_3 - \sigma'_1}$ ($\sigma'_1, \sigma'_2, \sigma'_3$ are respectively the maximum, intermediate and minimum principal deviatoric stress).

In many cases, the same face of a pebble was engraved by curving or crossing slickensides. In the first case their curvature is related to the tilting of the conglomerate beds during the tectonic phase; consequently, we only took in account the earlier part of the slickensides, provided that the beds are to be tilted back to the unfolded position, in order to find the maximum horizontal stress. Where we observed two successive sets of slickensides it has been possible, when the chronology between the two sets was clear, to sort out two groups of data and to calculate a stress state for each: these cases correspond to two diagrams for one site (Fig. 9, 10); the double circle is related to the younger data.

B - HIGH ATLAS NORTHERN BORDER

Despite of the fact that the studied sites are situated at the junction between the Middle and the High Atlas, the results appear quite homogeneous. The fact that some sites provided two compressive stress states (φ_1 and φ_2), is consistent with the geological data: indeed, it is known that a general unconformity separates the Mio-Pliocene from the Plio-Villafranchian or younger Pleistocene deposits and also that in many places the Plio-Villafranchian has been folded (Huvelin, 1973; Dutour and Ferrandini, 1985; Petit et al., 1985).

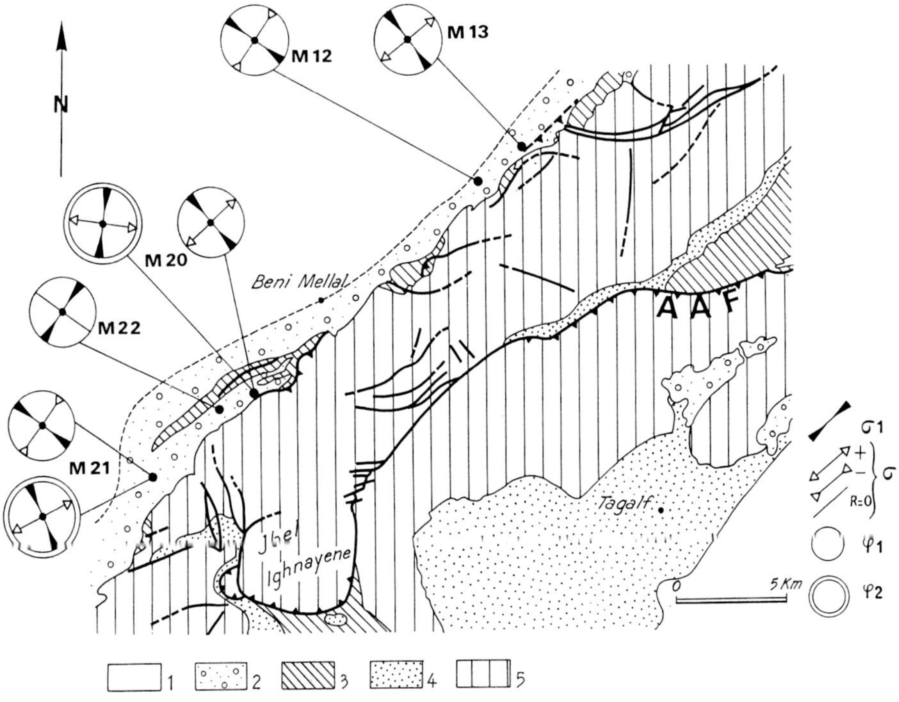

Fig. 9 - Microtectonic results. Region of Beni Mellal (for location see fig. 1) 1: Quaternary. 2: Neogene (conglomerates). 3: Eocene and Cretaceous. 4: Bathonian and Upper Jurassic. 5: Lower Mesozoic (Trias, Lias). $\sigma'1$: compressional stress. $\sigma'2$: intermediate stress (without arrow: value of ratio R=0). $\sigma'3$: extensional stress. $\varphi 1$: post-Miocene event (untilted bedding plane). $\varphi 2$: post-Plio-Villafranchian event

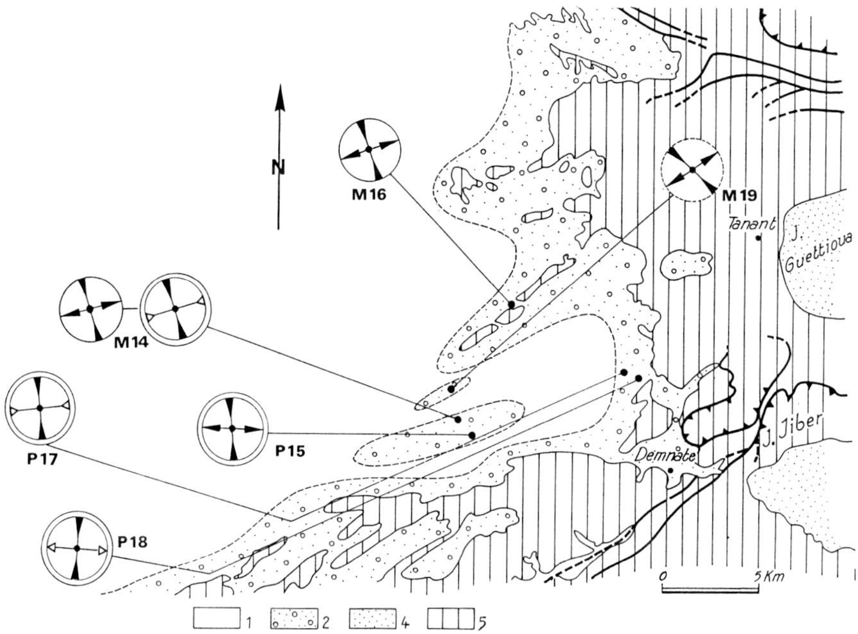

Fig. 10 - Microtectonic results. Region of Demnat (location fig.1 and legend fig. 9)

In the Beni Mellal region, the Plio-Villafranchian is not well defined; the Mio-Pliocene conglomerates and marls are folded, with overturned limbs, and overthrusted by the Mesozoic limestones. The data of sites M20 and M21 (Fig. 9) are the only proof for the successive compression registered in the Mio-Pliocene conglomerates.

In the region West of Demnate (Fig. 10), the sites P_{15}, P_{17}, P_{18}, obtained from Upper Pliocene beds, have given only one stress solution corresponding to the last compressive event (φ_2). In the Miocene conglomerates, the site M_{14} has given two solutions. The second one is in good agreement with the φ_2 compressive direction obtained from the sites "P". On the site M_{22} , the σ'_1 and σ'_2 are very near, so that the indicated shortening direction is not representative.

The shortening direction for the first event is the same as the directions (Fig. 1) reported by other authors (Dutour and Ferrandini, 1985; Petit et al., 1985; Huvelin, 1973) or deduced from fold axes in Mio-Pliocene formations. But in the Beni Mellal region (Fig. 9), this compression is notably different; this suggests that the fault which runs from the North side of the Jebilet into the High Atlas, North of Demnate, was active during the first event and caused a refraction of the stress direction. About the post-Plio-Villafranchian compression, it seems to have remained the same until now (Coisy and Frogneux, 1980).

C - HIGH ATLAS SOUTHERN BORDER

In the O. Dades region, the Pliocene Aït Seddrat formation provided three consistent shortening directions (Fig. 11; P_1, P_{IV}, P_V). Although we have shown that thrusts have developed before its deposition, the two other sites (in the Miocene and the Liassic) did not show a different stress state; it is possible that for both tectonic events the stress characteristics should be the same, thus undistinguished by the microtectonic analysis. As σ'_1 is horizontal, the value of R, near 1, indicates that thrusts or strike slip faults are equally able to act and actually both types of structures are encountered in this zone. The N150 direction of shortening reminds the one we have obtained in the Beni Mellal region, but the lack of stratigraphic correlations prevents us from any definitive conclusion.

In the Ouarzazate Basin, the Miocene has suffered only one compression, which was also registered in the Eocene. The shortening directions, somewhat dispersed, due to the near values for σ'_1 and σ'_2, are sub-meridian. As these directions are significantly different from those gathered near O. Dades, one could suggest that, here again, the border faults could have refracted the compression directions.

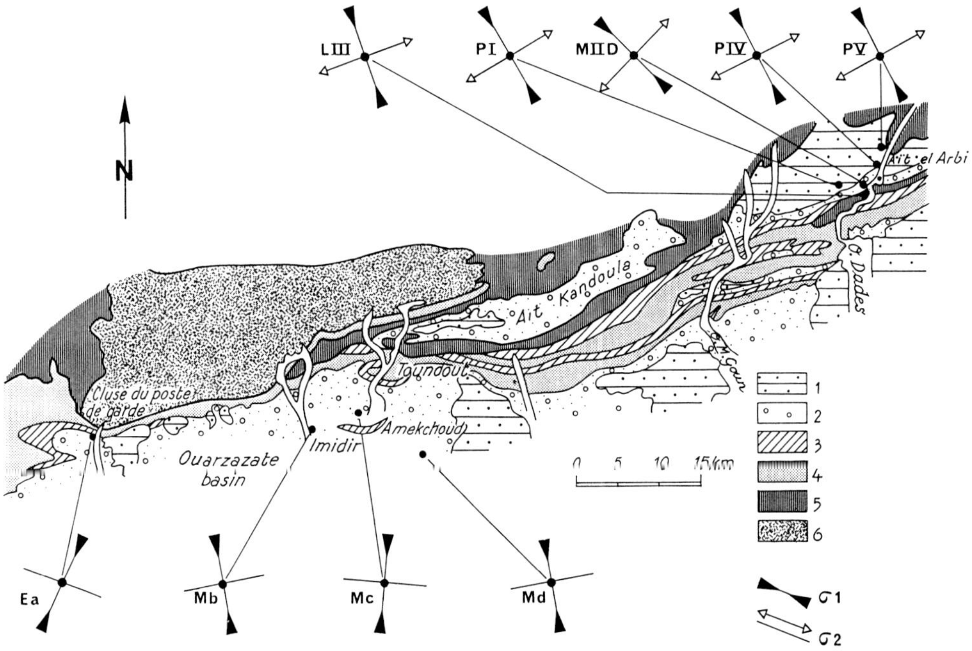

Fig. 11 - Microtectonic results. South High Atlas border (for location see fig. 1) after Laville's map. 1: Pliocene (?)(Aït Seddrat formation). 2: Mio-Pliocene continental formations. 3: Eocene and Cretaceous. 4: Lias. 5: Paleozoic. σ 1: compressional strain. σ 2: middle strain (without arrow: R=0).

IV - CONCLUSIONS

The preliminary results presented here indicate that, as a whole, the High Atlas was subjected since the Pliocene to compressional deformations. These cannot be directly related to the North-South relative motion between Africa and Europe during this recent period, because the inherited faults were probably reactivated and caused local changes in the shortening directions.

During Oligocene-Miocene times, the Southern part of the Central High Atlas suffered two compressive events well registered by the development of Oligocene-Lower Miocene detrital deposits; while in the North the deformations seem to be weak, and correlatively the Oligocene deposits are lacking.

These differences can be related to the important décollement of the Jurassic Carbonate cover which was initiated during the Late Cretaceous in the southern border of the Central High Atlas.

BIBLIOGRAPHY

BENZAQUEN M., 1963. La bordure septentrionale de l'Atlas de Beni-Mellal. Contribution à l'étude géologique de la région d'El Ksiba. Notes Serv. Géol. Maroc, 22, n°170, p.45-80.

CAREY E., 1976. Analyse numérique d'un modèle mécanique élémentaire appliqué à l'étude d'une population de failles: calcul d'un tenseur moyen des contraintes à partir des stries de glissement. Thèse 3 è cycle, Univ. Paris XI (ined.).

CHOUBERT G., 1946. Note préliminaire sur le Pontien au Maroc (Essai de synthèse orogénique du Maroc atlasique). Notes et Mem. Serv. Géol. Maroc, n°65 and Bull. Soc. Géol. France (5), 15, pp. 677-764.

CHOUBERT G. and FAURE-MURET A., 1960-62. Evolution du Domaine atlasique marocain depuis les temps paléozoïques. M.h.ser. Soc. Géol. France (Livre Mémoire, P. Fallot), t. 1, p. 447-527.

COISY P. and FROGNEUX M., 1980. Une étude de microséismicité.La région de Tillouguite (Haut Atlas de Beni-Mellal). Mines, Géol. et Energie, n°47, Rabat, pp. 93-102.

DRESNAY R. du, 1975. Influence de l'héritage structural tardi-hercynien et de la tectonique contemporaine sur la sédimentation jurassique, dans le sillon marin du Haut Atlas, Maroc. 9 è Congr. Int. Sédimentol., Nice, 4, 103-108.

DUTOUR A. and FERRANDINI J., 1985. Nouvelles observations néotectoniques dans le Haut Atlas de Marrakech et le Haouz Central (Maroc). Apports sur l'évolution récente d'un segment du bâti atlasique. Revue Géol. Dyn. Géog. phys., Vol. 26, Fasc. 5, p. 285-297.

GAUTHIER H., 1957. Contribution à l'étude géologique des formations post-liasiques des bassins du Dades et du Haut-Todra (Maroc méridional). Notes et Mém. Serv. Géol. Maroc, n°119, 212 p.

GÖRLER K. and ZUCHT M., 1986. Stratigraphie und Tektonik des Kontinentalen Neogens im Süden des Zentralen Hohen Atlas, Provinz Ouarzazate (Marokko). Berliner Geowiss. Abh., (A), 66, p. 471-497.

GRATIER J.P., 1981. Caractères des déformations successives dans l'Atlas de Béni-Mellal (Haut Atlas marocain). Notes Serv. Géol. Maroc, t. 42, n°308, 1981, pp. 195-216.

HELMDACH F.F., 1986. Zur Alterseinstufung des höchsten marinen Tertiärs im Bereich der Südatlas-Randzone (Provinz Ouarzazate, Marokko). Berliner Geowiss. Abh., (A), 66, pp. 495-502.

HERBIG H.G., 1986. Lithostratigraphisch-fazielle Untersuchungen im marinen Alter-tiär, südlich des Zentralen Hohen Atlas (Marokko). Berliner Geowiss. Abh., (A), 66, p. 343-380.

HINZ K., DORTMANN M. and FRITSCH J., 1982. The continental margin of Morocco: Seismic sequence, structural elements and geological development. In Geology of the Northwest African Continental Margin. U. von Rad, K. Hinz, M. Sarntheim and E. Seibold, ed., Springer Verlag, Berlin, p. 34-60.

HUVELIN P., 1973. Mouvements pré-atlasiques, atlasiques et récents dans les Jebilets et sur leur pourtour. Notes Serv. Géol. Maroc, t. 33, n°249, pp. 83-123.

JACOBSHAGEN V., 1986. Zur Baugeschichte des Zentralen Hohen Atlas (Marokko). Berliner Geowiss. Abh., (A), 66, pp. 455-470.

JENNY J., LE MARREC A. and MONBARON M., 1980. Les couches rouges du Jurassique moyen du Haut Atlas Central (Maroc): corrélations lithostratigraphiques, éléments de datation et cadre tectono-sédimentaire. Bull. Soc. Géol. France, (7), 23, p. 627-639.

LAVILLE E., 1975. Tectonique et microtectonique d'une partie du versant Sud du Haut Atlas marocain (Boutonnière de Skoura, nappe de Toundout). Thèse 3 è cycle, Univ. Sc. Tech. Languedoc, Montpellier.

LAVILLE E., 1978. Incidence des jeux successifs d'un accident synsédimentaire sur les structures plicatives du versant Nord du Haut Atlas Central (Maroc). Bull. Soc. Géol. France, (7), 20, p. 329-337.

LAVILLE E., 1981. Rôle des décrochements dans le mécanisme de formation des bassins d'effondrement du Haut Atlas marocain au cours des temps triasique et liasique. Bull. Soc. Géol. France, (7), 23, p. 303-312.

LAVILLE E., 1985. Evolution sédimentaire, tectonique et magmatique du Bassin jurassique du Haut Atlas (Maroc): modèle en relais multiples de dérochements. Thèse Doctorat d'Etat, Univ. Sc. Tech. Languedoc, Montpellier.

LAVILLE E., LESAGE J.L. et SEGURET M., 1977. Géométrie, cinématique (dynamique) de la tectonique atlasique sur le versant sud du Haut Atlas marocain. Aperçu sur les tectoniques hercyniennes et tardi-hercyniennes. Bull. Soc. Géol. France, (7), 19, pp. 527-539.

MATTAUER M. TAPPONNIER P. et PROUST F., 1977. Sur les mécanismes de formation des chaînes intracontinentales. L'exemple des chaînes atlasiques du Maroc. Bull. Soc. Géol. France, (7), 19, p. 521-526.

OLIVET J.L., BONIN J., BEUZART P. et AUZENDE J.M., 1984. Cinématique de l'Atlantique Nord et Central . Rapp. Centre National Exploitation des Océans, Paris, n°54, 108 p. 8 pl.

PETIT J.P., RAYNAUD S., CANTRU J.P., 1985. Microtectonique cassante lors du plissement d'un conglomérat (Mio-Pliocène du Haut Atlas, Maroc). Bull. Soc. Géol. France, (8), I, p. 415-421.

PROUST F., PETIT J.P. et TAPPONNIER P., 1977. L'accident du Tizi n'Test et le rôle des décrochements dans la tectonique du Haut Atlas occidental (Maroc). Bull. Soc. Géol. France, (7), 19, p. 541-551.

STETS J. and WURSTER P., 1982. Atlas and Atlantic-Structural Relations. In "Geology of the Northwest African Continental Margin", U. von Rad, K. Hinz, M. Sarntheim and F. Seibold (eds.), Springer Verlag, Berlin, p. 34-60.

RECENT DATA ON THE GEOLOGY OF THE MIDDLE-ATLAS (MOROCCO)

Renaud du Dresnay
Service de la Carte géologique,
Ministère de l'Energie et des Mines,
B.P.6208, Rabat-Instituts, Morocco

ABSTRACT

The Moroccan Atlas is subdivided into two SW-NE trending zones: a tabular one in the northwest ("Causse moyen-atlasique") and a folded one ("Moyen-Atlas plissé") in the southeast. Both show a Paleozoic, partly Carboniferous basement exposed at a few places. It is covered by thick sediments of mainly Mesozoic age. The lower Jurassic consists of carbonate rocks that were deposited on an extended platform. Middle Jurassic sequences are thin on top of the anticlines, but can be very thick in the cores of the synclines. Upper Jurassic deposits have not been clearly identified and a lack of deposition occurs up to the lower Cretaceous. Cretaceous rocks are confined to restricted areas in the center of some synclines as Lower Cretaceous or Maestrichtian beds, which unconformably overlie older rocks.

Perpendicular to the strike of the mountain range, a NW-SE trending Quarternary volcanism is spread over a huge area in the centre of the Middle-Atlas. The most obvious morphological features of the Middle-Atlas indicate the main tectogenesis to have happened during the Cenozoic ("Pyrenean" and "Alpine" phases according to traditional interpretations). Early tectonic movements started, however, as soon as the mid-Liassic and increased during mid-Jurassic times.

RESUME

Les deux régions essentielles du Moyen-Atlas marocain, de direction SW-NE, le Causse moyen-atlasique (ou Moyen-Atlas tabulaire) et le Moyen-Atlas plissé, montrent, au-dessus d'un socle paléozoique (en partie carbonifère) affleurant peu, une épaisse sédimentation essentiellement mésozoique, avec des faciès de plateforme assez uniformes pour le Jurassique inférieur (Lias) et des séquences du Jurassique moyen, relativement réduites sur le sommet des anticlinaux, mais pouvent être épaisses, voire très épaisses, dans les synclinaux. Des sédiments de Jurassique supérieur n'y ont pas été reconnus avec certitude, et il y a lacune des dépôts jusqu'au Crétacé: celui-ci occupe des superficies limitées au coeur des certains synclinaux, et est discordant sur les couches antérieures au niveau du Crétacé inférieur ou au niveau du Maestrichtien.

Un volcanisme quarternaire important, de direction perpendiculaire NW-SE recouvre une assez grande surface du centre du Moyen-Atlas. L'orogenèse du Moyen-Atlas s'est surtout manifestée, dans ses aspects morphologiques les plus visibles, pendant le cénozoique (interprétation traditionelle en phases "pyrénéenne" et "alpine"), mais certaines de ses phases prémonitoires ou précoces ont débute aussi tôt que le Lias moyen, et se sont multipliées au cours du Jurassique moyen.

1. Introduction

The Middle-Atlas, one of the four Moroccan mountain ranges, is situated between the Rif range to the north and the High-Atlas range to the south. Its name was introduced by the German geographer K. RITTER (1822), later made popular by the explorer C.DE FOUCAULT (1888, p.99), and taken up again by P. SCHNELL (1892, 1898,p.27). The latter author's works are of fundamental importance as far as the geomorphological terminology of Morocco is concerned and are still valid today: "The third important portion of the Moroccan Atlas, due to its altitude, is called by Foucault 'Middle-Atlas', a term which we want to adhere to because of the lack of a better one...".

The Middle-Atlas forms part of what is called the "Atlas Domain", a region situated to the north of the sub-atlas fault (lineament) extending from Agadir to Tunis. It is essentially a "maghrebian" or North-African chain forming part of the "Berbèrie", a geomorphological unit used mainly by geographers. It is more Mediterranean (Tethysian) than African in character.

2. Position within the North African Orogen

Amongst the North-African Atlas chains which form roughly parallelograms with W-E or SW-NE orientation, the Middle-Atlas represents clearly the SW-NE trend. It can be noted that this direction corresponds to that of the Atlantic coast of Morocco. One can thus talk about an "Atlantic" trend as opposed to the "Mediterranean" trend (KANES et al., 1973).

Along a north-south traverse, the Middle-Atlas is situated to the north of the High-Atlas (and thus well to the north of the African or Saharan Craton) but to the south of the Rif ranges which overrun it with thrusts, and from which it is separated by a zone of depressions called "southern Rif corridor".

Along a W-E longitudinal section, the Middle-Atlas is located between the High-Plateau Domain to the east (Algerian High Plateaus: Oran Meseta or Oran-Morocco Meseta) and the western Meseta (or Moroccan Meseta, Coastal Meseta, or Central Morocco) to the west.

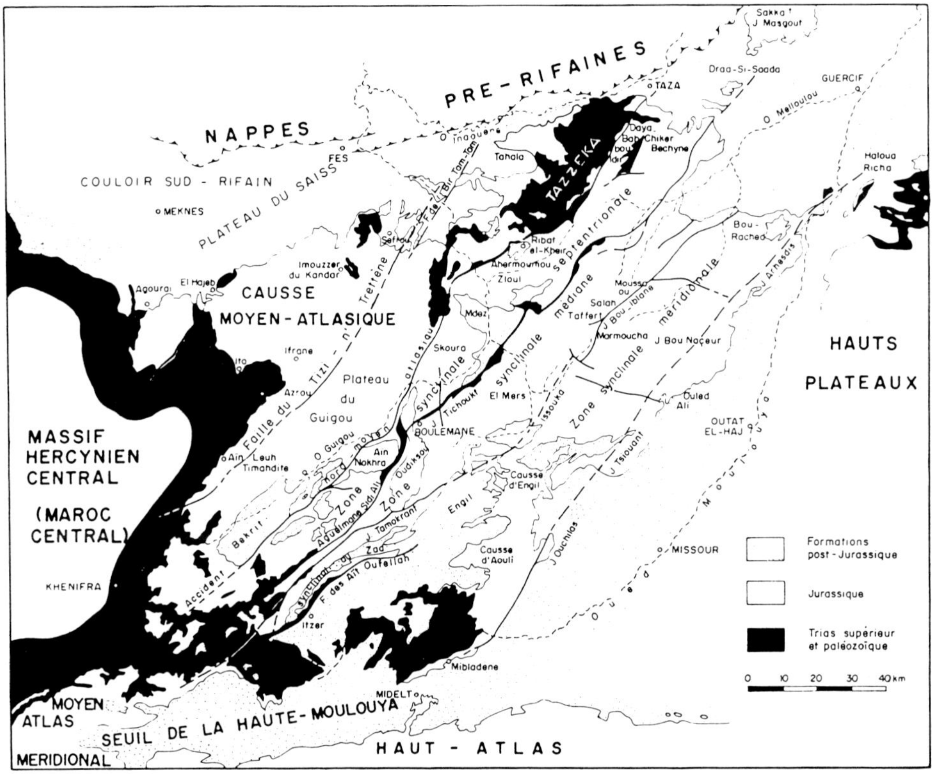

Fig. 1: Structural map of the Middle-Atlas showing main adjoining areas

2.1 Geomorphological Subdivisions of the Middle-Atlas

According to the geomorphological expression of the actual outcrops, the Middle-Atlas according to MARTIN (1981) can be subdivided into three (or four) different regions (Fig. 1):

-1) In the west, the Middle-Atlas "Causse" (a French term meaning "platform") or Tabular Middle-Atlas which overlaps onto the Paleozoic rocks of Central Morocco (Central Hercynian Massif) in a westerly direction, and in which mainly horizontal calcareous-dolomitic rocks of Jurassic and Cretaceous age can be observed.

-2) In the southeast and east, the Engil Rise Causse and the High-Moulouya Causse (region of Itzer and Mibladène)

-3) This unit could be extended farther to the southwest (although this attribution was, and is again, much disputed) to the borders of the High-Atlas near the boundary ranges of Tadla (Ksiba and Béni-Mellal areas: southern Middle-Atlas).

-4) Finally, the folded Middle-Atlas to the north and northeast, the most important unit of the mountain range, consisting of more marly, deeper, and at the same time thicker formations which are furthermore affected by folding. The folded Middle-Atlas is separated from the Middle-Atlas Causse by a zone of frontal faults called "Northern Middle-Atlas Fault". COLO (1960-62) has distinguished four anticlinal ridges (numbered 1 to 4):

1) the frontal Timhadite-Ribat el-Kheyr (Ahermoumou)-Bab Bou Idir ridge,
2) the Boulemane-Jebel Tichoukt-Jebel Ouarhirt ridge,
3) The Jebel Tamokrant-Issouka-Jebel Bou Iblane ridge,
4) the Mibladène-Jebel Tsiouant-Jebel Arhesdis ridge.

They are separated by three synclinal zones, the southern, central, and northern synclinal zones.

The folded Middle-Atlas forms the highest elevations of the range. The two highest points are the isolated and poorly accessible Jebel Bou Naceur (3340 m) in the eastern folded Middle-Atlas which overlooks the Moulouya valley, and the 3190 m high Moussa ou Salah in the Jebel Bou Iblane range.

3. History of Geological Studies

The geological studies carried out in the Middle-Atlas so far are essentially of a morphological, paleontological, stratigraphic, and sedimentological nature. Relatively few profound structural or tectonic studies are available.

Reconnaissance studies were published mainly by GENTIL (1916), BEAUGÉ (1920), and P.RUSSO (1930,1931,1933,1935,1936,1938).

During 1936 and 1938 the first general synoptic studies were carried out, represented in particular by the four volumes of H.TERMIER's (1936) thesis.

The main paleontological studies are given below together with the relevant fossils: MONESTIER (1930, ammonites), DUBAR (1936,1942, brachiopods), LE MAITRE (algae), LAMBERT (echinoderms), COLO (1961-64), PETITOT (1959, echinoderms), A.F. DE LAPPARENT (1955, dinosaurs), ROUSSELLE (1965-68, brachiopods), MONGIN (1967, lamellibranchs), GUEX (1973, ammonites), etc. Microfaunas and microfacies have been studied by REY & NOUET (1958), YACOUBI (1984) and SEPTFONTAINE (1984).

The morphological and geographical inventory of the southern Middle-Atlas has been described by MARTIN (1973,1981).

Subsequent to the studies by H.TERMIER, the stratigraphic basis of the Jurassic succession was established by COLO (1961-64), LORENCHET DE MONTJAMONT (1963-64), BENZAQUEN (1964), and DRESNAY (1963-75). It has recently been re-assessed by EL HATIMI & NASSILI (1977), LAADILA (1982), CHARRIERE (1985), and BEN SHILI & FEDAN (1984,1985).

The first structural interpretations, bold at their time, were given by GENTIL (1916) and thereafter by P.RUSSO (1927,1930 to 1938). In recent years only few tectonic or structural investigations have been carried out in the Middle-Atlas. The most recent ones are those of DUÉE et al. (1977), COLETTA (1977), ROBILLARD (1978,1981), and FEDAN (1977,1980 to 1986).

4. Stratigraphic and Paleontological Data

The Middle-Atlas consists essentially of a Mesozoic cover (Jurassic to Cretaceous) which overlies, with a thin intervening Triassic layer, a Paleozoic basement which crops out all around the Middle-Atlas and beyond. Within the Middle-Atlas these older rocks appear on the surface through certain windows: Tazekka in the north, B'Sabis and Imouzzer du Kandar in the centre, and the Aouli-Mibladène dome in the southeast. The rocks are all of Carboniferous age, their geological problems being different from those described here.

Above the Paleozoic basement, the sedimentary cover is made up of:
- beds attributed to the upper Triassic and the lowermost Liassic, consisting of brown or red pelitic rocks (locally with gypsum, salts, or conglomerates) and of basaltic effusives (COGNEY et al., 1971; MANSPEIZER et al.,1978; ROBILLARD,1978; ROBILLARD & PIQUÉ, 1981-83; OUARHACHE et al.,1985,1987).
- a Jurassic sequence, initially carbonatic, then more marly, belonging to the Liassic; thereafter terrigenous sediments of Dogger (mid-Jurassic) age. The latter can be up to several thousands of meters thick. The filling of the Middle-Atlas trench

is terminated by red brackish, locally conglomeratic sediments, their age straddling the Bathonian-Callovian boundary. Rocks of upper Jurassic age are not known with certainty.

- a Cretaceous sequence of reduced but highly variable thickness.
- a Paleocene sequence (starting with the Maestrichtian) unconformably overlying all previous formations (in itself an indication for orogenic movements) and dated by foraminifera (Koubbat Series) and by pulmonate gastropods (Clavator of the southern hammadas).
- the Oligocene is only sporadic; Miocene sediments, marine in the north, continental in the center and the south, follow unconformably.

	MOYEN-ATLAS W	MOYEN-ATLAS E	BASSIN DE GUERCIF Prolongement N-E du Moyen-Atlas	HAUTS-PLATEAUX Marche Est du Moyen-Atlas
Kimméridgien			Calcaire du Terni	
Oxfordien Callovien ?	? gypses de Skoura	? gypses de l'Arhesdis	Couches à gypse de Tamtlilit	
Bathonien	Formation d'El Mers et de Skoura	grès de Bou-Rached	grès de Bou-Rached	Marnes d'Aguelmous
Bajocien supérieur	Form. d'Ich-Timellaline Form. du J. Bou-Akrabène "calcaires-corniches"	Marnes à Rhynchonelles Marnes à Pholadomyes	Marno-calcaires de la Moulouya Marnes à Pholadomyes	Marno-calcaires de Guéfaït Marnes à Rhynchonelles Marnes à Pholadomyes
Bajocien sup.-moyen	Formation de R'cifa	"Calcaire-corniche"	"Calcaire-corniche"	"Calcaire - corniche"
Bajocien moyen Bajocien inférieur	Marnes de Boulemane		Marnes de Sakka	Marnes de Boulemane Dolomie récifale du J. Mechkakor / Dolomies et calcaires des Hauts-Plateaux
Aalénien	"Calcaires à Cancellophycus"		Calc. à "Cancellophycus"	Calcaires à "Cancellophycus"
Toarcien	Marnes de Taffert		Marnes de Bechyne	Couches rouges du J. Nador (ou de Debdou)
Pliensbachien: Domérien	Calcaires lités	Faciès rouges	Calcaires de l'Ouarirt	?
Pliensbachien: Carixien	Calcaires à silex Récifs à coraux	d'Itzer et de Mibladène	Calcaires d'El Ahmar	Dolomies
Sinémurien (Lias inférieur)	Dolomies à Paléodasycladacées	Récifs à éponges Dolomies	Dolomies stromatolithiques du J. Masgout	(Dolomies de Debdou) ?
TRIAS SUPERIEUR				

Fig.2 Lithostratigraphic formations of the Jurassic cover of the Middle-Atlas and its surroundings

4.1. Important Elements of the Mesozoic Paleontology and Chronology (Jurassic and Cretaceous)

The Jurassic of the Middle-Atlas, well developed and well exposed (Fig.1), presents some quite distinctive characteristics of great interest which merit consideration as classical examples. It consists of:

- a rather poorly dated, essentially massive dolomitic lower Liassic (Sinemurian), with algae (Paleodasycladus mediterraneus) which is terminated by a more bedded dolomitic section with special endemic brachiopods of the family Cancellothyridae, studied by DUBAR (1942). These Hesperithyris, the genotype of which is T.renierii CATULLO of the Italian Dolomites (Verona region), form separate species in Morocco with which one can establish, by correlation from one section to the next, a valuable local stratigraphic column. The facies represents a coastal carbonate platform.
- a mid-Liassic (Pliensbachian), carbonatic throughout but in well-separated layers, with flint-bearing limestones and thin beds of marly limestones which are well datable with the aid of ammonites. Ammonites older than the mid-Carixian have not been observed, the oldest ones belonging to the genus Tropidoceras (type species: T. mediterraneum GEMMELARO) and to Metaderoceras (ex-Crucilobiceras). At the base there are possibly some Uptonia, but this question is not yet resolved. Older ammonites mentioned, such as Arnioceras, Oxynoticeras, etc. are misidentifications.

The mid-Liassic ammonite faunas of the Middle-Atlas (MONESTIER,1930; H.TERMIER,1936; DUBAR,1938,1942; GARDET & GERARD, 1946; COLO,1961-64) are quite diagnostic of the Tethys realm, however, with certain northwestern European influences. One actually observes, in addition to the mid-Carixian Tropidoceras:

. Liparoceras (Becheiceras) of upper Carixian age (northwest European fauna);

. Hildoceratids (Fuciniceras and Protogrammoceras) of lower Domerian age (mediterranean fauna);

. numerous genera ("Coeloceras", Argutarpites, Arieticeras, Canavaria, Emaciaticeras, Taurомeniceras, etc.), initially described by Italian authors (FUCINI, amongst others). These genera show certain mediterranean affinities with Italy, and in particular with Sicily, but also with the Causse region of France and southern Spain (Andalusia).

. one can observe everywhere the contradictory, widespread, and locally even abundant presence of Amaltheus, considered as an indicator for a cold, so-called northwestern European fauna. This, amongst other indications, implies connections with the Aveyron of southern France which at the time was situated much farther to the northeast. So far no amaltheids have been observed in the High-Atlas and the southern limit of their distribution coincides with the border of the Middle-Atlas (DUBAR,1960-62). This observation poses a problem, not so much of climatology in the classical interpretation, but more of the paleogeographic connections between the Middle- and the High-Atlas. At the time there appears to have been a physical land barrier, separating these two trends of the Atlas range.

. the top of the mid-Liassic is characterized by the presence of Pleuroceras spinatum, the index amaltheid of this zone.

- In the Toarcian (and the upper Liassic) there is a complete rejuvenation of faunas and a reorientation of the paleogeographic currents. There is:

. a great abundance of classical and rather common faunas with, amongst others, Hildoceras (BEN SHILI & FEDAN,1984);

. also the sporadic arrival of oriental faunal elements of Arabian provenance, and even from Madagascar with e.g. Protogrammoceras madagascariense THEVENIN, some Bouleiceras sp. (NAJJIB, FEDAN, unpubl.) found at Issouka and on the Engil Plateau; one Nejdia aff. pseudogruneri (THEV.) by GUEX (1973); and some unusual or new forms such as Taffertia taffertensis GUEX and Harpoceratoides maghrebense GUEX.

The Toarcian is locally absent, especially to the southwest in the direction of the High-Moulouya paleo-rise, a ridge separating the Middle- and High-Atlas from each other. In contrast to this the Toarcian becomes more prevalent to the northeast.

- In the Aalenian, red facies ("ammonitico rosso") as well as transgressive ferruginous facies make their appearance, with Catulloceras, Erycites, Tmetoceras, hammatoceratids, etc., similar to the "Cap San Vigilio" fauna of the Lake Garda and to the ferruginous faunas of Lorrain and Luxembourg. These facies are known from Tizi Nehassa (BEN SHILI,1985), Tizi Issoulitène (Jebel Tichoukt), and Seloum at the source of the Oum-er-R'bia.

- The mid-Jurassic (Bajocian and Bathonian) is characterized by thick marly sediments interbedded with more calcareous layers ("calcaires-corniches" = coastal limestones).

The sediment thicknesses are largest in the synclines, locally on an almost inflationary scale: several thousands of meters alone for the upper Bajocian and the lower Bathonian in the central parts of the Middle-Atlas. In contrast to this the thicknesses are reduced, locally very extensively, on the anticlinal axes. From these observations the presence of a buried paleorelief can be assumed.

Certain levels have furnished important paleontological benchmarks (ammonites):

. in the lower Bajocian: Sonninia;

. in the mid-Bajocian: stephanoceratids, in particular St.humphriesianum (rather abundant) and Dorsetensia;

. in the upper Bajocian, after a hiatus in sedimentation, there is a renewed appearance of ammonites from the eastern part of the Tethys (Saudi-Arabia, Balouchistan, etc.) such as Ermoceras (Telermoceras), a fauna first decribed from Jebel Moghara on the Sinai Peninsula (DRESNAY,1964); as well as various Cadomites, Teloceras, Leptosphinctes, and Parkinsonia;

. these ammonites from the base of the upper Bajocian are succeeded by particular "Arabian" faunal elements such as "Thambites", as well as rarely Micromphalites,

. following these, in the lower Bathonian, Morphoceras, Ebrayiceras, as well as an endemic North-African genus: Oraniceras FLAMAND, corresponding actually to Parkinsonia wurttembergica;

. the succession changes to detritic and becomes more sandy to the top, but still contains some rare ammonites: Choffatia, Grossouvria, during the final stages of the Bathonian and during the Callovian.

The few younger ammonites mentioned by CHOUBERT & FAURE-MURET, 1967) are rather dubious and should be omitted from the literature. The question of the Callovian and the final stages of the upper Jurassic, stirred up by COLO (1961) and locally by CHOUBERT (but in contrast to this refuted by TERMIER & DUBAR) can no longer be sustained (DRESNAY,1975).

However, one has to take cognizance of a "continentalisation" of the upper, very thick portion of the succession: red facies, gypsum, lignites, additionally with lagoonal, euryhaline facies poor in species but each with abundant populations: Corbula,

Protocardia, Modiolus, Mytilus (MONGIN,1967; H.&G.TERMIER,1967). These are equivalents of the "Mytilus-Beds" of Switzerland and central Europe. There also is a report of a remarkably large sauropod dinosaur in Morocco (Bathonian): Cetiosaurus maghrebiensis LAPPARENT (1955).

- upper Jurassic: One has to go as far as Taza-Oujda to encounter carbonate rocks attributed with some certainty to the Oxfordian or Kimmeridgian, dated generally with the aid of microfossils (clypeins and ostracods) and some rare ammonites (HAMEL,1967). To the east, the transgressive Callovian is only encountered again to the south of Oujda where it is dated through ammonites (Reineckeia) which are much more numerous in Algeria (Oran region).

These facies represent a transition with those of the upper Jurassic of the Rif and Pre-Rif ("Ferrysch", WILDI,1983).

4.2. Cretaceous and Cenozoic (Eocene, Miocene) Cover

Cretaceous and Cenozoic terranes form only a thin surficial cover in the cores of the synclinal zones in the Middle-Atlas (synclines of Koubbat-Bou Angueur, Zad, Skoura, etc.).

The Cretaceous stratigraphy is highly variable from one outcrop to the next. It has recently been reviewed by ANDRIEUX & CHARRIÈRE (1985). The mid-Cretaceous has been dated by a few ammonites. The upper Cretaceous is mostly represented by sands containing fish teeth. The age of the core of the Marmoucha syncline, still under discussion, has been reported as Cretaceous by FEDAN (1984,1986). At the base of the Maestrichtian there is an important discontinuity with a clear unconformable relationship (Foum Kheneg), above which the Eocene sedimentation commences.

This is well represented by the marine limestones of Timhadite (CHOUBERT et al.,1952; RAHHALI,1970; SEGONZAC et al.,1985) which pass laterally into the gypsum beds of Ain Nokhra (synonymous with the "Feleddi Beds") and into the lacustrine limestones of Bou-Angueur with Bulimides (these "Clavator", known from the sub-Saharan hammadas and first described from the Aquitanian, are now assigned to the Eocene: TRUC et al.,1985).

The sedimentary sequences of the Middle-Atlas are terminated by conglomeratic formations (Jebel Hayan) attributed to the Aquitanian and, in the northern Middle-Atlas, by the lateral continental equivalents of the marine Eocene of the Pre-Rif zone.

5. Sedimentological Data

The Middle-Atlas during the Jurassic presents an interesting assembly of facies and sedimentary structures (DRESNAY,1972). One finds:

- coastal platform facies of the Middle-Atlas Causse area, represented essentially by secondary dolomites with hollow fossil imprints (brachiopods, lamellibranchs, gastropods, and a few isolated corals). There are furthermore dasycladacean algae together with foraminifera (SEPTFONTAINE,1984) such as Orbitopsella praecursor, and also some Calcarea (hydrozoans, BEAUVAIS 1980) as well as locally silicified fossil wood. These facies are accompanied by sedimentary structures typical of platform environments:
- "birds-eye" structures (cavernous sands);
- laminations on mm-scale, representing algal stromatolites, the vertical filaments of which are still discernible under the microscope;
- oolitic sands with cross-bedding;
- micro-atolls of crinoidal and coral-bearing sands;
- dolomitic pseudo-breccias, the origin of which, although still under dispute, should be sought in the diagenesis of the rocks.

These facies and structures suggest a submarine dune environment with algal and crinoidal meadows. At the lower/mid-Liassic boundary one finds furthermore the continued presence of tepee-structures (DRESNAY,1976; SEPTFONTAINE,1984) which suggest a transition into a sabkha facies.

- on the break of the slope between the coastal platform and the deepest basin of the folded Middle-Atlas there are various different types of reefs:

. massive carbonate facies, more or less in groups, with thickness in the m-range, and with portions containing silicified Trichites;

. a thick-bedded facies with large colonial lamellibranchs such as Durga, Lithioperna, Cochlearites, (resembling Lithiotis);

. bioherms ("lithoherms") consisting of sponges, especially well developed in the synclines bordering the Jebel Bou-Iblane (Taffert) chain;

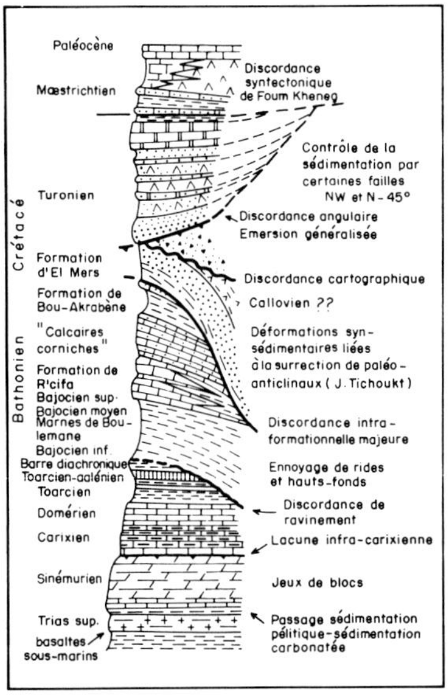

Fig.3 Stratigraphic section of the Mesozoic in the Middle-Atlas (partly after CHARRIÈRE)

. genuine reef domes on km-scale consisting of corals (DRESNAY,1971,1975-77; MORABET,1974), stacked in time (at least three episodes, ranging from mid-Carixian to the upper Domerian) and easily datable. The reef aprons of the coralline slope debris interdigitate with a deeper environment of bedded limestones with ammonites. The reef forming episode is terminated at the end of the mid-Liassic.

- The mid-Jurassic is equally rich in various types of sedimentary structures which, however, so far have been little recorded and would merit further studies. The thick mid-Jurassic of El Mers exhibits numerous shallow trace fossils such as U-shaped tubes of Thalassinoides, etc. (H.&G.TERMIER,1967). The facies are marine euryhaline passing into paralic and have been described by MONGIN (1967). The great thickness of these fairly continuously subsiding shallow-water sediments represents an especially suitable target for paleoecological investigations.

6. Tectonic and Structural data

So far only few structural investigations with tectonic and micro-tectonic data have been carried out in the Middle-Atlas chain.

- The controversy about the limits of the Middle-Atlas itself has not been settled. The following questions can be raised:

. Is the southwestern region (southern Middle-Atlas or Tadla and Béni-Mellal Atlas) really part of the Middle-Atlas? The first marginal folds of the Tadla Atlas exhibit the platform facies typical of the Middle-Atlas and one has to go as far as Tassemit, dominating the Béni-Mellal chain, to encounter the typical ammonite facies of the Sinemurian (BOUJO, unpubl.). DUBAR (1952-54) preferred a transition between the Middle- and the High-Atlas.

. Do the folds of the Paleozoic Jebilets farther to the southwest represent the eroded basement of the Middle-Atlas (P.RUSSO,1933,1934)? This interpretation is possible, taking recourse to the connection with the Atlas of Béni-Mellal, but the question has to be raised again, as it had already been posed at the time (1888) by C. DE FOUCAULT:"Where does this chain commence? Where does it end? One cannot say with any certainty. For us it commences to the north of Demnate at the Tessaout gap where its last slopes coalesce with those of the High-Atlas."

. Furthermore, how far does the Middle-Atlas extend to the north-east? Up to the Oujda hills (called the "horst region") or to the Béni-Snassène massif? Or is it cut off by the large E-W thrust still called the "Rabat-Taza Oujda Trend" (RAKUS,1979; COURBOULEIX et al.,1981)?

- The directional grid of the large folds and the important faults has already been likened for quite some time to a pattern consisting of rhombs or parallelograms associated with the SW-NE or Atlantic direction and the E-W or Mediterranean direction. This structural pattern according to E.SUESS is developed throughout the North-African mountain ranges from Agadir to Tunis. The SW-NE direction in particular forms the connection between the High- and the Middle-Atlas (Mibladène range) which itself extends under the overriding Rif zone (Bab Bou-Idir range and Jebel Masgout massif).

Another important structure ("lineament") is the SW-NE trending fault of Tizi n'Trettène, actually known from Khénifra (Hercynian Central Massif) up to, and possibly even beyond Sefrou and Bir Tamtam.

Other directions or lineaments can be noted by remote-sensing with the aid of satellite imagery which specifically exhibits the NW-SE alignment ("ougartian" direction). The ground truth of these observations should be verified by more detailed analyses.

- Finally, the configuration of the network of the folds and of the faults affecting them longitudinally and locally also transversally suggests an interpretation as "pull-apart basins" as proposed by LAVILLE (1979,1981,1982) and recently (1985) developed further by him. This interpretation would appear to be the best one, taking into account the field observations and the results of the studies of various folds of the Middle-Atlas.

7. Data on the Orogenic Processes in the Middle-Atlas

A paradoxical aspect of the geological studies in the Middle-Atlas is the large number of orogenic theories, based upon which these studies have been carried out, whereas the field observations are insufficient to allow such interpretations. And also the "questions of geological fashion" could potentially be scattered by the basic data.

The main orogenetic interpretations on the Middle-Atlas have been published by GENTIL (1916), STAUB (1926), RUSSO (1927,1935-36), FALLOT (1941), CHOUBERT & FAURE-MURET (1960-62), COLO (1961-64), DUÉE et al. (1977,1978), LAVILLE (1977,1979,1985). The majority of these interpretations emphasized the Tertiary phases of the Middle-Atlas orogenesis in which they distinguish a "pyrenean" and an "alpine" phase. Amongst the more extreme interpretations the question of "subduction"of the eastern border of the Middle-Atlas along the Moulouya valley has to be mentioned (DEWEY et al.,1973). This was probably prompted by the interpretation of the Triassic basalts of the Middle-Atlas as ophiolites.

- The question most actively debated in recent years is whether the Middle-Atlas orogenesis was "continuous" or "discontinuous" (Fig.3).

The traditional interpretation was to consider the essentially Cenozoic orogenesis as consisting of two phases, one of which is the pyrenean (Eocene) one. The second, or alpine (Helvetian) one, is morphologically more pronounced. It is clearly evident that the Eocene discontinuity of Foum Kheneg, an angular unconformity eroding the Liassic limestones under the Maestrichtian beds, implicates the presence of an "event" between the lower Jurassic and the upper Cretaceous. It should also be noted that the stratigraphic gaps, channelling, and discontinuities appear already during the mid-Liassic (DRESNAY,1962-64) and increase in frequency during the central portions of the mid-Jurassic. The phenomenon of reduction of the sedimentary thicknesses on the crests of anticlines, with variations and accentuations on the marginal slopes in the form of fans has been aptly described by H.TERMIER since 1936. This arrangement has been described by the morphogenetic term "progressive disconformity" by DUÉE et al. (1977) and FEDAN (1980). It should, however, rather be called "syntectonic disconformity" following RIBA (1976). Besides long periods without sedimentation (especially during the upper Jurassic to lower Cretaceous hiatus), it is evident that periods of quiescence followed periods of increased activity (during distension and then during compression). Each of the orogenic "phases" should be studied with its own typical expressions.

- Furthermore, the presence of gabbroic intrusions, so far not distinguished from the upper Triassic basalts, has recently been reported by FEDAN et al. (1985,1986) and compared with similar phenomena known for quite some time already from the High-Atlas. The ascent of this magma could be interpreted as the result of orogenic rearrangements in the mountain range.

- The magmatic activity appears to have continued up to very recent times and expresses itself in the important Quarternary volcanism of the Middle-Atlas. After the initial inventory by H.TERMIER (1936), its study has been continued by HARMAND (1984-85) and MOUKADIRI (1983,1986). According to these authors the volcanism is associated with compression, the main general directions of alignment of the Quarternary volcanoes being NW-SE, i.e. at right angles to the general trend of the chain.

- The rather important seismic activities in the Middle-Atlas have been studied by BEN SARI (1978) and RAMDANI & TADILI (1980). The results are naturally a function of the density of the seismographic network of the region (one station at Ifrane). One notable centre of activity appears to be situated at the edge of the Moulouya (Missour) valley and especially along the longitudinal SW-NE axis which passes through Morocco from Agadir to Nador.

Along this direction the strike slip fault of Tizi n'Trettène could play an important role. This SW-NE axis was substantiated again by AIT BRAHIM (1985). However, the tremor of Ain Leuh, near Azrou on January 17th,1979 with a magnitude of 4.7 (studied by RAMDANI & TADILI,1980) could be related to the SE-NW axis indicated by the alignment of the Quarternary volcanoes of the Middle-Atlas.

The seismicity of the Middle-Atlas, becoming more pronounced in the direction of the Melilla-Nador region, has left its marks on the neotectonic evolution of the chain itself and its surroundings. This has been analyzed for the Guercif basin by COLETTA (1977). Farther to the west, the descent of the Middle-Atlas platform under the Rif thrusts and below the lacustrine limestones of Saiss, as well as its relations with the southern Rif ranges, have been investigated in detail by AIT BRAHIM & CHOTTIN (1983). The study of the folds and faults has indicated a compressive episode of N-S orientation in the Plio-Quarternary limestones of the Saiss basin.

Acknowledgements:
I would like to thank Mr.B.Said, Director of Geology, and Mr.Dahmani, Divisional Head of General Geology, for the facilities made available at the Geological Service of Morocco. Messrs. Ramdani and Tadili furnished data on their observations on the seismicity of the Middle-Atlas. Their assistance is gratefully acknowledged as is that of B.Fedan and A.Charrière who freely gave information on their own studies.

REFERENCES

AIT BRAHIM, L. & CHOTTIN, P. (1983): Mise en évidence d'un épisode compressif dans le calcaires plio-quaternaires du bassin du Saiss, Rif, Maroc.- C.R. Acad. Sci. Paris, (II), 296, 1333-1336.

AIT BRAHIM, L. (1985): La zone de fracture transmarocaine Agadir-Nekor. Critères géophysiques, données de terrain et analyse de documents Landsat.- Bull. Inst. Sci. Rabat, 10, (sous presse).

ANDREU, B. & CHARRIERE, A. (1985): Données stratigraphiques et sédimentologiques nouvelles sur le Crétacé du synclinal de Tirhboula (Moyen-Atlas, Maroc).- Coll. Corrélation du Mésozoique et Cénozoique de l'Afrique de l'Ouest, Marrakech, (IGCP No. 183), p. 46.

ARSICAULT, G. (1965): Présence de bois silicifiés dans le Lias moyen du Maroc.- C.R. Acad. Sci. Paris, 260, 3720-3721.

- (1965): Le Domérien des Aguelmane Tatchas (Moyen-Atlas marocain).- C.R. somm. Soc. géol. France, 1965(2), p. 41.

- (1971): Socle schisteux et arkoses permo-triasiques d'Imi-Ourarn (Moyen-Atlas, Maroc).- C.R. somm. Soc. géol. France, 1971(1), p. 31.

- (1973): Socle schisteux et arkoses permo-triasiques d'Imi-Ourarn (Moyen-Atlas, Maroc).- Notes Serv. géol. Maroc, 34, 109-112.

BEAUGE, A. (1920): Une hypothèse sur la jonction du Moyen-Atlas Nord et du Haut-Atlas marocains.- Bull. Soc. géol. France, (4), 20, 271-279.

BEAUVAIS, L. (1977): Main characters of the Liassic coral fauna from Morocco.- 3rd intern. Coral Reef Symp., Miami, 2, 375-379.

- (1980): Les Calcarea (spongiaires) du Lias du Maroc.- Ann. Paléont. (Invertébrés), 66, 21-41.

BENBOUZIANE, A. (1984): Stratigraphie et sédimentologie du Lias et du Dogger inférieur du bassin de Guercif (Maroc oriental).- Thèse 3ème cycle Univ. Lyon 1, 150 p.

- (1985): Modèles de dolomitisation des formations liasiques (Lias inférieur) et aaléno-bajociennes du pourtour du bassin de Guercif (Est marocain). Comparaison avec les modèles disponibles.- Coll. Bassins sédimentaires du Maroc, Univ. Fes, p. 3.

BEN SARI, D. (1978): Connaissance géophysique du Maroc.- Thèse Univ. Grenoble, 1-262.

BEN SHILI, K. & FEDAN, B. (1984): Coupures fauniques et sédimentologiques dans le Lias-Dogger d'Issouka (Moyen-Atlas, Maroc).- 10ème Réun. ann. Sci. Terre, (Bordeaux), p. 476.

BEN SHILI, K. (1985): Biostratigraphie et sédimentologie de la série Lias-Bajocien du Tizi Nehassa (Moyen-Atlas oriental, Maroc).- Coll. Corrélation du Mésozoique et Cénozoique de l'Afrique de l'Ouest, Marrakech, (IGCP No. 183),p. 33.

BENTAYEB, A. & LECLERC, C. (1977): Le Moyen-Atlas: le Causse moyen-atlasique.- Notes et Mém. Serv. géol. Maroc, No. 231, (Ressources en eau du Maroc, 3),37-66.

BENZAQUEN, M. (1964): Sur la puissance exceptionnelle des formations jurassiques du J. Ouarirt (Moyen-Atlas septentrional, Maroc).- C.R. somm. Soc. géol. France, (1964), 5, 190-191.

BOURROUILH, R. (1966): Gastéropodes du Lias inférieur et moyen du domaine atlasique marocain.- Notes et Mém. Serv. géol. Maroc , 196, 25-67.

CANTAGREL, J.M. & HARMAND, C. (1984): Le volcanisme récent du Moyen-Atlas (Maroc): chronologie K/Ar et cadre géodynamique.- 10ème Réun. ann. Sci. Terre, (Bordeaux), p. 115.

CHARRIERE, A. (1985): Les grands traits de l'évolution mésozoique de la couverture du Moyen-Atlas central (Maroc).- 110ème Congr. Sociétés savantes, (Montpellier), 6, 257-268.

CHOUBERT, G. & FAURE-MURET, A. (1960-62): Evolution du domaine atlasique marocain depuis les temps paléozoiques.- Mém. h. sér. Soc. géol. France (Livre Mém. P. Fallot), 1, 447-527.

\- (1967): Etudes sur le Jurassique du Moyen-Atlas (Maroc): le Jurassique de la région d'El Mers - Skoura (Moyen-Atlas).- Notes et Mém. Serv. géol. Maroc, No. 200, 1-32.

CHOUBERT, G., SALVAN, H., TERMIER, H. & TERMIER, G. (1952): Sur l'âge des calcaires de Timhadite (Moyen-Atlas).- C.R. Acad. Sci. Paris, 234, 2090-2092.

COGNEY, G., TERMIER, H. & TERMIER, G. (1971): Sur la présence de "pillow-lavas" dans le basalte du Permo-Trias au Maroc central.- C.R. Acad. Sci. Paris, (D), 273, 446-449.

COGNEY, G., NORMAND, M., TERMIER, H. & TERMIER, G. (1974): Observations sur le basalte triasique de Romani-Maaziz (Maroc occidental).- Notes Serv. géol. Maroc, 36, 153-173.

COGNEY, G. & FAUGERES,J.C. (1975): Précisions sur la mise en place des épanchements basaltiques des formations triasiques de la bordure septentrionale du Maroc central.- Bull. Soc. géol. France, (7), 17, 721-733.

COLETTA, B. (1977): Evolution néotectonique de la partie méridionale du bassin de Guercif (Maroc oriental).- Thèse 3ème cycle, Univ. Grenoble, 136 p.

COLO, G. (1951): L'Aalénien et le Bajocien inférieur dans la région de Meghraoua (Moyen-Atlas septentrional).- Notes Serv. géol. Maroc, 4, 79-91.

\- (1952): Au sujet de Terebratula termieri.- Notes Serv. géol. Maroc, 6, 103-116.

\- (1953): Sur la faune de térébratules multiplissées d'un gisement du Lias moyen de la région d'Ifrane (Causse moyen-atlasique). - Notes Serv. géol. Maroc, 7, 145-171.

\- (1961-64): Contribution à l'étude du Jurassique du Moyen-Atlas septentrional.- Notes et Mém. Serv. géol. Maroc, No. 139,

No. 139 bis, 226 p., 8 cartes.

COLO, G. & MORIN, P. (1952): Sur l'extension du Miocène au Sud et à l'Ouest de Taza.- Notes Serv. géol. Maroc, 6, 27-31.

COURBOULEIX, S. (1977): Moyen-Atlas et Moyenne-Moulouya: étude bibliographique et paléogéographique.- Rapp. inéd.,Bur. Rech. Partic. Min., Rabat, 83 p.

COURBOULEIX, S., DELPONT, G. & DESTEUCQ, C. (1981): Un grand décrochement E-W au Nord du Maroc à l'origine des structures plissées atlasiques: arguments géologiques et expérimentaux.- Bull. Soc. géol. France, (7), 23, 1, 33-43.

DELANCE, J.H., LAURIN, B. & ROBILLARD, D. (1981): Présence de la Rhynchonelline Sulcirostra brusinai (Eichenbaum) dans le Carixien du Moyen-Atlas septentrional marocain.- Ann. Soc. géol. Nord, 101, 75-78.

DEWEY, J.F., PITMAN III, W.C., RYAN, W.B.F. & BONNIN, J. (1973): Plate tectonics and the evolution of the alpine system.- Geol. Soc. America, Bull., 84, 3137-3180.

DIOURI, M. (1965): Etat d'avancement des levês de la carte d'Azrou au 1/100 000.- Notes Serv. géol. Maroc, 24, 19-21.

DRESNAY, R. DU (1962-64): Les discontinuités de sédimentation pendant le Jurassique dans la partie orientale du domaine atlasique marocain, leurs conséquences stratigraphiques et leurs relations avec l'orogenèse atlasique.- C.R. et Mém., Coll. Jurassique, Luxembourg 1962, publ. Inst. grand-ducal, (1964), 899-912.

\- (1963): Données stratigraphiques complémentaires sur le Jurassique moyen des synclinaux d'El Mers et de Skoura (Moyen-Atlas, Maroc).- Bull. Soc. géol. France, (7), 5, 883-900.

\- (1964): Les découvertes, dans le Dogger du Maroc, de faunes d'Ammonites de la province arabique à Ermoceras: historique, localisations et répartition paléogéographique.- C.R. Acad. Sci. Paris, 259, 4754-4757.

DRESNAY, R. DU & KUNTZ, P. (1964): Un affleurement de Bajocien supérieur à faune d'Ermoceras dans la vallée de la Moyenne-Moulouya (Maroc).- C.R. somm. Soc. géol. France, (1964), 5, 211-212.

DRESNAY, R. DU (1966): Observations stratigraphiques au sujet des gisements de Gastéropodes liasiques du Causse moyen-atlasique et du Haut-Atlas oriental, étudiés par R. Bourrouilh.- Notes et Mém. Serv. géol. Maroc, 196, 5-23.

\- (1967-71): Relations paléogéographiques (Haut- et Moyen-Atlas) et plateforme des Hauts-Plateaux (Maroc oriental) pendant le Bajocien.- 2ème Coll. intern. Jurassique, Luxembourg 1967, Mém. Bur. Rech. géol. Min., 75,(1971), 147-161.

\- (1971): Extension et développement des phénomènes récifaux jurassiques dans le domaine atlasique marocain, particulièrement au Lias moyen.- Bull. Soc. géol. France, (7), 13, 46-56.

- (1972): Sédiments carbonatés intercotidaux et supra-cotidaux du Lias inférieur, dans le domaine atlasique marocain.- C.R. Acad. Sci. Paris, (D), 275, 341-344.

- (1975): Données topographiques, stratigraphiques et paléontologiques concernant une ammonite citée dans la formation d'El Mers (Moyen-Atlas, Maroc) et leurs conséquences sur l'âge attribué à cette formation.- Bull. Soc. géol. France, (7), 17, 1144-1146.

- (1975-77): Le milieu récifal fossile du Jurassique inférieur (Lias) dans le domaine des chaînes atlasiques du Maroc.- 2ème Symp. intern. coraux récifs et coraux fossiles, Paris 1975. Mém. Bur. Rech. géol. Min., 89, (1977), 296-312.

- (1976): Les structures en "tipis" liées aux faciès carbonatés intertidaux et supratidaux du Lias inférieur, dans le domaine des chaînes atlasiques du Maroc.- C.R. Acad. Sci. Paris, (D), 282, 2059-2062.

- (1976): Development and paleogeographic extent of buildups and organic reefs during Jurassic time in Atlas mountains of Morocco, Northwest Africa.- Bull. American Assoc. Petrol. Geol., 60, 4, p. 667; American Assoc. Petrol. Geol., Abstr. vol., New Orleans 1976, 60-61.

DRESNAY, R. DU, ENAY, R., MANGOLD, C. & RAKUS, M. (1985): Le Maroc, un lieu de rencontre entre faunes européennes et faunes d'origine arabe au Bajocien-Bathonien.- Coll. Corrélation du Mésozoique et Cénozoique de l'Afrique de l'Ouest, Marrakech, (IGCP No. 183), p. 39.

DUBAR, G. (1932): Faunes liasiques du Moyen-Atlas marocain.- Ann. Soc. géol. Nord, 57, 3, 181-182.

- (1938): Lias et Jurassique de Tamkroucht (Moyen-Atlas).- C.R. somm. Soc. géol. France, (1938), 6, 93-95.

- (1942): Etudes paléontologiques sur le Lias du Maroc.- Brachiopodes: Térébratules et Zeilleries multiplissées.- Notes et Mém. Serv. géol. Maroc, No. 57, 104 p.

- (1960-62): Notes sur la paléogéographie du Lias marocain (Domaine atlasique).- Mém. h. sér. Soc. géol. France, (Livre mém. P. Fallot, 1), 529-544.

DUBAR, G. & TERMIER, H. (1932): Sur les faciès du Lias inférieur au Toarcien dans le Moyen-Atlas marocain.- C.R. Acad. Sci. Paris, 195, 890-892.

- & - (1940): Notice explicative de la carte géologique provisoire du Moyen-Atlas septentrional au 1/200 000.- Notes et Mém. Serv. Mines Carte géol. Maroc, No. 24 bis, 74 p.

DUBAR, G. & LORENCHET DE MONTJAMONT, M. (1964): Le Lias des Djebel Haloua-Richa et Kebibicha.- C.R. somm. Soc. géol. France, 1964(7), 257-258.

DUEE, G., HERVOUET, Y., LAVILLE, E., DE LUCA, P. & ROBILLARD, D. (1977): Activité synsédimentaire d'un accident coulissant.- Articles, sommaires, analyses, correspondances. Dép. Géol. struct. Univ. Pierre et Marie Curie, Paris, 3, 7-8.

- (1978): L'accident nord-moyen-atlasique dans la région de Boulemane (Maroc): une zone de coulissement synsédimentaire.- Ann. Soc. géol. Nord, 98, 145-162.

DUEE, G. & KAZI-TANI, N. (1982): Cinématique des plaques et tectonique intraplaque: exemple des chaînes intracontinentales d'Afrique du Nord.- 9ème Réun. ann. Sci. Terre (Paris), p. 215.

EL ARABI, H., CANEROT, J. & CHARRIERE, A. (1985): Données stratigraphiques et sédimentologiques nouvelles sur le Lias du Causse moyen-atlasique (région d'Imouzzer-Ifrane).- Coll. Correlation du Mésozoique et Cénozoique de l'Afrique de l'Ouest, Marrakech, (IGCP No. 183), p. 31.

EL HATIMI, N., LAADILA, M. & NASSILI, M. (1977): Contribution à l'étude géologique des environs de Ribat-el-Kheir (Moyen-Atlas, Maroc). - DEA Toulouse, (inéd.), 43 p.

ELMI, S. (1984): Jurassic paleogeography of North Africa and its implications.- Intern. Symp. Jurassic Stratigr., Erlangen, 3, 630-639, (Geol. Surv. Denmark, Kopenhagen).

FALLOT, P. (1941): Sur le Moyen-Atlas.- C.R. Acad. Sci. Paris, 212, 708-710.

FEDAN, B. (1977): Etude structurale d'une portion de l'accident nord moyen-atlasique (Moyen-Atlas central de Boulemane, Maroc).- Bull. Inst. sci. Rabat, 2, 87-96.

- (1978): Etude structurale de l'accident sud-moyen-atlasique entre Enjil des Ikhatarn et Imouzzer des Marmoucha (Moyen-Atlas, Maroc).- Bull. Inst. sci. Rabat, 3, 169-184.

- (1980): Etude structurale du versant Sud du Moyen-Atlas central d'Enjil des Ikhatarn à Imouzzer des Marmoucha.- Thèse 3ème cycle, Rabat, 148 p.

FEDAN, B., DUEE, G. & LAVILLE, E. (1984): Mise en évidence de structures distensives toarciennes dans le Moyen-Atlas (Maroc).- 10ème Réun. ann. Sci. Terre, (Bordeaux), p. 219.

FEDAN, B. (1984): Le Lias-Dogger d'Issouka (Moyen-Atlas): biostratigraphie et sédimentologie.- Bull. Inst. sci. Rabat, 8, 67-76.

- (1984): Les formations carbonatées du plateau de Dwira-Enjil (Maroc): évolution sédimentologique et paléogéographique.- Bull. Inst. sci. Rabat, 8, 51-65.

- (1984): Découverte du Crétacé dans le synclinal des Marmoucha (Moyen-Atlas, Maroc).- Bull. Inst. sci. Rabat, 8, p. 93.

- (1984-85): Naissance et évolution d'une plateforme carbonatée: l'exemple du Moyen-Atlas (Maroc) au cours du Lias: comparaison avec les régions voisines.- Bull. Inst. sci. Rabat, 9, 43-66.

FEDAN, B., HARMAND, C. & LAVILLE, E. (1985): Données préliminaires sur la découverte d'intrusions du Jurassique moyen dans le domaine moyen-atlasique.- Journée géol. min. Maroc, inéd. résumé, p. 79.

FEDAN, B. (1985): Répercussions de la "crise tectonique" du Domérien-

Toarcien inférieur sur l'évolution du Moyen-Atlas au cours du Toarcien.- Coll. Bassins sédimentaires du Maroc, Univ. Fes, p. 2.

FEDAN, B., DELFAUD, J. & DUEE, G. (1986): L'évolution du Moyen-Atlas pendant le Lias-Dogger: prerifting, rifting, comblement, sénescence.- 11ème Réun. ann. Sci. Terre, (Clermont-Ferrand), p. 65.

FEDAN, B. (1986): Présence du Crétacé dans le Moyen-Atlas plissée à l'Est du méridien de Boulemane.- Bull. Inst. sci. Rabat, 10 (sous presse).

- (1986): Données préliminaires sur la découverte d'intrusions jurassiques moyen-atlasiques.- Bull. Inst. sci. Rabat, 10, (sous presse).

FOUCAULD, C. DE (1888, rééd. 1934, 1939, 1985): Reconnaissance au Maroc (1883-1884), 495 p.

GARDET, G. & GERARD, C. (1946): Contribution à l'étude paléontologique du Moyen-Atlas septentrional (Lias inférieur, Lias moyen, Lias supérieur, Aalénien, Bajocien, Bathonien).- Notes et Mém. Serv. géol. Maroc, No. 64, 88 p.

GAUTIER, E.F. (1925): Le Moyen-Atlas. Hesperis Rabat, 3ème trimestre, 25-64.

GENTIL, L. (1916): Sur la structure du Moyen-Atlas (Maroc central).- C.R. Acad. Sci. Paris, 162, 329-332.

- (1916): Note sur la région volcanique du Maroc central.- Bull. Soc. géol. France, (4), 16, 186-218.

GIRET, P. (1985): Histoire paléogéographique, pétrologique et structurale du district à fluorine de Taourirt (Maroc oriental).- Thèse, Univ. Orléans, 192 p.

GUBLER, J. (1935): Rapport géologique sur les travaux éxécutés dans le Moyen-Atlas en 1935 par W. Gelga, M. Gouskov et J. Gubler. Corrélations stratigraphiques et présentation par J. Gubler.- Rapp. inéd. Soc. chérif. pétroles, Rabat.

GUBLER, J., LAPPARENT, A.F. DE, TERMIER, H. (1940): Reptiles et poissons du Bathonien d'El Mers (Moyen-Atlas marocain).- C.R. Acad. Sci. Paris, 210, 768-770.

GUEX, J. (1973): Aperçu biostratigraphique sur le Toarcien inférieur du Moyen-Atlas marocain et discussions sur la zonation de ce sous-étage dans les séries méditerranénnes.- Eclogae geol. Helv., 66, 3, 493-523.

HAMEL, C. (1967): Etude géologique de la terminaison occidentale de la chaîne du Gareb (avant-pays du Rif oriental).- Notes et Mém. Serv. géol. Maroc, No. 199, 1-78.

HARMAND, C.E. (1984): Le complexe à Ijolite-Néphélinite de Talzast (Moyen-Atlas, Maroc).- 10ème Réun. ann. Sci. Terre, (Bordeaux), p. 283.

HARMAND, C.E. & CANTAGREL, J. (1984): Le volcanisme alcalin tertiaire et quaternaire du Moyen-Atlas (Maroc): chronologie K/Ar et cadre géodynamique.- Jour. African Earth Sci., 2, 1, 51-55.

HARMAND, C.E. (1985): Ijolite-nephelinite association at Talzast (Middle Atlas, Morocco): magmatic and metasomatic processes.- Jour. African Earth Sci. (sous presse).

HARMAND, C.E. & MOUKADIRI, A. (1986): Synchronisme entre tectonique compressive et volcanisme alcalin: exemple de la province quaternaire du Moyen-Atlas (Maroc).- Bull. Soc. géol. France, (8), 2, 595-603.

HOEPFFNER, C. (1978): Le massif paléozoique du Tazekka (Maroc): analyse des déformations liées à un linéament tectonique.- Sci. géol. Bull., 31, 33-44.

JAILLARD, L., FEDAN, B. & DRESNAY, R. DU (1984): Subdivisions des formations de la plateforme carbonatée du Lias de Dwira-Angil (Moyen-Atlas, Maroc).- 10ème Réun. ann. Sci. Terre, (Bordeaux), p. 303.

KANES, W.H., SAADI, M., EHRLICH, E. & ALEM, A. (1973): Moroccan crustal response to continental drift.- Science 180, No. 4089, 950-952.

LAADILA, M. (1982): Etude structurale du Moyen-Atlas septentrional (région d'El Aderj, Maroc).- Thèse 3ème cycle, Univ. Mohammed IV, Rabat, 173p.

LAADILA, M., FEDAN, B. & DUEE, G. (1984): Fonctionnement synsédimentaire miocène des accidents d'Al Adarj (Moyen-Atlas, Maroc).- 10ème Réun. ann. Sci. Terre, (Bordeaux), p. 322.

LAADILA, M. (1985): Evolution sédimentaire et structurale de la région d'El Aderj (Moyen-Atlas) au cours du Secondaire.- Bull. Sci. Terre Univ. Rabat, 1, 1, 88-95.

LAGAAY, R., ARENS, P.J., IMANDT, J. & SITTER, L.U. DE (1952): La série mésozoique des environs de Taourirt (Maroc oriental).- Geol. en Mijnb., N.S., 14, 8, 259-267.

LAPPARENT, A.F. DE (1955): Etude paléontologique des Vertébrés du Jurassique d'El Mers (Moyen-Atlas).- Notes et Mém. Serv. géol. Maroc, No. 124, 1-36.

LAVILLE, E. (1977): L'évolution sédimentaire et tectonique de la couverture jurassique de la région d'Imouzzer des Marmoucha (Moyen-Atlas, Maroc).- Bull. Soc. géol. France, (7), 19, 5, 1151-1158.

\- (1979): L'ouverture des bassins subsidents du Haut-Atlas central et du Moyen-Atlas (Maroc). Serait-elle le résultat de 2 directions d'allongement? - 7ème Réun. ann. Sci. Terre, (Lyon), p. 281.

\- (1981): Rôle des décrochements dans le mécanisme de formation des bassins d'effondrement du Haut-Atlas marocain au cours des temps triasiques et liasiques.- Bull. Soc. géol. France, (7), 23, 3, 303-312.

LAVILLE, E. & HARMAND, C.E. (1982): Le "pull-apart basin" haut-atlasique au Mésozoique (Maroc): initiation et fonctionnement.- Coll. rift anciens, Marseille St. Jérôme, Trav. Lab. Sci. Terre, (B), 19, 13-14.

LAVILLE, E. (1985): Evolution sédimentaire, tectonique et magmatique du bassin jurassique du Haut-Atlas (Maroc): modèle en relais multiples de décrochements.- Thèse, Univ. Languedoc, Montpellier, 166 p., 2 cartes.

LORENCHET DE MONTJAMONT, M. (1963): Le Bathonien et le Jurassique supérieur de la plaine de Guercif et de ses bordures (Maroc oriental).- C.R. somm. Soc. géol. France, 1963(3), 102-104.

- (1964): La Bajocien du Moyen-Atlas oriental (Maroc).- C.R. somm. Soc. géol. France, 1964(3), 110-112.

- (1964): La limite du Bajocien et du Bathonien dans le Moyen-Atlas oriental (Maroc).- C.R. somm. Soc. géol. France, 1964(5), 122-124.

MANSPEIZER, W., PUFFER, J.H. & COUSMINER, H.L. (1978): Separation of Morocco and eastern North America: a triassic-liassic stratigraphic record.- Geol. Soc. America, Bull., 89, 901-920.

MARTIN, J. (1973): L'évolution du bassin de Skoura (Moyen-Atlas) au cours du Néogène.- Rev. Géogr. Maroc, 14, 117-131.

- (1981): Le Moyen-Atlas central. Etude géomorphologique.- Notes et Mém. Serv. géol. Maroc, No. 228, No. 228 bis, 445 p., 5 cartes au 1/100 000.

MATTAUER, M., TAPPONIER, P. & PROUST, F. (1977): Sur les mécanismes de formation des chaînes intracontinentales. L'exemple des chaînes atlasiques du Maroc.- Bull. Soc. géol. France, (7), 19, 3, 521-526.

MICHARD, A., WESTPHAL, M., BOSSERT, A. & HAMZEH, R. (1975): Tectonique de blocs dans le socle atlaso-mesetien du Maroc: une nouvelle interprétation des données géologiques et paléomagnétiques.- Earth Planet. Sci. Lett., 24, 363-368.

MICHARD, A. (1976): Elements de Géologie marocaine.- Notes et Mém. Serv. géol. Maroc, No. 252, 420 p.

MONESTIER, J. (1930): Observations nouvelles sur les Ammonites liasiques du Moyen-Atlas marocain.- Notes et Mém. Serv. Mines Carte géol. Maroc, No. 8, 1-36.

MONGIN, D. (1967): Etudes sur le Jurassique du Moyen-Atlas (Maroc): Les mollusques du Bathonien saumâtre du Moyen-Atlas.- Notes et Mém. Serv. géol. Maroc, No. 200, 35-94.

MORABET, A.A. (1974): The Lias reefs of the Middle Atlas platform, Morocco.- M.S. Thesis, Univ. Southern Carolina, 52 p.

MOUKADIRI, A. (1983): Les enclaves ultrabasiques associées aux basaltes alcalins dans le district volcanique d'Azrou-Timahdit (Moyen-Atlas, Maroc).- Thèse 3ème cycle, Univ. Clermont-Ferrand II, 150 p.

MOUKADIRI, A. & KORNPROBST, J. (1984): Garnet and/or spinel bearing pyroxenites in alkali basalts near Azrou (Middle-Atlas, Morocco): mantle derived alumina-rich xenoliths related to the "ariegite-grospydite" trend.- In: KORNPROBST, J. (ed.), Kimberlites II, 179-189 (Elsevier).

MOUKADIRI, A. & HARMAND, C. (1986): Synchronisme entre tectonique compressive et volcanisme alcalin: exemple de la province quaternaire du Moyen-Atlas.- 11ème Réun. ann. Sci. Terre, (Clermont-Ferrand), p. 133.

NASSILI, M. (1982): Etude structurale de la terminaison nord-occidentale du Moyen-Atlas plissée (SW Meghraoua).- Thèse 3ème cycle, Univ. Mohammed V, Rabat, 114 p.

OUARACHE, D., CHALOT-PRAT, F. & CHARRIERE, A. (1985): Découverte d'un volcanisme explosif fini-triasique sur la bordure nord-occidentale du Moyen-Atlas (Maroc).- Coll. Corrélation du Mésozoique et Cénozoique de l'Afrique de l'Ouest, Marrakech, (IGCP No. 183), p. 27.

OUARACHE, D. (1987): Etude géologique dans le Paléozoique et le Trias de la bordure NW du Causse Moyen-Atlasique.- These 3ème cycle Toulouse, 137 p., 1 carte.

PETITOT, M.L. (1959-61): Contribution à l'étude des Echinides fossiles du Maroc (Jurassique et Crétacé).- Notes et Mém. Serv. géol. Maroc, No. 146, vol. Texte: 184 p., vol. Atlas: 74 p., 17 pl.

PIQUE, A. & ROBILLARD, D. (1983): Un golfe à chlorite dans les argilites permo-triasiques de Taza.- Coll. Permo-Trias Maroc, Bull. Fac. Sci. Marrakech, 1, 70-71.

PORTHAULT, B. & TIXIER, M. (1975): Evolution sédimentologique et géochimique du Jurassique inférieur du bassin de Guercif (Maroc oriental). Interprétation paléostructurale.- 9ème Congr. intern. Sédiment., Nice, 5, 2, 335-339.

RAHHALI, I. (1970): Foraminifères benthoniques et pélagiques du Crétacé supérieur du synclinal d'El-Koubbat (Moyen-Atlas, Maroc).- Notes Serv. géol. Maroc, 30, 51-86.

RAKUS, M. (1979): Evolution et position paléogéographique des Monts d'Oujda au cours du Mésozoique.- Mines, Géol. et Energie, 46, 75-78.

RAMDANI, M. & TADILI, B. (1980): Sismicité detaillée du Moyen-Atlas: sondages sismiques profonds dans le Maroc central.- Thèse 3ème cycle, Grenoble, 169 p.

RAYNAL, R. (1961): Plaines et piedmonts du bassin de la Mouloya (Maroc oriental). Etude géomorphologique.- Thèse Fac. lett., Paris; Rabat Ecole livre, 618 p.

REY, M. & NOUET, G. (1958): Microfacies de la région prérifaine et de la Moyenne Moulouya.- Intern. sédiment. petrogr. Ser., 3, 41 p., 97 pls. (Brill).

RIBA, O. (1976): Tectogènese et sédimentation: deux modèles de discordances syntectoniques pyrénéennes.- Bull. Bur. Rech. Géol. Min., (2è sér.), 1, 4, 383-401.

- (1976): Syntectonic unconformities of the Alto Cardener, Spanish Pyrenees: a genetic interpretation.- Sedimentology, 15, 213-233.

ROBILLARD, D. (1977): Données structurales nouvelles sur le Moyen-Atlas septentrional (région de Taza, Maroc).- Articles, sommaires, analyses, correspondances, Dep. Géol. struct. Univ. Pierre et Marie Curie, Paris, 3, 8-9.

- (1978): Etude pétrographique du complexe basaltique "permo-triasique" du Moyen-Atlas septentrional (région de Taza).- Ann. Soc. géol. Nord, 97, 2, 134-144.

- (1978-81): Etude structurale du Moyen-Atlas septentrional (région de Taza, Maroc).- Thèse 3ème cycle, Univ. Lille; Notes Serv. géol. Maroc, 42, (1981), 101-193.

ROBILLARD, D. (1979): Tectonique synsédimentaire du Moyen-Atlas septentrional.- Bull. Soc. géol. France, (7), 21, 4, 441-447.

ROBILLARD, D. & PIQUE, A. (1981-82): Epimétamorphisme du Permo-Trias dans le Moyen-Atlas septentrional (Maroc): Mise en évidence et zonation régionale.- Rev. Géol. dynam. Géogr. phys., 23, 4, 301-308.

- (1983): Mise en évidence par la cristallinité de l'illite de l'épimétamorphisme du Permo-Trias moyen-atlasique septentrional.- Coll. Permo-Trias Maroc, Bull. Fac. Sci. Marrakech, 1, p. 72.

ROBILLARD, D., FEDAN, B., DUEE, G., LAADILA, M. & DELFAUD, J. (1984): Evolution structuro-sédimentaire du Moyen-Atlas (Maroc) au cours du Jurassique.- 10ème Réun. ann. Sci. Terre, (Bordeaux), p. 476.

ROCH, E. (1950): Histoire stratigraphique du Maroc.- Notes et Mém. Serv. géol. Maroc, No. 80, 440 p.

ROUSSELLE, L. (1965): Rhynchonellidae, Terebratulidae et Zeilleridae du Dogger marocain (Moyen-Atlas septentrional, Hauts-Plateaux, Haut-Atlas).- Trav. Inst. sci. cherif., 13; Notes et Mém. Serv. géol. Maroc, No. 187, 186 p.

- (1966): Remarque sur la répartition verticale et l'écologie de Burmirhynchia? termierae R., seul brachiopode connu de la formation supérieure des synclinaux de Skoura et d'El Mers (Dogger du Moyen-Atlas septentrional).- C.R. somm. Soc. géol. France, 1966(9), 346-347.

ROUSSELLE, L. & BISCH, J. (1967-68): Deux nouvelles espèces de Tetrarhynchiinae (Rhynchonelles) dans le Lias moyen du Causse moyen-atlasique (Maroc).- Bull. Soc. géol. France, (7), 9, 777-783.

RUSSO, Ph. (1927): Les chaînes de l'Atlas. Essai d'une vue d'ensemble sur la structure des chaînes marocaines. Rev. géogr. Maroc, 6, 73-82.

- (1927): Sur la présence du Toarcien dans le massif du Maskedal (Moyen-Atlas septentrional).- C.R. somm. Soc. géol. France, 1927(2), p. 12.

- (1930): Le plongement du Moyen-Atlas sous la plaine de la Moulouya (Maroc septentrional).- C.R. Acad. Sci. Paris, 190, 1571-1572.

RUSSO, Ph. & RUSSO, L. (1930): La jonction de l'Atlas et du Tell dans le couloir Taza-Oujda (Maroc oriental).- C.R. somm. Soc. géol. France, 1930(14), 190-191; Bull. Soc. géol. France (4), 30, 1121-1159.

RUSSO, Ph. (1931): Le couloir de Taza à Oujda: géologie des chaînes nord-africaines.- Publ. Assoc. Etude géol. Méditerranée occidentale, 5, 1, 1-18; Barcelone.

- (1933): Géologie des systèmes montagneux du Moyen-Atlas marocain. Morphologie du groupe des Djebilet.- Publ. Assoc. Etude géol. Méditerranée occidentale, 5, 10, 1-16, Barcelone.

- (1933): Au-sujet des rapports des divers segments du Moyen-Atlas et du Grand-Atlas (Maroc).- C.R. somm. Soc. géol. France, 1933(15), p. 238.

- (1933): Remarques tectoniques sur le Moyen-Atlas et les Djebilet.- C.R. somm. Soc. géol. France, 1933(13), 195-196.

- (1934): Sur la géologie de la partie occidentale du Moyen-Atlas et sur les Djebilet.- Publ. Assoc. Etude géol. Méditerranée occidentale, 5, 3, 1-6, Barcelone.

- (1935): Sur la distinction entre le Moyen-Atlas central et le Moyen-Atlas septentrional (Maroc central).- C.R. somm. Soc. géol. France, 1935(5), 69-70.

- (1936): Sur le Lias de l'extrêmité septentrionale du Moyen-Atlas.- C.R. Acad. Sci. Paris, 202, 1866-1867.

- (1936): Géologie des systèmes montagneux du Moyen-Atlas. Rapports tectoniques généraux des reliefs moyen-atlasien.- Publ. Assoc. Etude géol. Méditerranée occidentale, 5, 10, 3-9, Barcelone.

- (1938): Observations nouvelles sur les rapports entre l'aire tectonique du Rif et celle du Moyen-Atlas au Maroc oriental.- Bull. Soc. géol. France, (5), 8, 653-664.

SAADI, M. (1980): Les grandes fractures du Maroc et leur relation avec l'évolution structurale du pays, la sismicité, le volcanisme et les gîtes minéraux.- Rapp. inéd. Ministère Energie Mines, Rabat, No. 1104, 187 p.

SCHMITT, J.M. (1976): Sédimentation, paléoaltération, géochimie et minéralisation en plomb de la série triasique de Zeida (Haute Moulouya, Maroc).- Thèse Ecole nat. Mines, Paris, 97 p.

SCHNELL, P. (1892): Das marokkanische Atlasgebirge.- Petermanns Mittl., 38; trad. franç. (1898) A. Bernard, 326 p., (LEROUX, Paris).

SEGONZAC, G. (1985): Algues chlorophycées des calcaires de Timahdite (Maroc) avec référence particulière à Halimeda.- Coll. Correlation du Mésozoique et Cénozoique de l'Afrique de l'Ouest, Marrakech, (IGCP No. 183), p. 51.

SEGONZAC, G., PEYBERNES, B. & RAHHALI, J. (1986): Les algues du "calcaire rosé de Timahdite" (Eocène inférieur) dans le Moyen-Atlas (Maroc): description d'Halimeda nana Pia, 1932, dans sa localité type.- Jour. African Earth Sci, 5, 501-507.

SEPTFONTAINE, M. (1984): Biozonation (à l'aide des foraminifères imperforés) de la plateforme interne carbonatée liasique du Haut-Atlas.- Rev. Micropaleont., 27, 209-229.

SITTER, L.U. DE & LAGAAIJ, R. (1948): Sur le mode de gisements des laves permo-triasiques dans le Maroc oriental.- C.R. somm. Soc. géol. France, 1948, 13/14, p. 279.

STAUB, R. (1926): Über Gliederung und Deutung der Gebirge Marokkos.- Eclogae geol. Helv., 20, 275-288.

TERMIER, H. (1931): Les discordances de la série mésozoique et cénozoique dans le Maroc central et le Moyen-Atlas.- C.R. Acad. Sci. Paris, 192, 688-690.

- (1936): Etudes géologiques sur le Maroc Central et le Moyen-Atlas septentrional: 1: les terrains primaires et le Permo-Trias; 2: les terrains posttriasiques; 3: Paléontologie, Pétrographie; 4: Atlas.- Notes et Mém. Serv. Mines Carte géol. Maroc, No. 33, 1566 p.

- (1938): Note préliminaire sur la moitié Est de la feuille au 1/100 000ème de Reggou (Maroc oriental).- C.R. somm. Soc. géol. France, 1938(5), 74-76.

TERMIER, H. & LECKWIJCK, W. van (1938): Observations géologiques sur la partie Sud-Ouest de la feuille Taza au 1/100 000ème (Maroc oriental).- C.R. somm. Soc. géol. France, 1938(6), 86-87.

TERMIER, H., GUBLER, J. & LAPPARENT, A.F. DE (1940): Reptiles et poissons du Bathonien d'El Mers (Moyen-Atlas marocain).- C.R. Acad. Sci. Paris, 210, 768-770.

TERMIER, H. & TERMIER, G. (1967): L'estran sablo-vaseux du Bathonien pendant l'exondation de la série d'El Mers (Moyen-Atlas marocain).- C.R. Acad. Sci. Paris, (D), 265, 947-949.

TRUC, G., ABDELJAOUAD, S. & BENSALAH, M. (1985): Unité lithologique et faunique de l'Eocène continental du Maghreb: sédiments détritiques encroûtés contenant une malacofaune à Bulimes.- Coll. Corrélation du Mésozoique et Cénozoique de l'Afrique de l'Ouest, Marrakech, (IGCP No. 183), p. 52.

WILDI, W. (1983): La chaîne tello-rifaine (Algérie, Maroc, Tunisie): Structure, stratigraphie et évolution du Trias au Miocène.- Rev. Géol. dynam. Géogr. phys., 24, 201-297.

YACOUBI, A.N. (1984): Etude micropaléontologique (Ostracodes et foraminiferes) des Marnes de Boulemane (Dogger du Moyen-Atlas marocain).- Thèse 3ème cycle, Univ. Lyon, 133 p.

- (1985): Microfaune (Foraminifères et ostracodes) des Marnes de Boulemane de la localité-type (Dogger du Moyen-Atlas marocain).- Coll. Corrélation du Mésozoique et Cénozoique de l'Afrique de l'Ouest, Marrakech (IGCP No. 183), p. 36; Coll. Bassins sédimentaires du Maroc, Univ. Fes, p. 5.

ZERYOUHI, I. (1970): Etude géologique et hydrogéologique de la partie Nord-Est du Moyen-Atlas plissée et de la Moyenne Moulouya inférieure.- Thèse 3ème cycle, Univ. Montpellier, 179 p.

- (1977): Le Moyen-Atlas plissé.- Notes et Mém. Serv. géol. Maroc, No. 231, (Ressources en eaux du Maroc, 3), 67-84.

SYNSEDIMENTARY TECTONICS IN THE NORTHERN MIDDLE ATLAS (MOROCCO) DURING THE LATE CRETACEOUS AND TERTIARY

Hans-Georg Herbig
Institut für Geologie der Freien Universität Berlin
Altensteinstrasse 34A, D-1000 Berlin 33

Abstract: A pre-Cenomanian to Tertiary sedimentary succession is preserved in five synclines in the central Northern Middle Atlas. The start of the post-Jurassic sedimentation, even within one syncline at different stratigraphic levels above an angular unconformity, is controlled by synsedimentary active, relief creating faults. Further indications of synsedimentary tectonics within the Cretaceous-Tertiary succession are progressive unconformities and conglomerates yielding reworked Jurassic limestones as well as brusque facies changes between the synclines. In the studied area the movements are connected to the Northern Middle Atlas Fault whose southern block is up-thrown and, less amply demonstrated, to the Middle Atlas Fault. The entire Northern Middle Atlas is subject to continuously acting synsedimentary strike-slip faulting since Liassic times.

Résumé: Une série stratigraphique s'étendant de l'anté-Cénomanien jusqu'au Tertiaire est préservée en cinq structures synclinales dans le Moyen-Atlas septentrional central. La sédimentation post-jurassique s'installe dans des niveaux stratigraphiques différents au-dessus d'une discordance angulaire voire dans un seul synclinal. Ce commençement est contrôlé par des accidents synsédimentaires formant des reliefs. La tectonique synsédimentaire dans la série post-jurassique est aussi indiquée par des discordances progressives, des bréches aux calcaires jurassiques et des changements brusques du faciès sédimentaire entre les synclinales. Dans la région étudiée, les mouvements sont liés à l'accident nord moyen-atlasique et, moins évidemment, liés à l'accident moyen-atlasique. Le Moyen- Atlas sepentrional total est affecté par des décrochements synsédimentaires continuant depuis du temps liassiques.

1. Geological setting

The term "Middle Atlas" is often used in a very wide sense. Herein, the terminology introduced by TERMIER (1936) and adopted by COLO (1961) is used which are among the most explicit studies of this mountain chain.

According to their terminology, the studied area is situated in the Northern Middle Atlas which stretches from the Permotriassic scarp of the Oued Serrou in the southwest to the Neogene covered plain of Guercif in the northeast. Two morphological and structural units are discerned in a northwest-southeast directed section. These are the "Causse moyen- atlasique" ("Moyen-Atlas subtabulaire") and the Middle Atlas s.s. ("folded" Middle Atlas). The units are separated from one another by the 40-50° N trending Northern Middle Atlas Fault. In the south, the Neogene basin of Moulouya separates the Middle Atlas s.s. from the east-west trending central High Atlas (Fig. 1).

The "Causse moyen-atlasique" constitutes an elevated plateau consisting of generally subhorizontally lying Jurassic limestones. It is subdivided into several blocks by northeast trending faults. The Middle Atlas s.s. displays Jurassic to Palaeogene, subordinate Permotriassic sediments. Quaternary alkali basalts cover important regions in the central part of both units. The "folded" Middle Atlas consists of four, approximately northeast striking, narrow anticlines ("rides anticlinales" of French authors) which are separated by wide synclines. The synclines may be subdivided by structural highs oriented oblique or normal to their strike. The northwestern flanks of the anticlines are accompanied by important faults with local overthrusting (COLO 1961).

The Upper Cretaceous/Tertiary sediment pile of the Northern Middle Atlas is conserved in five synclines in the central part of the mountain chain. These are from the northwest to the southeast (Fig. 1):

1. Syncline of Bekrit (Syncline of El Koubbat, RAHHALI 1970).
2. Syncline of Timahdit. Situated in the northeastern, striking prolongation of the Syncline of Bekrit, it is widely covered by Quaternary alcali basalts. Both synclines belong to the southernmost part of the "Causse moyen-atlasique".
3. Syncline of Bou Angueur. Separated from the Synclines of Bekrit and Timahdit by the Northern Middle Atlas Fault and the connected anticline ("première ride anticlinale", COLO 1961), it is also widely covered by Quaternary alcali basalts.
4. Syncline of Felledi (Syncline of Ain Nokra). It is situated in the northeastern, striking prolongation of the Syncline of Bou Angueur south of the Northern Middle Atlas Fault and the connected anticline. The synclines are separated by an east-west trending fault and connected anticline (Fault and Anticline of Felledi).
5. Syncline of Oudiksou. It is separated from the Syncline of Felledi by the Middle Atlas Fault (COLO 1961) and an accompanying anticline ("deuxième ride anticlinale", COLO 1961). Further to the west this structure corresponds to the Fault/ Fold of the Aguelmane Sidi Ali (TERMIER 1936, HARMAND & MOUKADIRI 1986). The northeastern prolongation of the Syncline of Oudiksou north of the Middle Atlas Fault is named Syncline of Tirhboula. It displays only Cretaceous sediments.

Synclines without marine Palaeogene sediments (Syncline of Zad, Syncline of El Mers, Syncline of Skoura) are outside of the considered area.

2. Stratigraphic framework

TERMIER (1936) was the first to establish a stratigraphic frame of the post-Jurassic sedimentation in the Northern Middle Atlas. According to him, limestones and marls of

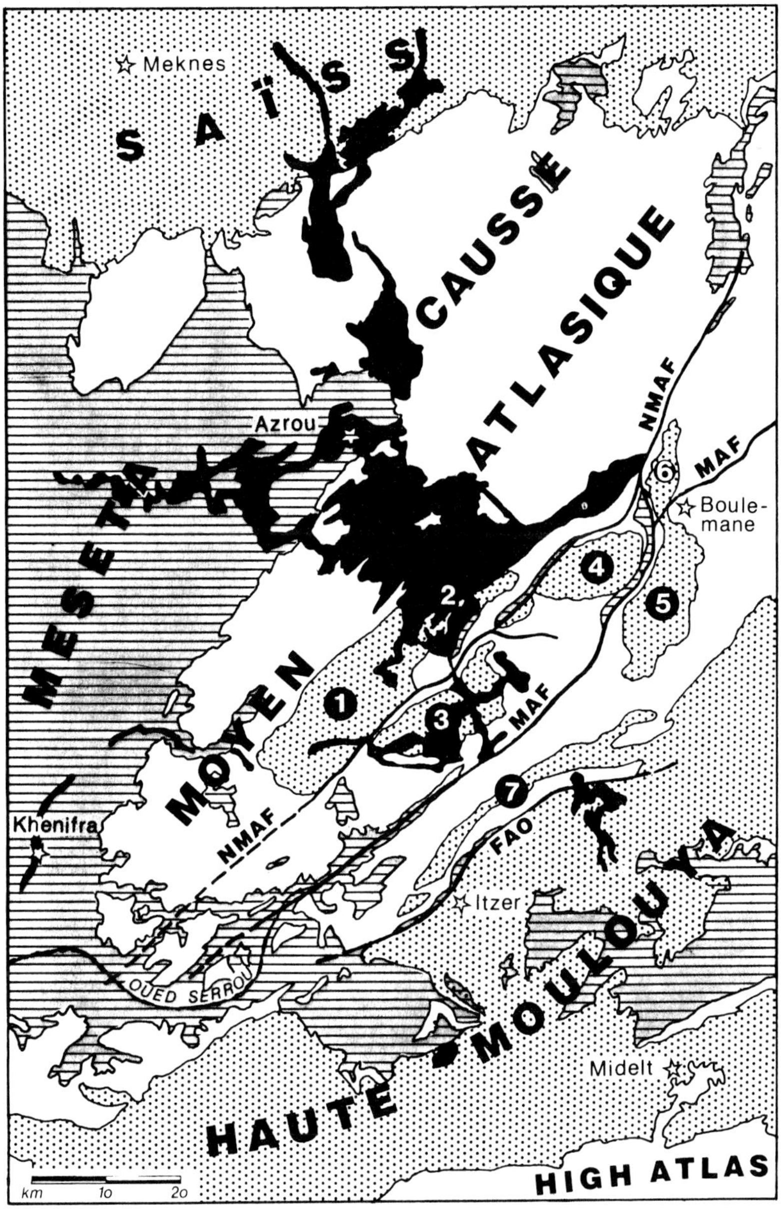

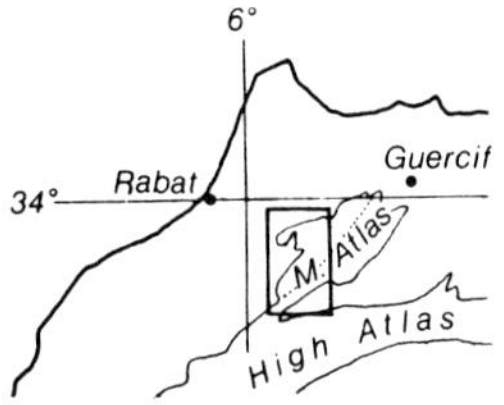

Fig. 1. Geological setting of the Northern Middle Atlas, southern and central part. Horizontally ruled: Palaeozoic-Permotriassic sediments and granites (Haute Moulouya); white: Jurassic sediments; stippled: post-Jurassic sediments; black: Quaternary alkali basalts. NMAF: Northern Middle Atlas Fault; MAF: Middle Atlas Fault; FAO: Fault of Ait Oufella. 1-5: Synclines conserving Cretaceous to Tertiary sediments (1: Syncline of Bekrit; 2: Syncline of Timahdit; 3: Syncline of Bou Angueur; 4: Syncline of Felledi; 5: Syncline of Oudiksou). 6-7: Synclines conserving Cretaceous sediments only (6: Syncline of Tirhboula; 7: Syncline of Zad).

Cenomanian-Turonian age are overlain by unfossiliferous marls and limestones atttributed to the Senonian. COLO (1961) and MARTIN (1981) first described pre-Cenomanian sediments in the Syncline of Tirhboula north of Boulemane and at the northwestern flank of the Syncline of Bekrit (MARTIN 1981). DIOURI (1965) generally stated the existence of pre-Cenomanian conglomerates in the central Northern Middle Atlas.

Later on, only RAHHALI (1970) and MARTIN (1981) paid attention to the litho- and biostratigraphy of Cretaceous and lowermost Tertiary sediments of the Northern Middle Atlas. In the Syncline of Bekrit, above marly limestones of Santonian age, RAHHALI (1970) introduced two Upper Cretaceous formations dated by foraminifers. These are the 90 m thick Campanian Oued Izem Formation composed mainly of marly, slightly bituminous limestones and intercalated oyster lumachelles. It is overlain by the 130 m thick El Koubbat Formation of Maastrichtian age. From the bottom to the top, it is composed of phosphatic limestones, compact marls, black shale, phosphatic marls and lithographic limestones.

The overlying Palaeocene Irbzer Formation consists of a lower, 70 m thick sandstone unit and an upper, 160 m thick unit composed of red mudstones and intercalated gypsum-beds (RAHHALI 1970).

The Irbzer Formation is overlain by the Lower to Middle Eocene Bekrit-Timahdit Formation consisting of 50 m of shallow-marine, mostly escarpment-forming limestones (PIA et al. 1932, CHOUBERT et al. 1952, SEGONZAC et al. 1986).

The overlying, exclusively continental sediments are poorly known. Redbeds, in part with thick gypsum lenses ("Couches rouges de Felledi", TERMIER 1936, CHOUBERT et al. 1952) are thought to be of Upper Eocene age. The overlying lacustrine limestones are attributed to the Upper Eocene and/or Oligocene (CHOUBERT et al. 1952, DU DRESNAY 1969). At the Jbel Hayane, they are overlain by another sequence of redbeds with intercalated lacustrine limestones. Above a slight angular unconformity a several hundred meters thick fining-upward sequence follows, starting at the base with conglomerates and ending, again above a slight angular unconformity, with lacustrine marls and limestones. The conglomerates are tentatively attributed to the Oligocene (DU DRESNAY 1969, MARTIN 1973, 1981), the lacustrine limestones and marls to the Miocene (DU DRESNAY 1969).

3. Synsedimentary tectonics

3.1. Previous studies

Only very few studies explicitly deal with synsedimentary movements in the Northern Middle Atlas. LAVILLE (1977) proved synsedimentary dextral strike-slip faulting during Early and Middle Jurassic times affecting the boundary between the Synclines of El Mers and Immouzzer de Marmoucha east of it. ROBILLARD (1979, 1981) described synsedimentary strike-slip faulting at the Northern Middle Atlas Fault and related faults south of Taza at the northeastern end of the Northern Middle Atlas. Proven from earliest Middle Liassic (Carixian) to early Bajocian times, the movements resulted in the differentiation of the sedimentary realm in a shoal at the eastern border of the "Causse moyen-atlasique" and into a basin east of it in the zone of the future "folded" Middle Atlas. Sedimentary gaps, progressive unconformities, synsedimentary faults, slumping and formation of breccias are related to the movements. Based on similar observations than ROBILLARD (1979, 1981), DUEE et al. (1979) concluded that the Northern Middle Atlas Fault, actually a fault zone in the region of Boulemane, acted as a strike-slip-fault continuously from Upper Liassic to Oligocene times. Strike-slip was dextral from Liassic to Bathonian and sinistral from Upper Jurassic to Oligocene. The vertical component movement controlled the sedimentation by the individualization of shoals and basins.

For Miocene times, LAADILA et al. (1984) described synsedimentary faulting near Al Adarj at the eastern end of the Neogene basin of Skoura. Finally, HARMAND & MOUKADIRI (1986) proved contemporaneous strike-slip faulting during the Middle Quaternary alkaline volcanic activity in the central part of the Northern Middle Atlas. Present day natural seismicity reveals the connection of seismic events with the first order, northeast trending faults of that region (references in HARMAND & MOUKADIRI 1986).

Further studies mention a variety of Neogene movements. MARTIN (1968, 1981) described movements of Miocene, probably Tortonian age and, moreover, movements at the transition from Miocene to Pliocene times and prior to the Villafranchian. Finally, he observed early Quaternary sinistral strike-slip faulting affecting conglomerates of Pliocene-Villafranchian age along the Southern Middle Atlas Fault. MARTIN (1981) stressed the fact that most anticlines and faults were active through all tectonic phases he discerned. The most recent faults affect Middle Quaternary basalts (DIOURI 1965).

Many other authors supposed synsedimentary movements during Mesozoic times or synsedimentary movements can be interpreted from their descriptions. Among those are TERMIER (1936), COLO (1961) - compare examples cited in DUEE (1979), MICHARD (1976) and MATTAUER (1977). In two important papers DU DRESNAY (1963, 1969) described an angular unconformity of Bajocian strata above Liassic limestones, thickness variations and intraformational unconformities in the Syncline of Skoura and the adjoining Syncline of El Mers which are also related to synsedimentary tectonics.

3.2. The base of post-Jurassic sedimentation

The outcrop of Cretaceous and Tertiary strata in the Northern Middle Atlas is restricted to an aerea approximately 65 km long and 25 km wide. In this very small area post-Jurassic sedimentation started at different levels between pre-Cenomanian and Upper Eocene times above an angular unconformity on top of Jurassic limestones (Fig. 2). This may be interpreted as the progressive levelling of a pronounced post-Jurassic relief (e.g. MARTIN 1981). In many cases, however, that is ruled out by the features of the transgressions ,which include:

- Low-angle unconformities, rarely exceeding 15° predominate.
- The direction of dip below and above the unconformities is allways consistent.
- In general, the transgression surfaces on top of the Jurassic strata are smooth. Neither a relief nor karstic phenomena point to a longer subaerial exposure.
- No continually developed coarse-grained (conglomeratic?) near-shore sediments or any other lateral facies differentiations pointing to a shore-line are encountered. Instead, strong facies differentiations are observed between different synclines (chapter 3.5.).

These characteristica do not point to a single Late Jurassic - Early Cretaceous tectonic phase forming a pronounced relief, but to a continually synsedimentary upward-movement of the Jurassic basement along faults and a more or less contemporaneous cover of the newly created relief by successively younger strata. The action of synsedimentary faults is underlined by the fact that the angle of the basal unconformities increases up to 45° (MARTIN 1981: 70°) in sections located near important faults, e.g. in sections near the Northern Middle Atlas Fault at the southwestern flank of the Syncline of Bekrit, at the western end of the Syncline of Bou Angueur and at the northwestern flank of the Syncline of Felledi. The consistent direction of dip below and above the basal unconformities, thereby, points to an unchanged tectonic stress tilting firstly the Jurassic and later, towards the same direction, sediments ranging from pre-Cenomanian to Upper Eocene age.

Connected to the pronounced angle of the basal unconformities, basal conglomerates are only developed in two regions. At the northern flank of the Syncline of Felledi a 5-15 m thick basal conglomerate fills up a palaeorelief. At the western end of the Syncline of Bou Angueur a basal limestone conglomerate is limited to a segment of less than one kilometer in length. There, the Lower - Middle Eocene Bekrit-Timahdit

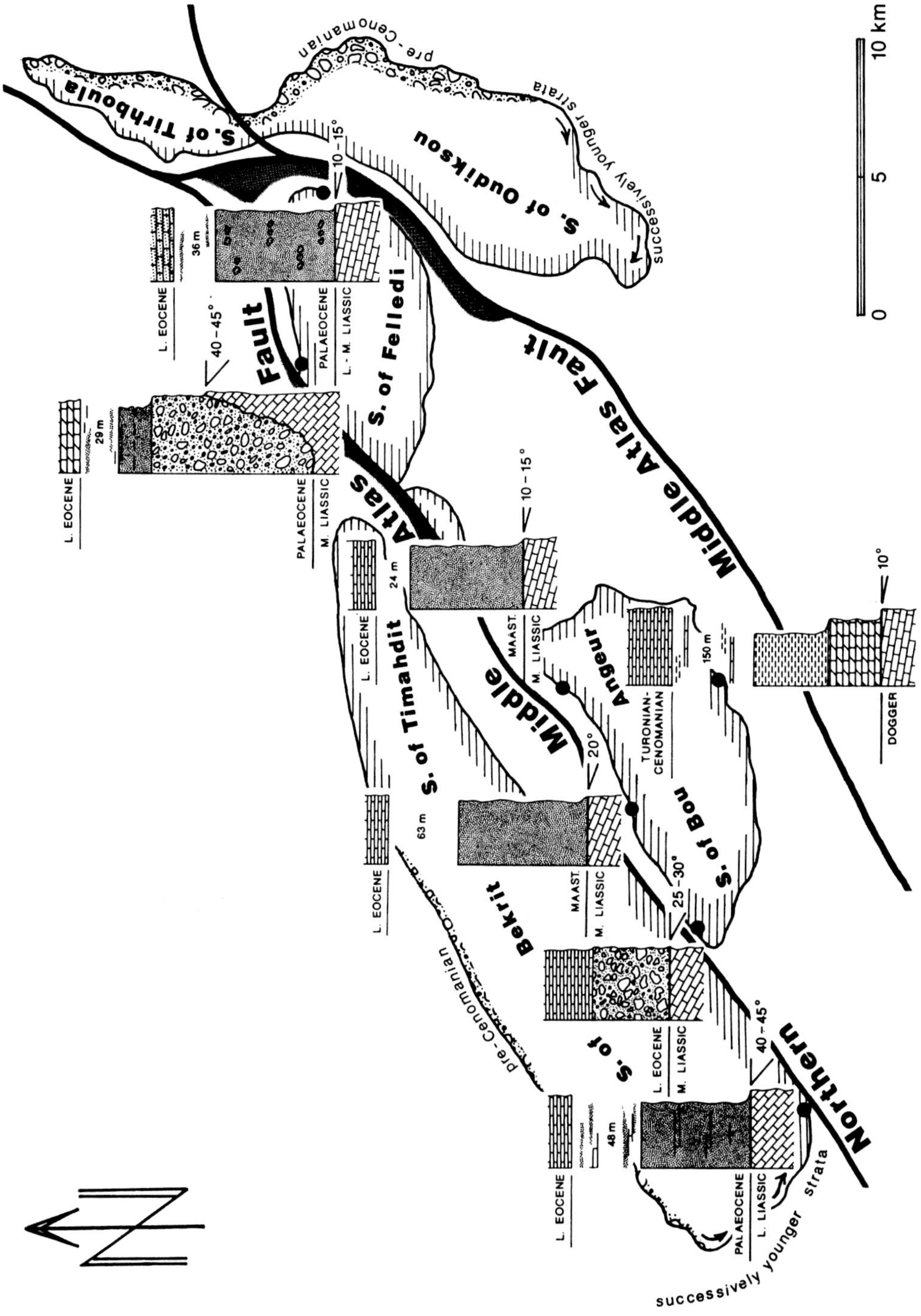

Fig. 2. Angular unconformities at the base of post-Jurassic strata. Grey: Permotriassic outcrops related to major fault movements.

Formation cuts in an angle of 10-20° through Middle Liassic limestones (Fig. 3b, Fig. 4) due to continuous, but local uplift of the Liassic along the Northern Middle Atlas Fault. Local uplift is demonstrated by the fact that all along the northwestern flank of the syncline the Middle Liassic is invariably overlain by the Maastrichtian (Fig. 3a), except in the segment described. Continuously lasting, but slow uplift is shown by the minimal erosion relief on top of the Middle Liassic, which rarely exceeds one meter (Fig. 3c). Further, a slow accumulation of the rather coarse conglomerate (average diameter of boulders: 0.15 m, Fig. 3d; maximum diameter: 1.6 m) is indicated by numerous borings of lithophageous lamellibranchs in the limestone boulders.

Post-Jurassic sedimentation started in pre-Cenomanian to Late Cretaceous times at the northwestern flank of the Syncline of Bekrit and at the southeastern flanks of the Synclines of Bou Angueur, Oudiksou and Tirhboula (CHOUBERT et al. 1956, COLO 1961, DUEE et al. 1979, MARTIN 1981). Opposed to that, it starts between Maastrichtian and Eocene times at the northern flanks of the Synclines of Bou Angueur and Felledi which are situated directly south of the Northern Middle Atlas Fault (Fig. 2). Combined with the conclusions drawn from the increased angle of the basal unconformity in sections close to that fault, the following conclusion is evident: Along the continuously upward moving southern block of the Northern Middle Atlas Fault, a relief could be maintained from the southern end of the Syncline of Bekrit to the eastern end of the Syncline of Felledi, i.e. over the studied distance of 55 km, from pre-Cenomanian to at least Upper Eocene - (?)Oligocene times. Further upward-movements lasting to Recent times are probably (chapters 3.3, 4). The strong upward movements at the Northern Middle Altas Fault are underlined by the fault contact or transgressive contact of "Oligocene" conglomerates with Permotriassic redbeds and associated basalts in the southeastern part of the Syncline of Felledi (CHOUBERT et al. 1956, DUEE et al. 1979).

Fig. 3. Basal unconformities of post-Jurassic strata at the northern limb of the Syncline of Bou Angueur.
a. Section Foum Kheneg. Maastrichtian (m) to Lower/Middle Eocene strata (e, escarpment-forming limestones of the Bekrit-Timahdit Formation) transgress with stratigraphic onlap over Middle Liassic (ml) limestones. View to the northeast.
b. The southeastern flank of the Syncline of Bekrit (left) is separated from the western end of the Syncline of Bou Angueur (right) by the Northern Middle Atlas Fault running through the valley up to the Tizi n'Afit. Within a short distance, the Lower/Middle Eocene Bekrit-Timahdit Formation (eo, light) cuts obliquely through Middle Liassic limestones (ml, dark). This was caused by secular, local movements at the fault. At the very right (dark) Upper Eocene-Oligocene continental strata of the southwestern slope of Jbel Hayane are visible. View to the north.
c. Smooth contact without any signs of an erosional relief between Middle Liassic limestones and the basal conglomerate of the Bekrit-Timahdit Formation, western end of the Syncline of Bou Angueur. Length of hammer: 32,5 cm.
d. Basal conglomerate of the Bekrit-Timahdit Formation. Location as c. Length of knife: 9 cm.

e
m
ml
a
Tizi n´Afit
eo
ml
eo
b
c
d

Analogous observations and conclusions concern the Middle Atlas Fault between the Synclines of Felledi and Oudiksou. The very late (Palaeocene) start of the post-Jurassic sedimentation at the southeastern flank of the Syncline of Felledi and the fault contact between the Palaeocene redbeds of the Irbzer Formation and tectonized Permotriassic redbeds and associated basalts further to the west (Fig. 2) (CHOUBERT et al. 1956, DUEE et al. 1979) reveal the maintenance of a palaeorelief lasting at least into Upper Eocene times (deposition of the "Couches rouges de Felledi", chapter 3.6.).

3.3. Unconformities within the Cretaceous-Tertiary succession

Angular unconformities within the post-Jurassic succession are especially noted at the northern flank of the Syncline of Bou Angeur, i.e. at the Jbel Hayane. Their common feature is the consistent direction of dip below and above an low-angle unconformity which rarely does exceed 15°. Therefore, these progressive unconformities reflect a repeated upward-movement of the southern fault-block of the Northern Middle Atlas Fault which is governed by an unchanged tectonic regime.

Above the basal Maastrichtian sandstones, locally an angular unconformity of up to 40° is present between Maastrichtian and Eocene rocks (MARTIN 1981). The often discussed unconformity between these stages at the section of Foum Kheneg could not be proved. The Upper Eocene -(?)Oligocene lacustrine limestones are separated by an angular unconformity from the Lower - Middle Eocene Bekrit-Timahdit Formation at the western end of the Syncline of Bou Angueur. The structural position of the 35 m thick intervening succession of redbeds (lateral equivalents of the "Couches rouges de Felledi") is not known due to bad exposure. A third angular unconformity is observed at the base of the "Oligocene" conglomerates constituting the cycle B of DU DRESNAY (1969). A last, faible unconformity separates the "Oligocene" cycle B from a "Miocene" cycle C constituted by lacustrine limestones and marls (DU DRESNAY 1969, MARTIN 1981). Still younger movements tilted also these "Miocene" sediments.

The angular unconformity at the base of the "Oligocene" conglomerates is also known in the Syncline of Felledi (DUEE et al. 1979, MARTIN 1981) and, therefore, represents a regional feature. A further angular unconformity within the post-Jurassic succession is known to separate two pre-Cenomanian, but post-Bajocian conglomerate series in the Syncline of Tirhboula (MARTIN 1981).

3.4. Conglomerates within the Cretaceous-Tertiary successsion

Conglomerates containing reworked Jurassic limestones give strong hints for tectonic pulses elevating Jurassic strata above erosion level. Such conglomeratic intercala-

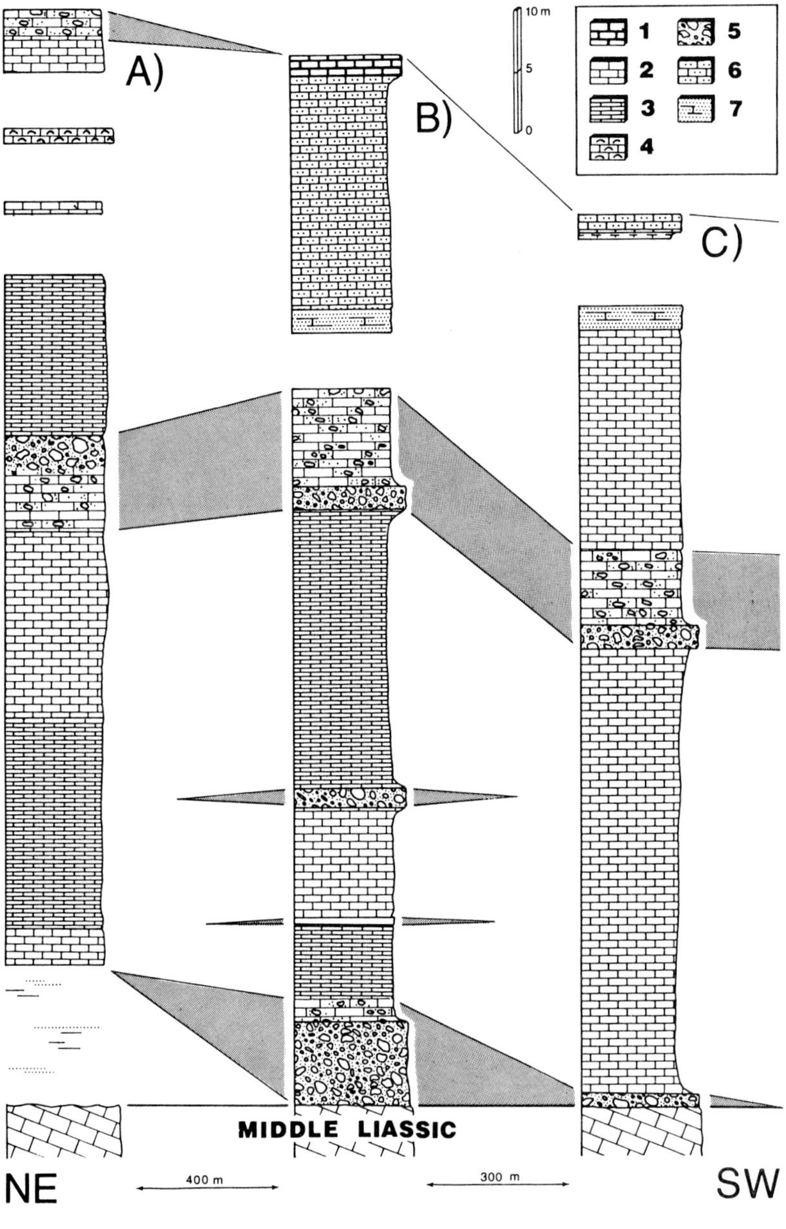

Fig. 4. Sections of the Bekrit-Timahdit Formation at the western end of the Syncline of Bou Angueur show the rapid outwedging of poorly exposed pre-Eocene strata between sections A and B, development of basal conglomerate outwedging to the southwest (sections B, C) and additional conglomerate horizons in higher levels of the Formation. The features are connected to the oblique cut of the Bekrit-Timahdit Formation through the Middle Liassic, caused by local movements of the Nor- thern High Atlas Fault.
1: Packstones, 2: whitish, bioclastic wackestones, 3: white, "chalky" mudstones, 4: oyster-boundstones, 5: conglomerate horizons, 6: sandy mud-/wackestones, 7: calcareous sandstones.

tions are locally encountered at different levels in the uppermost part of the Irbzer Formation, in the Bekrit-Timahdit Formation or its lateral siliciclastic equivalents (southern flank of the Syncline of Oudiksou) and in the succession of Upper Eocene - Oligocene? lacustrine limestones. The conglomerates cannot be correlated across the sections and record small local move- ments of tectonic blocks. Characteristically they are missing in tectonically undisturbed areas such as at the northwestern flank of the Syncline of Bekrit and in the Syncline of Timahdit. They are strongest developed near important faults, such as at the western end of the Syncline of Bou Angueur (Fig. 4). There, the conglomerates are intimately connected with the progressive erosion of the Middle Liassic by the Bekrit-Timahdit Formation (chapter 3.2.). Repeated upward-movements of the southeastern fault-block along the Northern Middle Atlas Fault are reflected by several conglomerate intercalations. According to the close position to the fault, the conglomerates consist exclusively of Middle Liassic limestones and cherts.

Polymict "Oligocene" conglomerates were encountered in the western part of the Syncline of Bekrit, in the Synclines of Bou Angueur and Felledi (DU DRESNAY 1969, MARTIN 1981) and at the Northern flank of the Syncline of Oudiksou. Their regional importance is underlined by an angular unconformity at their base which is known from the Synclines of Bou Angueur and Felledi (chapter 3.3.).

3.5. Facies differentiation

Remarkable facies differences exist between the synclines of the central Northern Middle Atlas from Cretaceous to Middle Eocene times (MARTIN 1981). In the area studied, they were especially noted between the Synclines of Bekrit and Bou Angueur. The brusque change of the sedimentary facies is considered to be the result of the synsedimentary active Northern Middle Atlas Fault which created a palaeorelief (Fig. 5).

In Maastrichtian times the northern, downthrown block (Synclines of Bekrit and Timahdit) constituted a shallow, but enclosed basin, in which the bituminous marls and the black shale of the El Koubbat Formation were deposited . The time-equivalent sediments on the upthrown southern block are phosphatic-calcareous beach sands (northern flank of the Syncline of Bou Angueur) and marly sediments, probably of lagoonal origin (southern flank of the Syncline of Bou Angueur). In Palaeocene times the northern downthrown block trapped the siliciclastic sediments of the Irbzer Formation. The red siltstones and intercalated gypsum-beds in the upper part of the Formation, which are of sabkha or playa origin, reveal the levelling of the sediment-trap. On the southern further upthrown block beach environments or shallow-marine, lagoonal environments with predominantly marly sedimentation still were present. In this way, in Maastrichtian and Palaeocene times 360 m of sediment were accumulated in the Syncline of Bekrit in contrast to 35 m at the northern flank of the Syncline of Bou Angueur.

Striking facies differences are also observed concerning the limestones of the Bekrit-Timahdit Formation. A brusque change is observed between the southeastern flank of the Syncline of Bekrit and the western end of the Syncline of Bou Angueur just across the valley leading to the Tizi n'Afit. In the Synclines of Bekrit and Timahdit rose-coloured and yellow bioclastic wackestones and bioclastic grain-/ packstones predominate. They yield abundant and relatively diversified biota and derive from high-energy environments. At the northern flank of the Syncline of Bou Angueur whitish microcrystalline carbonate mudstones and bioclastic wackestones predominate indicating a low-energy and probably lagoonal environment. Moreover, thicknesses are reduced and a further strong reduction is observed towards the east of the Syncline of Bou Angueur. At its southern flank the Bekrit-Timahdit Formation is present, but its thickness is not yet established due to a strongly deviating facies development.

A similar, but less conspicuous change in the facies development of the Bekrit-Timahdit Formation, obviously governed by a palaeorelief formed by the Middle Atlas Fault, is observed between the Syncline of Felledi and the northern flank of the Syncline of Oudiksou. In the first, especially at its southern flank, a monotonous succession dominated by white, in part very sandy carbonate mudstones is observed; they reflect the restricted environments encountered in the Syncline of Bou Angueur. At the northern flank of the Syncline of Oudiksou sandy bioclastic wackestones and yellow, sandy bioclastic grainstones are common reflecting a less restricted environment and elevated water-energy. A palaeorelief, formed by synsedimentary upward-movements directly north of the Middle Atlas Fault is also suggested by enormous thickness variations of the "Couches rouges de Felledi" in the Syncline of Felledi: 200-300 m thick in the north, they wedge out towards the south, where at the Jbel Irhennba the Upper Eocene - (?)Oligocene lacustrine limestones directly overly the Bekrit-Timahdit Formation (MARTIN 1981).

3.6 Interpretation of post-Middle Eocene continental sedimentation

After the retreat of the Lower/Middle Eocene sea, red siltstones with intercalated gypsum lenses ("Couches rouges de Felledi"), a thick sequence of pure lacustrine limestones and, known from the Jbel Hayane only, a new series of red siltstones with intercalated lacustrine limestones were deposited. They are unconformably overlain by polymict conglomerates of "Oligocene" age. The cyclic change between redbed and lacustrine sedimentation at the Jbel Hayane may reflect climatic variations but tec-

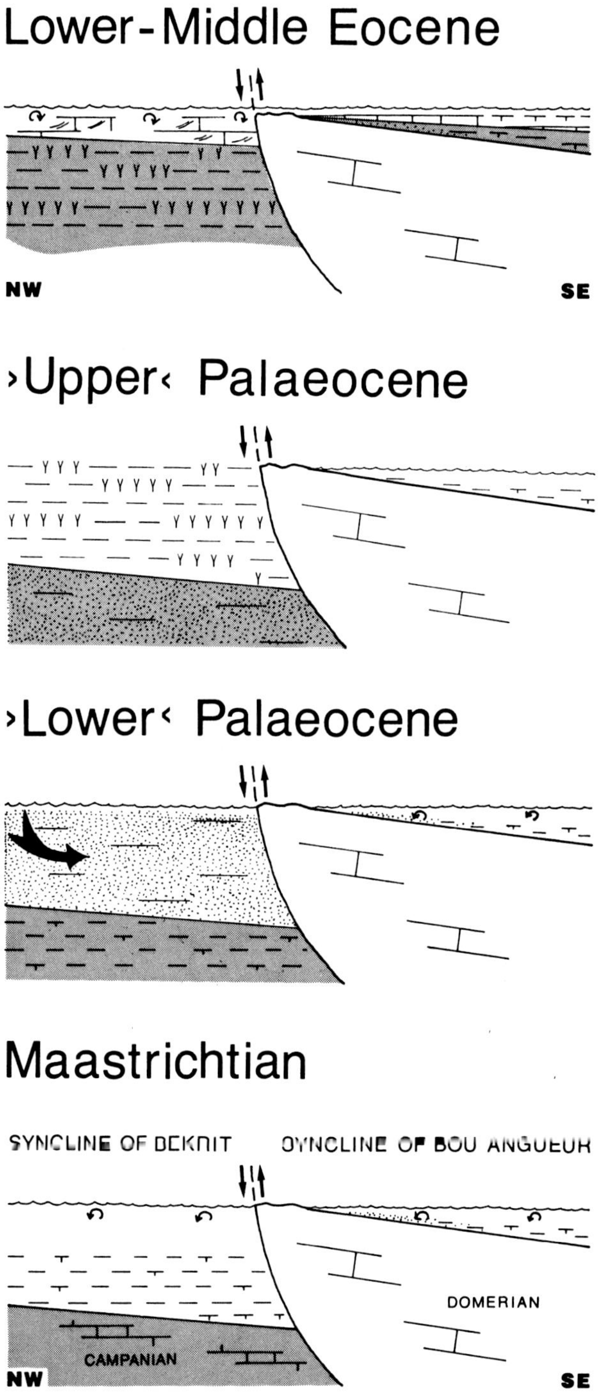

Fig. 5. Facies differentiations at the Northern High Atlas Fault due to synsedimentary movements. See text for further explanation. Not to scale.

tonic pulses as well which might be responsible for ponding effects. Considering the importance of synsedimentary movements in Upper Cretaceous and Lower Tertiary times, a tectonic control of the lacustrine sedimentation is envisaged.

4. Conclusions

Many small-scale, in part local tectonic movements could be demonstrated in the post-Jurassic (pre-Cenomanian - "Oligocene") strata of the central Northern Middle Atlas. Summing up, they result in continuously acting, synsedimentary fault tectonics. In the studied area the movements are related to the Northern Middle Atlas Fault, the southern block of which is upthrown and, not as amply demonstrated, to the Middle Atlas Fault. The Northern Middle Atlas Fault displays a strong synsedimentary sinistral strike-slip component movement at the south-western flank of the Syncline of Bekrit. This can be deduced from the geological map of CHOUBERT et al. (1956) using the method of 'synclinal hooks' (ARTHAUD et al. 1977, compare also DUEE et al. 1979). According to this method, progressively younger strata of the downthrown block (i.e. the Syncline of Bekrit) get in fault contact in direction of the strike-slip, when the view is directed to the opposed upthrown block. These results confirm the study of DUEE et al. (1979), who described continuosly active sinistral strike-slip faulting at these faults in the region of Boulemane from Upper Jurassic to Oligocene times. Both faults are considered to be reactivated late Variscan structures, like other strike-slip faults of the Atlas domain (MATTAUER et al. 1977).

Considering the studies of DUEE et al. (1979), HARMAND & MOUKADIRI (1986), LAADILA et al. (1984), LAVILLE (1977), MARTIN (1968, 1973, 1981), ROBILLARD (1979) and the presented results, the Northern Middle Atlas was subject to synsedimentary faulting latest since Liassic time. Strike-slip movements are of prime importance. Seismic data reveal Recent movements. Last not least, the mountain range from the Jbel Hayane (2409 m) to the Jbel Ben Ij (2426 m) and further to the Jbel Tadjda (2305 m), situated directly south of the Northern Middle Atlas Fault, offers some of the most culminating points of the investigated area; for that, also very young movements may be responsible.

The motor of the synsedimentary movements, at least since the late Jurassic - early Cretaceous folding phase of the Atlas System is the convergence of the European and the African plates (MATTAUER et al. 1977, HARMAND & MOUKADIRI 1986). The responsibility of a continuously acting, unchanged tectonic process is demonstrated by the unchanged tectonic stress in the post-Jurassic sequence (chapter 3.2, 3.3). The connection between plate tectonic collision processes in the western Mediterranean, i.e. orogenic movements in the Betic Cordillera and the Rif Mountains and the Oligocene to

Neogene tectonics of the central High Atlas, situated still further to the south, was also stressed by JACOBSHAGEN (1986).

Acknowledgements

The study is a contribution of the reseach group "Mobility of active continental margins", section "Sedimentary cover of the Atlas System"; it was generously financed by the "Deutsche Forschungsgemeinschaft" and the "Freie Universität Berlin". The friendly support of the "Ministère de l'Energie et des Mines, Division de la Géologie", Rabat, and especially, of Dr. M. Bensaid and M. Dahmani is greatly acknowledged. I am obliged to Prof. V. Jacobshagen, Berlin, who improved the manuscript by a careful review. S. Huhn carried out the photographic work.

Literature

ARTHAUD, F., MEGARD, F. & SEGURET, M. (1977): Cadre tectonique de quelques bassins sédimentaires. - Bull. Centres Rech. Explor.-Prod. Elf-Aquitaine, 1, 1, 147-188.

CHOUBERT, G., FALLOT, P., LEVY, R., MARÇAIS, J., SUTER, G. & TILLOY, R. (1956): Carte géologique du Maroc 1/500 000, Sheet Rabat. - Notes et Mém. Serv. géol. Maroc, N° 70.

CHOUBERT, G., SALVAN, H., TERMIER, H. & TERMIER, G. (1952): Sur l'âge des calcaires de Timhadit (Moyen-Atlas). - C.R. Acad. Sci. Paris, **234**, 2090-2092.

COLO, G. (1961): Contribution à l'étude du Jurassique du Moyen Atlas septentrional. - Notes et Mém. Serv. géol. Maroc, N° 139, N° 139 bis, 226 p.

DIOURI, M. (1965): Etat d'avancement des levés de la Carte d'Azrou au 1/100 000. - Notes Serv. géol. Maroc, **24**, 19-21, Rabat.

DRESNAY, R. DU (1963): Données stratigraphiques complémentaires sur le Jurassique moyen des synclinaux d'El Mers et de Skoura (Moyen Atlas, Maroc). - Bull. Soc. géol. France, (7), 5, 883-900.

DRESNAY, R., DU (1969): Discussions stratigraphiques sur les conditions de gisements de bois fossiles mésozoiques du Maroc, étudiés par Mlles. Y. Attims, F. Crémier et F. Gazeau. - Notes et Mém Serv. géol. Maroc, N° 210, 121-178.

DUEE, G., HERVOUET, Y., LAVILLE, E., DE LUCA, P. & ROBILLARD, D. (1979): L'accident nord moyen-atlasique dans la région de Boulemane (Maroc): une zone de coulissement synsédimentaire. - Ann. Soc. géol. Nord, **98**, 145-162.

HARMAND, C. & MOUKADIRI, A. (1986): Synchronisme entre tectonique compressive et volcanisme alcalin: exemple de la province quaternaire du Moyen Atlas (Maroc). - Bull. Soc. géol. France, (8), **2**, 4, 595-603.

JACOBSHAGEN, V. (1986): Zur Bau-Geschichte des zentralen Hohen Atlas (Marokko). - Berliner geowiss. Abh., (A), **66**, 455-470.

LAADILA, M., FEDAN, B. & DUEE, G. (1984): Fonctionnement synsedimentaire miocene des accidents d'Al Adarj (Moyen Atlas, Maroc). - 10ème Reun. ann. Sci. Terre, (Bordeaux), p. 322.

LAVILLE, E. (1977) L'évolution sédimentaire et tectonique de la couverture jurassique de la région d'Immouzzer de Marmoucha (Moyen Atlas - Maroc). - Bull. Soc. géol. France, (7), **19**, 5, 1151-1158.

MARTIN, J. (1968): L'évolution du bassin de Skoura (Moyen-Atlas) au cours du Quaternaire. - Rev. Géogr. Maroc, **14**, 117-132.

MARTIN, J. (1973): Carte géomorphologique du Moyen-Atlas central, au 1/100 000 avec un schéma structural au 1/500 000. - Notes et Mém. Serv. géol. Maroc, N° 258 (5 sheets).

MARTIN, J. (1981): Le Moyen-Atlas central. Etude géomorphologique. - Notes et Mém. Serv. géol. Maroc, N°258 bis, 446 p.

MATTAUER, M., TAPPONNIER, P. & PROUST, F. (1977): Sur les mécanismes de formation des chaînes intracontinentales. L'exemple des chaînes atlasiques du Maroc. - Bull. Soc. géol. France, (7), **19**, 3, 521-526.

MICHARD, A. (1976): Eléments de géologie marocaine. - Notes et Mém. Serv. géol. Maroc, N° 252, 408 p..

PIA, J., PFENDER, J. & TERMIER, H. (1932): Etudes géologiques sur les calcaires de Bekrit et de Timhadit (Moyen Atlas). - Notes et Mém Serv. Mines Carte géol. Maroc, N° 20, 19 p..

RAHHALI, I. (1970): Foraminifères benthoniques et pélagiques du Crétacé supérieur du synclinal d'El-Koubbat (Moyen Atlas - Maroc). - Notes Serv. géol. Maroc, **30**, 51-98.

ROBILLARD, D. (1979): Tectonique synsédimentaire du Moyen-Atlas septentrional au Sud de Taza (Maroc). - Bull. Soc. géol. France, (7), **21**, 4, 441-447.

ROBILLARD, D. (1981): Etude stratigraphique et structurale du Moyen Atlas septentrional (région de Taza, Maroc). - Notes Serv. géol. Maroc, **42**, 101-193.

SEGONZAC, G, PEYBERNES, B. & RAHHALI, I. (1986): Les algues du "Calcaire rosé de Timahdite" (Eocène inférieur) dans le Moyen-Atlas (Maroc): description d'*Halimeda nana* Pia, 1932, dans sa localité-type et son paléoenvironment. - Jour. African Earth Sci., **5**, 501-507.

TERMIER, H. (1936): Etudes géologiques sur le Maroc central et le Moyen Atlas septentrional. Tome II. Les terrains post-triasiques. Tome IV. Atlas des figures et tableaux hors-texte. - Notes et Mém. Serv. Mines Carte géol. Maroc, N° 33.

GEOPHYSICAL STUDIES OF THE EARTH'S CRUST AND UPPER MANTLE IN THE ATLAS SYSTEM OF MOROCCO

Gerhard Schwarz & Peter J. Wigger
Institut für Geophysikalische Wissenschaften
Freie Universität Berlin
Rheinbabenallee 49, D-1000 Berlin 33

Abstract

The cause and mechanism that led to the uplift of the Atlas Mountains of Morocco is unknown. In some aspect it may be seen as a singular phenomenon within the frame of the new global tectonics. The key for a better understanding is the knowledge of the deep structure of this orogene. This was the reason to perform seismolgical observations, deep seismic-, magnetotelluric-, and geomagnetic depth soundings. The field measurements were carried out along a profile from Erfoud in the Anti-Atlas to Azrou in the Middle Atlas in 1983.

Seismicity in Morocco is concentrated to the Rif, the Middle and the High Atlas. Most earthquakes occur at shallow depth, some of them have intermediate depth.

The High Atlas region shows a distinct intracrustal discontinuity at 15-21 km depth, crustal thickness itself varies between 30 and 37 km, with its maximum near the southern border and the contact zone towards the Middle Atlas. The lower crust is characterized by a low velocity layer of different thickness and extension.

Electrical conductivity is high at mid crustal depth from the inner High Atlas towards the Middle Atlas. Further to the south, towards the Anti Atlas normal resistivities were found in crust and mantle.

The anomalous conductivity structures in the middle crust as well as the seismic low velocity layer may point towards partial melt at this depth range, and a higher fluid pressure, most likely indicating planes of overthrusting.

1. Introduction

The Atlas Mountains of Morocco are bordering the northern and north-western African continent. Whereas their structure and development have been described by a number of authors, e.g. MICHARD (1976), STETS & WURSTER (1981), very little is known about the reason of the uplift of the Atlas Mountains. The geology shows very different processes: collision zones at the northern margin at the Mediterranean Sea, and a passive margin at the Atlantic coast. The Rif-Atlas, the northernmost Atlassique chain, is dominated by wide spanned overthrusting with the typical features of Mediterranean orogenes. In comparison, Middle and High Atlas do not show alpine tectonics, as crustal shortening or large overthrustings. Nevertheless, with an elevation of more than 4000 m a.s.l., the High Atlas Mountains belong to the high mountain systems of the world. The southernmost mountain range, the Anti-Atlas, already belongs to the African craton, a more or less stable platform since the beginning of the Mesozoic.

The reasons are unknown, which led to the formation of the Middle and High Atlas. With the early Mesozoic the region has undergone - like a rifting zone - subsidence and sedimentation, whereas in the Eocene an inversion of this movement is believed to have started: Middle and High Atlas developed to high mountains. An inversion of such large scale is unknown in plate tectonics.

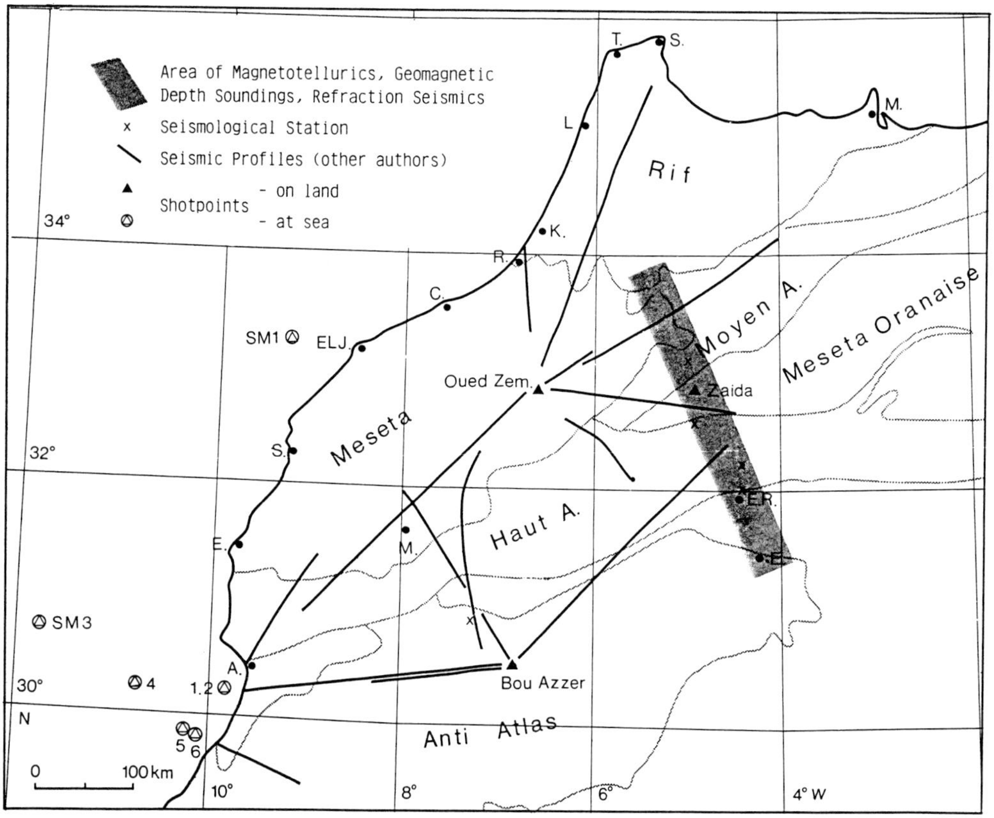

Fig. 1: Location map of magnetotelluric, seismic (inside shaded area), and seismological recording sites in the Atlas system of 1983 field campaign. Straight lines represent seismic refraction profiles from other authors discussed in text.

As already mentioned, the surface geology of the Atlas system is roughly known, whereas physical parameters are still missing to describe the state of the crust and upper mantle. Geophysical investigations in the Atlas mountains have been restricted to a few methods and regions. Most of the work has been refraction seismics (e.g. MAKRIS et al. 1985, BEN SARI 1978) in the Moroccan Meseta and the western High Atlas. Gravimetric investigations and reflection seismics have been carried out in the Rif-Atlas (e.g. DEMNATI 1972). The gravity field (VAN DEN BOSCH 1971) and the magnetic field (ETUDE AEROMAGNETOMETRIQUE 1974) have been investigated on regional scale. Nevertheless, these data have not yet been sufficient to construct a profile through the crust and Upper Mantle, crossing all structural units of the Atlas system.

For the first time the distribution of the electrical conductivity was studied in the Atlas system by the magnetotelluric method. The electrical conductivity depends on the physical state, as well as on the material. Of special interest is the relation between the content of fluids in the earth s crust and electrical conductivity (e.g. SCHWARZ et al. 1986a).

One of the aims of our investigations is to collect uniform physical parameters for the crust and upper mantle, on a profile from the Anti-Atlas, on the African platform, towards the Rif, with its Mediterranean structural style. These investigations covered seismology, refraction seismics, magnetotellurics and geomagnetic depth soundings. The first profile section has been measured in 1983 (see fig. 1), the second one with seismics quite recently late in 1986. In the following we will summarize the existing data, present our own results and try to infer a geodynamic model from these data.

2. Gravity

A Bouguer map of Morocco was published by VAN DEN BOSCH (1971). The bouguer gravity of this region shows mainly negative values except in a narrow strip at the coast of the Alboran sea, the western part of the Moroccan Meseta and its southern prolongation. The most interesting features are two negative anomalies with values down to - 150 mgal. The northern one correlates to the Rif mountains. The larger one covers the central High and the Middle Atlas. The north-western border of these mountain ranges is situated between the -50 and -100 mgal lines. Also the Meseta Oranaise which lies in the east between the Middle and the High Atlas shows a minimum of less than -100 mgal. The lowest values (<-150 mgal) indicate some local intramountaneous structures at the southern border of the central High Atlas. The gradient of Bouguer gravity is high at the north-western border of the Atlas chains and low at the southern border towards the Anti-Atlas.

Along our section a minimum of -120 mgal is reached at the Haute Moulouya, a high plateau placed between Middle and High Atlas. North of the Middle Atlas as well as south of the High Atlas towards the Anti Atlas values of about -50 mgal can be found. Two local heights are superposed. One with an amplitude of about 10 mgal at the southern border of the Middle Atlas and the second one in the range of the High Atlas with an amplitude of about 20 mgal.

MAKRIS et al. (1985), QURESHI (1986) and TADILI et al. (1986) calculated some models of crustal structure basing on the gravity field.

3. Seismicity

The Atlas system in northwestern Africa is situated in the tensional field associated with the convergent movements of the African and European plates and of the long transform faults reaching from the Mid-Atlantic ridge to the Azores and farther into the street of Gibraltar. A description of the regional seismicity and the tensional field is given by UDIAS (1982). The seismicity of the Gibraltar arc and adjacent regions was described by HATZFELD et al. (1977) and HATZFELD & FROGNEUX (1981). Fig. 2 shows the epicenter map compiled by these authors. They relate shallow earthquakes to the transform fault, but there are intermediate events, too, under the Gibraltar arc and Alboran sea. Furthermore there are two very deep events with a focal depth greater than 600 km below southern Spain.

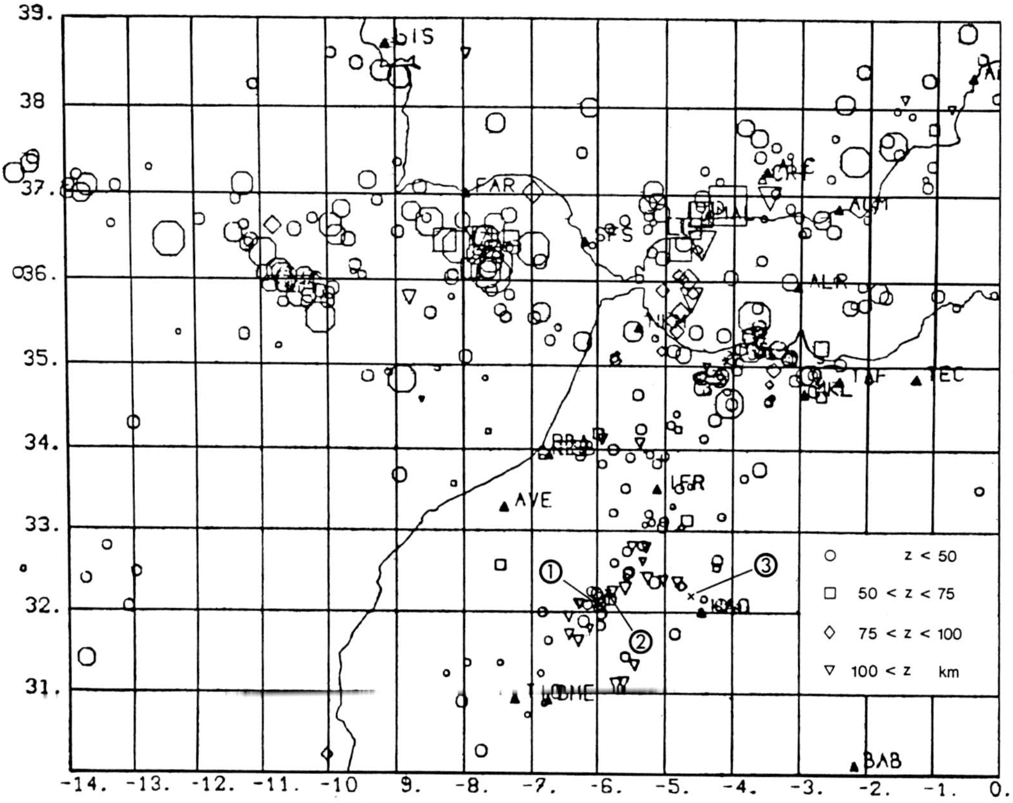

Fig. 2: Distribution of epicenters for NW-Africa, the arc of Gibraltar and adjacent regions after HATZFELD & FROGNEUX (1981). The number 1 to 3 indicate intermediate deep earthquakes which were recorded during our field work in 1983.

The seismicity of the Atlas system is bound to distinct structures: To the arc-like southern border of the Rif mountains on one hand and to a strip of the Middle,

High and Anti-Atlas on the other hand which are characterized by Bouguer gravity values of less than -80 mgal. The area with the highest seismicity inside this zone is the contact between High and Middle Atlas. The magnitudes of the Moroccan earthquakes are moderate. Strong events (M>6) are not known from Morocco, although quakes with catastrophic consequences may occur, e.g. the earthquake close to Agadir in the western part of the High Atlas on February 29th, 1960 (M = 5.7). Most of the Moroccan earthquakes are shallow ones, the rest are intermediate events with depths down to 180-200 km. Most of these happen under the central High Atlas. It should be mentioned, that the depth range between 30 and 100 km behaves relatively aseismic.

During our field work in 1983 we operated five short period seismic stations. The recording sites are given in fig. 1. Three of the recording stations (PCM-systems) were equipped with an event trigger, and the data were stored on magnetic tape, two stations recorded continuously on paper. More than 100 local and several far distance events could be observed. The program "HYPOPT" (BUNESS et al., 1986) was used for hypocenter localization. The velocity-depth values which are given in tab. 1 represent a reasonable model basing on a great number of seismic profiles from different authors (see next chapter: crustal seismics). Processing and interpretation of the data have not yet been completed, but some features can be mentioned here. Most of the localized events are shallow, except three of them for which an intermediate depth has been calculated. The epicenters are indicated in fig. 2 by the number 1, 2 and 3. The calculated focal depths are 185, 130 and 175 km, respectively. These hypocenters as well as the great number of intermediate events determined by HATZFELD & FROGNEUX (1981) are located in the High Atlas between 4.5° and 6.5° W. The northern limit of their occurrence is the junction between High and Middle Atlas. In absence of a known recent subduction dipping to the south which could explain these earthquakes, we have to look for another explanation for this intermediate seismicity in the depth range of 80-200 km confined to such a small zone in the upper mantle. This phenomenon may lead to the assumption that not only the general stress field of northwestern Africa is responsable for these earthquakes, but also a regional anomaly in the upper mantle may effect or trigger this seismicity.

Table 1: Velocity-depth model for P- and S-waves used for hypocenter localization.

Depth (KM)	VP (KM/S)
0.0	5.00
7.5	6.00
12.0	6.20
18.0	6.80
18.0	6.40
28.0	6.40
30.0	7.20
32.0	8.10
220.0	8.10
22o.0	8.55

Depth (KM)	VS (KM/S)
0.0	2.88
7.5	3.46
12.0	3.57
18.0	3.92
18.0	3.69
28.0	3.69
30.0	4.15
32.0	4.67
220.0	4.70
220.0	4.77

4. Crustal seismics

There exist a large number of crustal seismic profiles for Morocco, the northern adjacent to the Alboran Sea and the continental margin offshore to the Atlantic ocean. Descriptions of these profiles and their interpretation as well as the record sections can be found for northern Morocco and the Alboran sea in DEMNATI (1972), HATZFELD & BEN SARI (1977) and BEN SARI (1978), for the Atlantic continental margin in HINZ et al. (1982), WEIGEL et al. (1982), and THIESSEN (1979), for central Morocco and the Anti-Atlas in RAMDANI & TADILI (1980), HECK (1982), and MAKRIS et al. (1985). The recording lines related to the Atlas system are drawn in fig. 1. Summarizing the results of the above mentioned authors it can be stated that the Moroccan crust is a typical continental crust with a thickness of 30 km. It thickens a little under the Atlas Mountains but not comparable e.g. with the root of the Alps. The values of crustal thickness under the Atlas chains differ from region to region and from author to author between 30 and 40 km. An intracrustal discontinuity, which is generally well developed, can be found in depths between 15 and 23 km. As indicated by Bouguer gravity, the shape of the crust under the High Atlas mountains is asymmetric. From the northern border of the High Atlas, towards the Moroccan Meseta crustal thickness changes more rapidly to lower values than from the southern border towards the Anti-Atlas. Contrary to the above mentioned authors WIGGER & HARDER (1986) derived from the published record sections a slight low velocity zone in the lower crust under the Atlas chains.

The present knowledge of the crustal structure beneath the Atlas systems does not include the region of the eastern part of the High Atlas and the Middle Atlas (only few recording sites have been installed there). To fill that gap a seismic as well as a magnetotelluric survey were carried out on a NNW-SSE running line: Meknes - Azrou - Midelt - Errachidia - Erfoud. Due to limited fundings special explosions could not be arranged, the explosions in the lead mine of Zaida - situated in the center of the mentioned profile (fig. 1) - served as signal source for the seismic recordings. Details of field work and data processing are given in WIGGER & HARDER (1986).

Two profiles have been measured: Zaida to NNW and Zaida to SSE. The resulting two record sections are given in fig. 3. The generation of seismic wave energy by explosions in the lead mine Zaida was unfortunately very small, too small to produce reliable wave groups which have penetrated the lower crust and the crust/mantle boundary. Therefore no recordings were taken at distances greater than 120 km away from Zaida. Keeping further in mind a probable lateral inhomogeneous structure under the High Atlas and that these profiles are not reversed we have to look on the results very carefully.

Figure 4 shows the calculated one-dimensional velocity-depth function for the record section Zaida to NNW crossing the Middle Atlas. Up to 50 km distance a pro-

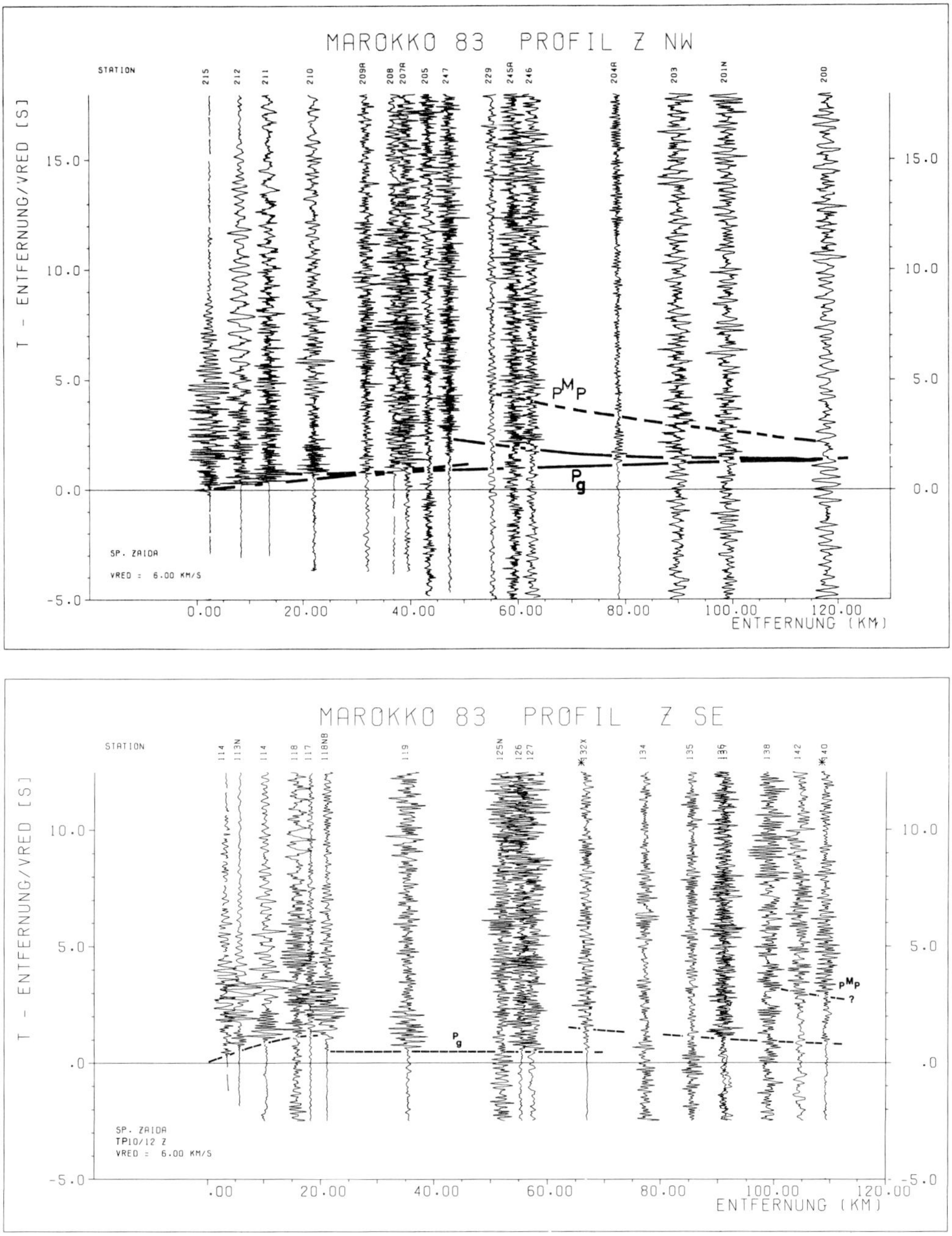

Fig. 3: Record sections with P-wave correlation (reduction velocity: 6 km/s).
Top: profile Zaida to NNW; bottom: Profile Zaida to SSE.

grade phase which is related to a sedimentary cover of a thickness of about 2.3 km and an increasing velocity from 4.7 to 5.3 km/s is observed. The P_g-wave develops from the adjacent retrograde phase with a velocity increase to 5.7 km/s. It can be correlated up to 120 km, the velocity increases to 5.9 km/s and the penetration depth is 10.5 km. Without time delay a retrograde phase is observed which reaches an apparent velocity of 7.1 km/s in 17 km depth. The further dashed drawn retrograde correlation between 120 and 60 km with a time delay of about 1 s could indicate the P^MP-phase. If this correlation is correct,crustal thickness under the Middle Atlas (Azrou) is 29 km. In this case the lower crust has a low velocity zone with 6.2 km/s.

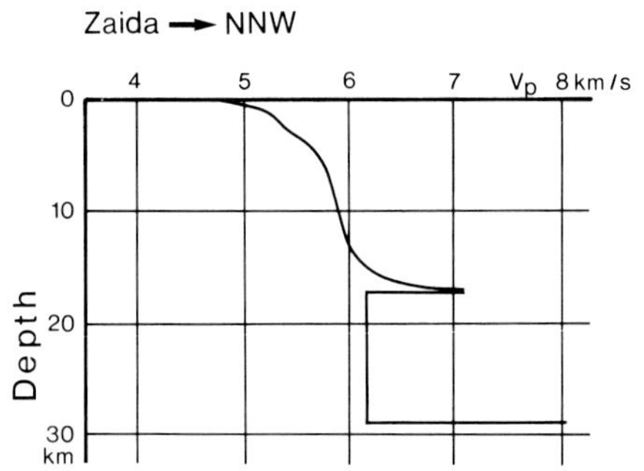

Fig. 4: Velocity-depth function for the line Zaida to NNW. Calculated on the base of the correlation shown in fig. 3.

The record section Zaida to SSE (fig. 3, bottom) shows clear arrivals up to 67 km and some single seismograms with good onsets. The correlation starts with a velocity under 3 km/s. This phase can only be followed to a distance of 20 km where the apparent velocity reaches 5.1 km/s. A local structure should be the reason for this wave group. In greater distances the Pg-wave is identified up to 70 km, with v = 6 km/s. The next phase is of retrograde character. It leads to a discontinuity in 14-15 km depth, where the velocity increases to 6.5 km/s. With certain reservations the marked late arrivals of the last two seismograms (105-115 km) could be part of the P^MP-reflection at the critical distance. With an assumption of 7.8 km/s at the Moho, crustal thickness under the High Atlas could be 35 km. In respect to the relatively late arrival a low velocity zone can be expected there.

The resulting velocities and depths as well as some extrapolated values from the next neighbouring profiles are combined with the results of the magnetotelluric measurements to a crustal section (fig. 9). As mentioned above, seismic results have to be looked upon very carefully and the length of the error bars, especially for the depth of the Moho, may reach 6 km. Open questions are e.g. the crustal structure under the Haute Moulouya between Middle and High Atlas, an asymmetric crustal structure beneath the High Atlas in this section. One of the most important questions is the velocity as well as the structure in the uppermost mantle.

To find adequate answers to these questions, crustal seismic investigation with special blasts have been carried out late in 1986. Data evaluation of this field work has been terminated now. Calculations and interpretation will follow and detailed information will be given in a later paper.

5. Magnetotelluric and Geomagnetic Depth Soundings

The electrical conductivity measurements were carried out parallel to the seismic profile starting in the Anti-Atlas near Erfoud, traversing the High Atlas and the Haute Moulouya, and ending near Azrou in the Middle Atlas for the moment (fig. 1). The continuation of this profile towards the North, the Rif, is planned to be carried out early in 1988. The field measurements of the 1983 campaign have been described in an earlier paper (see SCHWARZ et al. 1986b). In fig. 5 we show an example of the data analysis for site BUG, situated near the southern border of the Haute Moulouya - towards the High Atlas. Apparent resistivities have been calculated in a NS-EW coordinate system. Typical for all sites from the inner High Atlas towards the north, is a more or less strong anisotropy of apparent resistivities thus indicating strong lateral changes of apparent resistivity in the neighbourhood of each site. At site BUG the anisotropy is moderate, the electrical field is polarized in the direction of 60° E, e.g. near to the geological strike of the Haute Moulouya.

The strong anisotropy of apparent resistivity curves from the High to the Middle Atlas indicates two- or even three-dimensional conductivity structures. The influence of topography in particular on the measured data for B-polarization has to be considered as well. We assume that these two-dimensional structures are due to near-surface structures. In this case we can interprete the sounding curves of the E-polarization by one-dimensional models as a first approximation. E-polarization is characterized by an electrical field parallel to the conductivity structures, with a magnetic field perpendicular to it. Due to the strong effect of anisotropy the distinction between E- and B-polarization is most essential for the interpretation of measured data. This question, of how to decide between both cases has been discussed e.g. by SCHWARZ et al. (1984) for data from the Andes. We will follow their arguments and therefore we will rely on the magnetic induction arrows as an indicator for lateral changes in conductivity.

In fig. 6 the magnetic induction arrows for 4 periods are represented for the complete section from the Anti- towards the Middle Atlas. From their orientation as well as from their length one can infer a strong change in conductivity at the southern border of the High Atlas. The induction arrows reach their maximum in length for periods between 1000 and 4000 s, thus indicating structures of extreme high electrical conduc-

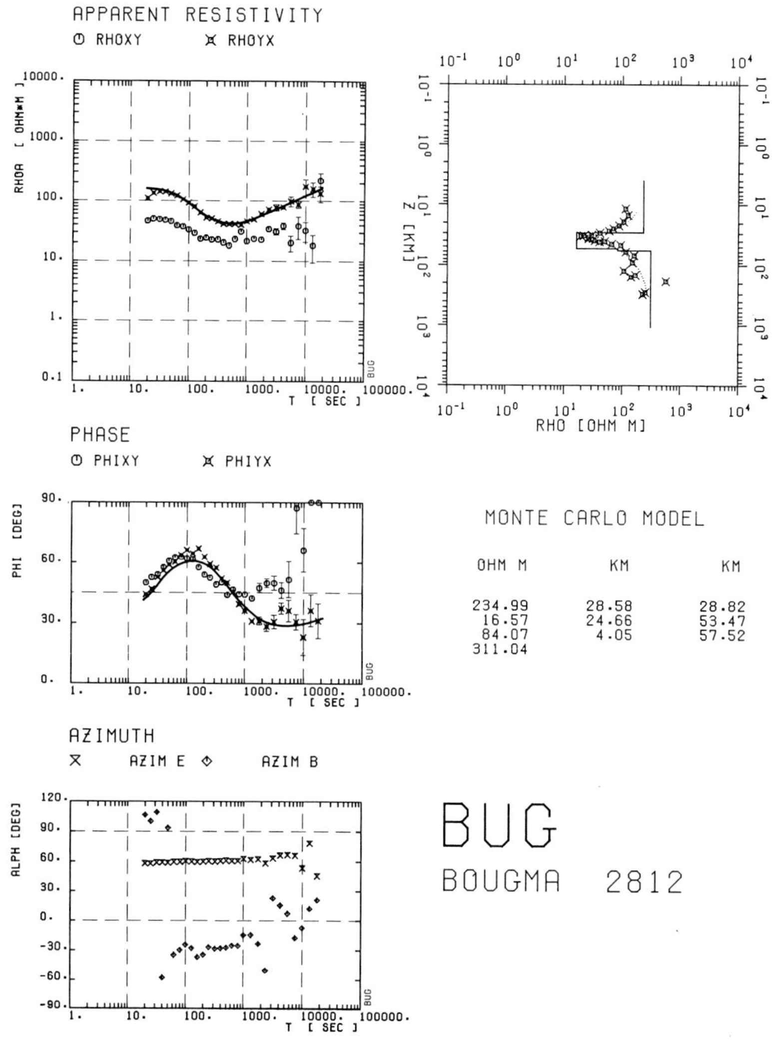

Fig. 5: Magnetotelluric results from site BUG in the Haute Moulouya. Left panels from top to bottom: Apparent resistivity, phase between electrical and magnetic field, and azimuth (polarization) of E- and B-field. Apparent resistivity and phase calculated from the electric EW-component and the corresponding magnetic NS-component have been used for resistivity model inversion: Upper right panel shows 4-layer resistivity model, deduced from Monte Carlo Inversion (MARTINEZ 1987), as well as resistivity $\rho^*(z^*)$, (SCHMUCKER 1971), marked by crosses. Drawn lines in left panels represent model curves.

tivity or at extreme crustal depth, respectively. The conductivity anomaly at the southern border of the High Atlas is striking under ENE-WSW, e.g. the geological strike of the mountain range. In the northern High Atlas and in the Haute Moulouya the induction arrows seem to represent a superposition of induction anomalies. The main strike of anomalies has N-S direction, only for a period of 4000 s the arrows point to the large scale anomaly, which is striking ENE-WSW. In the Middle Atlas the arrows display anomalous structures which strike approximately E-W. Due to the large amplitudes of magnetic transfer functions and due to the fact that they never change their sign along the profile one has to assume very complicated conductivity structures. It seems that their centre has not yet been crossed.

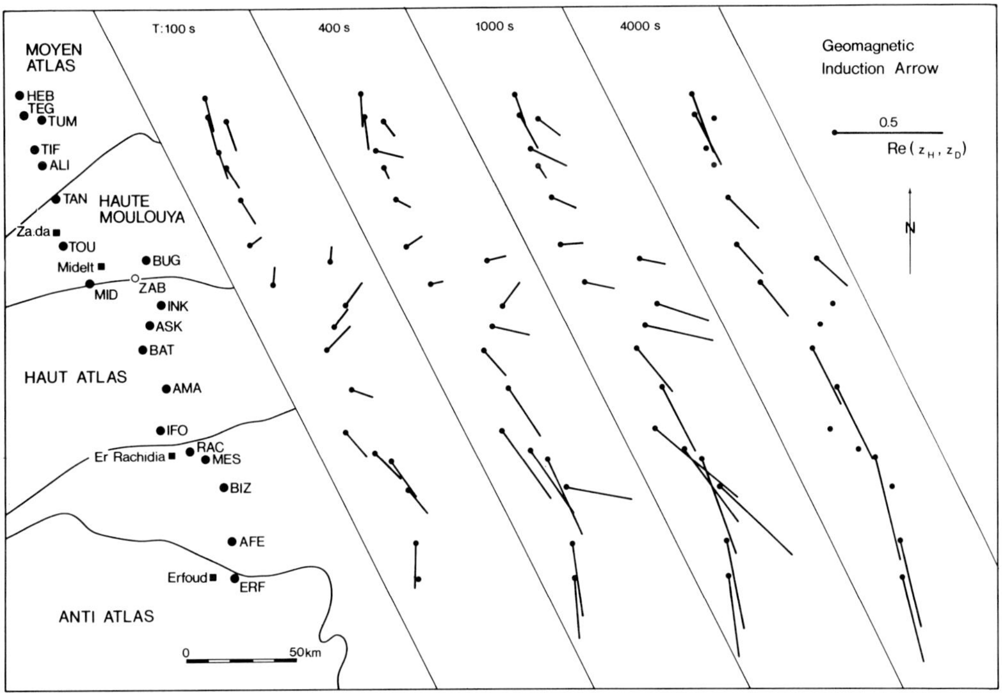

Fig. 6: Geomagnetic induction arrows along the profile from the Anti- towards the Middle-Atlas for periods of 100, 400, 1000, 4000 s. Only the real part of the complex magnetic transfer function (SCHMUCKER 1970) is shown. The arrow is pointing away from well conducting structures and reaches its maximum in length where the contrast of the electrical conductivity is greatest, e.g. the southern border of the High Atlas. Part of the sites were equipped with induction coil magnetometer. That is the reason for some data gaps at 4000 s period.

The situation shown by the magnetotelluric impedance along the profile is as well a complex one, due to static distortion effects of the electrical field from site to site. The preference directions of the E-field do not correlate with the main strike of the structural units, but seem to be controlled by small electrical inhomogeneities. The calculated transfer functions of the magnetic field along the section are

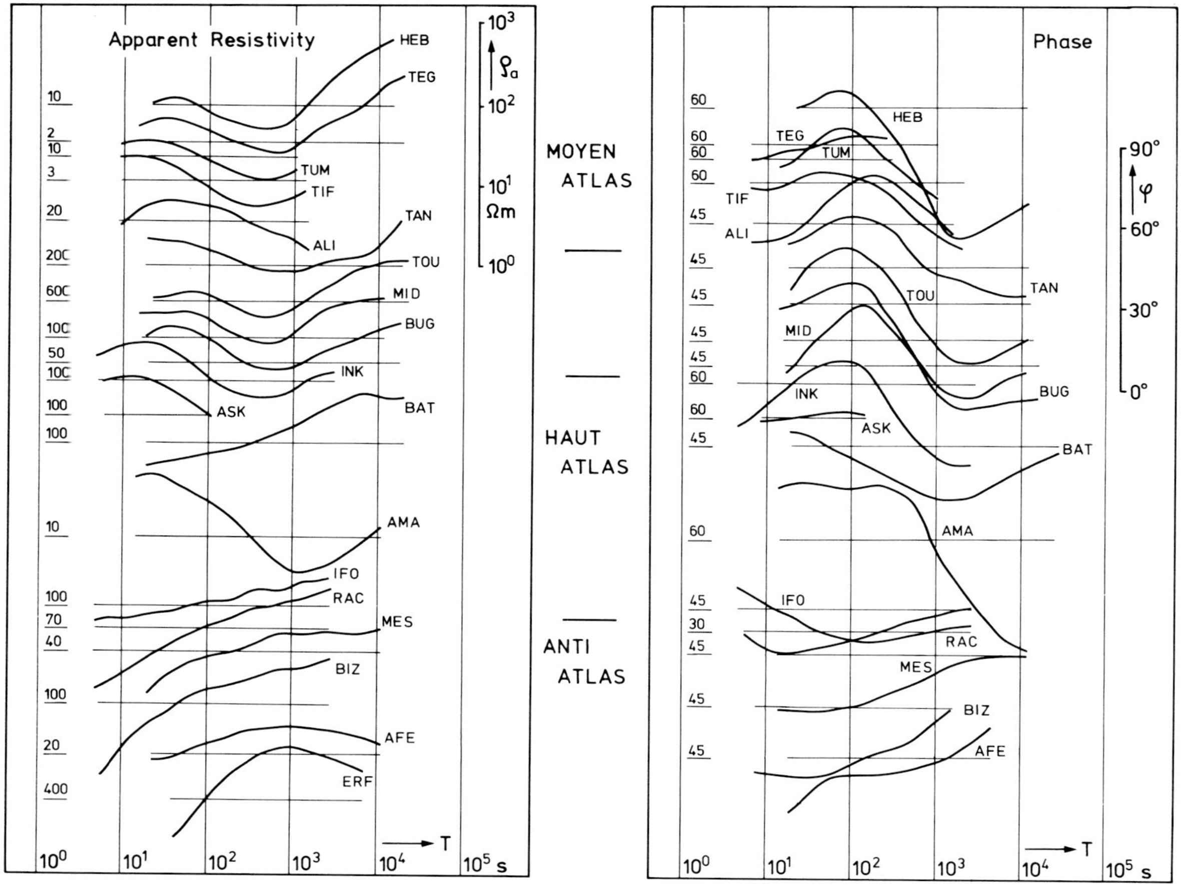

Fig. 7: Apparent resistivity and phase for E-polarization calculated from electrical field in E-W direction and inducing N-S component of magnetic field mounted for all 19 sites. Underlined numbers on the left of each panel indicate base lines of individual sounding curves (in ohm m or degrees).

smooth functions at each site, except from the Haute Moulouya. Therefore we have taken the direction perpendicular to the magnetic induction arrows as that one of the E-polarization, that is the E-W direction. That means we look on magnetotelluric impedances in a fixed, unrotated NS-EW coordinate system.

Figure 7 displays apparent resistivities and phases for E-polarization along the profiles from the Anti- to the Middle Atlas. Two groups of sounding curves can be distinguished by their similar shape: In the northern profile section, extending to the northern High Atlas, resistivity curves are characterized by a prominent minimum between 100 and 1000 s period, but their level is different from site to site. We have to discuss this fact later on. The other group of electrical resistivity curves, from the center of the High Atlas towards the south is characterized by a slight increase in resistivity with period, without any other prominent feature.

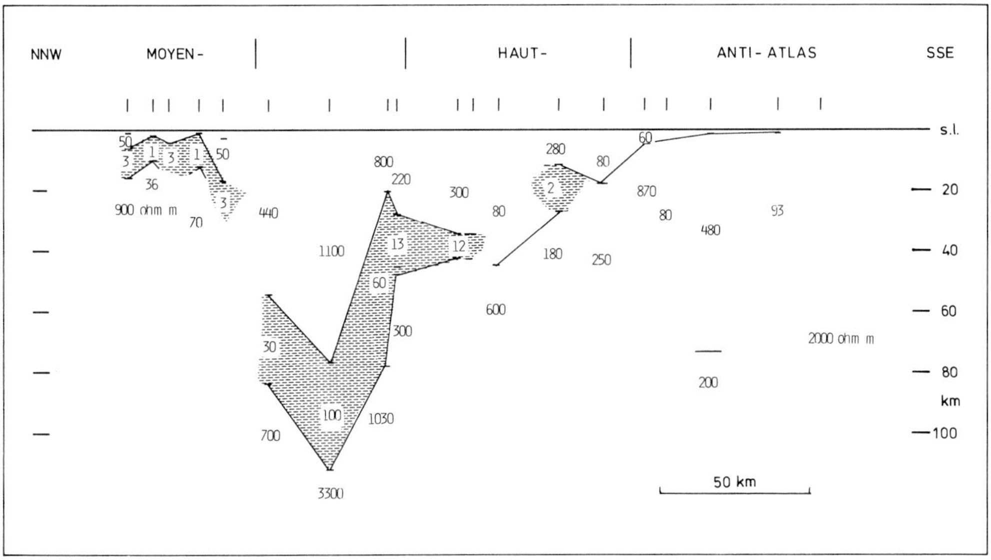

Fig. 8: Pseudo two-dimensional section of electrical resistivity versus depth, compiled from 1D-models, and unshifted. Dashed lines represent low resistive layer.

The data from fig. 7 have been used to calculate resistivity depth models, as shown in fig. 4, upper left panel. Electrical resistivity $\rho*$ versus depth $z*$ (SCHMUCKER 1971) has been computed as well as layered resistivity models using a Monte Carlo inversion technique (MARTINEZ 1987). In the latter case we looked for those models which could approximate measured curves with a minimum number of layers, in most cases this have been three layer-models (fig. 8). From the Middle- towards the northern High Atlas the resistivity models display a prominent low resistivity layer at different depth range, whereas the resistivity structure further to S, towards the Anti-Atlas is more homogeneous. In the northern profile section one can observe strong lateral changes in resistivity at about 100 km depth, with differences of more than one decade. For distances of about 10-20 km from site to site one should expect a much more homogenous resistivity structure beneath the Atlas, e.g. like that in the earth's mantle beneath the African platform, that means values of about 100 ohm m. This indicates a distortion of magnetotelluric impedances by near surface inhomogeneities. This is corroborated by the shapes of the phase curves (refer to fig. 7) as well, which are much more consistent from site to site versus period. Another justification for a distortion of resistivity curves at different sites could be a constant shift of resistivity values versus period along straight lines for same periods. That means the penetration depth z* can be transformed for same periods by a factor to a constant depth. Under these assumptions we multiplied the magnetotelluric transfer functions at the sites in the Haute Moulouya and the High Atlas with a constant distortion factor. This factor has been chosen in that way that the electrical resistivity at 100 km depth reaches 100 ohm m. The value of 100 ohm m is a "typical one" for the mantle depth of a "global" resistivity model. These shifted one-dimensional resistivity models were compiled to a pseudo two-dimensional model given in fig. 9 and will be discussed in detail in the next chapter.

6. Discussion

Geophysical investigations in the Atlas Mountains of Morocco up to now focussed on their western and northern part. Only very few information was available for the northeastern profile section from the Anti- towards the Middle Atlas, although this traverse offers the unique possibility to cross all structural units of the Atlas system. Geologists from Free University of Berlin (e.g. JACOBSHAGEN 1986) decided to concentrate their future activities on this section. That was the reason to concentrate our own geophysical activities on this profile as well, offering the chance to combine geological knowledge of subsurface structures with geophysical data, giving information on deep structures.

In fig. 9 we compiled the data from geology and geophysics in a section from the Anti- towards the Middle Atlas. As can be seen in the geological cross section (BREDE & HEINITZ 1987, CARTE GÉOLOGIQUE DU MAROC 1985) structures on the whole profile are layered, overthrusting of "large scale" does not occur. South of the High Atlas we find Jurassic flat layers covering a Paleozoic basin which crops out farther to the south. The tectonic setting inside the High Atlas is dominated by deep reaching faults, which again can be found in the Middle Atlas. In the northern Middle Atlas we have to assume dike-like young volcanic intrusions, in the Haute Moulouya Paleozoic granites of large extent are known. Crustal thickness, calculated from seismics, can be assumed to reach 30 km beneath the Middle Atlas, and beneath the High Atlas to reach about 35 km. Under both mountain ranges a low velocity zone has been found by refraction seismics for lower crustal depth.

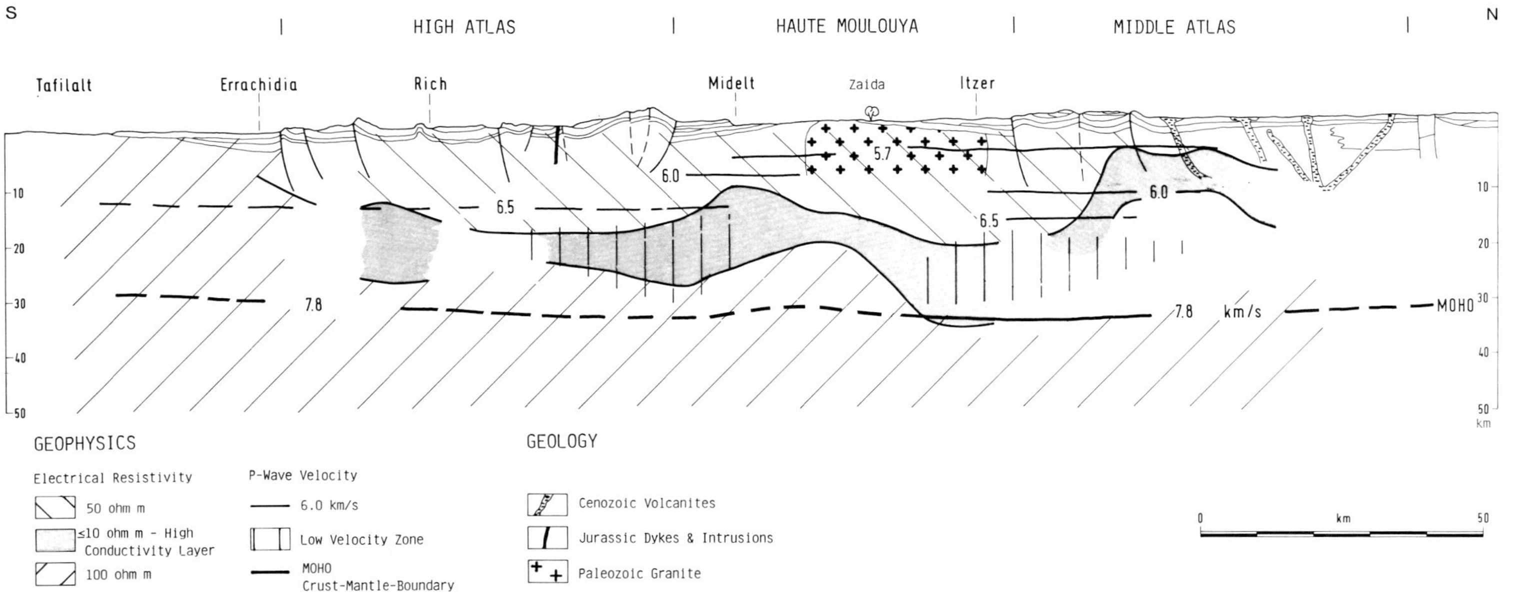

Fig. 9: Crustal section for the profile from the Middle Atlas towards S with pseudo 2D-model of electrical resistivity and lines of equal seismic velocity. Geology after BREDE & HEINITZ (1987) and CARTE GÉOLOGIQUE DU MAROC 1985).

The dominant result of the magnetotelluric survey is a low resistivity layer at mid crustal depth beneath Middle and High Atlas. The pseudo two-dimensional resistivity model (fig. 9) in this form does not explain the results from geomagnetic depth soundings (refer to fig. 6), which imply a much stronger lateral variation of the resistivity structure. Comparable anomalies have been found by GDS at the southern border of the High Atlas, W of Quarzazate (MENVIELLE & LE MOUEL 1985) and at the southern border of the Rif (LEGSP & DMGM 1977, MENVIELLE & ROSSIGNOL 1982). A characteristic feature of anomalous magnetic variations in these three Atlassique zones is a ratio of vertical to horizontal magnetic field of about 0.5 at periods of about 1 hour. This common feature could point towards comparable sources of the observed GDS-anomalies, e.g. such as deep reaching fault systems. Such a fault system has been found at the southern border of the High Atlas (MATTAUER et al. 1977, MICHARD et al. 1975). This lineament was interpreted as a large transform fault. We know from recent crustal seismic experiments that lateral fault systems of large extent could reach in their vertical extension even Upper Mantle depth (e.g. BREWER & SMYTE 1984). Such vertical fault systems can explain induction anomalies, if they contain fluids of high salinity. Of course, the origin of these fluids remains an open question, as well as the mechanism to keep fault systems open under high lithospheric pressure and prevent the fluids from escaping through the crust.

To explain the results of GDS in total one has to assume a complex fault system running along ENE-WSW through the High and even the Middle Atlas. Such a bundle of faults in NE-SW direction is indeed not only known from the western High Atlas (AMBROGGI & NELTNER 1952, PROUST et al. 1977), but also in nearly the whole Atlas system (e.g. JACOBSHAGEN 1986). The problem still stands how to combine GDS- and magnetotelluric results into one consistent resistivity model. A single large scale anomaly of electrical resistivity cannot explain the observed data. One has to ask, whether the vertical fault systems could be connected electrically to the crustal conductor. We will discuss this question with the help of geological data.

The geological development of the Atlas system is not very well known. Whereas MATTAUER et al. (1977) postulate an overthrusting under lateral shortening, STETS & WURSTER (1981) concluded that the Atlas has undergone prevailingly vertical movements. Most likely the High and Middle Atlas originated from a rifting process, which has to be seen in context with the opening of the Atlantic Ocean. The rifting occurred together with extensional tectonics and with transform faults of the NW-margin of North Africa. Such a rifting is normally initiated by mantle diapirism, which always is connected with a sinking of the crust involved (e.g. MIDDLETON 1984). Which physical process led to the inversion of movements, e.g. the development of the Atlas high mountain range? The wide spanned uplift of the mountain ranges to be seen in topography and to be inferred from the large scale gravity low indicates a deep source for vertical movements. Seismic activity at 120-200 km depth reinforces this assumption. Can one assume ascending basaltic melts and its accretion to the crust/mantle-boundary? The extended volcanism in the Middle Atlas and the Meseta Oranaise display that basaltic melts have been activated. Another hint for basic intrusions are the numerous small anomalies in the total magnetic field (ETUDE AEROMAGNETIQUE 1974). Finally one can observe secondary processes related to the rise and accretion of magma: crustal fluids have been generated by volcanic and subvolcanic activities, indicated by numerous hydrothermal deposits in the Meseta Oranaise. High fluid pressures within the crust could have led to a decoupling of lower and upper crust. In such a situation crustal blocks could have been moved upwards under lateral shortening, e.g. under the compressive stress of convergent plate motions.

The origin of induction anomalies in the Atlas mountains can thus be seen in small amounts of melt in the middle and lower crust. These melts could explain also the observed low velocity zones at crustal depth. Fluid phases liberated in a subsequent process have a high electrical conductivity and could produce induction anomalies while ascending through vertical fault systems to the upper crust.

Acknowledgements: The geophysical investigations in the Atlas Mountains were carried out in cooperation with the Ministry of Energy and Mines, Division of Geology, Rabat. We wish to thank all co-workers, in particular Drs. M. Bensaid, M. Hamuda, and A. Bellot, as well as the members of Institut Scientifique, Service Physique du Globe, and CNR, Rabat. We are indebted to SODIM Mining Cooperation, Zaida, for their cooperation in explosion seismology. Field assistance has been given by Sigrid Harder, Volker Rath, Peter Rüdinger, and Ulf Seidel, all from FU Berlin. Furthermore we would like to acknowledge instrumental support given by Drs. Martin Beblo, Alfred Berktold, München, and Hartmut Jödicke, Münster. We appreciate the discussions we have had throughout this research with Drs. Peter Giese, and Volker Haak. This research was supported by "Deutsche Forschungsgemeinschaft" under grant Gi 31/45-3, /51-1+3, and by "Freie Universität Berlin".

References

Ambroggi, R., Neltner, L. (1952): Haut Atlas et Anti-Atlas occidentaux. Généralités. IXX Congr. géol. intern. Alger 1952, livre-guide, Maroc, 12.

Ben Sari, D. (1978): Connaissance Géophysique du Maroc. Thèse Dr. Sc., Université Grenoble.

Bosch van den, J.W.H. (1971): Carte Gravimétrique du Maroc, au 1:500.000. Notes et Memoires du Service Géologique du Maroc, 234.

Brede, R., Heinitz, W. (1987): Priv. Communication.

Brewer, J.A., Smythe, D. (1984): MOIST and the continuity of crustal reflector geometry along the Caledonian-Appalachian orogen. J. Geol. Soc. London, 141, 105-120

Buness, F., Wetzig, E. , Wigger, P. (1986): Seismologische Studien in den Zentralen Anden. Berl. Geowiss. Abh., (A), 66, 5-30.

Carte Géologique du Maroc, Échelle 1:1000000 (1985): Ministère de l´Energie et des Mines, Direction de la Géologie, Rabat.

Demnati, A. (1972): Krustenstruktur im Rif-Bereich von Nord-Marokko aus gravimetrischen und aeromagnetischen Regionalmessungen. Boll. Geofis. Teor. Appl., XIV, 203-214.

Etude Aéromagnétométrique du Moyen Atlas, de la Haute Moulouya et du Haut Atlas Oriental. Carte en courbes isogammes du champ total. (1974) - Comp. Afric. Geophys., Rabat

Hatzfeld, D., Frogneux, M., Giradin, N. (1977): Etude de sismicité dans la région de l´arc de Gibraltar et l´Algérie du Nord. Bull. Soc. géol. France, (7), XIX, 741-747.

Hatzfeld, D., Ben Sari, D. (1977): Grands profils sismiques dans la région de l´arc de Gibraltar. Bull Soc. Géol. France, 1977, (7), t. XIX, 749-756.

Hatzfeld, D., Frogneux, M. (1981): Intermediate depth seismicity in the western Mediterranean unrelated to subduction of oceanic lithosphere. Nature, 292, 443-445.

Heck, C. (1982): Refraktionsseismische Untersuchungen in Westmarokko. Unpublished Diploma Thesis, Institut für Geophysik der Universität Hamburg.

Hinz, K., Dostmann, H., Fritsch, H. (1982): The continental margin of Morocco: Seismic sequences, structural elements and geological development. In: Geology of the Northwest African Continental Margin, Eds. U. von Rad, K. Hinz, M. Sarnthein, E. Seibold, Springer Verlag Heidelberg, 35-60.

Jacobshagen, V. (1986): Zur Baugeschichte des zentralen Hohen Atlas. Berl. Geowiss. Abh. (A), 66, 455-470

LEGSP, DMGM (1977): Premiers résultats d´une campagne d´enregistrements des variations géomagnétiques dans le Nord du Maroc. Ann. Géophys., 33, 225-236

Makris, J., Demnati, A. (1978): Deep seismic soundings in Marocco and a crust and upper mantle model deduced from seismic and gravity data. Inst. Geophys. Hamburg, 15 S., unveröffentlicht

Makris, J., Demnati, A., Klußmann, J. (1985): Deep seismic soundings in Morocco and a crust and uppermantle model deduced from seismic and gravita data. Ann. Geophysicae, 3, 369-380.

Martinez, M. (1987): Entwicklung neuerer Inversionsmethoden für die Magnetotellurik und ihre Anwendung auf Messungen in den Anden, Marokko und KTB Oberpfalz. Paper given at 47th annual DGG-meeting, Clausthal-Zellerfeld.

Mattauer, M., Tapponier, P., Proust, F. (1977): Sur les mécanismes de formation des chaines intracontinentales. L´example des chaînes atlasique du Maroc. Bull. Soc. géol. France, 19, 521-526

Menvielle, M., Rossignol, J.C. (1982): Conséquences tectoniques de l´existence d´une anomalie de conductivité électrique au Nord du Maroc. Can. J. Earth Sci., 19, 1507-1517

Menvielle, M., Le Mouël, J.L. (1985): Existence d´une anomalie de conductivité dans le Haut Atlas marocain et concentration des courants telluriques à l´échelle régionale. Bull. Soc. géol. France, 553-558

Michard, A, Westphal, M., Bossert, A., Hamzeh, R. (1975): Tectonique de blocs, dans le socle atlasomésétien du Maroc: une nouvelle interprétation des données géologiques et paléomagnétiques. Earth Planet. Sci. Lett., 24, 363-368

Michard, A. (1976): Elément de géologie marocaine. Notes et Mémoires du Service Géologique du Maroc, 252.

Middleton, M.F. (1984): Formation of rifted continental margins: a discussion of mechanisms. First Break, p. 9-14

Proust, F., Petit,J.P., Tapponier, P. (1977): L'accident du Tizi-n'Test et le rôle des décrochements dans la tectonique du Haut Atlas occidental (Maroc). Bull. Soc. géol. France, 19, 541-551

Qureshi, S.N. (1986): Crustal studies of Morocco from gravity and seismic data. Dissertation, Fachbereich Geowissenschaften, Universität Hamburg.

Ramdani, M., Tadili, B.A. (1980): Sismicité detaillée du Moyen-Atlas, sondages sismiques profonds dans le Maroc Central. Thèse Dr. de 3e cycl, Université Scientifique, Grenoble.

Schmucker, U. (1970): Anomalies of geomagnetic variations in the southwestern United States. Bull. Scripps Inst. Ocean. Univ. Calif., 13.

Schmucker, U. (1971): Neue Rechenmethoden zur Tiefensondierung mit langperiodischen Variationen. In: Weidelt, P. (ed.): Prot. Koll. Erdmagn. Tiefensond., Rothenberge, S. 1-39

Schwarz, G., Haak, V., Martinez, E., Bannister, J. (1984): The electrical conductivity of the Andean crust in northern Chile and southern Bolivia as inferred from magnetotelluric measurements. J. Geophys., 55, 169-178.

Schwarz, G., Martinez, E., Bannister, J. (1986a): Untersuchungen zur elektrischen Leitfähigkeit in den Zentralen Anden. Berl. Geowiss. Abh., (A), 66, 49-72.

Schwarz, G., Rath, V., Haak, V. (1986b): Magnetotellurische und erdmagnetische Sondierungen im marokkanischen Atlas. Berl. Geowiss. Abh., (A), 66, 289-300.

Stets, G., Wurster, P. (1981): Zur Strukturgeschichte des Hohen Atlas in Marokko. Geol. Rundschau, 70, 801-841.

Tadili, B., Ramdani, M., Ben Sari, D., Chapochnikov, K., Bellot, A. (1986): Structure de la croûte dans le nord du Maroc. Ann. Geophysicae, 4, 99-104.

Thiessen, J. (1979): Seismische und gravimetrische Untersuchungen der Krustenstruktur am marokkanischen Kontinentalrand. Unpublished diploma thesis, Institut für Geophysik der Universität Hamburg.

Udias, A. (1982): Seismicity and seismotectonic stress field in the Alpine-Mediterranean region. In: Berckhemer, Hsü (eds.): Alpine-Mediterranean Geodynamic. Geodynamic Series, 7, 75-82, A.G.U. Washington D.C.

Weigel, W., Wissmann, G., Goldflam, P. (1982): Deep seismic structure (Mauretania and Central Morocco). Geology of the Northwest African Continental Margin. Eds.: U. van Rad, K. Hinz, M. Sarnthein, E. Seibold, Springer Verlag Heidelberg, 133-159.

Wigger, P., Harder, S. (1986): Seismologische und krustenseismische Untersuchungen im Atlas-System/Marokko. Berliner geowiss. Abh., (A), 6, 273-288.

Cenozoic Basins

THE UPLIFT OF THE CENTRAL HIGH ATLAS AS DEDUCED FROM NEOGENE CONTINENTAL SEDIMENTS OF THE OUARZAZATE PROVINCE, MOROCCO.

Konrad Görler[1], Friedrich-Franz Helmdach[1], Pieter Gaemers[2], Kurt Heißig[3], Winfried Hinsch[4], Karl Mädler[5], Werner Schwarzhans[6] and Monika Zucht[7]

[1]Institut für Geologie, Freie Universität Berlin, Altensteinstr.34 A, D 1000 Berlin 33; [2]Rijksmuseum van Geologie en Mineralogie, Hooglandse Kerkgracht 17, NL-2312 HS Leiden; [3]Universitäts-Institut und Staatssammlung für Paläontologie und Historische Geologie, Richard-Wagner Straße 10, D-8000 München 2; [4]Geologisches Landesamt Schleswig-Holstein, Mercatorstr. 7, D-2300 Kiel; [5]Hamsunstr. 33F, D-3000 Hannover; [6]Stammannstr. 13, 2000 Hamburg 60; [7]Gesellschaft für Hydrogeologie und Umweltschutz mbH, Bredtschneiderstr.11, D 1000 Berlin 19

ABSTRACT

South of the central High Atlas, continental Neogene sediments crop out in the following areas: in a palaeo-drainage system in the Anti-Atlas, in a southern foredeep of the central High Atlas (Ouarzazate basin), and in the intensively folded South Atlas Marginal Zone. The latter two are characterized at the base and the top of the Neogene by thick alluvial-fan sequences and gravitational-glide slabs. They are correlated to episodes of rapid uplift of the axial zone of the High Atlas during Oligocene/Early Miocene and Late Pliocene/Early Pleistocene time. The intermediate sedimentation comprises widespread lake deposits. Charophytes, mollusks, ostracodes, fishes and small mammals from the lacustrine sediments have been studied and their biostratigraphic position (Mellalian/Vallesian to Late Pliocene) has been established. Lithology and the faunas of ostracodes and fishes demonstrate a frequent change of salinity during the lacustrine sedimentation.

1. Introduction

Motive of these investigations on the continental Neogene south of the central High Atlas has been a research project of the Freie Universität Berlin named "mobility of active continental margins". One of the objects of study has been the intracratonic Atlas system relating to the mobilization of continental crust several hundred kilometres away from the active continental margin. Its build up was contemporaneous with the last climax of the orogenic processes in the Mediterranean region in Neogene time.

A source of information reflecting the mobilization of continental crust can be found in contemporaneous sediments that document tectonic movements by the pattern and environment of their deposition and by the history of their deformation. Key parameters are facies, stratigraphy and structural geology.

As marine Neogene sediments are missing in the area of investigation, the following study concentrated on continental Neogene deposits, which consist mainly of red beds.

1.1 Geological setting and working area

The Moroccan Atlas system is composed of four different mountain-ranges: Rif, Middle Atlas, High Atlas and Anti-Atlas (fig. 1). The Rif is part of the Mediterranean Alpine orogene-system, whereas Middle, High and Anti-Atlas have an intracratonic origin. The Anti-Atlas owes its elevation of more than 3000 m to post-Eocene upwarping of the Precambrian and Palaeozoic of the Sahara platform and some associated volcanism. On the other hand the Middle and High Atlas are the result of a structural inversion of an earlier rift system in an area already affected by Variscan orogeny and consequently include thick sequences of Mesozoic and Tertiary sediments and volcanics now exposed to erosion.

The High Atlas can be divided in a western, central and eastern portion. Between the western and central High Atlas the Mesozoic sediments are interrupted by a region of Precambrian and Palaeozoic rocks, the "massif ancien". The Jbel Toubkhal (4 175 m), the highest elevation of the Atlas-system is situated there. In the past the "massif ancien" separated the Mesozoic sedimentary realm of the Atlantic-connected Essaouira gulf in the west from a more complex Tethys-connected rift graben system and epicontinental embayment in the east known as the Atlas gulf that later evolved to Middle Atlas and the central and eastern portions of the High Atlas.

The central High Atlas ("calcareous Atlas") is mainly constructed of Lower and Middle Jurassic marine limestones which are underlain by Triassic red beds. Precambrian and Palaeozoic rocks affected by the Variscan orogeny are exposed especially in the southwest near the "massif ancien".

The "South Atlas Marginal Zone" of the central High Atlas in its western section is characterized by folds and thrusts of Cretaceous and Palaeogene sediments with southern vergencies and small areas of continental Neogene sediments in between. Locally the South Atlas Marginal Zone is overridden by large slabs of Jurassic limestones, as for example the "nappe of Toundout" in the study area. In Ait Kandoula region continental Neogene has been deposited on top of this nappe. The South Atlas Marginal Zone borders and partly overthrusts towards the south the Neogene - Quaternary basin of Ouarzazate. In this region the continuous huge fault postulated by many authors the "accident sud-atlasien" could not be observed (ZYLKA 1986, 1988).

South of the South Atlas Marginal Zone a foredeep filled with continental Neogene, the Ouarzazate basin, is situated. The investigated area comprises Neogene sediments of the South Atlas Marginal Zone and the Ouarzazate basin. More detailed investigations have been carried out in the area between Toundout and Boumalne de Dadès and especially in the Ait Kandoula region. Furthermore some small outcrops of continental Neogene sediments in the Anti-Atlas have been worked on.

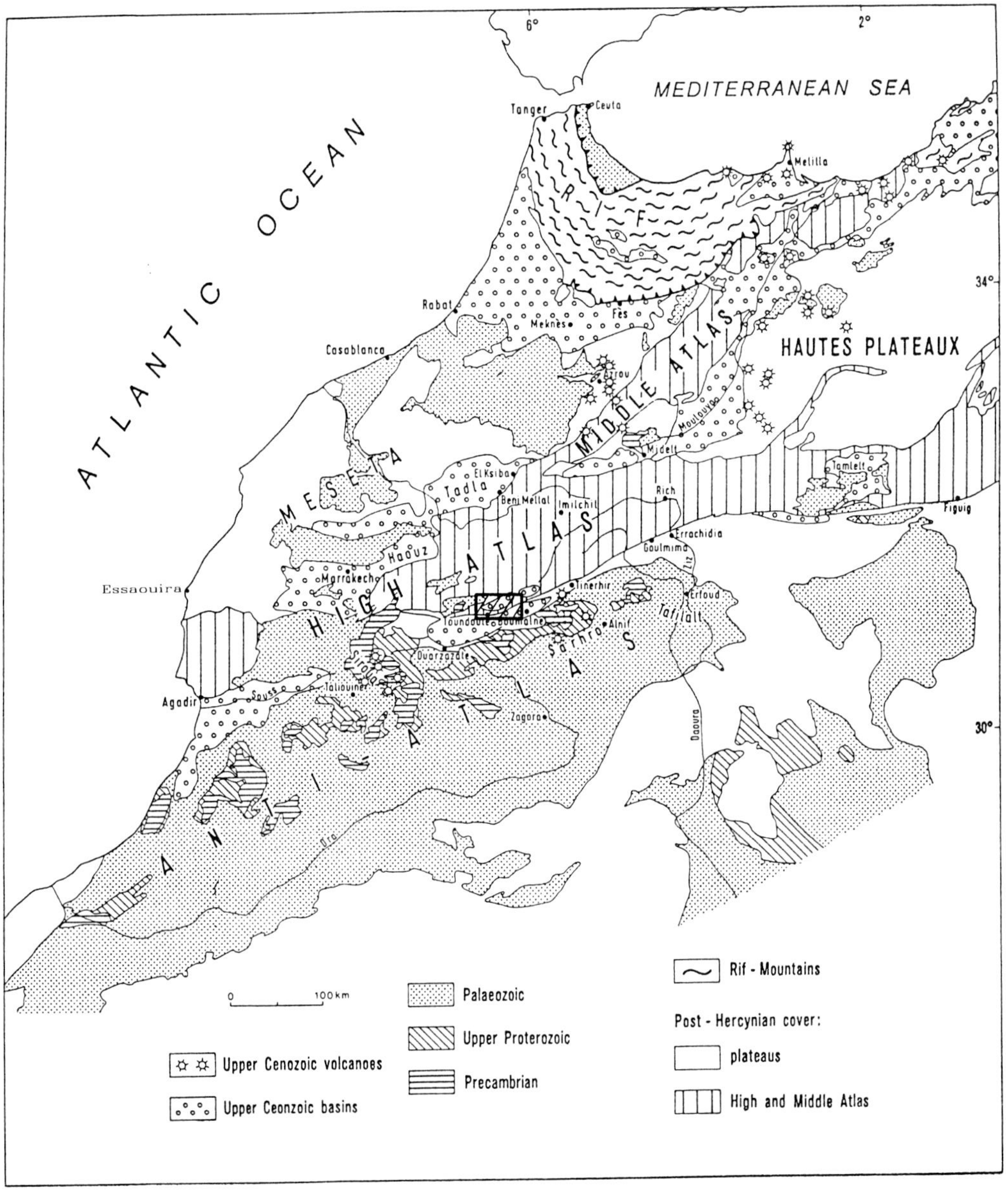

Fig. 1 Geotectonic sketch-map of the Atlas system (based upon MICHARD, 1976, from JACOBSHAGEN et al. this vol.) and main study area.

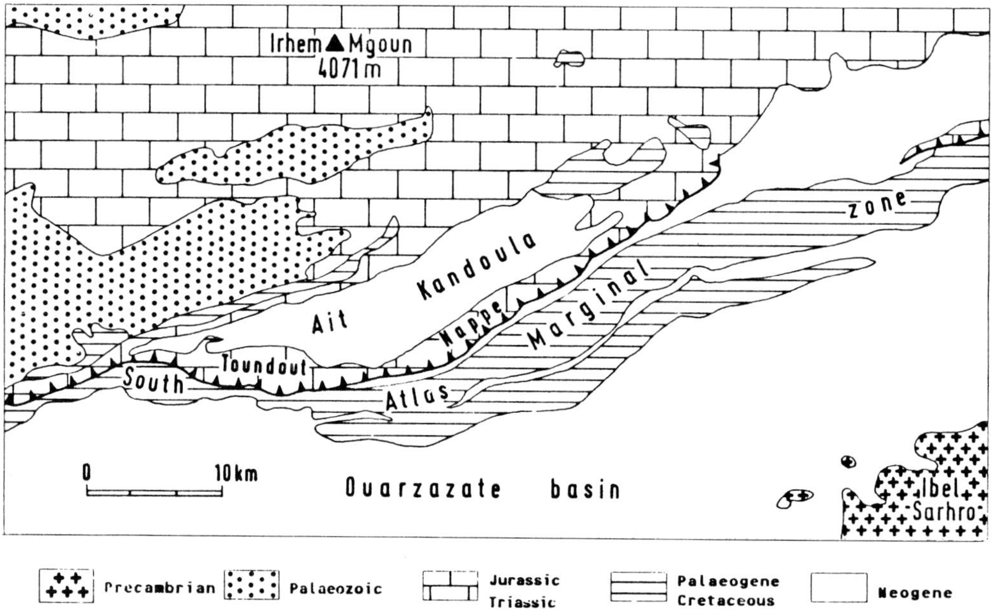

Fig. 2 Generalized geological map of the southern rim of the central High Atlas (modified from the sheet Ouarzazate of the geological map of Morocco 1 : 500.000)

1.2 Present knowledge

Although there is little controversy as to the depositional environment and the degree and style of the structural deformation of the above mentioned sediments the timing of the tectonic events has remained poorly understood.

In the past during the investigation of the High Atlas and the adjacent areas the continental Neogene was somewhat neglected. Especially the stratigraphic correlation varies widely between the different authors. For example the same sediments of the Ouarzazate basin were attributed to the Oligo-/Miocene (ROCH 1939), the Pontian (CHOUBERT 1945) and the Mio-/Pliocene (GAUTHIER 1957). ROCH (1939) and GAUTHIER (1953) found gastropodes which according to JODOT (1953, 1954) are indicative for Pliocene. JAEGER 1977 discovered microvertebrates at a locality in the Ait Kandoula region indicative of a Middle to Late Vallesian age.

LAVILLE (1980) distinguishes, mainly on the base of the tectonic position, within the continental Neogene "éocène superieur", "molasses post-éocène", "mio-pliocène 1" and "mio-pliocène 2".

Since reliable index fossils are rare, previous workers tended to define their respective stratigraphic units of continental Tertiary sediments with the help of unconformities or regressive events observed elsewhere unaware of the chronostratigraphic implications such a procedure might provoke. Therefore the main purpose of our investigations was to define the stratigraphic position of the synorogenic continental Tertiary sediments to obtain a more reliable timetable of the tectonic processes within the central High Atlas and its southern borderland.

2. Outcrop areas and observed thickness of the continental Neogene

In Morocco south of the central High Atlas continental sediments of Neogene age have been deposited in a large (probably more than 5.000 km^2) basin with internal drainage between the rising High Atlas mountains and the former source region of terrigenous material, the northern rim of the Sahara platform, i. e. Anti-Atlas and Jbel Sarhro. Due to tectonics and erosion this basin is actually divided in three different outcrop areas (fig. 1):

a. The Anti-Atlas
b. The Ouarzazate basin south of the central High Atlas
c. The South Atlas Marginal zone

2.1 Anti-Atlas

Within the northern part of the Anti-Atlas and west of the Ouarzazate basin continental Neogene deposits crop out in a fossil valley-system, as already observed by CHOUBERT (1945) and GAUTHIER (1957). These palaeo-valleys are partly connected with the Ouarzazate basin and cut into Precambrian and Palaeozoic rocks within the Anti-Atlas. West of the Ouarzazate basin, Eocene limestones were also eroded by the palaeo-drainage system indicating the drainage to have taken place during post-Eocene time while the palaeo-valleys filled with continental red beds and lacustrine sediments during the Neogene. The drainage pattern was clearly different from today's and the thickness of sediments within it varies widely. Profiling and geoelectrical measurements yielded a maximum value of 120 m for the two kilometres broad palaeo-valley west of Anzel.

2.2 The Ouarzazate basin

The Ouarzazate basin with a length of 140 km is situated directly south of the central High Atlas. Near its western end it has its maximum width of 35 km and towards the east its borders converge a few kilometres west of the village of Tinerhir. Near its western and southern margins "islands" of older rocks sticking out through the continental Neogene cover are indicative of only moderate thicknesses of the Neogene sediments and their deposition over a palaeo-relief. Precambrian andesites beneath Tertiary red beds were encountered below 316 m in a well drilled near Taferghoust and below 535 m in a second well 5 km further to the north.

Near the northern border of the Ouarzazate basin Tertiary continental red beds crop out in the core of the Amekchoud anticline reaching a thickness of more than 1.200 m. These observations indicate that the Ouarzazate basin is pronouncedly asymmetric with strongly increasing thickness towards the north.

2.3 South Atlas Marginal Zone

Within the South Atlas Marginal Zone, which is built up mainly by folded and faulted Jurassic, Cretaceous and Palaeogene sediments, narrow elongated outcrops of continental Tertiary sediments occur (fig. 2). One of them, the Ait Kandoula region, was studied in more detail, because it exhibits pronounced lateral and vertical facies changes and is locally rich in fossils.

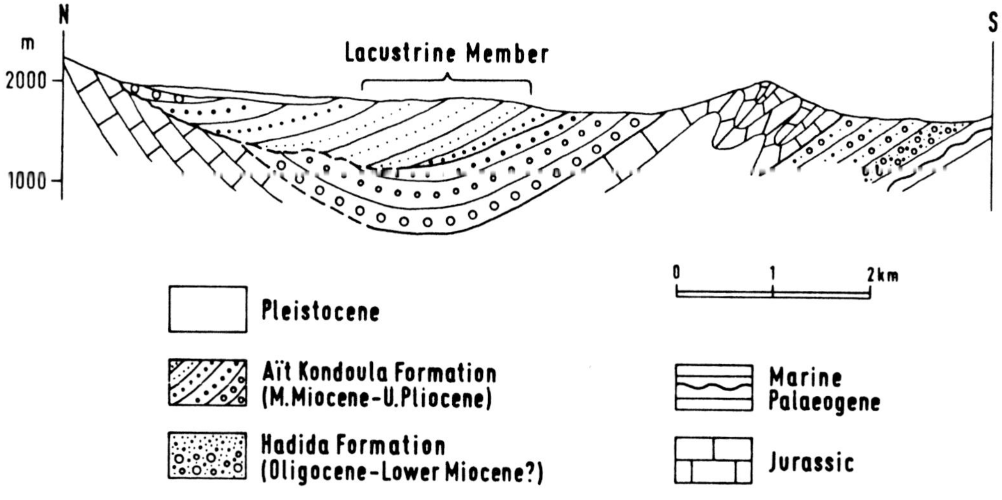

Fig. 3 Schematic section across the Ait Kandoula region and the Nappe of Toundout

The term "Ait Kandoula" or "Ait Kantoula" referring to a Berber tribe was introduced as it shows up on modern topographic and geologic maps contrary to the name "Tiniguert" used by ROCH (1939) and GAUTHIER (1957).
The Ait Kandoula Neogene outcrop has a length of about 50 km and a maximum width of 6 km. Basin analysis, however, suggests that during Miocene it must have been part of a single large basin, also comprising the present Ouarzazate basin and other outcrop areas of continental Tertiary along the South Atlas Marginal Zone. In the Ait Kandoula region continental Neogene is mainly deposited on top of Liassic limestones of the "nappe of Toundout". The leading edge of this nappe lies on top or is intercalated within continental Tertiary sequences (fig. 2) and there is good evidence that the emplacement of this nappe took place during the deposition of the continental Neogene strata (see 3.2.2.1 and fig. 3). In agreement with LAVILLE, 1980 the range of overthrust is thought to be in the order of 10 - 15 km.

On top of the Toundout nappe maximum thickness of continental Tertiary (Lower Miocene - Lower Pleistocene) amounts to about 1.700 m with more than 1.200 m of red beds beneath, which are Oligocene to Early Miocene in age. The region of the southern front of the Toundout nappe is therefore considered as the area with the maximum sedimentation of continental Tertiary south of the central High Atlas. This view is supported by the fact, that one of the most pronounced negative Bouguer gravity anomalies (less than -150 mgal) of Morocco is situated in this region (fig. 4). The increasing thickness of continental Tertiary sediments towards the north of the Ouarzazate basin seems therefore to represent a trend which continues even into the High Atlas mountains where the Neogene is often overridden by slabs of older sediments. Although towards the north there seems to be a pronounced change over a short distance from the area of maximum sedimentation to an area of erosion during the Neogene, there is no sign at the northern limit of the Ait Kandoula region of a fault between source and sedimentation area (fig. 3).

3.0 Lithology, content of fossils and biostratigraphy, environment of deposition and palaeoclimatology

In all of the three outcrop areas the environment of deposition is very similar, even though the total thickness of sedimentation is quite different. Neogene fossils have been found in all three regions, but it was only possible to establish stratigraphic correlations based on these fossils in the northern part of the Ouarzazate basin and especially in the central portion of the Ait Kandoula region. As mentioned before there is strong evidence that the Ouarzazate basin and the Ait Kandoula region once formed a single area of deposition, therefore they will be dealt within one chapter.

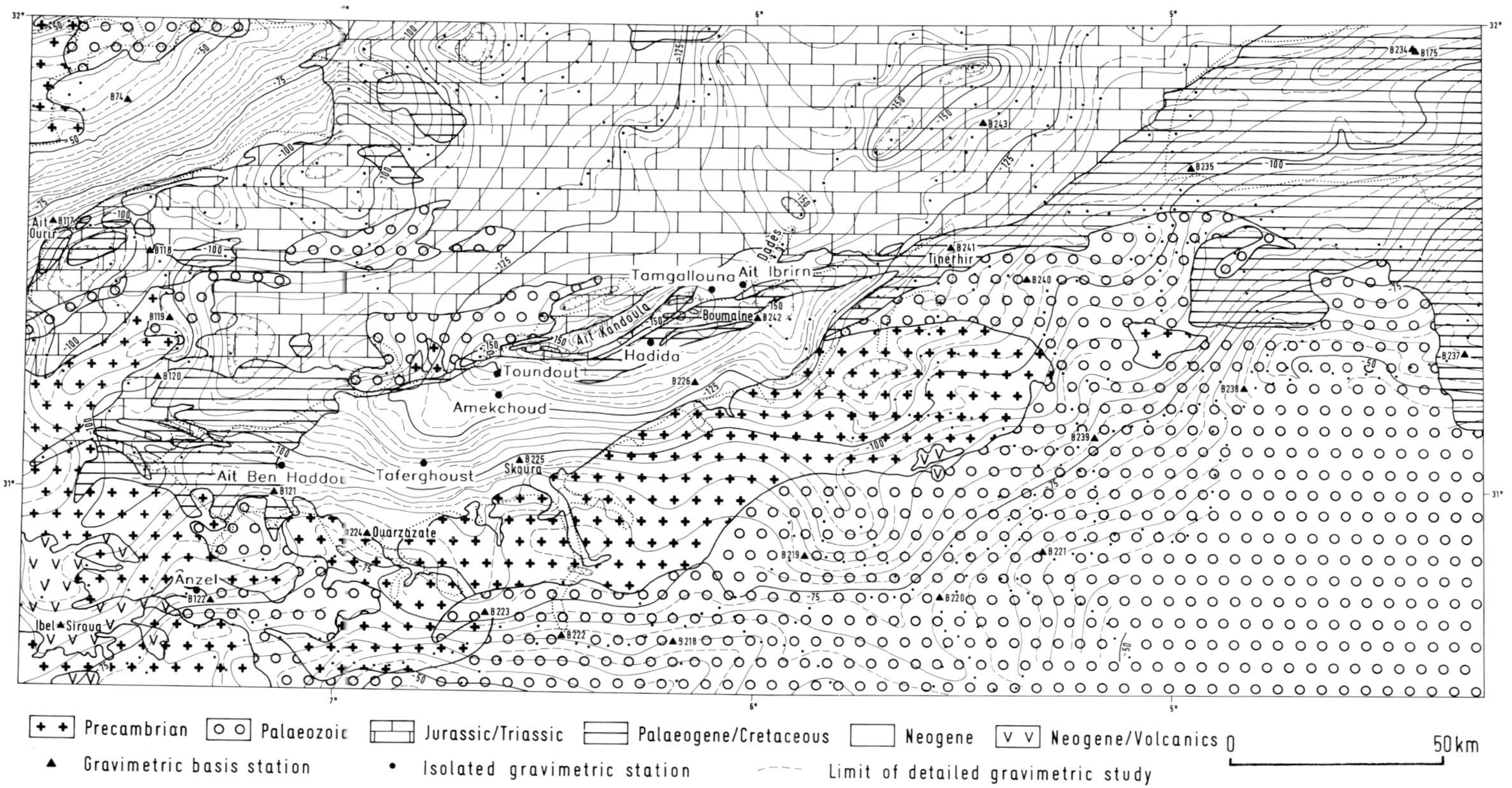

Fig. 4 Geological and Bouguer gravity map of the southern central High Atlas and adjacent areas, according to published maps 1 : 500 000

3.1 Anti-Atlas

The terrigenous continental Neogene deposits within the palaeo-valleys are dominantly coarse grained. Marls and lacustrine limestones are rare, except for near the western end of the Ouarzazate basin, where the alternation of marls and gypsum, deposited in a lacustrine environment continues into the palaeo-drainage system. Further to the west, and to the south and east of Ouarzazate, sediments are exposed at the base of the palaeo-valleys that are built up of lenticular orthoconglomerates and sandstones which in many instances display the tabular or trough cross-bedding. Their thickness rarely exceeds one metre. The conglomerates contain pebbles of exotic material not cropping out in the neighbourhood. These orthoconglomerates and sandstone are interpreted as stream-channel deposits. The bulk of the sediments within the palaeo-valleys are quite monotonous. They consist of a sequence of not very distinctive layers each in the order of decimetres to metres of homogeneous paraconglomerates grading towards its top into sandy material with parallel bedding. The bedding planes between the layers are somewhat irregular, but in most cases non-erosive. The source area was most likely nearby, because abundant angular to subangular fragments of rocks cropping out in the same area occur in the palaeo-valleys. These paraconglomerates are interpreted as mudflow deposits. Intercalated single cross-bedded layers of orthoconglomerates overlain by small sets of cross-bedded sandstone are probably the result of occasional sheet floods. Red massive sandy mudstones displaying greenish, mostly vertical, root-structures (?) of up to several metres in length are interpreted as palaeosoils. In some places they contain calcitic nodules or calcite pseudomorphs replacing selenite. Especially the conglomerates are well cemented by calcite. Within the Anti-Atlas lacustrine bituminous marls and limestones have been found at three locations only. Their exposed thicknesses do not exceed more than a few metres. They contain:

Charophytes:

Charites molassica (STRAUB) HORN AF RANTZIEN

Mollusks:

Pisidium sp.
Bithyynia gaudryi (PALLARY)
Stagnicola chouberti (JODOT)
Anisus leucostomus (MILLET) or *A. falsani* (LOCARD)
Galba cf. *truncatula* (MÜLLER)
Planorbarius cf. *doumergouei* (PALLARY)
Radix cf. *subperegra* (PALLARY)
Radix cf. *thomasi* (TOURNOUVER)
Cochlicella conoidea (DRAPARNAUD)

Cochlicopa cf. *lubrica* (MÜLLER)
Columella cf. *aspera* WALDEN
Oxyloma cf. *elegans* (RISSO)
Rumina atlantica PALLARY
Vallonia cf. *pulchella* (MÜLLER)
Vertigo (*V.*) *antivertigo* (DRAPARNAUD)
Vertigo (*V.*) *pygmaea* (DRAPARNAUD)
Vertigo (*Vertilla*) *angustior* JEFFREYS
Xeroclausa cf. *meda* (PORRO)

Ostracodes:
Candona compressa (KOCH 1837)
Candona aff. *marchica* HARTWIG 1899
Candona neglecta SARS 1887
Candona pratensis HARTWIG 1901
Cypridopsis sp.
Cypridopsis vidua (O. F. MUELLER 1776)
Darwinula sp.
Herpetocypris reptans (BAIRD 1835)
Herpetocypris brevicaudata KAUFMANN 1900
Heterocypris cf.*parva* (LIENENKLAUS 1905)
Ilyocypris bradyi (SARS 1890)
Ilyocypris inermis KAUFMANN 1900
Ilyocypris sp..
Potamocypris sp..
Potamocypris villosa (JURINE 1820)

The ostracodes indicate a late Pliocene to Pleistocene age and a freshwater to oligohaline environment. The sediments are locally interfingering with the volcanics of the Jbel Siroua region, which according to CHOUBERT et al. (1968) and unpublished reports cited in HERNANDEZ and BELLON (1985) yielded radiometric ages between 5 to 10 m. y., i. e. Late Miocene. A Late Miocene to early Pleistocene age for the sedimentation of the continental Neogene in the Anti-Atlas region seems therefore most probable.

3.2 Ouarzazate basin and South Atlas Marginal Zone

The continental Tertiary sediments throughout most of the Ouarzazate basin are deposited unconformable upon older rocks, ranging from Precambrian magmatic rocks to marine Eocene sediments. Only in a limited area along the northernmost part of the

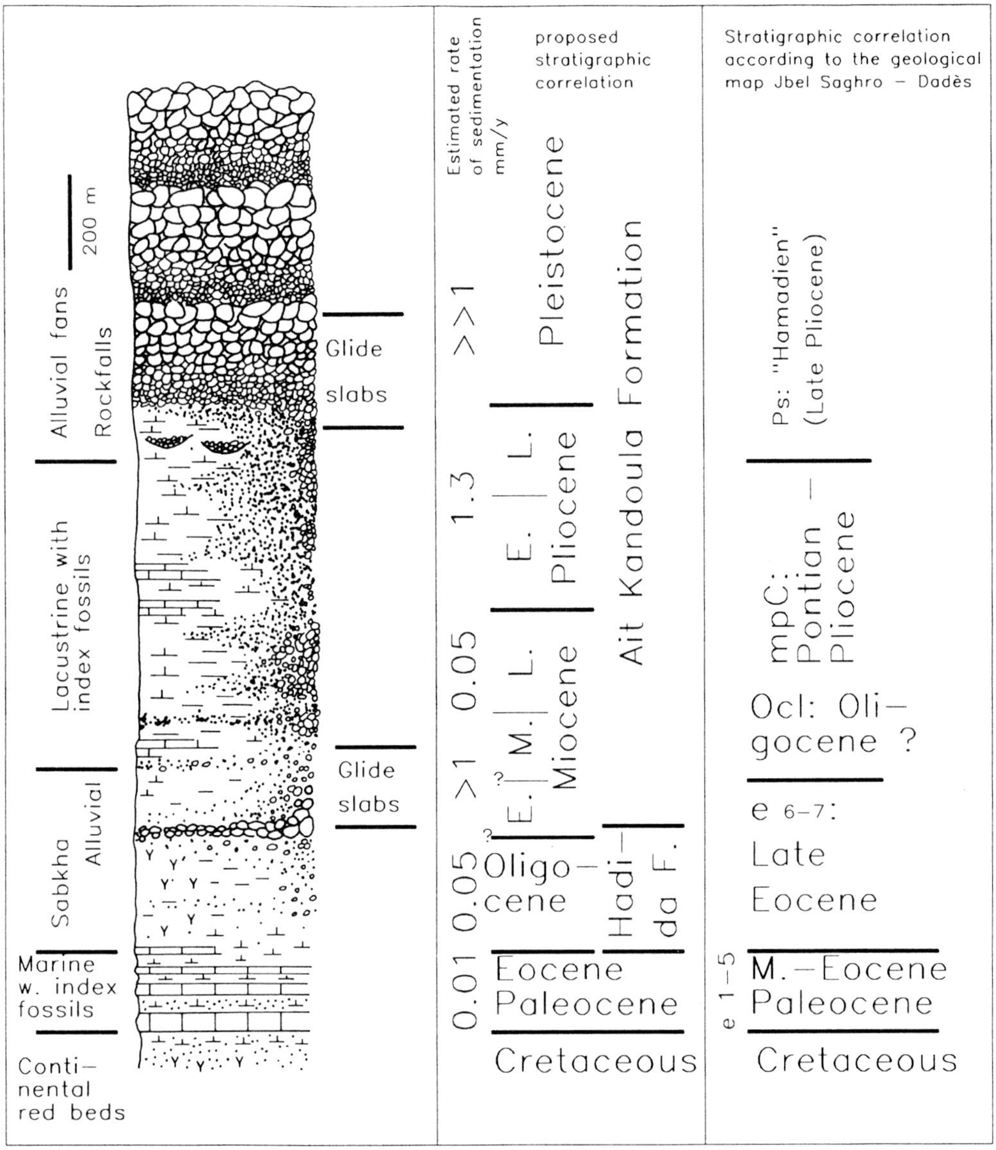

Fig. 5 Schematic stratigraphic section of the continental Tertiary of the northern Ouarzazate basin and Ait Kandoula region. The rate of sedimentation is calculated for the Mgoun valley and the Ait Kandoula region.

Ouarzazate basin and in the South Atlas Marginal Zone is a conformable transition observed from the shallow marine Eocene limestones and marls into the continental red beds of the Hadida-Formation (GAUTHIER, 1957; HERBIG, 1986; GÖRLER and ZUCHT 1986). A schematic stratigraphic section of the continental Tertiary is given in fig. 5.

3.2.1 Hadida Formation (Oligocene - Lower Miocene ?)

In a shoaling-upwards cycle shallow marine to littoral limestones and marls (MORET, 1938; ROCH, 1939; GAUTHIER, 1957; HERBIG, 1986) are succeeded by red mudstones and silts with layers of crystalline gypsum and gypsum nodules, interpreted as sabkha sediments. Fine to medium grained sandstones with large-scale cross-bedding are intercalated; hence an eolian origin seems very probable. In the upper portions of the Hadida Formation some of the sedimentary deposits are interpreted as proximal braided rivers and alluvial fans (couches des Ait Arbi of GAUTHIER, 1957). These crudely bedded conglomerates with irregularly intercalated cross-bedded sandstones fit the Scott type of Miall (1978).

The Hadida Formation reaches a total thickness of more than 600 m near the village of Hadida. But the average thickness is much less. The pebbles laid down within this formation are composed of Precambrian and Palaeozoic rocks which infer a source area in the south, i. e. the Sahara Platform.
On regional geological maps (1 : 200.000 and 1 : 500.000) the base of the continental Tertiary sequence is considered as "eocène supérieur", following the Lutetian marine marls and limestones (GAUTHIER 1957, HERBIG 1986). However, ostracode analyses of the uppermost portion of the marine sequence at several locations have revealed *Limburgina moosi* BASSIOUNI 1969, *Cytherella jonesiana* (BOSQUET 1852) and species of the genus *Buntonia* indicative of latest Eocene/Early Oligocene (HELMDACH 1986, GÖRLER and ZUCHT, 1986). The onset of continental sedimentation may therefore be slightly younger during Early Oligocene. The Hadida Formation thus is in our opinion of Oligocene to Early Miocene (?) age, but similar to the overlying Alluvial Base Member of the Ait Kandoula Formation no index fossils have been found, and thus an entirely Oligocene age of the Hadida Formation can not be ruled out.

3.2.2 Ait Kandoula Formation (? Lower/Middle Miocene - Upper Pliocene/? Lower Pleistocene)

The Ait Kandoula Formation can generally be subdivided into three lithostratigraphic units:

3. a second conglomerate sequence at the top, the "Alluvial Top Member"
2. a middle portion characterized by silts, gypsum, bituminous marls and limestones, the "Lacustrine Member"
1. breccias, conglomerates and sandstones at the base, the "Alluvial Base Member"

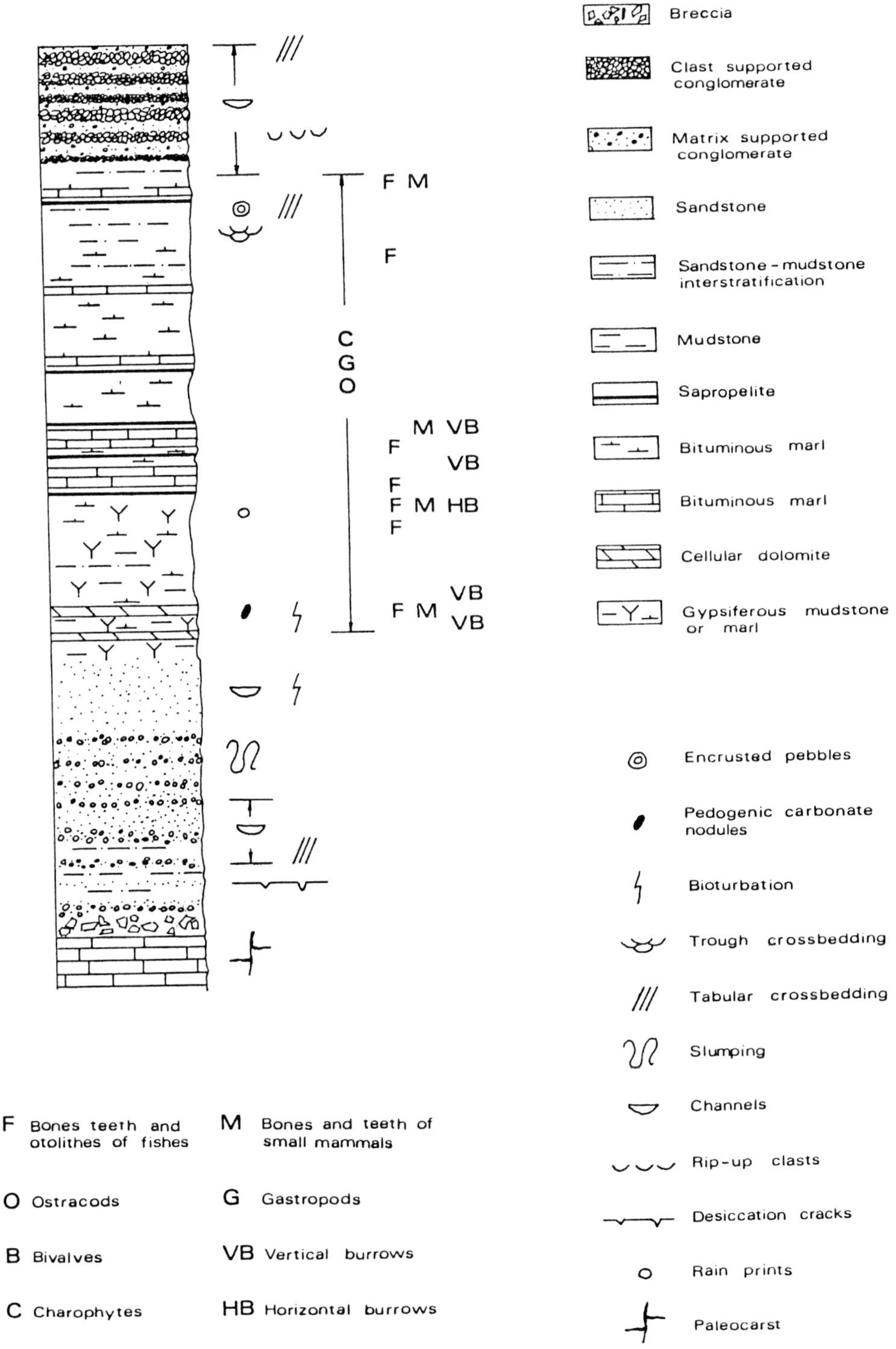

Fig. 6 Stratigraphic section of the Ait Kandoula Formation in the area with maximal thickness of Neogene lacustrine sediments

Towards the northern border of the sedimentary basin the whole section grades into a sequence constructed of alluvial-fan deposits with a middle portion characterized by smaller grain sizes.

3.2.2.1 Alluvial Base Member (?Lower Miocene - Middle Miocene)

According to the definition by GÖRLER and ZUCHT (1986) the main criterion distinguishing the Ait Kandoula Formation from the Hadida Formation is the presence of material derived from the High Atlas region as pebbles composed of Mesozoic or Tertiary sediments or volcanics. The first appearance of such material is indicative of the beginning of the uplift of the High Atlas. Following this definition the onset of the Ait Kandoula Formation varies distinctly in its sedimentary structure from one location to another. In some places a coarse breccia or coarse conglomerate of poorly rounded boulders up to 1 m in diameter is deposited as a disconformity or even an angular unconformity on top of the Hadida Formation or older sediments. Here and there the base of the Ait Kandoula Formation is only represented by the first appearance of pebbles from Mesozoic or Tertiary rock units without any other change in sedimentary pattern. In this case the lithostratigraphic unit "eocène superieur" of the regional geological maps (synonymous with the "couches rouges de Boumalne", CHOUBERT 1946 and "couches des Ait Arbi", GAUTHIER 1957) must be split as follows into two parts: A lower part, void of pebbles derived from High Atlas material - the Hadida Formation - and a upper part equal to the base of the Ait Kandoula Formation. The upper part is still dominantly composed of pebbles derived from the Precambrian and Palaeozoic of the Sahara Platform, but it also contains pebbles from Eocene limestones, and other reworked rocks from High Atlas material.

Generally the higher portions of the Alluvial Base Member of the Ait Kandoula Formation increasingly contain in its conglomeratic parts pebbles composed of conglomerates and of sandstones of the Hadida Formation and also pebbles of Palaeogene limestones, Cretaceous sandstones and small amounts of Jurassic limestones, probably representing the uppermost strata deposited in the axial zone of the later High Atlas. The first appearance of High Atlas material is thought to mark the beginning of the uplift of the axial zone of the central High Atlas during Early Miocene time. Even in regions with a more gradational transition from the Hadida to the Ait Kandoula Formation the sudden appearance of similarly composed very coarse breccias or conglomerates is a diagnostic characteristic. The breccia is most likely related to a first phase of gravitational gliding of large slabs of Cretaceous and Tertiary sediments caused by a sharp uplift of the axial zone of the central High Atlas (s. 4.0). In many places, as can be seen 2 km northwest of Toundout, this breccia is overlain by Eocene limestones. At other locations a conglomeratic sedimentation continued on top of the breccia without the intercalation of slabs of older sediments.

Matrix supported debris flow deposits and intercalated crudely bedded orthoconglomerates dominate. The base of the orthoconglomerates is characterized by scours cut into the underlying debris flow deposits. This lithofacies assemblage is typical for proximal alluvial-fan deposits and represents the Trolheim type of fluvial deposition as described by MIALL 1978, 1982. Geometry of cross-bedding, imbrication and direction of channel fill structures (fig. 10) indicate that the fans originated from the uprising High Atlas in the north. In this upward coarsening sequence large slabs of Eocene and Jurassic limestones are intercalated. This applies in particular to a large area at the southern rim of the nappe of Toundout, where megabreccias lie between the conglomerate and the base of the nappe. Neogene sedimentation then continued with coarse conglomerates, probably after a short interruption, as palaeokarst features developed in the Jurassic limestones on top of the intercalated Jurassic limestones of the Toundout nappe. It is therefore assumed, that the emplacement of the Toundout nappe took place during the deposition of the Ait Kandoula Formation. This is best documented near the western end of the Toundout nappe, where it splits into several slabs of Jurassic limestones isolated from each other by Neogene continental sediments. They have sedimentary megabreccias or coarse conglomerates at the leading edge and at their base. In front of these slabs, as is the case of the stratigraphic deepest slab grain size diminishes rapidly. There the time of emplacement of the slab was determined (Late Mellalian/Early Vallesian) with teeth of small mammals from that level (s. 3.2.2.2).

Further away from the southern edge of the Toundout nappe e. g. northwest of Amekchoud, within the Ouarzazate basin, similar sediments were deposited without intercalated slabs of older sediments. Also within the Soputh Atlas Maginal Zone, e. g. in front of the eastern part of the Tondout nappe in the Mgoun valley north of Tamgallouna the alluvial-fan sedimentation continued without the intercalation of slabs of older sediments.

The coarse conglomerates on top of the Toundout nappe are overlain by upward fining alluvial-fan deposits. They pass over to alluvial plain deposits with numerous channel fill structures and finally to paludal and lacustrine sediments forming the middle portion of the Ait Kandoula Formation.

3.2.2.2 Lacustrine Member (Middle Miocene - Upper Pliocene)

On top of the Toundout nappe the lacustrine sediments mentioned above reach a maximum thickness of 800 m, but in a lateral distance of few kilometres they are in many cases replaced by large alluvial fans that project from the north. Very similar lacustrine sediments of the same age have been found at the northern rim of the Ouarzazate basin within an area that is more than 120 km long between Ait Ben Haddou and Boumalne. The "oligocène présumé" série des Ait Ibrirene of GAUTHIER 1957) shown on the geological map 1 : 200 000 sheet "Jbel Sagrho-Dadès" corresponds in lithology and fossil content with the deeper portions of the described Lacustrine Member and is therefore considered a part of the Ait Kandoula Formation (Görler and ZUCHT 1986).

The stratigraphic section shown in fig. 6 and the following description is related to zones with maximal thickness of Neogene lacustrine sediments on top of the Toundout nappe, if not otherwise noted.

The paludal and lacustrine section is lithologicaly characterized by a sequence of mudstones with rare sandy or conglomeratic intercalations, bituminous marls, laminated bituminous limestones, sapropelites, travertines, cellular limestones and gypsum which show both a pronounced vertical and lateral alternation. Lithology, sedimentary structures and the fossil content indicate several changes of the environmental conditions. The bulk sequence of the Lacustrine Member corresponds to recent examples of hard-water marl lakes as described by ALLEN and COLLINSON, 1986. Rocks typical for the various facies of the Littlefield Lake in Michigan (MURPHY and WILKINSON, 1980) are common:

- limestone encrusted stems of larger plants (littoral or paludal environment)
- marls and micritic limestones very rich in gyrogonites and stems of charophytes (photic zone down to a depth of about 10 m)
- marls and micritic limestones rich in gastropodes and few gyrogonites (lower bench slope facies)
- bituminous marls and micritic, laminated limestones with ostracodes and only few or no gyrogonites and gastropodes (deepest portions of the lake beneath the photic zone).

Brown Sapropelic limestone laminites with intercalated black sapropelic coals contain only trace fossils such as up to 10 cm deep vertical burrows of about 1 cm diameter and branching burrows with a diameter of less than 1 mm. These sediments are interpreted as having been deposited in even greater depth. The layers of sapropelic coal are up to 30 cm thick and contain up to 80 % organic matter, which is composed of inertinite, liptinite, ulminite and euulminite. In contrast to the Littlefield Lake model we observed additional rocks as palaeosoils with calcareous concretions and rootlets, coated pebbles, stromatolites, oncoids, and scattered selenite within mudstones or marls and layered gypsum beds within the lacustrine/paludal section of the Ait Kandoula Formation. It therefore seems likely, that in the central area the sedimentary environment changed several times between "open freshwater lake" and "perennial saline lake" conditions as described by EUGSTER and KELTS (1983). This is confirmed by the vertical distribution of ostracodes and the induced palaeosalinities. Only in the westernmost part of the Ouarzazate basin between Ouarzazate and Ait Ben Haddou is the Lacustrine Member completely built up by marls and mudstones with abundant layered and nodular gypsum, representing an almost permanent saline lake environment.

An open question is whether the Lacustrine Member was deposited within one giant lake larger than 1.000 km^2 or in a number of separated lakes along the southern border of the uprising High Atlas. The latter case is supported by the inferred regional trend of increasing salinity in the N - S and E - W directions. In addition the repeated reappearance of endemic freshwater species throughout the 10 m. y. years of deposition is interpreted in favour of the separate lakes hypothesis. Perhaps they inhabited "eastern" lakes containing freshwater throughout this whole time span, whereas lakes more westward and less elevated displayed repeated variations in salinity as can be deducted from the changes or even disappearance of the ostracode- and fish faunas. During intervals of high precipitation the overflowing "eastern"freshwater lakes would permit a new spreading of their floras and faunas.

Another explanation of the observed regional and temporal changes in salinity on the one hand, and the persistence of many, partly endemic, species on the other, could be the existence of one large lake during times of high lake level, which disintegrated into several small lakes during low-level intervals. According to their position their salinities could be quite different and freshwater dependent taxa could survive within rivers coming from the uprising High Atlas.

The observed very low sedimentation rates on the other hand, suggest the existence of one large lake at least during the Late Miocene. The 800 m of lacustrine sediments of the main section were deposited within about 10 m. y. at rate of 0,08 mm/y. But the sedimentation rate was not constant. For the Middle and Late Miocene portion of the section the average rate was 0,05 mm/y, whereas for the Pliocene section, it was 1,3 mm/y. Further to the south, away from the uprising axial zone of the High Atlas, the sedimentation rate is markedly lower. Two biostratigraphically correlated sections between Amekchoud and Toundout show sedimentation rates of only 0,01 mm/y and 0,005 mm/y for the Late Miocene.

Only few fossils without stratigraphic value have been found in the westernmost Ouarzazate basin with its evaporitic sedimentation. In contrast the Lacustrine Member of the Ait Kandoula Formation further in the east has yielded great numbers of fossils from different taxa:

A. Diatoms are present but have not been studied.

B. Stromatolites and oncoids have been found at three locations in the northern Ouarzazate basin and at two locations in the Ait Kandoula region at the top or at the base of marls with charophytes and gastropodes in connection with red silts. The oncoids have a diameter of up to 5 cm and many display irregular shapes. It is probable that crusts on pebbles within conglomeratic channel fillings also originated as blue-green algae mats.

C. Unspecified remainders of larger plants are preserved in travertines. In the marls various calcified seed-pods have been found, which are not yet identified.

Even marls and limestones rich in organic material did not yield palynomorphs. This may be due to the climate in the southern central High Atlas which causes considerable temperature variations and frequent fluctuations between aridity and heavy rainfalls.

D. Charophytes: Stems and gyrogonites of charophytes are very common. At many locations they are the main calcareous component of limestones and marls. The following genera and species have been identified (in alphabetical order):

Amblyochara anatolica MÄDLER
Amblyochara cf. *anatolica* MÄDLER
Amblyochara concava MÄDLER
Amblyochara ovalis MÄDLER
Amblyochara tortonica MÄDLER
Charites cylindrica MÄDLER
Charites densospirae MÄDLER
Charites elongata MÄDLER
Charites cf. *elongata* MÄDLER
Charites minutissima (MÄDLER) HORN AF RANTZIEN
Charites molassica (STRAUB) HORN AF RANTZIEN
Charites multispirae (PAPP) MÄDLER
Charites nannocorpo SCHWARZ
Charites postconica MÄDLER
Charites sadleri (UNGER) HORN AF RANTZIEN
Charites sadleri cf. (UNGER) HORN AF RANTZIEN
Charites straubii MÄDLER
Croftiella escheri (A. BRAUN ex UNGER 1850) HORN AF RANTZIEN
Croftiella rochettiana (HEER) HORN AF RANTZIEN
Grambastichara sp. (STRAUB)
Grambastichara cylindrica (MÄDLER) HORN AF RANTZIEN
Grambastichara tornata (REID and GROVES) HORN AF RANTZIEN
Maedlerisphaera ulmensis (STRAUB) HORN AF RANTZIEN
Microchara ovalis MÄDLER
Sphaerochara sp. MÄDLER
Sphaerochara granulifera (HEER) MÄDLER
Tectochara globula (MÄDLER) HORN AF RANTZIEN
Tectochara meriani (A.BRAUN) GRAMBAST
Tectochara meriani globula MÄDLER
Tectochara cf. *meriani helvetica* MÄDLER
Tectochara meriani meriani (PAPP) GRAMBAST
Tectochara meriani octospirae MÄDLER

Divergences from species in Europe and Anatolia are frequent. At present the charophytes are not of great stratigraphic value in the Ait Kandoula Formation. Most species occur in stratigraphic levels throughout the lacustrine sediments, which due to findings of ostracodes and teeth of small mammals range at least from Serravallian to Upper Pliocene. Species as *Croftiella rochettiana* (HEER) HORN AF RANTZIEN und *Grambastichara tornata* (REID and GROVES) HORN AF RANTZIEN, which in Anatolia are restricted to Palaeogene and Lower Miocene (MÄDLER and STAESCHE, 1979) still occur in the Ait Kandoula Formation in levels of Upper Miocene and Pliocene age. There seems to be a general tendency of charophyte species in Morocco to have survived for a longer time span due to a warmer climate. The diversity of the flora differs markedly between the northernmost lacustrine outcrops in the Ait Kandoula region and the Ouarzazate basin. In the later the number of species is reduced to less than the half.

E. Mollusks: Particularly gastropodes are very common too and occur at various levels throughout the Lacustrine Member. The first gastropodes of the working area have been described by JODOT, 1953, 1954, 1955. From our samples up to now the following mollusks have been determined:

Bivalvia:
Pisidium cf. *casertanum* (POLI)
Pisidium sp.
Sphaerium sp.

Archaeogastropoda:
Theodoxus (Theodoxus) sp.
Theodoxus (Calvertia) sp.

Mesogastropoda:
Bithynia cf. *gaudryi* PALLARY
Bithynia cf. *glabra* (ZIETEN)
Bithynia sp. (*Opercula*)
Cincinnatia (?Neohoratia) atlasica n. sp.
Hydrobia (Ventrosia) iberica JODOT
Hydrobia (Ventrosia) dubia SCHLOSSER
Limnidia n. sp.
Melanoides tuberculata (MÜLLER)
Melanopsis (Lyrcaea) acuminata PALLARY
Melanopsis sp.
Nystia sp.
Prososthenia sparsistriata n. sp.
Tryonia kandoulae n. sp.
Basommatophora:
Anisus cf. *leucostomus* (MILLET) + *falsani* (LOCARD)
Anisus mariae (MICHAUD)

Armiger cf. *crista* (L.)
Bulinus (Diastropha) cf. *contortus* (MICHAUD)
Galba cf.*truncatula* (MÜLLER)
Gyraulus cf. *riparius* (WESTERLUND)
Gyraulus cf. *zoebeleini* SCHLICKUM and STRAUCH
Hippeutis sp.
Lymnaea cirtana (PALLARY)
Lymnaea cf. *jobae* TOURNOUVER
Planorbarius cf. *doumerguei* (PALLARY)
Planorbarius cf. *belnensis* (TOURNOUVER) + ?*philippei*(LOCARD)
Planorbis sp.
Physa sp.
Radix cf. *subperegra* (PALLARY) = *deydieri* (FONTANNES)
Radix cf. *thomasi* (TOURNOUVER) = *larteti* (NOULET)
Segmentina cf. *nitida* (MÜLLER)
Seminolina n. sp.
Stagnicola chouberti (JODOT)

Stylommatophora (terrestrial):
Cochlicella conoidea (DRAPARNAUD)
Cochlicopa cf. *lubrica* (MÜLLER)
Columella cf. *aspera* WALDEN
Ferussacia sp.
Keraea bertholiana (PFEIFFER)
Oxyloma cf. *elegans* (RISSO) = *primaeva* (MATHERON)
Rumina atlantica PALLARY
Sucinella sp.
Vallonia cf. *pulchella* (MÜLLER)
Vertigo (V.) antivertigo (DRAPARNAUD) + *callosa* (REUSS)
Vertigo (V.) pygmaea (DRAPARNAUD)
Vertigo (Vertilla) angustior JEFFREYS
Xeroclausa cf. *meda* (PORRO)
Xeroplexa cf. *setubalensis* (PFEIFFER)

Like charophytes, mollusks do not seem to be of great stratigraphic value in the continental Neogene deposits of Morocco. Most species occur throughout the whole lacustrine section and are related to well-described Algerian faunas, which according to JODOT (1955) have a Pliocene age. But JAEGER, 1977 (page 7) points out, on the basis of rodents, found in the same areas, that the "Pliocene" gastropodes described by JODOT in the various regions of the Maghreb should have in fact a Miocene age.

The three new species *Tryonia kandoulae n. sp.*, *Cincinnatia (? Neohoratia) atlasica n. sp.* and *Seminolina n. sp.* show closer relationships to North American than to European genera. *Tryonia kandoulae n. sp.* has been found only in sample MA 529 and within strata directly beneath it. This species is probably a descendant of the Burdigalian *Tryonia bleicheri* (PALLARY) {? *Melania.*}. Based on the content of teeth of small mammals sample MA 529 yielded an uppermost Miocene age (Upper Turolian). The two other forms related to North America occur throughout the whole Lacustrine Member from Middle Miocene to Upper Pliocene. The Miocene/Pliocene-boundary by mollusks is characterized by the last occurrence (LOD) of *Tryonia kandoulae* n. sp. in Late Miocene (MA 529) and the first appearance (FAD) of *Melanopsis acuminata* PALLARY in Early Pliocene time (MA 541).

Like charophytes the gastropodes too show a marked difference in the number of species occurring in the Ouarzazate basin and in the Ait Kandoula region. In the former less than half could be detected of the amount found in the Ait Kandoula region.

A final taxonomic revision of some groups of mollusks, especially Planorbidae and some Hydrobiidae was not yet possible due to lacking old literature. There may be new species in these groups too, and some more index fossils are expected useful for a further subdivision of the Ait Kandoula Formation .

F. Ostracodes: The third group of fossils very common throughout the Lacustrine Member are ostracodes, which are mostly well preserved. Their ecology and stratigraphic distribution will be discussed and a systematical description will be given separately (HELMDACH, this vol.).

Due to the large number of ostracodes picked, only the section of maximum thickness within the Ait Kandoula region will be presented in detail. Lithology, fish assemblages (see next chapter) and the ostracode faunas indicate a frequent change in salinity of the aquatic environment from freshwater to oligohaline and mesohaline (fig. 7). A freshwater to oligohaline environment has been dominant. Most of the species of charophytes and several species of the ostracode genera *Darwinula, Limnocythere, Candona, Cypris, Ilyocypris, Heterocypris, Cyprinotus and Zonocypris* indicate a salinity of 0.005 % - 0.05 %. Some species of *Candona* and *Ilyocypris* are known from the oligohaline zone: 0.05 - 0.5 % (Venice System). The nearly monofaunal appearance of species of the genus *Cyprideis* finally suggests a salinity of 0.3 - 1.8 % for at least 4 distinct periods (fig. 7). Most species of this genus have been reported from a wide range of salinities, but are most common in brackish water with a salinity between 0.2 - 1.65 % (WAGNER 1957). Some species of *Cyprideis* occur even in hypersaline environments with up to 6 % salinity (HARTMANN 1954). For example in samples from a shore lake in Saudi Arabia with a salinity of 6 % shells of *Cyprideis* have been found. Short episodes with higher than mesohaline salinity during lacustrine deposition of the Ait Kandoula Formation may well be possible.

Changes in salinity can occur in very short intervals within few metres of thickness. This is reflected by pronounced changes of faunal composition with low salinity-tolerance. Due to their low tolerance of salinity, changes of the faunal composition have occurred often and rapidly. Periods of freshwater alternate nearly rhythmically with those of oligohaline and mesohaline environment. Based on 44 sam-

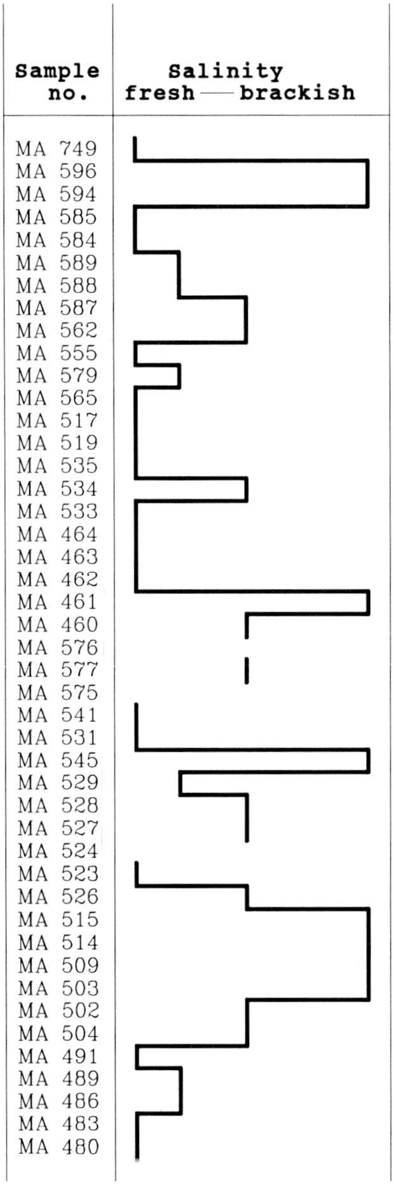

Fig. 7 Palaeosalinities inferred from ostracode faunal composition

ples containing ostracodes taken from the 800 metres of the fossiliferous main section, this could result in a subdivision into several oecostratigraphic zones each characterized by their faunal composition. Denser sampling probably would result in an even more detailed picture of salinity changes. More detailed investigations are planned.

The lowermost part of the Lacustrine Member in the studied section contains a fauna of fresh water ostracodes as follows:

Limnocythere rostrata STRAUB 1952
Darwinula aff. *cylindrica* STRAUB 1952
Cyprinotus unicornutus n.sp.952
Heterocypris aff. *parva* (LIENENKLAUS 1905)

This assemblage is nearly identical with the Southern German fauna from the Swabian Obere Süßwassermolasse (Kirchberg- und Sylvana- Schichten) described by STRAUB 1952. There, a Langhian to Lower Serravallian age has been analysed and therefore the same age is assumed for the base of the Lacustrine Member in the main section.

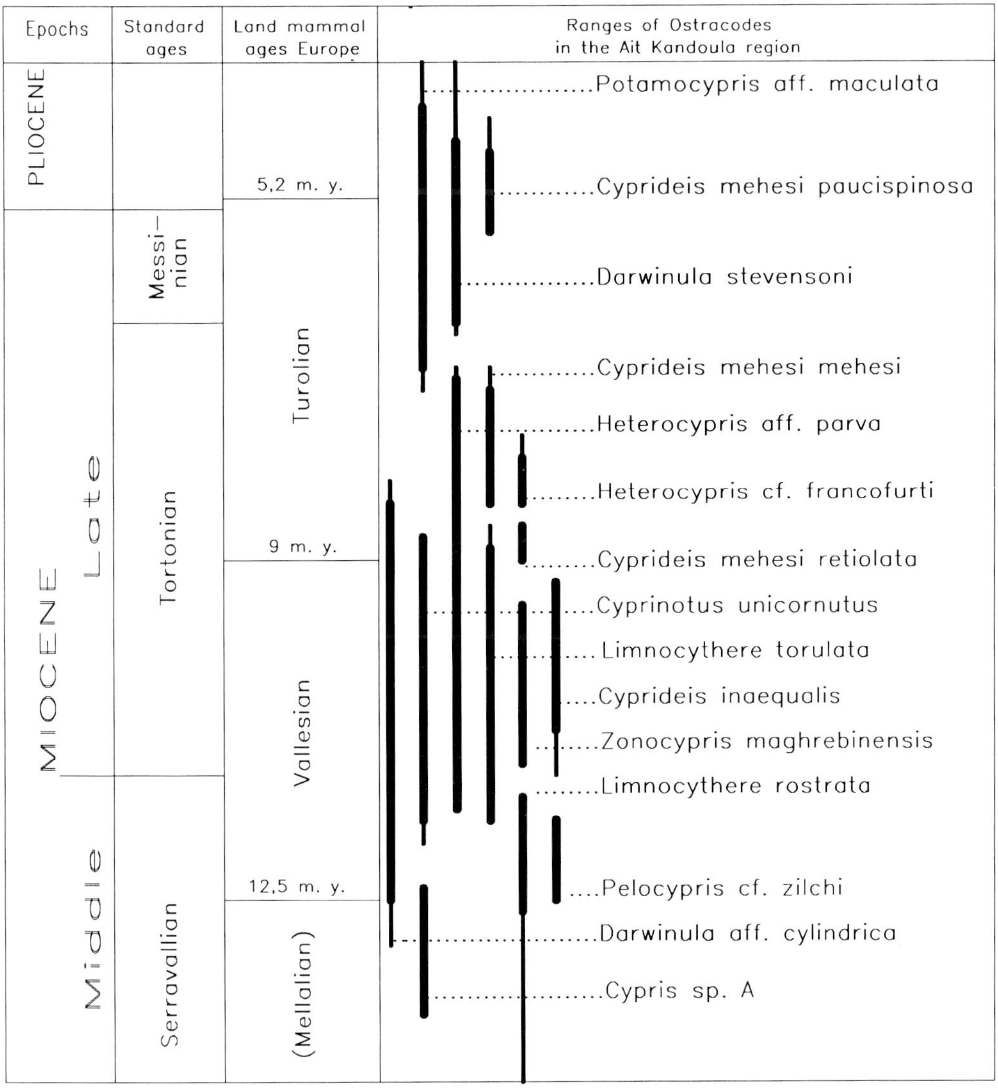

Fig. 8 Distributional range of selected ostracode species in the Ait Kandoula region. Stratigraphic correlation and ages according to BERGGREN et al. (1985). "Mellalian" is an African land mammal age.

sian with a circum-Mediterranean distribution (*Progonomys cathalai* replaced in the early Turolian by *Progonomys woelferi*). The Turolian is characterized by the appearance of Protatera - the first Gerbillid. The higher Turolian is indicated by the first *Paraethomys*. During the Pliocene the Murids dominated. Myocricetodonts have not survived.

The earliest fauna (sample MA 486) of lowermost Vallesian age is situated in the Ait Kandoula region only 5 m above the base of the Lacustrine Member. The sample MA 682 of comparable age, taken 23 km further to the west lies 60 m above the base of the Lacustrine Member and may thus indicate an earlier onset of the lacustrine sedimentation there. A probably even older fossiliferous lacustrine section exist in the northern Ouarzazate basin two kilometres west of Amekchoud. Beneath the sample A 97 that is of an Early to Middle Vallesian age another 400 m of fossiliferous lacustrine sediments crop out. They have not been studied in detail and have not yet supplied any index fossils.

The youngest sample with teeth usable for small mammal stratigraphy, MA 589 was situated about 65 m beneath the highest fossiliferous lacustrine layer and yielded an age of Late Pliocene. There is no indication of younger faunas. All samples of which mammal teeth have been obtained are depicted in fig. 9. relative to their biostratigraphic range analysis.

3.2.2.3 Alluvial Top Member (?Upper Pliocene/Lower Pleistocene)

In the uppermost portions of the Lacustrine Member intercalations of sandstone occur and red siltstone with numerous conglomeratic channel fill structures prevail. These channel fills have a width in the order between a few m and up to 40 m and a thickness of up to 3 m. Even in this level small intercalations of fossiliferous marls and stromatolites occur. This portion of the section is interpreted as a transition between an intralittoral and an alluvial plain environment. Further up a mixture of clast supported and matrix supported conglomerates occur. They build up a large upward coarsening and thickening megasequence. Channel-fill streamflow deposits are very common as are sieve sediments, very poor in matrix and with open framework.

Pebbles of Cretaceous or Palaeogene material occur only in the westernmost portion of the Ait Kandoula region; otherwise there are only Jurassic limestones, Triassic sandstones and volcanics, and a few Palaeozoic pebbles. In several sections a sudden change in the composition has been observed from a distinct predominance of Jurassic pebbles to a clear prevalence of older material.

Channel orientation and measurements of the imbrication in the streamflow deposits yielded a semicircular radial pattern of palaeocurrent directions, indicating a general transport direction from N towards S (fig. 10). The sequence is interpreted as a result of alluvial-fan sedimentation triggered by a renewed uplift phase of the

High Atlas in Late Pliocene (?) and/or Early Pleistocene times. This view is supported by large megabreccias with blocks of diameters up to more than 10 m -probably rockfall debris- within the sequence near the northern contact to the underlying Jurassic limestones. In the western part of the study area even glide-slabs of Jurassic limestones in the order of kilometres in length are intercalated in the Alluvial Top Member of the Ait Kandoula Formation.

The apexes of the reconstructed alluvial fans coincide -probably not by chance- with the outlets of the major actual gorges into the plain of the Ait Kandoula region (fig. 10) although during and after the buildup of the alluvial fans the whole region has been affected by folding and faulting.

The palaeocurrent patterns of these alluvial fans in the Ait Kandoula region indicate fan widths (E - W) of up to 15 km. Today, the outcrop area of Neogene sediments in the Ait Kandoula region on top of the Toundout nappe has a maximum "axial" length of 6 km. However recent alluvial fans are normally longer than wide, and therefore a greater former extent of the fans in the N - S direction has to be assumed. In addition there occur sieve sediments even near the southern margin of the outcrop area. In modern alluvial fans such sediments are near the apex of a fan. These occurrences near today's southern margin of the outcrop area hint at originally much larger fans within the Ait Kandoula region. The southern prolongation of these fans probably is disrupted or missing because of tectonics and/or erosion. The narrow, elongated outcrops of continental Neogene between the Ait Kandoula region and the Ouarzazate basin may be considered as a prolongation of the fans in the Ait Kandoula region. They too, display N - S palaeocurrent directions but their northward prolongation can not be determined, as they are overridden from the north by slabs of Mesozoic or Palaeogene rocks. But as midfan or even outer fan sediments are in some places overridden, a northward prolongation beneath the overlying older sediments must be assumed. A prolongation of the alluvial fans of the Alluvial Top Member into the Ouarzazate basin seems unlikely, because the conglomerates of the Ouarzazate basin contain at least locally large amounts of pebbles of Cretaceous or Eocene age, which are missing in the conglomerates of this level within the Ait Kandoula region.

In the western Ouarzazate basin the uppermost portion of the Ait Kandoula formation (or oldest reg) is built up by conglomerates and coarse sandstones which show trough- and planar cross-bedding. This sequence is typical for braided river deposits of the Donjek type (MIALL 1978, 1981) and is deposited across the Ouarzazate basin. The base of the sequence is cut into the older sediments of the Lacustrine Member. Because pebbles of these conglomerates are far better rounded than in the alluvial-fan sequences, and also because boulders with diameters up to 50 cm of High Atlas material show up even near the southern border of the Ouarzazate basin, long-range cross-basinal fluvial transport must be assumed. The transport of these exceptional large boulders for at least 40 km, from the High Atlas across the whole Ouarzazate basin, therefore could only have happened when the basin was filled by sediments. At that time the modern drainage pattern of the Ouarzazate had to be established, and the "Ouarzazate lake" suddenly drained off towards the south. This interpretation is confirmed by the determination of palaeocurrent directions, namely few measurements of imbrication and cross-bedding that point to the northern end of the Dra gorge cutting through the Anti-Atlas as the outlet from the Ouarzazate basin.

Within the Alluvial Top Member no fossils have been detected. The youngest index fossils found in the Lacustrine Member indicate an Upper Pliocene age for a level about 65 m beneath the base of the Alluvial Top Member. The Alluvial Top Member is therefore considered to be of latest Pliocene, or more likely, Early Pleistocene age. The actual drainage system of the Ouarzazate basin via the river Dra towards the south seems to have first been active during this time.

3.2.3 Quaternary

Besides the Alluvial Top Member of the Ait Kandoula Formation, Quaternary sediments are widely distributed exhibiting an angular unconformity to the older sediments. According to published geological maps 1 : 200.000, the Quaternary sequence is six-fold divided in regs and terraces of different elevations. Their respective ages are mainly determined by archaeological artifacts. No detailed investigations within Quaternary sediments have been carried out by us. Only at one location (MA 606) in the Ait Kandoula region lacustrine Quaternary sediments with the following gastropodes were investigated:

Galba cf.*truncatula* (MÜLLER)
Radix cf. *subperegra* (PALLARY)
Stagnicola chouberti (JODOT)
Keraea bertholdiana (PFEIFFER)
Succinella sp.
Vertigo (V.) antivertigo (DRAPARNAUD)
Vallonia cf.pulchella (MÜLLER).

3.3 Evolution of depositional environment and climate

During the evolution of the High Atlas the depositional environment oscillated several times between continental and marine sedimentation of different water depths. As mentioned in 3.2.1 the latest transition from marine to nonmarine sedimentation probably took place in latest Eocene/earliest Oligocene time. Then shallow marine to littoral conditions were followed by a sabkha environment influenced by dunes and subsequently by a first period of alluvial-fan and braided stream sedimentation. During the Middle Miocene a paludal and lacustrine environment was established, but alluvial-fan sedimentation from the uprising axial zone of the High Atlas continued. Lacustrine sedimentation persisted up to the Late Pliocene, but salinity and the area and shape of the lake surface changed repeatedly. Afterward, during a second phase of rapid uplift of the High Atlas, alluvial-fan sedimentation was re-established. Most likely the closed basin began to drain towards the south in Early Pleistocene time and since then erosion prevailed not only in the Anti-Atlas, but also in the Ouarzazate basin and the Ait Kandoula region.

The history of the climate in this region is rather monotonous. According to reconstructions of the continental displacements since the Triassic (SMITH et al. 1981; OWEN, 1983) the latidude of the study area changed from about 10^{0} N to 31^{0} N today. During the Mesozoic enormous amounts of palaeozoic rocks must have been eroded in the region south of the High Atlas (s. next chapter), but contemporaneous coarse-grained sediments are scarcely known from the High Atlas region. There silts, limestones and marls prevail. A warm and humid climate, therefore, seems likely for this time span.

During Eocene a warm and dry desert climate can be assumed due to the microflora found in littoral sediments (MOHR and FECHNER, 1986). Similar conditions seem to have persisted into the Oligocene as documented by sabkha sedimentation and dunes. Later on, braided river deposits, the sedimentation of thick alluvial-fan sequences, the absence of laterite and the dominance of limestone pebbles suggests more semiarid climatic conditions during Early Miocene times. During the subsequent lacustrine episode the climatic conditions probably were equivalent to that of today in the southern Mediterranean area. JODOT (1955) assumed, from analyses of the Vallesian gastropode fauna of the Ait Kandoula region that a warm Mediterranean climate prevailed similar to that of recent northern Algeria. Based on the occurrence of Middle and Late Miocene rodents found at various localities in Algeria and Morocco JAEGER (1977) infers a "steppe" environment of changing humidity: during the Mellalian and Early Vallesian time more drier conditions prevailed, followed by more humid conditions from Middle Vallesian to Early Turolian and beginning in Late Turolian time again by increasingly dry conditions. This point of view is supported by the changing lithology within the Lacustrine Member of the Ait Kandoula Formation. Gypsum layers and scattered gypsum crystals occur particularly in the deepest portion of the main section (Lower Vallesian) and directly beneath sample MA 529 (Upper Turolian). Above sample MA 529 gypsum is absent suggesting an increased humidity during Pliocene time. A much more complicated picture arises when evaluating the ostracode oecozones, which indicate a frequent change of palaeosalinities during the time represented by the Lacustrine Member. Such variations have even been detected at intervals of less than one metre thickness. The curve given in fig. 7 is only a first approximation, and dense linear sampling would certainly provide a much more complicated curve of inferred palaeosalinity, perhaps even relatable to the MILANKOVITCH curve.

The retreat and the advance of the alluvial fans at the beginning and the end of the lacustrine sedimentation is probably more related to tectonic movements than to changes of the climatic conditions. Nevertheless the periods of rapid uplift of the High Atlas might have also caused higher precipitation rates as well.

A further alternation of more arid and more pluvial episodes during the Pleistocene is well documented by the presence of various terraces and glacis during that time.

4.0 Uplift, deformation and basin evolution at the southern rim of the the central High Atlas during the Cainozoic.

The High Atlas is an intracratonic mountain range in an area previously affected by the Variscan orogeny. Sedimentation started after an initial rifting phase during the Triassic, which was probably related to the early history of the Atlantic Ocean Rift. The Mesozoic sediments of the Western High Atlas were deposited in an Atlantic-embayment known as the gulf of Essaouira extending towards the east as far as the "massif ancien". In contrast to the western High Atlas, the eastern and central High Atlas and the Middle Atlas were connected to the Tethys (Atlas Gulf) and the epicontinental sea of Jurassic age of the Arabian plate (CHOUBERT and FAURE-MURET, 1962; DU DRESNAY, 1971; ADAMS, 1979; HÜSSNER, 1985; ENAY et al. 1987). From Triassic to Palaeogene time the central High Atlas was a region of more or less continuous subsidence and sedimentation although the type of sedimentation changed several times from continental to marine. This is documented by the stratigraphic record and the almost persistent concordant deposition of the respective sediments. Further west and north local unconformities occur within the Late Cretaceous to Early Tertiary sequence (ROCH, 1939; LAVILLE, 1980; STETS and WURSTER, 1981; FROITZHEIM, 1984;) indicating an initial uplift of the High Atlas region. In the study area itself however the first obvious evidence of an uplift as a structural inversion from a formerly subsiding basin is the occurrence of rock debris and pebbles derived from the High Atlas at the base of the Ait Kandoula Formation. An Early Miocene age of the first uplift of the High Atlas in this particular area may therefore be more likely.

Closely related to the uplift of the High Atlas is the beginning of the subsidence of the Ouarzazate basin which is confined between the uprising axial zone of the High Atlas and the then still active source area at the northern rim of the Sahara platform. The evaporitic sedimentation at the western end of the Ouarzazate basin indicates that it was a closed basin at least since Middle Miocene time. In this aspect it is different from other foredeeps e. g. the Alpine Molasse basin, which includes marine sequences, too. The subsidence of the Ouarzazate basin is, therefore, only relative to the High Atlas and the Anti-Atlas, but not in respect to the sea-level. As marine Palaeogene strata outcropping in, and marginal to, the Ouarzazate basin today have an elevation of more than 1 000 m above the sea-level, an uplift of the Ouarzazate basin area in respect to sea-level during the Neogene has to be assumed.

An earliest angular unconformity in the study area was observed in the Dadès valley (ZYLKA and JACOBSHAGEN, 1986; ZYLKA, 1988) between the basal parts of the Alluvial Base Member and the higher portions of the Ait Kandoula Formation. It proves the earliest phase of folding of Cretaceous and Palaeogene sediments. As the lower parts of the Alluvial Base Member of the Ait Kandoula Formation are already affected by this folding, folding must have started slightly later than the first erosion of uplifted Mesozoic and Palaeogene sequences in the High Atlas realm. An age of early Middle Miocene seems most probable for this folding, as the uppermost portion of the Alluvial Base Member and the oldest (Lower Vallesian) parts of the fossiliferous marls and limestones of the Lacustrine Member of the Ait Kandoula Formation are overlying steeply dipping sediments of the folded Cretaceous to Lower Miocene sequence.

A still unresolved question is whether endogenic compressional forces or gravity sliding of the uppermost sedimentary cover from the uplifted axial zone of the High Atlas caused this folding or a combination of both. As mentioned in chapter 3.2.2.1, southward movements of Cretaceous and Palaeogene slabs and the intercalation of megabreccias are indications of gravity tectonics during the deposition of the Alluvial Base Member. The following observations favour a complete detachment at the lower Cretaceous red bed level and a southward gliding and folding of the detached sedimentary cover (Cenomanian/Turonian limestones, Upper Cretaceous red beds, Paleocene/Eocene marls and limestones, Hadida Formation):

1. Palinspastic unravelment of the South Atlas Marginal Zone and the occurrence of southward transported pebbles of Cretaceous and Palaeogene sedimentary rocks in the Alluvial Base Member of the Ait Kandoula Formation demonstrate that once such a continuous sedimentary cover existed on top of the Jurassic limestones. In the Ait Kandoula region the two higher Members, deposited after the assumed sliding of the Upper Cretaceous and Palaeogene cover, are void of pebbles of these rocks.

2. The sharp contrast of deformational style between the broad folds of Jurassic limestones in the High Atlas of this region and the intense steep folding and thrusting of the Cretaceous and Palaeogene in the South Atlas Marginal Zone is a result of a different amount of lateral shortening, only explained by some kind of detachment. Within the South Atlas Marginal Zone, wide synclines alternate with narrow, south vergent anticlines and thrust faults. The intensity of deformation increases from north towards south and the Palaeogene/Cretaceous sequence at many places overrides the Neogene rocks of the Ouarzazate basin. This style of deformation fits well the model of MERLE (1984) for "viscous gliding accomplished by low-viscosity basal strata".

3. Gravitational sliding is favoured by the fact that the most pronounced folding is adjacent to the barely folded region of the central High Atlas around the Irhem Mgoun (4 071 m), precisely the region where the greatest amount of uplift of the sedimentary cover occurred. Over a distance of 30 km the difference of uplift of the

Palaeogene strata between the High Atlas and the Ouarzazate basin is estimated to have been about 5.000 m (GÖRLER and ZUCHT 1985).

In the "calcareous" Atlas there is a clear tendency of increasing endogenic compressional tectonics towards the east, manifested within the Jurassic sediments (JACOBSHAGEN et al. this vol.). On the other hand there is a clear decrease in the amount of deformation of Cretaceous and Palaeogene sediments within the South Atlas Marginal Zone. East of Tinerhir there is even a gradational transition between the folded High Atlas and the South Atlas Marginal Zone. There the style of deformation due to compressional tectonics in both of them is the same and the only detectable difference is the occurrence of Cretaceous sediments in the latter one (ZYLKA 1988). Unfortunately Neogene or even Palaeogene sediments are missing there and the exact age of this endogenic folding, therefore, remains unknown. For the study area a combined deformation due to gravitational and endogenic tectonic seems most likely.

The next datable event which is probably related to the uplift of the High Atlas is the emplacement of large slabs of Jurassic limestones within the Ait Kandoula Formation. The largest Jurassic slab with Triassic at its base and locally with Lower Cretaceous red beds on its top is the nappe of Toundout. Near its western end it splits up into several stacked klippen isolated from each other by the sediments of the Ait Kandoula Formation. The level of the deepest stacked klippe was dated with the help of small mammal teeth (sample MA 682) indicating an Upper Mellalian/Lower Vallesian age for the emplacement of the Toundout nappe. The oldest datable sediments overlying this nappe have an age of lowermost Vallesian, whereas in the Dadès valley north of Ait Ibrirn a large slab of Jurassic limestone is emplaced on top of age equivalent lacustrine marls. The emplacement of the different slabs of Jurassic limestones thus has not occurred simultaneously, although mainly during the Middle Miocene. The lateral displacement of the nappe of Toundout is estimated to be in the order of 10 - 15 km (LAVILLE, 1980). The overthrust of the large klippe of Jurassic limestone in the Dadès valley north of Ibrirn could be in the same order. As mentioned above it is emplaced on top of lacustrine marls that were not deposited in the direct neighbourhood of the uplifting axial zone from where this klippe is thought to have originated. Because the Ait Kandoula region is characterized by negative gravity anomalies, a pronounced negative Bouguer gravity anomaly (less than -150 mgal) about 15 km farther north in the Dadès valley, where only Jurassic rock crops out, might suggest an extension of the thick Neogene sediments from the Ait Kandoula region beneath the overriding Jurassic sediments (fig. 4).

The nappe of Toundout and the klippe of Ibrirn must have been emplaced without significant internal deformation into a nearly horizontal position, because the Neogene strata on top of them are deposited nearly always concordantly with the Jurassic limestones, but over a relief marked by a deep reaching palaeokarst.

The alluvial-fan deposits on top of the nappe show a fining upwards tendency and grade into the Lacustrine Member of the Ait Kandoula Formation. The time interval of about 10 m. y. from late Middle Miocene to Late Pliocene after the emplacement of the

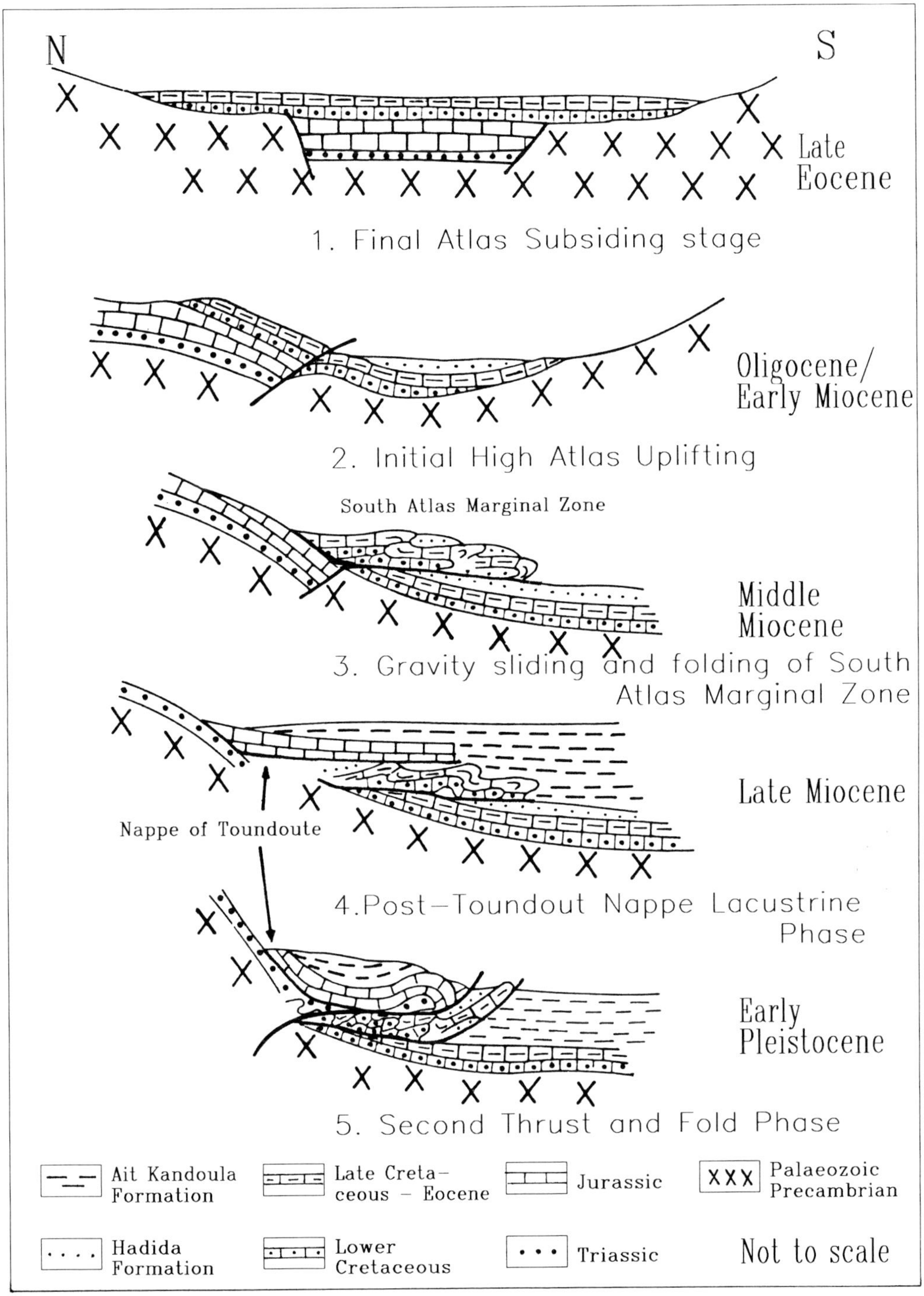

Fig. 11 Structural timing of the southern central High Atlas during Tertiary time

Toundout nappe, is thought to have been tectonically more stable because of a continuous lacustrine sedimentation. This could be especially true for the Late Miocene with its extremely low lacustrine sedimentation rates. Still, progressive unconformities, slumping structures and the local continuation of the alluvial-fan sedimentation throughout this time, suggest some small-scale vertical movements.

A second period of rapid uplift in the High Atlas is documented by the upwards coarsening alluvial-fan sedimentation on top of the Lacustrine Member, containing megabreccias and a second generation of Jurassic and even Palaeozoic glide-slabs (LAVILLE, 1980; SULZER, 1987).

As the youngest sample with index fossils of the Lacustrine Member 60 m beneath the base of the Alluvial Top Member yielded a Late Pliocene age, this renewed period of rapid uplift of the High Atlas is thought to have started in latest Pliocene or Early Pleistocene time. It was effective throughout the Pleistocene, as documented by the different elevation of glacis and terraces (see published geological maps of the region 1 : 200.000 and 1 : 500.000). High elevation and morphology are signs for a still active recent uplift of the central High Atlas. In the north of the Ait Kandoula region the lower two members of the Ait Kandoula Formation are missing and only the slightly tilted conglomerates of the Alluvial Top Member overlie the steeply dipping Jurassic limestones with an angular unconformity. The deeper portions of the Ait Kandoula Formation were probably eroded during the uplift. This uplift also caused a tilting of the northern part of the nappe of Toundout. The vertical movements are locally related to folding of previously emplaced slabs of Jurassic limestone. In particular the nappe of Toundout is deformed in its southern part together with the concordantly overlying lower portions of the Ait Kandoula Formation. At some locations this deformation led even to an upright or overturned attitude. In addition, at the northern rim of the Ouarzazate basin between Tamesoult and Tamassine conglomerates of a probable Pliocene age are in an upright position near southward overthrusted Eocene limestones. A compressional deformation seems therefore likely synsedimentary with the deposition of the higher levels of the Ait Kandoula Formation.

Signs of an even younger compressional deformation can be observed near Amekchoud. There, at a distance of a few kilometres from the High Atlas, the strata of the Ouarzazate basin form a large anticline overlain by Pleistocene strata (q 4, "Reg ancien" of the geological map 1 : 200 000) and separated from them by an angular unconformity . Near the core of the anticline these Pleistocene sediments are still tilted and incline with up to 25^{0} (GÖRLER and ZUCHT 1986). However, based on the available information a diapiric origin of this anticline cannot be excluded.

In contrast to the first phase of uplift during Early Miocene/early Middle Miocene the second phase affected not only the axial zone of the High Atlas but also the Ouarzazate basin and the Anti-Atlas, because there too, the older Pleistocene glacis and terraces are situated well above the younger ones. As the stratigraphic position of the Neogene sediments in the Anti-Atlas is not well known, it is too speculative at this stage to reconstruct the history of its uplift. During the Mesozoic and Palaeogene the region of today's Anti-Atlas and Jbel Sarhro once was a source area of

Continental Neogene in the southern High Atlas		
Stratigraphic position	Facies interpretation	Structural and basin evolution
Pleistocene	Erosion predominant	Uplift of High Atlas, Ouarzate basin and Anti-Atlas
Early Pleistocene Late Pliocene -	Upwards-coarsening alluvial fans, landslides	Rapid uplift of the High Atlas, local gravity sliding, second phase of folding. Start of the drainage of the Ouarzazate basin via river Dra
Pliocene	Lacustrine/paludal/alluvial in the Ait Kandoula region, alluvial in the Ouarzazate b.	Accelerated uplift and separation between Ait Kandoula region and Ouarzazate basin
Late Miocene	Lacustrine sedimentation at a very low rate	Slow to vanishing uplift of the High Atlas. Lakes and swamps in a closed basin between Anti-Atlas and High A.
Middle Miocene	Lacustrine/paludal/pedogenic/alluvial	Gravity sliding and emplacement of nappe of Toundout
Middle Miocene Early Miocene -	Fining-upwards alluvial fans/braided streams/slumping	Uplift of the axialzone of the High Atlas, large scale gravity sliding of Cretaceous and Palaeogene rocks, emplacement and first folding of the South Atlas Marginal Zone
Early Miocene	Alluvial fan sedimentation	Start of the uplift of the High Atlas
Oligocene	Sabkha, Braided stream	Renewed uplift of the Sahara platf.
Early Oligocene latest Eocene -	End of epicontinental marine sedimentation	Slow to vanishing subsidence in the High Atlas sedimentation area

Fig. 12 Environmental and structural timing in the southern High Atlas

the sediments shed into the pre-High Atlas depression and, therefore, a much longer uplift history must be assumed for the Anti-Atlas. Especially in the area bordering the Ouarzazate basin the Palaeozoic cover over the Precambrian is probably eroded during that early time. Only 50 - 100 km to the south in the Jbel Bani region and the Tindouf basin Palaeozoic strata reach a thick-ness of 10 kilometres. But during the sedimentation of the Palaeocene and Eocene shallow marine to littoral sequence in the High Atlas realm the type (limestones and marls) and the low rate (0.005 - 0.01 mm/y) of sedimentation indicate low relief and low erosion rates in the bordering Sahara platform. The coarse grained material of the Hadida Formation derived from there and its one order of magnitude greater sedimentation rates are signs of a renewed uplift and erosion in this area in Oligocene to Lower Miocene time.

Afterwards only in the southern portion of the Ouarzazate basin the Sahara platform remained the source area of terrigenous material. It is therefore possible to assume that the uplift of the northern rim of the Sahara platform ceased during Lower Miocene time, permitting sedimentation to take place in the palaeo-drainage system (see 3.1) during the Late Miocene and Pliocene. But the assumption of a continuous, slow, uplift of the Anti-Atlas is not contradictory to these observations, because the palaeo-drainage system which was filled up with sediments, was connected only to the "closed" Ouarzazate basin. Therefore conclusions about vertical movements of the Anti-Atlas with respect to sea-level are not possible.

The start of the erosion of the Miocene/Pliocene sediments.during the Pleistocene could therefore be due to a renewed uplift of the Anti-Atlas, but could also be caused only by the beginning of the drainage of the Ouarzazate basin *via* the river Dra.

In summary the history of the Ouarzazate basin seems to be a good example for a "loading" basin, as proposed by PRICE (1973): Tectonic thickening of supracrustal rocks causes isostatic subsidence which reaches beyond the actual limits of the supracrustal load due to flexural rigidity in the lithosphere. In this way the increasing thickness of the Ouarzazate basin towards the High Atlas (see 2.2) can be explained. This model is also in good agreement with the observation that the extent of the Ouarzazate basin is limited to that section of the southern High Atlas characterized by south-vergent overthrusts. Loading by thrust sheets and filling of the adjacent moat with sediments seems sufficient to explain the origin of the Ouarzazate basin. A lithospheric flow as proposed by BALLY and SNELSON (1980) as an additional process for the generation of foredeeps is not probable in this case. If this model were applicable, a pronounced foredeep would be expected to exist in front (southward) of the Jbel Toubkhal, the highest elevation of the High Atlas, with basement uplifted to more than 4 000 m above sea-level. But instead of a subsiding basin, there, in the southern foreland of the High Atlas, marine Eocene strata have been uplifted to an elevation of more than 2000 m above sea-level (GÖRLER and ZUCHT 1986).

ACKNOWLEDGMENTS

We thank Deutsche Forschungsgemeinschaft and Freie Universität Berlin for the financial support of the research project. This manuscript was reviewed by V. Jacobshagen, D. A. Warnke and H. Ibbeken. R. Zylka made useful comments on the tectonics of the South Atlas Marginal Zone. The English of the manuscript was revised by G. Lindemann. Essential technical assistance was provided by M. Bryant, E. Gerolymatos, S. Köhler, W. Jung, B. Schindzielorz and D. Salomon. The field work in Morocco would have been impossible without the friendly support by the Division de la Géologie of the Ministère de l'Energie et des Mines, Rabat and especially M. Bensaid, M. Dahmani and M. Khalek.

REFERENCES

ADAMS, A. E. (1979): Sedimentary environments and paleogeography of the western High Atlas, Morocco, during the Middle and Late Jurassic. - Palaeogeography, Palaeoclimatology, Palaeoecology, **28**: 185 - 196.

ALLEN, P. A. and COLLINSON, J. D. (1986): Lakes.- In READING, H. G. (ed.) Sedimentary Environments and Facies.- Blackwells, Oxford: 63 - 94.

BALLY, A. W. and SNELSON, S. (1980): Realms of Subsidence.- Canad. Petroleum Geol. Mem. **6**: 9 - 94.

BERGGREN, W. A.; KENT, D. V.; FLYNN, J. V. and VAN COUVERING, J. A. (1985): Cenozoic geochronology. - Geol. Soc. Amer. Bulletin, **96**: 1407 - 1418.

BERGGREN, W. A. and VAN COUVERING, J. A. (1974): The late Neogene. Biostratigraphy, geochronology and paleoclimatology of the last 15 million years in marine and continental sequences.- Palaeogeography, Palaeoclimatology, Palaeoecology, **16**: 1 - 216.

BERNASCONI, R. (1983): Géologie du Haut Atlas de Rich.- Thèse de l'Université de Neuchâtel, 1 - 107, Neuchâtel.

BREDE, R. (1987): Strukturelle Entwicklung des zentralen Hohen Atlas nordwestlich von Errachidia (Marokko).- Berl. geowiss. Abh., (A), **85**: 1 - 65.

BREDE, R. and HEINITZ, W. (1986): Zum Bau des Hohen Atlas zwischen Errachidia und Midelt (Marokko).- Berl. geowiss. Abh., (A), **66**: 433 - 354.

CANT, D. J. (1982): Fluvial Facies Models- in SCHOLLE, A. and SPEARING, D. (eds.): Sandstone Depositional Environments: 115 - 138, Amer. Ass. Petroleum Geol., Tulsa.

CHARISIUS, M. (1985): Stratigraphische und strukturgeologische Untersuchungen in der Südatlas-Randzone bei Boumalne du Dadès (Hoher Atlas, Marokko).- unpubl. Diplomarbeit FU Berlin,: 1 - 61, Berlin.

CHOUBERT, G. (1945): Note prélimininaire sur le Pontien au Maroc (Essai de synthèse orogénique du Maroc atlasique).- Notes et Mem. Serv. Géol. Maroc **65** and Bull. Soc. Géol. France (5) **15**: 677 - 764.

CHOUBERT, G. CHARLOT, R., FAURE-MURET, A., HOTTINGER, L., MARCAIS, J., TISSERAND, D and VIDAL, PH. (1968): Note prélimininaire sur le volcanisme messinien-pontien au Maroc. - C. R. Acad. Sci.(D) **266**: 197 - 199.

CHOUBERT, G. and FAURE-MURET, A. (1962): Evolution du domaine atlasique marocain depuis les temps paléozoiques.- Livre Mém. P. Fallot, **1** (Soc. géol. France): 447 - 528.

CONRAD, G. and CONRAD, J. (1982): La zone de jonction des cratons de l'Ouest-Africain et du Sahara Central au Cènozoique: évolution sédimentaire et structurale de l'Ahnet et de confins Erg Chech-Tanezrouft. - Memoirs géologiques de l'ùniversité de Dijon, **7**: 391 - 400.

COURBOULEIX, S., DELPONT, G. and DESTEUCQ, C. (1981): Un grand décrochement est-ouest au Nord du Maroc à l'origine des structures plissées atlasiques. Arguments géologiques et expérimentaux.- Bull. Soc. géol. France (VII), **23**, 1: 33 - 43.

COUVREUR, G. (1973): Etagement de formes et néotectonique sur la versant sud du Haut Atlas. - Rev. Géomorph. dyn., **22**, 3: 108 - 124.

DOMEYER, S. (1985): Geologische Untersuchungen in der Südatlas-Randzone nordwestlich von Boumalne du Dadés (Südmarokko).- Unpubl. Diplomarbeit FU Berlin, 1 - 61, Berlin.

DRESNAY, R. DU (1962): Les discontinuités de sédimentation pendant le Jurassique dans la parte orientale du domaine Atlasique Marocain, leurs conséquences stratigraphiques et leurs relations avec l'orogénèse Atlasique.- Coll. Jurassique, 899 - 912, Luxembourg.

DRESNAY, R. DU (1972a): Les phénomènes de bordure des constructions carbonatées du Lias moyen du Haut-Atlas oriental (Maroc).- C. R. Acad. Sci. Paris (D), **275**: 535 - 537.

DUTOUR, A. and FERRANDINI, J. (1985): Nouvelles observations néotectoniques dans Le Haut Atlas de Marrakech et le Haouz central (Maroc). Apports sur l'évolution récente d'un segment du bâti atlasique.- Rev. Géol. dyn. Géogr. phys., **26**, 5: 285 - 297.

ENAY, R., MANGOLD, C., DRESNAY, DU R. and RAKUS, M. (1987): Arrivals of Arabian origin among the ammonite faunas of Morocco during the Bajocian-Bathonian. - Palaeogeography, Palaeoclimatology, Palaeoecology, **61**: 107 - 120.

EUGSTER, H. P. and KELTS, K. (1983): Lacustrine chemical sediments.- In: GOUDIE, A. S. and PYE, K. (eds.) Chemical Sediments and Geomorphology, Academic Press, London: 321 - 368.

EVANS, I., KENDALL, C. G. St. C. and WARME, J. F. (1974): Jurassic sedimentation in the High Atlas mountains of Morocco during early rifting between Africa and North America.- Geology, **2**, 6: 295 - 296.

FAURE-MURET, A. and CHOUBERT, G. (1971): Le Maroc. Domain rifain et atlasique.- Tectonique de l'Afrique: UNESCO, Sciences de la Terre, **6**: 17 - 46, Paris.

FERRANDINI, J. and LE MARREC, A. (1982): La couverture jurassique à paléogène du Haut Atlas du Marrakech est allochtone dans la "zone de cuvettes" d'Ait Ourir (Maroc). - C. R. Acad. Sc., **295**: 813 - 816.

FOUCH, T. D. and DEAN, W. E. (1982): Lacustrine Environments- in SCHOLLE, A. and SPEARING, D. (eds.): Sandstone Depositional Environments: 87 - 114, The Amer. Ass. Petroleum Geol., Tulsa.

FREYTET, P.: Les sédiments lacustres carbonatés et leurs transformations par émersion et pédogenèse. Importance de leur identification pour le reconstitutions paléogéographiques. - Bull. Centres Rech. Explor.-Prod. Elf Aquitaine, **8**,1: 1 - 289.

FREYTET, P. and PLAZIAT, J.-C. (1982): Continental Carbonate Sedimentation and Pedogenesis - Late Cretaceous and Early Tertiary of Southern France.- Contrib. Sed. **12**: 1 - 213

FROITZHEIM, N. (1984): Oberkretazische Vertikaltektonik im Hohen Atlas SW' von Marrakech/Marokko - Rekonstruktion eines Bewegungsablaufs im Frühstadium der Atlas-Orogenese.- Neues Jb. Geol. Paläont. Mh., 1984: 463 - 471.

GAUTHIER, H. (1953): Découverte d'une faunule de mollusques continentaux pliocènes dans le synclinal des khelas (Maroc meridional). - C. R. Acad. Sc., **236**: 99 - 106.

GAUTHIER, H. (1957): Contribution à l'étude géologique des formations post-liasiques des bassin du Dadès et du Haut-Todra (Maroc méridional).- Not. Mém. Serv. géol. Maroc, **119**: 1 - 212.

GÖRLER, K. and ZUCHT, M. (1986): Stratigraphie und Tektonik des kontinentalen Neogens im Süden des zentralen Hohen Atlas, Provinz Ouarzazate (Marokko).- Berl. geowiss. Abh., (A), **66**: 471 - 494.

HAQ, B. U., HARDENBOL, J. and VAIL, P. R. (1987): Chronology of fluctuating sea levels since the Triassic.- Am. Assoc. Adv. Sci., **235**: 1156 - 1169.

HARTMANN, G. (1954): Ostracodes des Étangs Méditerranéens. - Vie et Milieu, **4**,4: 707 - 712.

HELMDACH, F.-F. (1986): Zur Alterseinstufung des höchsten marinen Tertiärs im Bereich der Südatlas-Randzone (Provinz Ouarzazate, Marokko).- Berl. geowiss. Abh., (A), **66**: 495 - 502.

HERBIG, H.-G. (1986): Lithostratigraphisch-fazielle Untersuchungen im marinen Altter-tiär südlich des zentralen Hohen Atlas (Marokko).- Berl. geowiss. Abh., (A), **66**: 343 - 380.

HERNANDEZ, J. and BELLON, H. (1985): Chronologie K - Ar du volcanisme miocène du Rif oriental (Maroc): implications tectoniques et magmatoliques. - Revue Géol. dynam. Geogr. phys., **26**: 85 - 94.

JACOBSHAGEN, V. (1986 a): Zur Bau-Geschichte des zentralen Hohen Atlas.- Berl. Geowiss. Abh., (A), **66**: 455 - 470.

JACOBSHAGEN, V. (1986 b): Untersuchungen zur geodynamischen Entwicklung Südmarokkos. Eine Einführung.- Berl. geowiss. Abh., (A), **66**: 265 - 271.

JACOBSHAGEN, V., BREDE, R., HAUPTMANN, M., HEINITZ, W. and ZYLKA, R. (1987): Structure and post-Palaeozoic Evolution of the Central High Atlas.- In JACOBSHAGEN (Ed.): Studies of the Geodynamic Evolution of the Atlas System (Morocco).- this volume

JAEGER, J. J. (1977): Les rongeurs du Miocène moyen et superieur du Maghreb.- Palaeovertebrata **8**,1: 1 - 166.

JAMES, N. P. (1984): Shallowing Upward Sequences in Carbonates.- in WALKER, R. G. (ED): Facies Models.- Geoscience Canada Reprint Ser. **1**: 213 - 228.

JENNY, J. (1983): Les décrochements de l'Atlas de Demnat (Haut Atlas central, Maroc): Prolongation orientale de la Zone de décrochement du tizi-n-test et clef de la compréhension de la tectonique atlasique.- Eclogae geol. Helv., **76**, 1: 243-251.

JODOT, P. (1953): Faune Pliocène de gastéropodes terrestres et d'eau douce de l'oued Imasssine (Sud du Haut Atlas marocain).- Notes Serv. Géol. Maroc, (7) **117**: 241 - 249.

JODOT, P. (1953): Gastéropodes continentaux du Pliocène inférieur de l'òued Assermo n'Ait Zarhar (Sud du Haut Atlas marocain).- Notes Serv. Géol. Maroc, (8) **120**: 21 - 30.

JODOT, P. (1955): Les subdivisions du Pliocène dans l'Afrique (Algérie - Maroc) d'après les faunes de mollusques continentaux.- Notes Mèm. Serv. Géol. Maroc **121**, 1 - 112.

KENDALL, A. C.(1984): Evaporites.- in WALKER, R. G. (Ed.): Facies Models.- Geoscience Canada Reprint Ser. **1**: 259 - 296.

LAVILLE, E. (1980): Tectonique et microtectonique d'une partie du versant sud du Haut Atlas marocain (Boutonnière de Skoura, nappe de Toundoute).- Not. Serv. géol. Maroc., **41**: 81-184.

LAVILLE, E. (1981): Róle des décrochements dans le mecanisme de formation des bassins d'effondrement du Haut Atlas marocain au cours des temps triasique et liasique.- Bull. Soc. géol. France (VII), **23**, 3: 303-312.

LAVILLE, E. and HARMAND, C. (1982): Evolution magmatique et tectonique du bassin intracontinental mésozoique du Haut Atlas (Maroc): un modèle de mise en place synsedimentaire des massifs "anorogéniques" liés à des décrochement.- Bull. Soc. géol. France (VII), **24**, 2: 213-227.

MERLE, O. (1984): Déplacement et déformation des nappes superficielles.- Revue géol. dynam. geogr. phys. **25**: 3 - 17.

MIALL, A. D. (1978): Lithofacies types and vertical profile models in braided rivers: a summary.- in MIALL, A. D. (ed.): Fluvial Sedimentology.- Canad. Soc. Petrol. Geol. Mem. **5**: 597 - 604.

MIALL, A. D. (1982): Analysis of fluvial depositional systems.- AAPG Education Course Note Ser. **20**: 1 - 75.

MICHARD, A. (1976): Eléments de géologie Marocaine. Not. Mém. Serv. géol. Maroc, **252**: 1 - 408.

MOHR, B. and FECHNER, G. (1986): Eine eozäne Mikrofauna (Sporomorphe und Dinoflagellaten-Zysten) aus der Südatlas-Randzone westlich Boumalne du Dadès (Marokko).- Berl. geowiss. Abh., (A), **66**: 381 - 414.

MONBARON, M. (1981): Sédimentation, tectonique synsédimentaire et magmatisme basique: l'évolution paléogéographique et structurale de l'Atlas de Beni Mellal (Maroc) au cours du Mésozoique; ses incidences sur la tectonique tertiaire.- Eclogae geol. Helv., **74**, 3: 625-638.

MURPHY, D. H. and WILKINSON, B. H. (1980): Carbonate deposition and facies distribution in a central Michigan marl lake.- Sedimentology, **27**: 123 - 135.

NILSEN, T. H. (1982): Alluvial Fan Deposits.- in SCHOLLE, A. and SPEARING, D. (eds.): Sandstone Depositional Environments: 49 - 86, The Amer. Ass. Petroleum Geol., Tulsa.

OWEN, H. G. (1983): Atlas of continental displacement, 200 million years to the present. - 1 - 159, Cambridge University Press, Cambridge.

PRICE, R. A. (1973): Large-Scale Gravitational Flow of Supracrustal Rocks, Southern Canadian Rockies. - In: JONG, K. A. and SCHOLTEN, R. (eds): Gravity and Tectonics: 491 - 502 - John Wiley & Sons, New York, London, Sydney, Toronto.

ROCH, E. (1939): Description géologique des montagnes à L'Est de Marrakech - Not. Mém. Serv. géol. Maroc, **51**: 1 - 438.

ROCH, E. (1950): Histoire stratigraphique du Maroc.- Not. Mém. Serv. géol. Maroc, **80**: 1 - 437.

RUST, B. R. and KOSTER, E. H. (1984): Coarse Alluvial Deposits.- in WALKER, R. G. (ED): Facies Models.- Geoscience Canada Reprint Ser. **1**: 53 - 70.

SCHERMERHORN, L. J. G. and GROTHE, CHR. (1986): Vulkanismus im Anti-Atlas (Marokko).- Berl. geowiss. Abh., (A), **66**: 333-342.

SCHMIDT, K.-H. (1966): Strukturbestimmte Relieftypen und Tektonik im Grenzbereich zwischen Hohem Atlas und Anti-Atlas.- Berl. geowiss. Abh., (A), **66**: 503 - 514.

SEUFERT, G. (1986): Zur strukturellen Entwicklung des zentralen Hohen Atlas: Die Region zwischen Boumalne du Dadès, Tinerhir und Imilchil (Süd-Marokko).- Diss. Freie Univ. Berlin, 1 - 143 (microfiches), Berlin.

SMITH, A. G., HURLEY, A. M. and BRIDEN, J. C. (1981): Phanerozoic paleocontinental world maps.- 1 - 102, Cambridge University Press, Cambridge.

STETS, J. and WURSTER, P. (1981): Zur Strukturgeschichte des Hohen Atlas in Marokko.- Geol. Rdsch., **70**, 2: 801-841.

STRAUB, E. W. (1952): Mikropaläontologische Untersuchungen im Tertiär zwischen Ehingen und Ulm a. d. Donau.- Geol. Jb. **66**: 433 - 523.

SULZER, H. (1986): Geologische Untersuchungen und Kartierung im westlichen Teil der Decke von Toundoute - Provinz Ouarzazate - (Südliches Marokko).- unpubl. Diplomarbeit FU Berlin, 1 - 65., Berlin.

TIERCELIN, J.- J., VINCENS, A., BARTON, C. E., CARBONEL, P., CASANOVA, J. DELIBRIAS, G., GASSE, F., GROSDIDIER, E., HERBI, J.-P., HUC, A.Y., ARDINE, S., LE FOURNIER, J., MELIERES, F.,OWEN, R. B., PAGE, P., PALACIOS,CH., PAQUET, H., PENIGUEL, G., PEYPOUQUET, J.-P., RAYNAUD, J.-F., RENAUT, R. W., RENEVILLE, P. de, RICHERT, J.-P., RIF, R., ROBERT, P., SEYVE, CH., VANDENBROUCKE, M. and VIDAL, G. (1987): Le demi-graben de Baringo-Bogoria, rift Gregory, Kenya 30 000 ans d'histoire hydrologique et sedimentaire. - Bull. Centres Rech. Explor.-Prod. Elf Aquitaine **11**, 2: 249 - 540.

TURNER, P.(1980): Continental red beds.-Developments in Sedimentology **29**: 1 - 562, Elsevier, Amsterdam, Oxford, New York.

WALKER, R. G. and KOSTER, E. H. (1984): Sandy Fluvial Systems.- in WALKER, R. G. (ed.): Facies Models.- Geoscience Canada Reprint Ser. **1**: 71 - 90.

WIGGER, P. and HARDER, S. (1986): Seismologische und krustenseismische Untersuchungen im Atlas-System/Marokko.- Berl. geowiss. Abh., (A), **66**: 273-288.

ZYLKA, R.(1988): Geologische und photogeologische Untersuchungen am Südrand des Hohen Atlas zwischen Toundout und Goulmima (Marokko). Diss. FU Berlin, in press.

ZYLKA, R. and JACOBSHAGEN, V. (1986): Zur strukturellen Entwicklung am Südrand des zentralen Hohen Atlas (Marokko).- Berl. geowiss. Abh. (A), **66**: 415-432.

GEOLOGICAL MAPS

All maps used are edited by the Ministère de l'Energie et des Mines, Direction de la Géologie, Rabat. They are issued as Not. Mém. Serv. géol. Maroc

<u>1 : 200 000:</u>

No. 55 Carte géologique provisoire des régions de Demnat et de Télouet, au 1/200 000, par J. BOURCART & H. TERMIER, 1941.

No. 138 Carte géologique de l'Anti-Atlas central et de la zone synclinale de Ouarzazate: feuilles Ouarzazate - Alougoum et Telout Sud, par G. CHOUBERT, J. DESTOMBES, A. FAURE-MURET, H. GAUTHIER, J. HINDERMEYER, H. HOLLARD, G. JOURAVSKY, au 1/200 000, 1971.

No. 161 Carte géologique du Maroc Jbel Saghro-Dadès (Haut Atlas central, sillon Sud-Atlasique et Anti-Atlas oriental), par G. CHOUBERT, J. DESTOMBES, H. GAUTHIER et J. HINDERMEYER, au 1/200 000, 1980.

1 : 500 000:

No. 70 Carte géologique du Maroc au 1/500 000 par G. CHOUBERT, 1959
Feuille: Ouarzazate (Réimpression en 1978).

No. 234 Carte gravimetrique du Maroc au 1/500 000 par J. W. H. VAN
Feuille: Ouarzazate DEN BOSCH, 1969.

1 : 2 000 000:

No. 278 Carte structurale du Maroc (Provinces du Nord) au 1/2 000 000
par H.M. SAADI, , 1982.

THE OSTRACODE FAUNA OF THE AIT KANDOULA REGION, SYSTEMATIC DESCRIPTION

Friedrich-Franz Helmdach

Freie Universität Berlin, D - 1000 Berlin 33, Institut für Geologie, Altensteinstr. 34 A

ABSTRACT

The stratigraphically important ostracodes of the non-marine sediments of the Ait Kandoula Basin (Morocco) are described, six of which are new. Age: Langhian to late Tortonian.

Introduction

This paper concentrates on the systematic description of the ostracodes from the Neogene strata south of the central High Atlas mountains mainly the Ait Kandoula sequence. Ecology, distribution and stratigraphic range of the fauna have already been discussed in the article by GÖRLER et al. "The Uplift of the Central High Atlas as deduced from Neogene Continental Sediments of the Ouarzazate Province, Morocco" (this volume). The present paper should be considered in connection with that article.

Remarks

Two groups of ostracodes will not be included in the systematic descriptions for the following reasons:

a. The species are well known, sufficiently described and figured in numerous publications.

b. The ostracodes are badly preserved and/or only few specimens could be collected.

Representatives of both groups will be listed below in alphabetical order:

First group: *Candona angulata* G. W. MUELLER 1900
Candona neglecta SARS 1887
Candonopsis kingsleii (BRADY and ROBERTSON 1870)
Cypris pubera O. F. MUELLER 1776
Cypridopsis vidua (O. F. MUELLER 1776)
Darwinula stevensoni (BRADY and ROBERTSON 1870)
Ilyocypris bradyi SARS 1890
Ilyocypris gibba (RAMDOHR 1808)

Second group: *Candona* aff. *fabaeformis* (FISCHER 1851)
Candona cf. *parallela* G. W. MUELLER 1900
Candona praecox STRAUB 1952
Candona suevica STRAUB 1952
Candona (Typhlocypris) aff. *beogradica* KRSTIC
Herpetocypris cf. *reptans* (BAIRD 1835)
Heterocypris cf. *incongruens* (RAMDOHR 1808)
Potamocypris aff. *maculata* ALM 1914
?*Stenocypris major* (BAIRD 1859)
?*Strandesia* sp.

Systematic description

Suborder Podocopina SARS 1866
Family Darwinulidae BRADY & NORMAN 1889
Genus *Darwinula* BRADY & ROBERTSON 1885 (in JONES 1885)
Type species *Polycheles stevensoni* BRADY & ROBERTSON 1870

Darwinula aff. *cylindrica* STRAUB 1952
Plate I, fig. 1 and 3

1952 *Darwinula cylindrica* n. sp.STRAUB , p 497-498, fig. 19 - 20

Material: About 100 valves
Measurements: Length: 0,47 - 051 mm
Heigth: 0,21 - 0,24 mm

Remarks:
Our specimens agree well with STRAUB's figures in shape, size and the weakly narrowed lateral parts about the middle in dorsal view.Only the shape of the RV in the back part of the ventral margin is different : most specimen having a small lump near the curvature towards the posterior margin. A close relationship seems to be obvious.

Occurrence:Rarely in the Chattian and Aquitanian, more frequent in the Serravallian (Sylvana Schichten) from Southern Germany.
The species occurs in Morocco (Ait Kandoula region) in the layers:
MA 480-489, 502, 545, 460

Family Cyprididae
Subfamily Cypridinae BAIRD 1845
Genus *Cypris* O.F. MUELLER 1776

Cypris sp. A
Plate I, figs. 2 and 4

Material: 5 adult and 15 juvenile values, about 30 larger fragments.

Remarks:
This species is characterized by distinct and regular pits on the lateral surfaces and a well rounded antero-dorsal outline, allowing a clear separation from other species.Because of the few well preserved specimens the erection of a new taxonomic unit must be omitted.

Occurrence and age:
Cypris sp. A could be found only in the lowermost layers (MA 480 and 483) of the Ait Kandoula sequence.
Age: Presumably Langhian to Lower Serravallian

Genus *Cyprinotus* BRADY 1886
Type-species: *Cyprinotus cingalensis* BRADY 1886

Remarks: The typical hump within the group of *Cyprinotus*-species will be regarded according to MALZ 1976 as a genetically fixed feature, allowing a distinct separation from the genus *Heterocypris*.

Cyprinotus unicornutus n.sp.
Plate I, figs. 7 - 9

1952 *Cypridarum* Gen. et sp. inc. Nr. 4 STRAUB, p 496, Taf. B, figs. 53 - 54

1965 Gen. et. sp. inc. Nr. 1 LUTZ, p. 315 - 316, fig. 30

Name: unus (lat.) =single, cornutus (lat.) = horned, after the horn-shaped prolongation of the RV
Holotypus: RV of a ?male, plate I, fig. 7 (cat. no. REM II, MA 489-1)
Paratypes: More than 40 most slightly broken valves and carapaces (cat. no.: 489-2, Ma 489-3, Ma-St-489-1 Ma-St-489-2)
Locus typicus: Ait Kandoula Basin, Morocco.
Stratum typicum: Lowermost layers of the Ait Kandoula profile. Middle Miocene.
Diagnosis: A species of *Cyprinotus* distinguished by a horn to spine shaped wing at the dorsal margin of the right valve.

Description: Overlap not clearly recognizable because of an at least slight deformation of all carapaces. LV overlaps RV at the anterior, and posterior margin and presumably very wide in the ventral region. Distinct differences in shape between RV and LV except in the wide, regular half-circled anterior margin.
RV: Ventral margin in its frontal third slightly convex, in the middle concave, turning modestly narrow rounded into the steeep ascending posterior margin. Dorsal margin little inclined and nearly straight . Posterior dorsal angle faintly marked. The wing like dorsal protuberance starts in front at one third of the length of the valve, forming a lump like elevation of about 1/6 of the height of the valve. It ends in front of the posterior third with a spine being directed towards the posterior end.
LV: Greatest height after one third of the valve length. Dorsal margin descending to the posterior dorsal angle in straight to slightly concave line. Posterior margin slightly oblique, half circled.It passes continuously into the widely convex bent ventral margin
Dorsal view: Carapaces regularly lens-shaped, pointed anteriorly and posteriorly . The dorsal spine-like prolongation of RV being directed obliquely backwards with an angle of 45 .
Inner lamella: Moderately wide. A small vestibulum could be observed. Lateral pore canals not recognizable. Normal pore canals moderately frequent.
Hinge: RV with a long, slightly bent groove.
Ornamentation:Lateral surfaces very finely but distinctly pitted with exception of the median part and the spine.

Muscle scars: Scar arrangement corresponds with the pattern of the Cyprididae. Upper scar longish and narrow with a small rounded scar slightly above of its posterior limitation. Two irregular oval formed scars below the frontal one being more longish. Another small longish scar appears about underneath the gap between these two scars.Two additional slim and longish scars are located below. They form a row with a small gap in between, the posterior one underneath the scar above. The scar arrangement in total is inclined towards the back.
Sexual dimorphism: Valves with a wider rounded posterior margin and more obliquely ascending anterior dorsal margin are presumably female ones.
Affinities: *Cyprinotus unicornutus* seems to be related with *Cyprinotus scholiosus* (SOHN & MORRIS 1963). Both species agree quite well in development of the hinge, muscle scars, overlap, ornamentation and in its tendency of forming a dorsal protuberance in the RV. *C. unicornutus* differs from other species by its more longish shape in lateral view and the dorsal protuberance. There are little doubts that large number of broken spines found by STRAUB 1952 in the Kirchberg- and Sylvana-Schichten in southern Germany belong to *C. unicornutus*.They agree well in structure, shape and size with the numerous broken isolated spines found in the Ait-Kandoula layers.
Measurements: Length: 1 - 1,14 mm, height: 0,56-0,64 mm
Occurrence: Southern Germany: Upper Kirchberg - Schichten, and Sylvana Schichten. Morocco: Ait-Kandoula Basin (layers 489, 502, 523, 545), and in layer 11 of the Ouarzazate Basin (GÖRLER & ZUCHT 1986)
Age: Serravallian to Early Tortonian

Subfamily Cypridopsinae
Genus *Zonocypris* G.W. MÜLLER 1898
Type species: *Zonocypris madagascarensis* G.W. MÜLLER 1898

Zonocypris maghrebinensis n. sp.
Table II, figs.: 1-4, 6 and 8

Name: maghreb (arab.) = West, name of the western Arabic countries.
Holotype: LV of a ? male, plate II, fig.2 (REM II 523 - 4) -
Paratypes: More than 100 valves and carapaces (M 523-2, 523 ...)
Locus typicus: Basin of Ait Kandoula,Morocco.
Stratum typicum: Layer 523 of the Ait Kandoula profile, Morocco

Diagnosis: A species of *Zonocypris* distinguished by a pattern of fine ribs on the surfaces and a marginal crenulation of the RV.
Description: LV overlaps RV anteriorly and posteriorly of dorsal margin. Valves in lateral view almost triangular.Transition from anterior to ventral margin in a narrow rounding Ventral margin concave in its middle, in external view partially covered by the inflated lower parts of the valves. Posterior ventral margin and posterior margin forming about a half circle, turning continuously into the straight to widely rounded dorsal margin (see sexual dimorphism)
Lateral surfaces finely ribbed. These ribs are connected by regular but not always distinctly built bars, forming a pattern of small pits in a row-like arrangement.
Inner lamella: Rather wide anteriorly, with a large anterior vestibulum and a smaller posterior one.
Marginal zone narrow,radial pore canals presumably numerous. Selvages of RV with small denticulations. Normal pores modestly numerous,ending on the external ribs. (see plate II,fig.8)
Sexual dimorphism not pronounced. Males with distinct dorsal angles caused by the nearly straight formed dorsal margin. Females in dorsal view wide and evenly rounded,the valves meet anteriorly and posteriorly with an angle of about 90°. Greatest width about in the middle. Males in dorsal view acute angled,greatest width behind the middle.
Affinities: *Zonocypris madagascarensis* G.W.MUELLER 1898 has a similar outline and rib-pattern. The ribs are not connected by bars,more irregularly arranged and run less close to each other.Females longer then males.

Measurements:	Females,	length:	0,65 - 0,69 mm
		height:	0,39 - 0,42 mm
	Males,	length:	0,67 - 0,71 mm
		height:	0,35 - 0,38 mm

Occurrence: *Zonocypris maghrebinensis* n.sp. could be found only in the layer 523 of the Ait Kandoula sequence. The species also occurs in the Ouarzazate Basin (layer A 103)
Age: Serravallian

Genus Type - species: *Pelocypris* KLIE 1939

Type - species: *Pelocypris lenzi* KLIE 1939

Pelocypris cf. *zilchi* TRIEBEL 1953
Plate I ,figs.: 5 and 6

Material: One complete RV, several broken valves.
Remarks: In shape, construction of inner lamella, muscle scars and marginal denticulations similar to *P. zilchi*. Differences exist in details of the shape: especially the posterior margin is more irregular and narrower rounded, in a finer marginal denticulation and in the absence of marginal spines.
The poorly preserved material did not allow a separation as a new species. A close relationship however with *Pelocypris zilchi* T. seems to be evident. *P. zilchi* is known from Pleistocene deposits of El Salvador. The Moroccan specimes will be regarded as a related, earlier occuring form.
Occurrence: Morocco, Ait Kandoula Basin, Layers: 480,483 and 489.
Age: Middle Serravallian

Genus *Heterocypris* CLAUS 1893
Type-species *Cypris incongruens* RAMDOHR 1808

Heterocypris aff. *parva* (LIENENKLAUS 1905).
Plate II, figs.: 5 and 6

1950 *Cyprinotus* cf. *francofurti* (LIENENKLAUS 1905)
STRAUB, p. 491 - 492, Taf. B, Fig. 41, Abb. 11

Material: More then 60 valves and carapaces.

Remarks: The specimens figured by STRAUB 1950 aggree well with the Moroccan individuals. Because of similarities in shape and marginal denticulations a close relationship with *Heterocypris* parva can be assumed.
H. francofurti differs from our specimens by a less pronounced dorsal break and a more rounded anterior dorsal margin in lateral view. Both forms differ also in the weak embayment of the ventral outline being localized in the Moroccan specimens about in the first third of the length and not in the middle as in the individuals figured by MALZ 1976.
The flat and faintly ornamented small depression running parallel to the anterior margin could only be observed in the Moroccan specimens. Because of the small number of adult valves a separate taxonomic unit could not be established.
Measurements: Length (presumably females): 0,93 - 0,97 mm
Height: 0,56 - 0,58 mm

Occurrence: In the lower part of the Ait Kandoula sequence. Layers: 489, 503 541 and 535.
Age: Late Serravallian to Middle Tortonian

Heterocypris cf. *francofurti* (LIENENKLAUS 1905)
Plate VI, fig.5

cf.1976 *Heterocypris francofurti* (LIENENKLAUS 1905).- Malz, Taf.2, Fig. 9 - 11

Material: About 12 mostly juvenile valves.

Remarks: The Moroccan specimens resemble *H. francofurti*.The dorsal margin however seems to be more flat and the dorsal break is less strongly developed. Because of the few preserved individuals a separation as a new taxonomic unit must be omitted.
Occurrence: *H.*cf. *francofurti* could be recorded in the layer 577 of the Ait Kandoula sequence.
Age: Tortonian

Family	Cytheridae BAIRD 1850
Subfamily	Cytherideinae SARS 1925
Genus	*Cyprideis* JONES 1857
Subgenus	*Cyprideis (Cyprideis)* JONES 1856
Type-species	*Candona torosa* JONES 1850

Remarks: The location and numerical designation of nodes occuring on the surfaces of several species of the genus *Cyprideis* will be used according to SANDBERG (1964). Position of node 1 : Upper part of the valves, behind the middle, about adjacent to the sulcal area. Position of node 2 and 2a : Near the latero-ventral bending, about in middle of the valve length. If both nodes are present, they appear as a short ridge. Position of node 3 : In front of the middle, about in the muscle scar area.
According to SANDBERG (1964, p.37) the presence or absence of marginal denticulations will be used in the following for specific and subspecific differentiations. SANDBERG stated a constancy in the development of marginal spines within several species of *Cyprideis*: "Furthermore,

if denticulations are present, their number and general appearance constitute stable diagnostic characters for species discrimination".

Cyprideis (Cyprideis) inaequalis n.sp.
Plate IV, figs.:3, 5, 8, 9; Plate 5, fig. 7

Name: inaequalis (lat.) = different, after the differences within the groove pattern of the surfaces.
Holotype: Left valve of a female. Plate 5, fig. 7 (cat. no.: REM II-514 - 5)
Paratypes: More then 200 valves and carapaces. (cat. no. 514)
Locus typicus: Ait Kandoula Basin, Morocco
Stratum typicum: Lower part of the Ait Kandoula sequence, Late Serravallian.

Diagnosis:
A species of *Cyprideis (Cyprideis)*, distinguished by relatively large grooves in the central regions of the valve surfaces and on the swellings. Their size decreases towards the margins. Posterior margin in both sexes steeply and nearly straight sloping, extending widely downwards. Without denticulations at the anterior margin..

Description:
Females in lateral view truncated trapezoid, males longish trapezoid. Anterior margin obliquely rounded: rather wide in the upper third, narrower below. Transition into dorsal margin widely bent. Dorsal outline with a slight embayment in its middle.
Pronounced posterior dorsal angles in both sexes.
Posterior margin of females weakly bent and steeply inclined, in males more oblique and nearly straight sloping. Transition into the ventral margin narrow rounded.
Ventral outline sigmoidal with a distinct embayment near the anterior margin and a more weak one posteriorly. The submedian convex parts being less pronounced in males.
Sculpture: A flat and slightly bent sulcus extends from the dorsal embayment to an about longitudinal bulge in the median ventro-lateral region. There is a second weak and short sulcus below the anterior dorsal angle.
RV with one or two spines in the postero-ventral corner being longer and less blunt in males.

Node 3 (SANDBERG - scheme) nearly always developed and a weak longitudinal bulge presumably formed by node 2a occurs frequently. Female valves with an extensive vertical inflation about in the posterior third.
Ornamentation: Surfaces covered with grooves of irregular size except the areas below the anterior dorsal angle and the sulcal regions where only isolated pits occur. The largest grooves appear on the inflated parts of the valves. They become smaller towards the margins and occur in a row arrangement near the free margin.
Hinge: Anterior hinge element of RV with 12 - 13 teeth.Median hinge consists in a short crenulated groove anteriorly and in a long finely crenulated bar behind ascending continuously and nearly straight into the faintly arched posterior hinge element.
The posterior hinge can be distinguished from the median one by its 5 - 6 large and longish teeth. Hinge elements of LV correspond with those of RV.
Marginal zone: Anterior marginal area moderately wide.About 30 radial pore canals anteriorly being straight to slightly curved,rarely branching and little widened about in the middle.
Muscle scars: Upper mandibular scar obliquely elongated lower one small,rounded to slightly oval. Adductor scars according to the subgenus.
Sexual dimorphism: Strongly developed. Males in dorsal view distinctly slimmer than females with faint narrowings in the sulcal area. Females distinctly inflated posteriorly. Greatest width at 3/4 of the length. Outlines of the valves meet posteriorly with an angle of about 110 o. Node 3 and narrowings in the sulcal areas clearly recognizable.
Ontogeny: Juvenile specimens with straight sloping dorsal margin posteriorly. Nodes 1, 3, 2 and 2a in most individuals present.

Measurements:	Females,	length:	1,01 - 1,03 mm
		height:	0,57 - 0,60 mm
	Males,	length:	1,07 - 1,11 mm
		height:	0,53 - 0,57 mm

Remarks: By the steep and nearly straight sloping posterior margin and the large grooves on the lateral swellings differing from other species. *Cyprideis (C.) maxima* CARBONNEL 1969 differs by a more constant pit pattern,by a straight to concave ventral margin and by a narrower rounding of the anterior margin.
Occurrence: Ait Kandoula Basin, Morocco, in the layers 480, 503, 509, 514, 515, 526, 528, 529.

Age: Serravallian to Early Tortonian

Cyprideis (Cyprideis) mehesi SISSINGH 1972

Diagnosis: see KOLLMANN 1960, p. 161
Extended diagnosis: Dorsal margin of LV of both sexes sloping towards the back. RV with distinct spine posteroventrally and denticulated anterior margin.

Remarks: Because of primary homonymy of *Cytheridea tuberculata* TERQUEM 1878 and *Cytheridea pannonica* n. sp. var. *tuberculata* MEHES 1908 the species has been redesignated by SISSINGH 1972as *Cyprideis mehesi* nom. nov.
Subspecies: *Cyprideis (C.) mehesi mehesi* SISSINGH 1972. Plate 3, figs.1 - 6. *Cyprideis (C.) mehesi retiolata* n. ssp. Plate 5, figs. 1 - 5. *Cyprideis mehesi paucispinosa* n. ssp. Plate 4, figs. 1 - 2, 4 and 6 - 7 .

Occurrence and age: *C. (C.) mehesi mehesi* occurs in Morocco and in different areas of the Paratethys. Age: Tortonian to Messinian. *C. (C.) mehesi retiolata* n.ssp. seems to be restricted on the Early Tortonian of Morocco and *C.(C.) mehesi paucispinosa* could be collected from Moroccan sediments of presumably Messinian to Pliocene age.

Cyprideis (Cyprideis) mehesi mehesi SISSINGH 1972
Plate IV, figs.: 1 - 6

1960 *Cyprideis tuberculata* (MEHES 1908)._
KOLLMANN, 161 - 163, Taf. 13, Fig. 7 - 14
1979 *Cyprideis (Cyprideis) mehesi* SISSINGH 1972 ._
BASSIOUNI, 102 - 103, Taf. 2, Fig.11 - 15 ; Taf. 7, Fig. 13 (with complete synonomy)
1985 *Cyprideis tuberculata* (MEHES 1908)._
JIRICEK, p.397, Taf. 53, Fig. 10 - 12

Diagnosis and description: See KOLLMANN 1960, p.161 - 163
Material: Material: More than 300 valves and carapaces (cat.no.:Ma
Remarks: The Moroccan specimens diverge insignificantly in size, shape and ornamentation from the individuals figured by BASSIOUNI,KOLLMANN and JIRICEK . Nodes could not be observed. Marginal denticulations

consist mostly in one small spine of the RV situated at the anterior margin in about half of the height.Rarely occur 3 - 4 small spines in the lower half of the anterior margin.
KOLLMANN (1960) mentioned 6 - 7 marginal teeth of the RV and a faint denticulation of the anterior margin of the LV. JIRICEK (1985) described a fine and faint marginal denticulation of the RV only.

Measurements: Females,	length:	0,88 - 0,96 mm
	height:	0,51 - 0,56 mm
Males,	length:	0,94 - 1,05 mm
	height:	0,48 - 0,55 mm

Affinities: Because of its relatively small size and the steep sloping posterior margin of RV *C.(C.) mehesi mehesi* differs distinctly from other species. The specimens figured by CARBONNEL (1969) seem to be little related to *C. (C.) mehesi mehesi*. They differ especially in dorsal view by an edged outline.

Occurrence and age: *C. (C.) mehesi mehesi* occurs in the Early and Middle Pannonian in different areas of the Paratethys. The specimens of Messinian age described by DECIMA (1964) most probably do not belong to this species. The Moroccan material seems to have a Middle to Late Tortonian age.
Within the Ait Kandoula sequence the species has been recorded from the layers 545, 461, 462, 533, 534 and 519 .

Cyprideis (Cyprideis) mehesi retiolata n.ssp.
Plate V, figs.:1-5

Name: retiolum (lat.) = little net, after the pattern of fine ribs on the anterior surfaces.
Holotype: LV of a female, plate 5, fig.2 (cat.no.: Ma 545-1)
Paratypes: More than 150 valves and carapaces (cat. no.: 545 I - III)
Locus typicus: Ait Kandoula Basin, Morocco
Stratum typicum: Layer 545, Early Tortonian

Diagnosis: A subspecies of *Cyprideis (C.) mehesi* SISSINGH 1972 distinguished by a pattern of fine ribs near the frontal margin and frequently in the first third of the surface of the valves. RV with a short spine in the posterior ventral region and 3 - 4 marginal teeth anteriorly.

Description: Females in lateral view truncated oval. Males more elongate. Shape in the posterior half triangular. Anterior margin about half circled. Transition into the dorsal margin widely bent with a weak concavity in front of the anterior dorsal angle.
Dorsal outline of the LV of females slightly embayed in front of the middle. Posterior dorsal margin passing about constantly bent into the posterior margin.
Ventral margin slightly convex with a weak embayment in the anterior half.
Transition from posterior margin into the ventral one in males distinctly narrower curved than in females. About straight formed posterior dorsal margin and posterior margin in male specimens.
Posterior dorsal angle moderately well developed.
RV of both sexes with 4 small spines in the lower half of the anterior margin and a short pointed one in the posterior corner.
Dorsal view: Females about hemispherical posteriorly, converging from the sulcal area widely bent towards the front.
Males constantly wide rounded with exception of the faint constricted sulcal areas. Valves meeting with acute angles anteriorly and posteriorly.
Ornamentation: Surfaces decorated with rather uniform small pits and a pattern of fine ribs near the anterior margin. These ribs form a rhomboid pattern of meshes being arranged in their longitudinal extension parallel to the outline of the anterior margin.The rib-pattern covers in most adult specimens the first third of the surfaces.
Hinge:RV with about 14 teeth anteriorly and 6 - 7 somewhat larger teeth posteriorly. Median hinge element with a short depression in front forming afterwards a crenulated ridge passing continuously into the posterior hinge element. Teeth of posterior hinge element differ from the crenulated median ridge by being distinctly larger.
Sculpture: The nodes 1,2 and 3 in a triangular arrangement could only be observed in juvenile specimens. Only node 3 occurs frequently in adult specimens. Both valves are constricted by an arched sulcus starting in front of the middle and ending near the lower limitation of node 3. A smaller and shorter sulcus can be observed behind the anterior dorsal angle. Within the area between the sulci irregular formed pits and ridges appear.
Marginal zone: Inner lamella moderately broad with about 30 radial pore canals. These pore canals are almost straight, rarely branching and little widened in their middle.
Muscle scars: Upper mandibular scar longish, irregular oval shaped becoming narrower posteriorly.

Measurements: Females, length: 1,00 - 1,04 mm
height: 0,58 - 0,61 mm
Males, length: 1,01 - 1,06 mm
height: 0,50 - 0,56 mm

Affinities: This subspecies can be distinguished from *mehesi mehesi* by its rib-pattern anteriorly ,by an indistinct posterior dorsal angle and by its larger size.

Remarks: Because of the stratigraphically earlier occurrence *mehesi retiolata* will be regarded as a precursor of *mehesi mehesi*.

Occurrence and age: *C.(C.) mehesi retiolata* n.ssp. could be collected only from layer 545 of the Ait Kandoula sequence. Age: Most probably Early Tortonian.

Cyprideis (Cyprideis) mehesi paucispinosa n.ssp.
Plate IV, figs.:1, 2, 4, 6, and 7 .

Name: After the poorly developed posterior spine of the RV.
Holotype: LV of a female, pl. 4, fig. 6 (cat.no. REM II 596 - 4)
Locus typicus: Ait Kandoula Basin, Morocco
Stratum typicum: Uppermost part of the Ait Kandoula sequence, layers 594 - 596.
Paratypes: More than 300 valves and carapaces (cat. no. Ma 596 - PZ I and II, FZ 1 .)

Diagnosis: A subspecies of *Cyprideis (C.) mehesi mehesi* distinguished by an ornamentation consisting of pits of different size, by about parallel dorsal and ventral margins and by a faintly developed spine in the postero-ventral corner.

Description:LV: Shape in lateral view truncated oval in females. Dorsal and ventral margins slightly convex about parallel to each other, with a weak embayment in the antero-ventral region. Anterior margin oblique ,half-circled. Transition into dorsal margin widely bent. Dorsal margin of LV about straight in both sexes. Posterior dorsal angle poorly developed. Posterior margin in males about straight in upper half, less steeply inclined than in females.Lower half of posterior margin passes with an even rounding into ventral margin. The rounding being distinctly narrower than in females.
Ventral outline in males sinuate, in females only embayed anteriorly.
RV: Greatest height in both sexes in the middle.Dorsal outline consists of an even and wide rounding which slopes in male valves without

interruption to the narrow rounding of the postero-ventral corner. Dorsal outline of females about straight sloping moderately inclined into posterior margin.Posterior dorsal angle in females well rounded in males not recognizable. Postero-ventral corner in females half-circled.

Dorsal view: Females pear-shaped with indistinct edges, distinctly constricted in the sulcal areas. Valves meeting posteriorly with an angle of 90 degrees. Males slim,pointed anteriorly and posteriorly, with weak sulcal constrictions.

Sculpture and ornamentation:

Surface of valves decorated with pits being distinctly larger on the inflated area of the posterior region. Pits are lacking only in a small area below the anterior dorsal angle. A flat and bent sulcus starts dorsally in front of the middle, enlarging downwards ending at the lower limitation of node 3. Nodes occur rarely also in juvenile specimens.Only node 3 can always be observed even if it is faintly developed.

Inner lamella: modestly broad. Radial pore canals about straight, rarely branching, little enlarged in their median part.

Hinge: RV anteriorly with 12 - 14 teeth. Median hinge depressed in its frontal region than forming a long denticulated ridge. Posterior hinge element with 4 - 6 teeth. LV with corresponding hinge elements.

Muscle scars: Central muscle scars in accordance with the genus. Upper mandibular scar longish oval.

Affinities: *C (C.) mehesi mehesi* can be distinguished from *mehesi paucispinosa* by its larger size and its distinctly developed postero-ventral spine. Dorsal outline in LV of females of *mehesi paucispinosa* about straight in its posterior portion running almost parallel to the ventral margin. In *mehesi mehesi* the posterior dorsal outline slopes distinctly in LV of female specimens.

Measurements:	Females,	length:	0,93 - 0,98 mm
		height:	0,55 - 0,58 mm
	Males,	length:	1,02 - 1,07 mm
		height:	0,53 - 0,56 mm

Occurrence and age: *C.(C.) mehesi paucispinosa* n.ssp. occurs in the uppermost layers (MA 594 and 596) of the Ait Kandoula sequence. Age: presumably Messinian to Pliocenian.

Subfamily Limnocytherinae KLIE 1938

Genus *Limnocythere* BRADY 1867

Type - species *Cythere inopinata* BAIRD 1843

Limnocythere rostrata STRAUB 1952
Plate V, figs. 6, 8 and 9

1952 *Limnocythere rostrata* n.sp.- STRAUB, p.499 - 500, Abb. 23, Taf. C, Fig. 55 - 56
1965 *Limnocythere rostrata* STRAUB,- LUTZ, p. 310 - 311, Abb. 26

Material: More than 50 mostly broken or deformed isolated valves.

Remarks: In outline, size and sculpture the Moroccan specimens agree well with the individuals figured by STRAUB (1952).
Completely preserved male specimens could not be collected. Some broken valves however seem to be males. They differ distinctly by their slim appearance and more narrow rounded posterior margin from the more truncated presumably female valves.
In some Moroccan specimens the pit-ornamentation seems to be lacking.
Occurrence and age: Rare to abundant in the lower part of the Ait Kandoula sequence. Layers: MA 480, 483, 486, 489, 491, 502 .The species occurs in Southern Germany in the Swabian and Frankonian Miocene and in the Ouarzazate Basin (Morocco) in the layers A 103 and A 117 (GÖRLER & ZUCHT 1986)
Age: Late Langhian to Serravallian

Limnocythere torulata n.sp.
Plate VI, figs.:1 - 4

Name: torulus (lat.) = braid, after the braid-like ventro-lateral logitudinal ridge.
Holotype: LV of a female, pl.VI, fig. 3, (cat.no. REM I - 486-1)
Paratypes: 19 valves ,cat. no. MA 486 - St. 2, MA 486 - Pz 1
Locus typicus: Ait Kandoula Basin, Morocco
Stratum typicum: Layer 486 of the Ait Kandoula sequence.

Diagnosis: A species of *Limnocythere* distinguished by a longitudinal ridge along the ventro-lateral bending.

Description: Dorsal margin faintly concave with distinct dorsal angles. Anterior margin widely rounded. Transition into ventral margin modestly narrow bent. Ventral outline concave, deepest embayment in

the middle. Posterior ventral margin and posterior margin slightly irregular half-circled.
Both valves distinctly sculptured with a broad postero-median inflation and a long antero-median sigmoid sulcus extending from the dorsal margin to the ventro-lateral ridge.
Two smooth node-like knobs are located on both flanks of the upper sulcal area and a third ornamented one in the muscle scar region. Adjacent to the anterior dorsal angle occurs an eye-ridge like node being continued without distinct interruption by a ridge running parallel to the anterior margin.
Sculpture: Anterior lateral surface with a vertical oval inflation between anterior marginal ridge and node in the muscle scar area. This inflation becomes narrower towards the venter, turns backwards and runs continuously into the widely extended longitudinal ventro-lateral ridge. The ventro-lateral ridge passes at its ending - slightly turned upwards - continuously into the lateral suface, being almost flat near the posterior margin.
Ornamentation: Lateral surfaces covered with coarse pits with exception of some parts of the nodes and ridges. The flattened venter is decorated with a mesh-like pattern of fine ribs.
Inner lamella: Modestly broad. Radial pore canals straight, presumably 8 - 10 at the anterior margin.
Muscle scars: 4 scars in a slightly bent row. The median ones being longish, the upper and lower one irregularly rounded oval.
(see pl.6, fig. 2)
Hinge: LV with oval sockets anteriorly and posteriorly. Median hinge consists in a row of roundish teeth being somewhat larger anteriorly. RV with corresponding longish teeth and a crenulated rim .(see pl.6, fig.1)
Sexual dimorphism: Males distinctly slimmer and less high than females. Dorsal and ventral margins in males about parallel to each other. They converge in females towards posterior end and are less distinct concave.
Measurements: Females, length: 0,56 - 0,60 mm
height: 0,34 mm
Males, length: 0,56 - 0,62 mm
height: 0,26 - 0,29 mm
Remarks: Because of shell morphology and muscle scars an affiliation to *Limnocythere* seems to be assured. Crenulations within the hinge elements have been recorded also from some species. The anterior and posterior teeth of *Limnocythere staplini* GUTENTAG & BENSON 1962 are crenulated and G.W.MUELLER (1900) figured a specimen of *Limnocythere*

inopinata (BAIRD 1843) where at least the posterior hinge element seems to be crenulated.

Affinities:

Limnicythere luculenta LIVENTAL 1929 is larger, has a bow-formed ventro-lateral ridge and seems not to be ornamented. *Limnocythere acquensis* CARBONNEL 1969 can be distinguished by a more concave ventral outline and an even formed posterior margin. *Limnocythere ? williamsi* SWAIN 1947 is similar in sculpture amd ornamentation but its shape is triangular because of the converging dorsal and ventral margins.

Occurrence and age: *Limnocythere torulata* could be collected from layer 486 of the Ait Kandoula sequence and from layer Z 7 of the Ouarzazate Basin (GÖRLER & ZUCHT 1986)
Age: Early Serravallian

All figured material is kept in the collections of:
Institut für Geologie der Freien Universität Berlin.

Sincerest thanks for the SEM photography which was carried out by Mrs.Freitag and Mrs.Strauß from the Bundesanstalt für Materialprüfung, Berlin and by Dr.Ghobarkar from the Institut für Mineralogie der Freien Universität Berlin.

REFERENCES

BASSIOUNI, M. A. 1979: Brackische und marine Ostrakoden (Cytherideinae, Hemicytherinae, Trachyleberidinae) aus dem Oligozän und Neogen der Türkei.-Geol. Jb. B 31,3 - 195.

CARBONNEL, G. 1969: Les Ostracodes du Miocène Rhodanien.- Thèse, Fac. Sci. Lyon no.32, 1 - 469.

DECIMA, A. 1964: Ostracodi del Gen.*Cyprideis* JONES del Neogene e del Quaternario italiani.- Palaeontographica Italiana 62, 81 - 133.

ELLIS, B. F. & MESSINA, A. R.: Catalogue of Ostracoda.- The American Museum of Natural History

GÖRLER, K. & ZUCHT, M. 1986: Stratigraphie und Tektonik des kontinentalen Neogens im Süden des zentralen Hohen Atlas, Provinz Ouarzazate (Marokko).- Berliner geowiss. Abh. (A), 66, 471 - 494.

JIRICEK, R. 1985: Die Ostracoden des Pannonien.-Chronostratigraphie und Neostratotypen. Miozän der Zentralen Paratethys, .VII, 378 - 425

KOLLMANN, K. 1960: Cytherideinae und Schulerideinae n. subfam. (Ostracoda) aus dem Neogen des östlichen Oesterreich.-Mitt. Geol. Ges. Wien, 51 (1958)

KRSTIC, N. 1985: Ostracoden im Pannonien der Umgebung von Belgrad.- Chronostratigraphie und Neostratotypen. Miozän der Zentralen Paratethys VII, 103 - 143

LUTZ, A. K. 1965: Jungtertiäre Süßwasser-Ostracoden aus Süddeutschland.- Geol.Jb. 82, 271 - 330

MALZ, H. 1976: *Heterocypris* vel *Cyprinotus* ? - Ist die Morphologie des Gehäuses entscheidend für die Bestimmung rezenter Ostracoden-Gattungen ? .- Senckenberg.lethaea 57, 185 - 199

MALZ, H. & MOAYEDPOUR, E. 1973: Miozäne Süßwasserostracoden aus der Rhön.- Senckenberg.lethaea 54, 281 - 309

MÜLLER, G. W. 1900: Deutschlands Süßwasserostracoden. Zoologica 12 (30), 1 - 112

SANDBERG,P. A. 1964: The Ostracod Genus *Cyprideis* in the Americas. Stockholm Contrib.Geol. XII, 1 - 178

SISSINGH,W. 1972: Late Cenozoic Ostracoda of the South Aegean Island Arc.- Utrecht Micropal.Bull. 6, 1-187

SOHN,I. G. & MORRIS, R. W. 1963: *Cheikella* a new freshwater Ostracode genus, and *Telekia* a new name for a homonym.-Micropal. 9 (3), 327 - 331

STRAUB, E. W. 1952: Mikropaläontologische Untersuchungen im Tertiär zwischen Ehingen und Ulm a.d.Donau.- Geol. Jb. 66, 433 - 523

TRIEBEL,E. 1953: Eine fossile *Pelocypris* (Crust. , Ostr.) aus El Salvador.- Senckenbergiana 34, 1 - 4

PLATES

Plate I

Fig.1 *Darwinula* aff. *cylindrica* STRAUB
RV external view, length: 0,47 mm
Fig.2 *Cypris* sp. A
RV external view, length: 1,11 mm
Fig.3 *Darwinula* aff. *cylindrica* STRAUB
LV external view, length: 0,5 mm
Fig.4 *Cypris* sp. A
RV internal view, length: 1,08 mm
Fig.5 *Pelocypris* cf. *zilchi* TRIEBEL
RV external view, length: 1,5 mm
Fig.6 *Pelocypris* cf. *zilchi* TRIEBEL
RV internal view, length 0,94 mm
Fig.7 *Cyprinotus unicornutus* n.sp. Holotype
RV of a ? male in external view, length: 1mm
Fig.8 *Cyprinotus unicornutus* n.sp.
Central muscle scars of RV, internal view, length of the complete muscle scar area: 0,16 mm
Fig.9 *Cyprinotus unicornutus* n.sp.
LV, external view, length: 1 mm

Plate II

Fig.1 *Zonocypris maghrebinensis* n. sp.
RV of a female, external view, length: 0,66 mm
Fig.2 *Zonocypris maghrebinensis* n.sp. Holotype
LV of a ? male, external view, length: 0,69 mm
Fig.3 *Zonocypris maghrebinensis* n. sp.
LV of a female, external view, length: 0,64 mm
Fig.4 *Zonocypris maghrebinensis* n.sp.
Female carapace in dorsal view. Length: 0,66 mm
Fig.5 *Heterocypris* aff. *parva* (LIENENKLAUS)
RV, internal view, length: 0,84 mm
Fig.6 *Zonocypris maghrebinensis* n.sp.
Central area in external view
Fig.7 *Heterocypris* aff. *parva* (LIENENKLAUS)
LV, external view, length: 0,92 mm
Fig.8 *Zonocypris maghrebinensis* n.sp.
LV of a ? female, internal view, Length: 0,64 mm

Plate III

Fig.1 *Cyprideis (Cyprideis) mehesi mehesi* SISSINGH
LV of a female, external view, length: 0,93 mm
Fig.2 *Cyprideis (C.) mehesi mehesi* S.
Right side of a female carapace. Length: 0,94 mm
Fig.3 *Cyprideis (C.) mehesi mehesi* S.
RV of a female, external view, length: 0,91 mm
Fig.4 *Cyprideis (C.) mehesi mehesi* S.
Rigth side of a male carapace. Length: 1,01 mm
Fig.5 *Cyprideis (C.) mehesi mehesi* S.
Female carapace in dorsal view. Length: 0,98 mm

Fig.6 *Cyprideis (C.) mehesi mehesi* S.
Male carapace in dorsal view. Length: 1,01 mm
Fig.7 *Cyprideis (Cyprideis) mehesi paucispinosa* n.ssp.
Female carapace in dorsal view. Length: 1,11 mm
Fig.8 *Cyprideis (C.) mehesi paucispinosa* n.ssp.
RV of a male (praeadult stage), internal view, legth: 1,03 mm
Fig.9 *Cyprideis (C.) mehesi paucispinosa* n.ssp.
RV of a male, external view, length: 1,05 mm

Plate IV

Fig.1 *Cyprideis (Cyprideis) mehesi paucispinosa* n.ssp.
LV of a male, external view, length: 1,03 mm
Fig.2 *Cyprideis (C.) mehesi paucispinosa* n.ssp.
LV of a male, external view, length: 1,07 mm
Fig.3 *Cyprideis (Cyprideis) inaequalis* n.sp.
Female carapace in dorsal view. Length: 0,97 mm
Fig.4 *Cyprideis (C.) mehesi paucispinosa* n.ssp.
Male carapace in dorsal view. Length: 1,06 mm
Fig.5 *Cyprideis (C.) inaequalis* n.sp.
Male carapace in dorsal view. Length: 1,07 mm
Fig.6 *Cyprideis (C.) mehesi paucispinosa* n.ssp.
Holotype, LV of a female,external view, length: 0,97 mm
Fig.7 *Cyprideis (C.) mehesi paucispinosa* n.ssp.
Carapace of a female, right side, length: 0,96 mm
Fig.8 *Cyprideis (C.) inaequalis* n.sp.
RV of a male, internal view, length: 1,02 mm
Fig.9 *Cyprideis (C.) inaequalis* n.sp.
LV of a male, external view, length: 1,1 mm

Plate V

Fig.1 *Cyprideis (C.) mehesi retiolata* n.ssp.
LV of a female, praeadult stage, external view, length: 0,97 mm
Fig.2 *Cyprideis (C.) mehesi retiolata* n.ssp. Holotype,
LV of a female, external view, length: 1,02 mm
Fig.3 *Cyprideis (C.) mehesi retiolata* n.ssp.
LV of a male, external view, length: 1,10 mm
Fig.4 *Cyprideis (C.) mehesi retiolata* n.ssp.
RV of a male, external view, length: 1,06 mm
Fig.5 *Cyprideis (C.) mehesi retiolata* n.ssp.
RV of a female, external view, length: 0,91 mm
Fig.6 *Limnocythere rostrata* STRAUB
LV of a male, external view, length: 0,57 mm
Fig.7 *Cyprideis (C.) inaequalis* n.sp. Holotype
LV of a female, external view, length: 1,05 mm
Fig.8 *Limnocythere rostrata* STRAUB
RV of a female, external view, length: 0,51 mm
Fig.9 *Limnocythere rostrata* STRAUB
LV of a female, external view, length: 0,53 mm

Plate VI

Fig.1 *Limnocythere torulata* n.sp.
incomplete LV, internal view, length:0,5 mm
Fig.2 *Limnocythere torulata* n.sp.
Central muscle scars of LV, internal view.

Fig.3 *Limnocythere torulata* n.sp. Holotype
LV, external view, length: 0,58 mm

Fig.4 *Limnocythere torulata* n.sp.
LV, external view, length: 0,53 mm

Fig.5 *Heterocypris* cf.*francofurti* (LIENENKLAUS)
RV, external view, length: 0,74 mm

Plate I

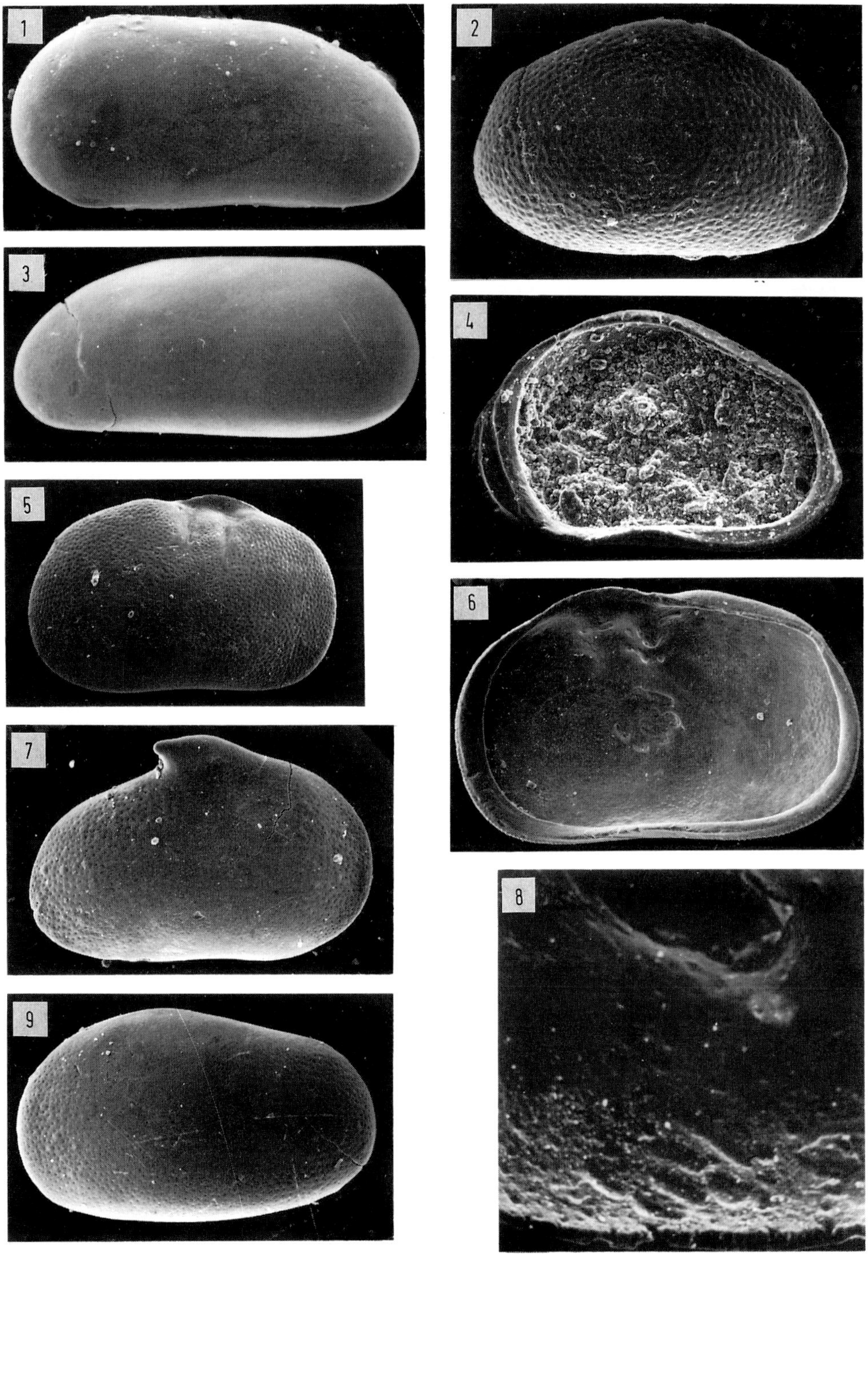

Plate II

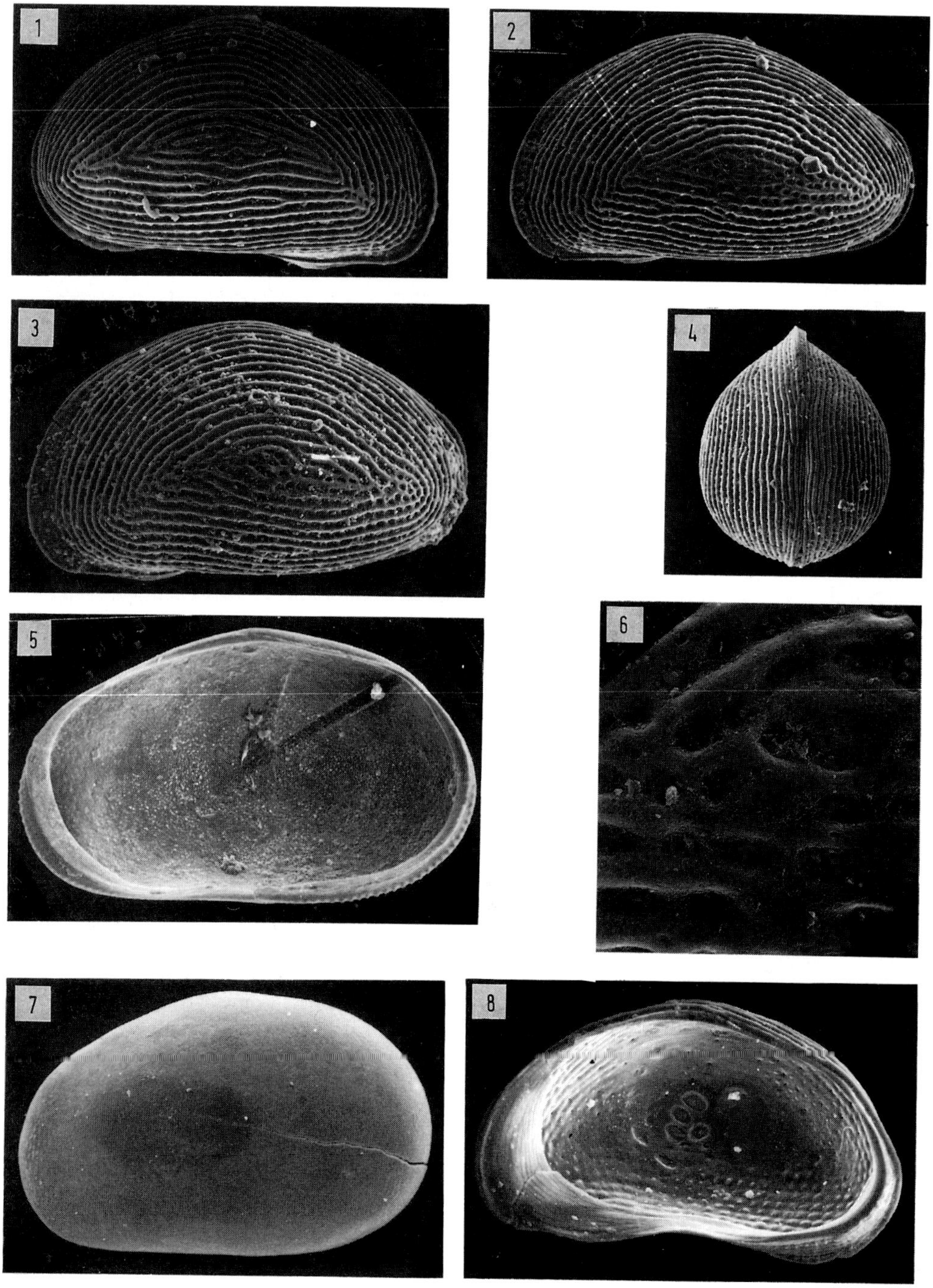

Plate III

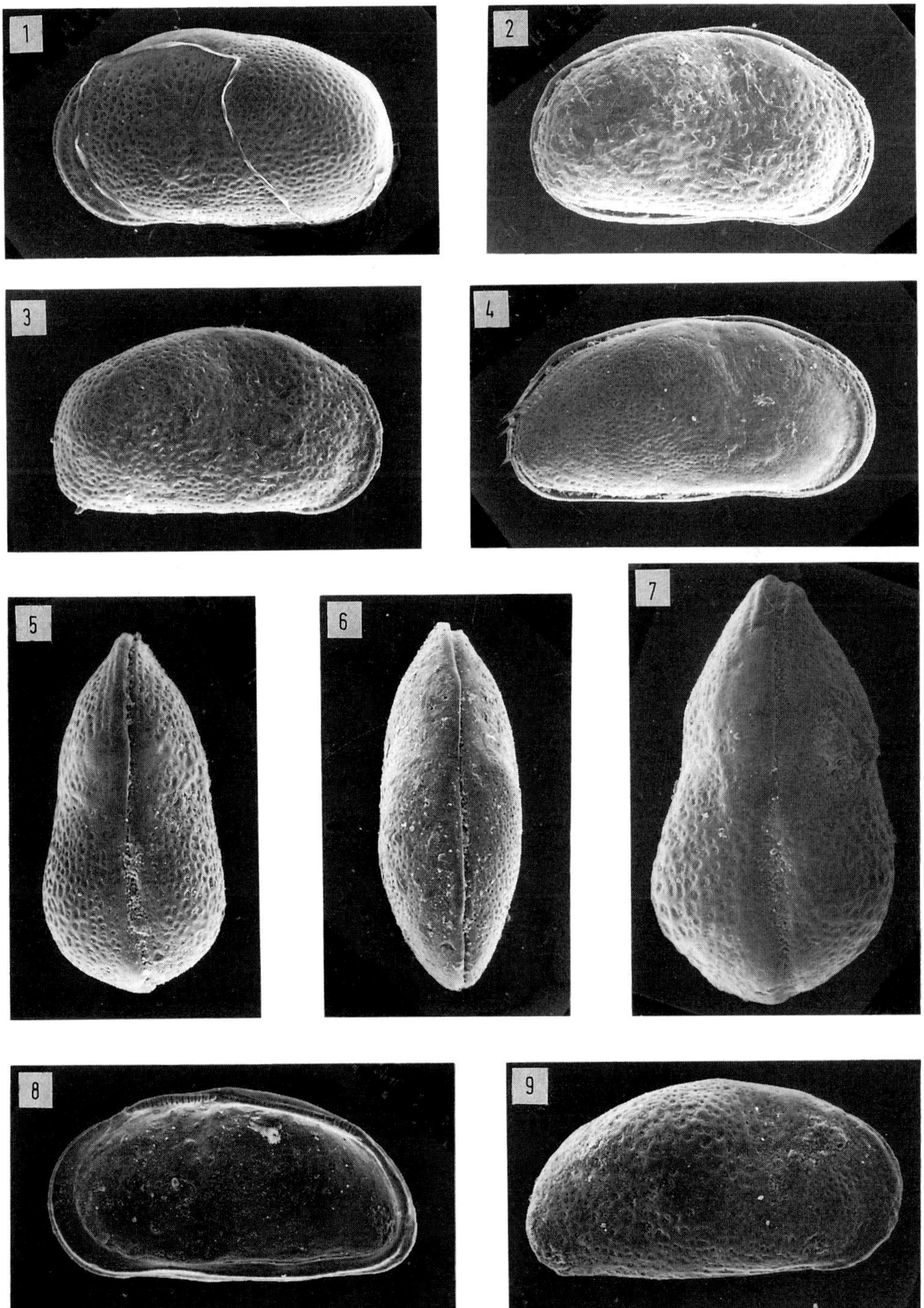

Plate IV

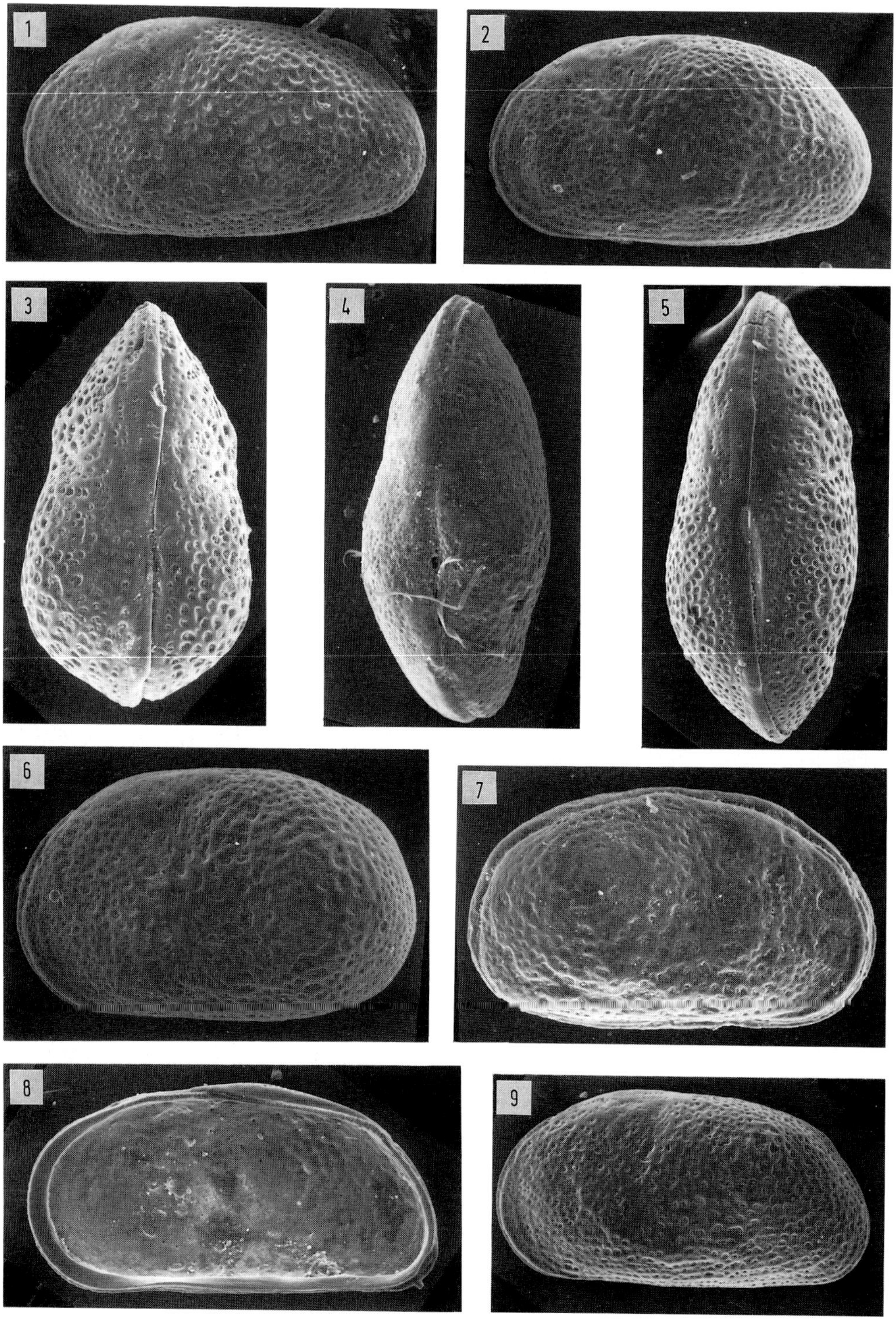

Plate V

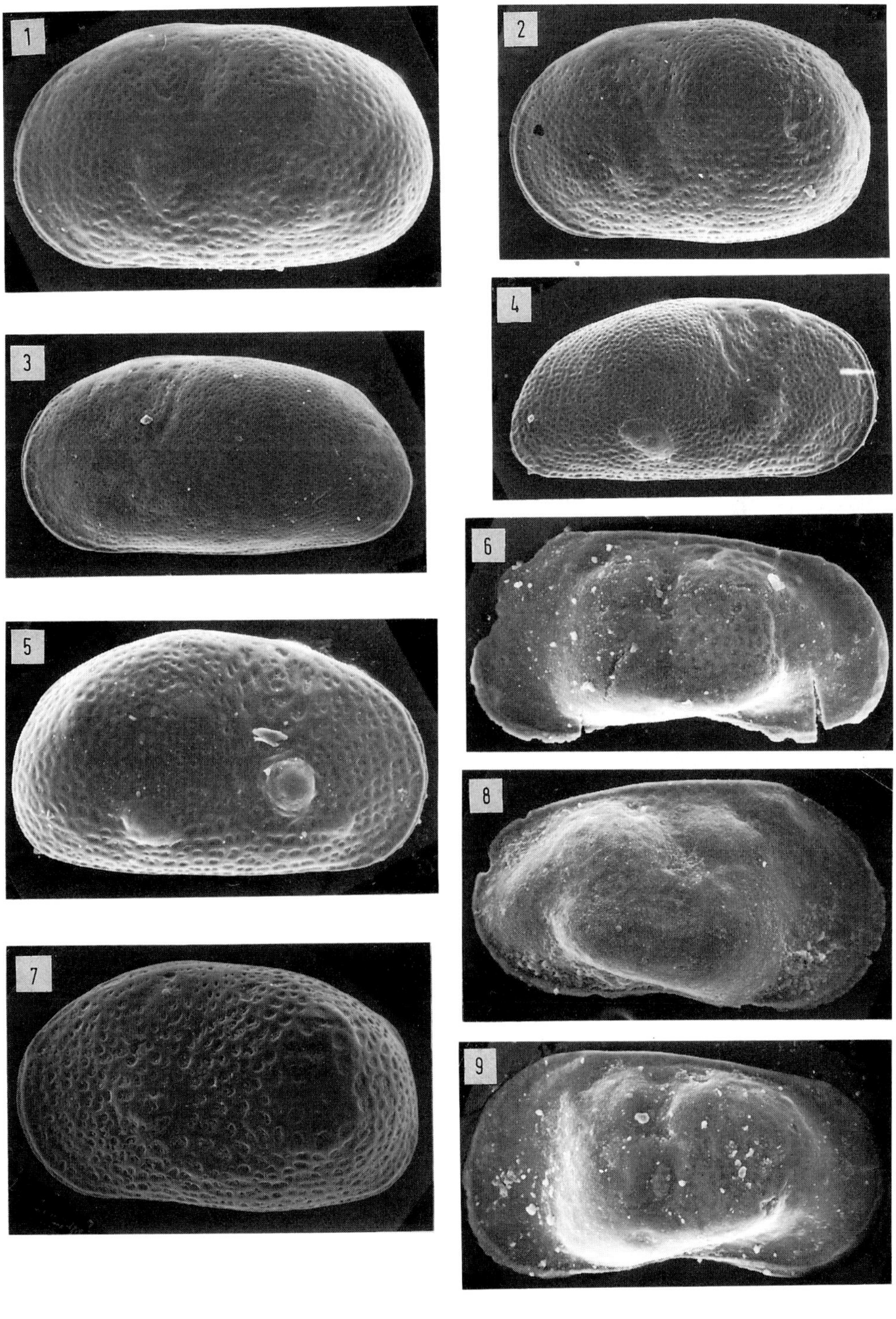

Plate VI

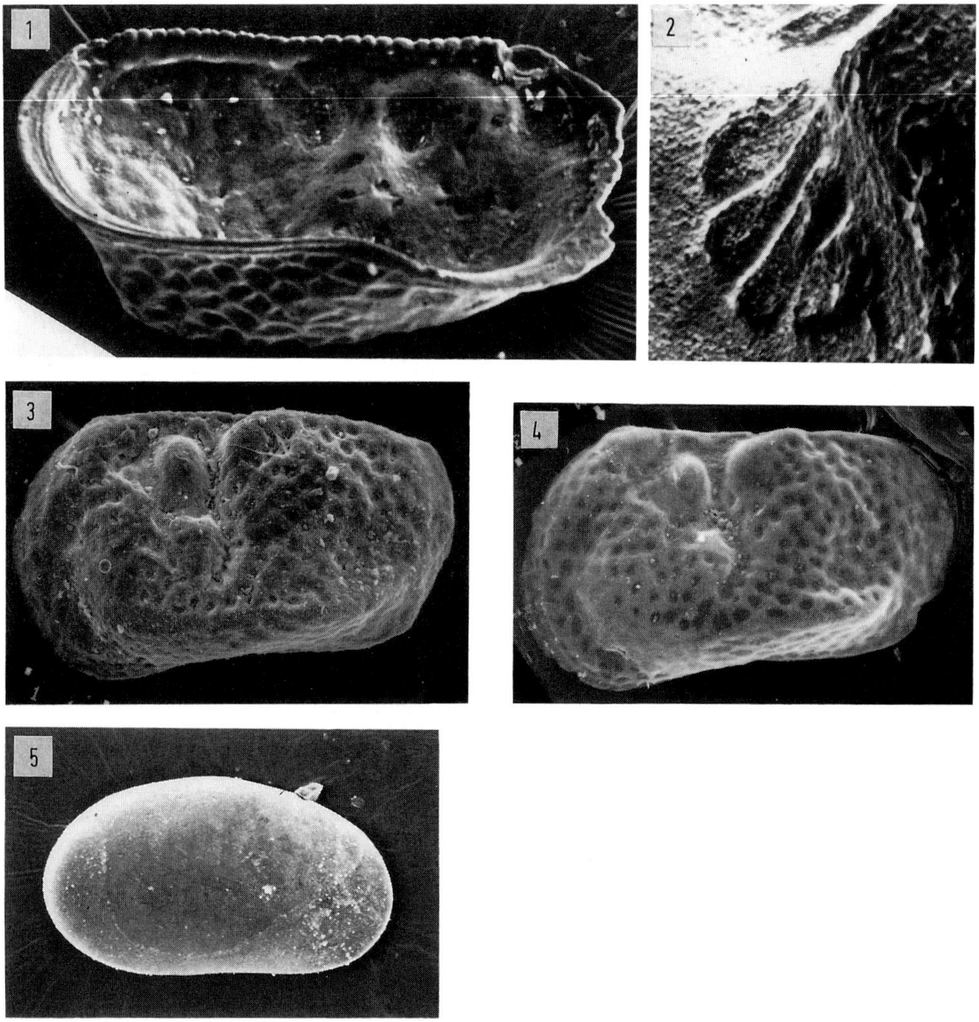

GEOMORPHOLOGICAL ASPECTS OF THE QUATERNARY EVOLUTION OF THE OUARZAZATE BASIN, SOUTHERN MOROCCO

Gerhard Stäblein

Physiogeographie, Universität Bremen
Postfach 330440, D-2800 Bremen 33

Abstract: The relief forms of the Ouarzazate Basin show a zonation with tectonic patterns. Only the levels and glacis-sediments of the northern marginal zone are disturbed tectonically in Quaternary times with folded structurs and anticlines. The inner part of the basin has predominantly horizontal layers in the underground with a bloc mosaic created by a grid of fractures, which are locally subrecently active with subsidence, especially in the southern part and in the Skoura alluvial plain section. Above the southern edge of the basin against the Anti Atlas there exist fossiliated older remnants of a stage, in which the basin on a higher level reached more southwards. The drainage development and the pronounced erosion of the Dra River through the Anti Atlas created the characteristic mesa shaped glacis hills ("khelas") of the basin. The recent erosion rate is not accelerated by neotectonics.

The relief forms in southern Morocco were mainly genetically explained under a climato-genetic aspect. The Quaternary with the climatic periods of cold ages and changing geomorphodynamic conditions, especially connected with shifting of the altitudinal belts, are the frame for detailed models about relief generations. DRESCH (1954), JOLY (1962), RAYNAL (1965), MENSCHING (1955), WICHE (1953), BÜDEL (1977: 166 ff), ANDRES (1977) and others are authors related to that approach. The development of periglacial and partly glacial altitudinal belts in the mountains, in the High Atlas and the Anti Atlas, and of plains and pediments in the foreland were interpreted as a consequence of the Quaternary exogenesis (MENSCHING & RAYNAL 1954, CHOUBERT 1955, 1961, WICHE 1955). The endogenetic conditions of the orogeny with uplift of the mountains and subsidence of the basin in their foreland were only mentioned as initial preconditions. The new geomorphic investigations of French colleagues, COUVREUR (1973, 1981) and RISER (1978, 1981), consider the regional differentiated geological conditions more precisely.

The transition from the zone with several mountain massifs in the northwest of Africa to the Saharian platform (MICHARD 1976) is characterized by a chain of elongated basins, the Presaharian marginal furrow, starting at the Golf of Agadir from the Atlantic coast with the plain of Souss to northeast, continuing with the Schott Dajas, and in southern Tunesia arrives the Mediterranean sea finally at the Golf of Gabes (JOLY 1952, CHAMAYOU & RUHARD 1977, DEPOIS & RAYNAL 1967). This is a typical geomorphological structure in the megarelief on the margin of continental plates. During the regional development, with collision, strain and stress periods, changing effects of stretching and compression the pattern of real relief forms was modified and differentiated.

In the frame of the project "mobility of active continental margins" (cf. GIESE 1986, JACOBSHAGEN 1986 a) we looked for traces and effects of neotectonics in the relief. In the first step we have analysed the area of the basin of Ouarzazate and its margins. The field studies give the following results which could be summarized in a section through the basin as a generalized geomorphological model (Fig.1).

1. The interior tectogenetic pattern of the basin

The interior pattern of the basin, as it was clearly recognized on satellite images and was mapped geomorphologically, is connected to the geoecological environment and the adaption of the vegetational formations. The geoecological division (EMBERGER 1951, ZILLBACH 1984) and distribution of the vegetational formation types in the basin are essentially determined by the tectonic structure and are correlated to a block faulting mosaic with bedrock of different lithology (cf. MÜLLER-HOHENSTEIN 1978). The borders of the units are frequently fault lines which were partly drawn by the hydrographic network in the basin. Locally the faultlines are morphologically pronounced as cuesta scarps (cf. SCHMIDT 1886). The mixed sliding nappes are characteristic preconditions for the relief forms changing in short distances at the southern mountain slope of the High Atlas (LAVILLE et al. 1977). This

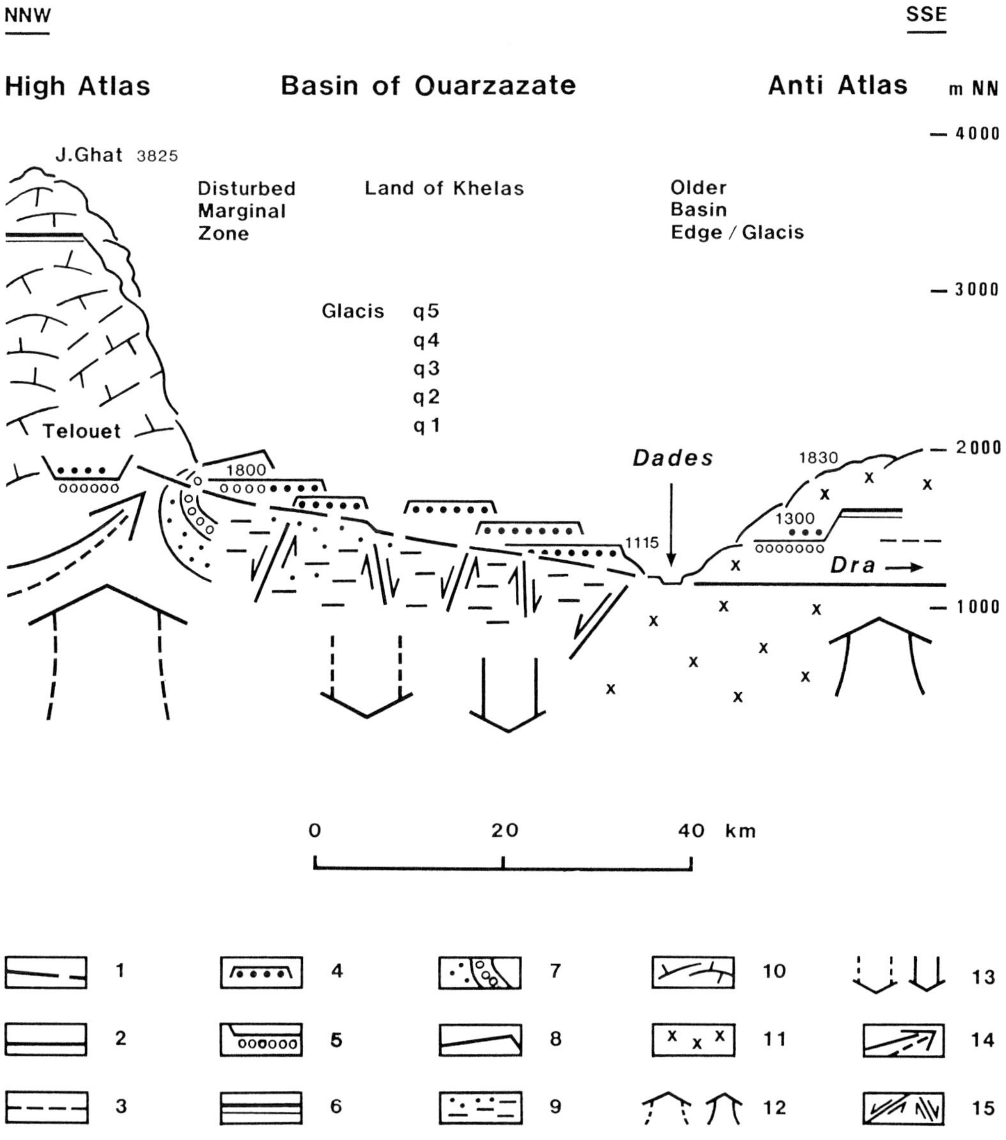

Figure 1: Generalized cross section of the Ouarzazate Basin, geomorphological model of the geomorphogenetic pattern.
1 = wadi, oued, episodic fluvial drainageways; 2 = Dades and Dra valley; 3 = Pleistocene terraces of Dra River; 4 = Pleistocene glacis levels; 5 = basin margin of upper Tertiary and intamountain basins; 6 = Tertiary planation surfaces in the mountains; 7 = pushed and overthrusted Cretaceous and Tertiary layers; 8 = Cuesta scarps and hogbacks; 9 = Teritary layers of continental, lacustrine and marine sediments; 10 = Mesozoic, locally overthrusted marginal layers and massifs; 11 = crystalline basement and Precambrian;
12 = older and younger uplift of the montain areas; 13 = older and younger subsidence; 14 = compression and overthrusting; 15 = subrecent fault lines.

structure is caused by a compressional deformation since the Oligocene-Miocene period (JACOBSHAGEN 1986 b).

2. The recent rate of erosion

The present-day average rate of erosion was computed to be 310 m^3 per year and km^2 or 0,3 mm per year from the sedimentation in the reservoir of the Dades River near Ouarzazate. This is equal to values typical for this zone (cf. SAUNDERS & YOUNG 1983) and therefore in the tributary area of the basin recent accelerated erosion and valley incision by neotectonic movements cannot be deduced. The erodibility, the disposition for linear and areal erosion, is differentiated in the basin area in connection with the tectonic pattern. - These aspects were proved by field experiments simulating erosion by water (WAGNER 1984).

3. Evidences for Quaternary tectonic activities

Unbalanced knickpoints in the gradient of the Wadi beds and dislocations of the Quaternary gravel sediment layers, especially in the transition zone of the northern border of the basin, locally show young movements at fault lines of the margins of the blocs (ZYLKA & JACOBSHAGEN 1986). At some places in the northern marginal zone of the basin with folded or tilted gravel layers of the older Pleistocene periods are exposed sites, e.g. near Amekchoud in a anticline with inclinations more than 20 degrees. In the inner part of the basin, the layers are predominantly horizontal and only disturbed with bloc movements. Therefore concordant or flat discordant layering were normally found. The discontinuity of Quaternary sedimentation is marked by missing and regionally eroded layers. But such periods of erosion are not found in the same position over all interior basin blocs of the area. Therefore we have to consider a regional differentiated tectonic development. A subsidence of one bloc initiated sometimes linear erosion with valley cutting in the neighboring blocs, which go down relatively more slowly.

The broad alluvial plain of Skoura is a segment of the basin with a pronounced subsidence. At several sites of the up to 600 m wide Oued

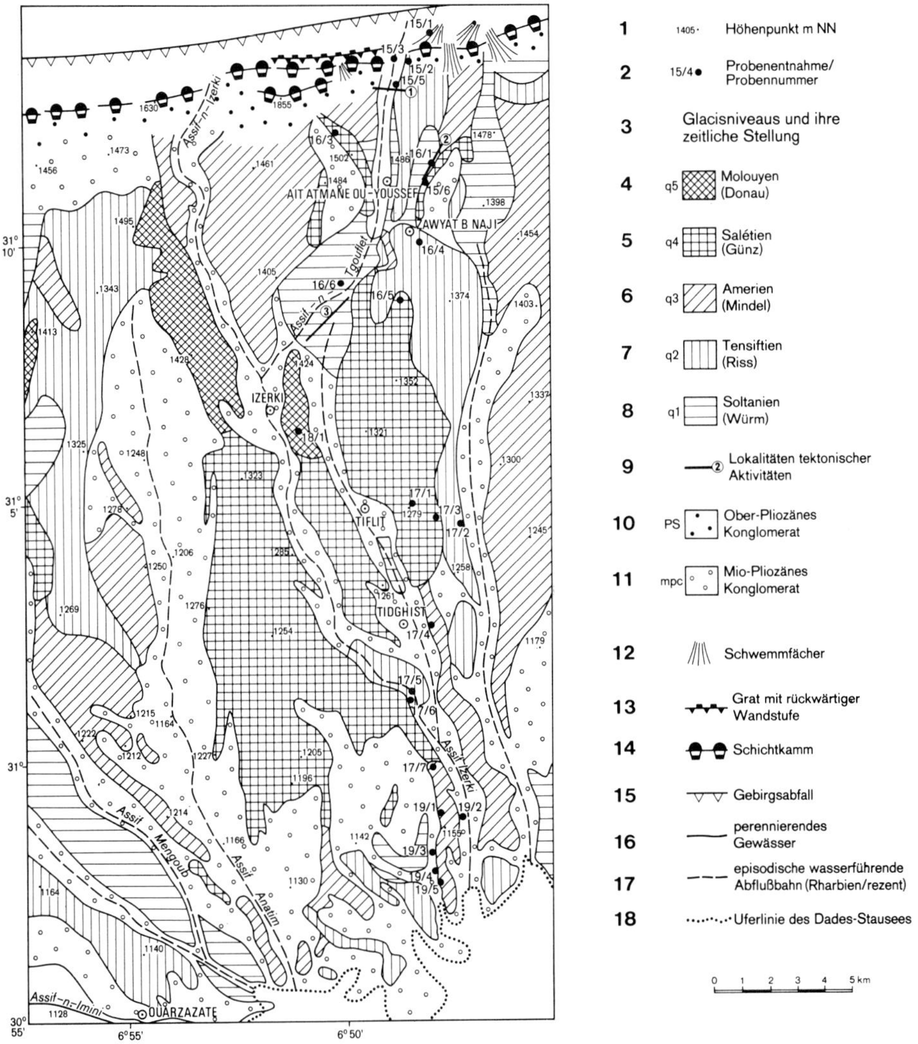

Figure 2: Geomorphological map of the western part of the Ouarzazate Basin with characteristic glacis levels.
1 = heights in m above sea level, 2 = points of sediment samples; 3 = glacis levels and their correlation to pleistocene periods; 4 = "oldest Reg" more than 150 m above the oueds, 5 = "upper old Reg" 80 to 100 m above Oueds, 6 = "lower old Reg"; 7 = "middle Reg"; 8 = "young Reg"; 9 = sites with traces of tectonic activities; 10 = Upper Pliocene conglomerates; 11 = Miopliocene conglomerates; 12 = recent alluvial fans; 13 = crest with backward steep slope; 14 = hogback; 15 = mountain slope of High Atlas; 16 = perennial streams; 17 = episodic drainage ways; 18 = coastline of the storage lake of Dadès River.

floors crossing the basin from north to south rotational segments are tilted towards the east (see site 1 in Fig.2). This should be explained by active north-south faults. At these lines the foreland blocs are moved relatively to another in a different way. By the same germanotype neotectonic the higher glacis surfaces of the marginal zone are tilted slowly to the east, e.g. west of Ait Atmane Ou Youseff (see site 2 in Fig.2) the q4-level is incident to east. There is a layer of conglomerates several meters thick, which is exposed on a glacis slope at a southwest-northeast striking line dipping with 20 degrees to the north. On northeast-southwest striking fault lines fluvial drainageways are locally changed in their former direction (see site 3 in fig.2). The q5-level was there divided of the mountain edge by a lower marginal bloc with q3-level relicts.

South of Imidi (12 km eastwards of the site 3 in fig.2) a southwest-northeast striking fault exists in folded Tertiary sediments. The Mio-Pliocene layers (mpc) dip locally with 10 to 12 degrees to north and they are unconformably overlain by q2-fanglomerates. The steep fault with 50 degrees towards the north is exposed at the slope of the Wadi. The southern bloc has subsided 20 m or more, whereas the northern bloc was uplifted. If the q2-level belongs to the Tensiftian period (cf. CHOUBERT 1951, 1961) the fault tectonic activity in the basin could be followed at this site up to the Younger Quaternary.

In the surrounding mountains several erosion surfaces exist in different heights (cf. COUVREUR 1981) and intramountain basins (cf. AMBOS 1977), e.g. the basin of Telouet. These phenomena must be correlated with tectonic stabiltity in some periods of Tertiary, but without sediments for dating it is not possible to identify stages of mountain uplift.

4. The levels of glacis

The pronounced table shaped mountains of pediment levels with only thin and compacted gravel layers of the glacis (2-5 m conglomerates) can only be grouped topographically by their absolute and relative heights into five levels (see Fig.2). Their usual correlation to the cold ages could not be proved by exact dating. Coarse sediment analysis (cf. STÄBLEIN 1970, MÖLLER et al. 1983) with many samples shows that, with respect to composition, there is, over the whole area, not any characteristic throughout one level. The regional differences in one sediment sequence are greater than the differences between samples of different sequences of different times. In the higher glacis gravels there is a large amount of rocks from the Anti- Atlas. Therefore and with respect to the measurements of imbrication of the glacis gravels, it can be shown that the older glacis sedimentation which reached far into the basin area, came from the south.

5. The fluvial development

The incision of valleys deeper than 100 m is primarily caused by the tectonically controlled development of the hydrographic base level at the Dadès River. A narrowing of the basin area occurred since the Pliocene period. Especially at the southern edge of the basin to the Anti-Atlas, Pliocene sediments and pediments, as remnants of an older basin stage, e.g. south of Skoura and Ouarzazate, reach heights of 1 300 m above sea level, 100 up to 200 m above the Dadès River. The former border of the basin is morphologically marked by high and steep scarps. Corresponding remnants of an erosion level continue to the south as flat and broad high valley bottoms without gravels high above the narrow canyon of the recent Dra River . The axis of subsidence of the basin shifted to the south during the Quaternary. This process increased the area of the High Atlas sedimentation.

The older flat drainage to the west, which did not induce a very active erosion, was proved by fillings of channels with Upper Tertiary sediments (indicated by GAUTHIER 1957, GÖRLER & ZUCHT 1986). A younger uplift in the massive of the Djebl Siroua stopped the outlet of the basin to the west. From the comparison of the glacis we assume that this occurred in the old Pleistocene (Amirien resp. Mindel), but we cannot date this event until now. For the younger phase of relief development, the incision of the breaching of the Dra River through the Anti-Atlas is of morphological importance. In the narrow canyon between Ouarzazate and Agdz there are no fluvial terraces at all. This is an argument that the diversion to the south of the main channel of the basin must have occurred in younger times. From Agdz to Zagora many remnants of terraces in heights of 30 m and more above the river level show that this part of the Dra valley must be much older.

We deduce that a young uplift of the Anti-Atlas effected a steepening of the valley gradient and therefore a stronger headward erosion capacity of the Dra River. This led to the capture of the Dadès River system. In this process the former tributary valley of Dadès River partly suffered a reversal of the flow direction. The break through of the Dra River is the cause for the deep valley incision into flat layered Tertiary lacustrine and terrestric and also marine basin sediments of up to 2 000 m thickness. These sediments prove that the basin as a whole initiated during the lower Tertiary. The opening and narrowing of the basin modified the zonal geomorphogenesis of the relief, which is controlled by the general tendency of subsidence with a regionally differentiated pattern of horsts, anticlines and bloc systems.

Several phenomena of the tectonic and morphologic scenery of the basin of Ouarzazate are comparable with the relief development in other depressions and rift systems (STETS & WURSTER 1981, WURSTER 1981), e.g. at the upper Rhine valley (cf. STÄBLEIN 1968). The relief is the result of the endogenetic tectonic and exogenetic erosional development. The erosion is stimulated by the changing climatic conditions of the glaciual periods, which created several geomorphodynamic sequences of arid and humid periods of activity respectively passivity of relief forming (cf.ANDRES 1980, SABELBERG & ROHDENBURG 1979).

References

AMBOS, R. 1977: Untersuchungen zur pleistozänen Reliefentwicklung im Oberen Todrhagebiet am Südrand des Hohen Atlas (SE-Marokko). - Mainzer Geogr. Studien, 12: 1-140, Mainz.

ANDRES, W. 1977: Studien zur jungquartären Reliefentwicklung des südwestlichen Anti-Atlas und seines saharischen Vorlandes (Marokko). - Mainzer Geogr. Studien, 9: 1-147, Mainz.

ANDRES, W. 1980: On the paleoclimatic significance of erosion and deposition in arid regions. - Z. Geomorph. N.F., Suppl. 36: 113-122, Berlin, Stuttgart.

BEAUDET, G. 1971: Le Quaternaire marocain: état des études. - Rev. Géogr. Maroc. Rabat, 20 (1971): 3-56, Rabat.

BIROT, P. & JOLY, F. 1951: Observations sur la morphologie de l'Anti-Atlas oriental. - Bull. Soc. Nat. Maroc Rabat, 1951: 209-229, Rabat.

BÜDEL, J. 1977: Klima-Geomorphologie. - 1-304, Stuttgart.

CHADENSON, L. 1952: Essai Morphologique et Tectonique sur le Pliocène Supérieur en Afrique du Nord et dans les Steppes Nord-Sahariennes. - Trav. Inst. Rech. Sahar., 8: 49-69.

CHAMAYOU, J. & RUHARD, J.P. 1977: Sillon préafricain à l'est du Siroua: les bassins de Ouarzazate et de Errachidia (Ksar-Es-Souk) - Boudenib. - Notes et Mémoires du Service Géologique, 231 (1977): 224-242, Rabat.

CHOUBERT, G. 1941: Sur les déformations de la surface hamadienne et sur les mouvements atlasiques récents dans l`Anti-Atlas. - C.R. Acad. Sci., 213: 586-588, Paris.

CHOUBERT, G. 1951: Note sur la géologie de l'Anti-Atlas. C.R. Congr. Géol. Intern. de Londres 1948, 14: 29-44, Paris.

CHOUBERT, G. 1955: Sur le mouvements tectoniques quaternaires au Maroc. - Geol. Rundschau, 43 (1): 203 ff, Stuttgart.

CHOUBERT, G. 1961: Quaternaire au Maroc. - Biuletyn Peryglacialny, 10: 9-28, Lodz.

CHOUBERT, G. & FAURE-.MURET, A. 1965: Manifestations tectoniques au cours du Quaternaire dans le sillon préafricain (Maroc). - Notes Serv. Géol. Maroc, 25: 57-63, Rabat.

COUVREUR, G. 1973: Étagement de formes et néotectonique sur le versant sud du Haut Atlas. - Rev. Géomorph. Dyn., 22 (3): 109-124, Strasbourg.

COUVREUR, G. 1981: Essai sur l'évolution morphologique du Haut Atlas Central Calcaire (Maroc). - Thèse presentée devant l'Université de Strasbourg le 3.2.1978: 1-877, Lille, Paris.

DEPOIS, J. & RAYNAL, R. 1967: Géographie de l'Afrique du Nord-ouest. - 1-570, Paris.

DRESCH, J. 1954: Formes et limites climatiques et paléoclimatiques en Afrique du Nord. - A. Géogr., 335: 56-59, Paris.

DRESCH, J. 1984: L'étagement des formes du relief dans le haut Atlas Central (Maroc). - Studia Geomorphologica Carpatho - Balcania, 18: 65-72, Krakow.

DRESCH, J. & RAYNAL, R. 1953: Note sur les formes glaciaires et periglaciaires du Moyen Atlas, du bassin de la Moulouya et du Haut Atlas oriental. - Notes Serv. Géol. Maroc, 7: 111-121, Rabat.

EMBERGER, L. 1951: Rapport sur les régions arides et semiarides de l'Afrique du Nord (les bases écologiques de la régenérations de la végétation des zones arides). - Union Intern. des Sci. Biol., Ser. B. 9: 50-61, Paris.

GAUTHIER, H. 1950: Sur le Pliocène et le Quaternaire de la région d'Ouarzazate (versant sud du Haut-Atlas marocain). - C.R. Acad. Sci, 231(25): 1519-21, Paris.

GAUTHIER, H. 1957: Contribution à l'étude géologique des formations post-liassiques des bassins du Dadès et du Haut Fodra (Maroc méridional). - Notes et Mémoires du Service Géologiquet, 119: 1-212, Rabat.

GIESE, P. 1986: Einleitung. Teil II Forschungsberichte aus den zentralen Anden (21-25oS) und aus dem Atlas-System (Marokko) 1981-1985, Forschergruppe Mobilität aktiver Kontinentalränder. - Berliner Geowiss. Abh., A 66: 1-2, Berlin.

GIGOUT, M. & RAYNAL, R. 1957: Corrélation des phénomènes marines et continentaux dans let Quatérnaire marocain. - C.R. Acad. Sci. Paris, 244 (1957): 2528-31, Paris.

GÖRLER, G. & ZUCHT, M. 1986: Stratigraphie und Tektonik des kontinentalen Neogens im Süden des zentralen Hohen Atlas, Provinz Ouarzazate (Marokko). - Berliner Geowiss. Abh., A 66: 471-494, Berlin.

JACOBSHAGEN, V. 1986 a: Untersuchungen zur geodynamischen Entwicklung des Atlas-Systems in Süd-Marokko; eine Einführung. - Berliner Geowiss. Abh., A 66: 265-272, Berlin.

JACOBSHAGEN, V. 1986 b: Zur Baugeschichte des zentralen Hohen Atlas. - Berliner Geowiss. Abh., A 66: 455-470, Berlin.

JENNY, J. 1983: Les décrochements de l'Atlas de Demnat (Haut Atlas central, Maroc): prolongation orientale de la zone de décrochement du Tizi-n-Test et clef de la compréhension de la textonique atlasique. - Eclogai geol. Helo., 76: 243-251, Basel.

JOLY, F. 1952: Le sillon-sud-atlasique. - Notes et Mém. Serv. Géol. du Maroc, 96: 93-105, Rabat.

JOLY, F. 1962: Etudes sur le relief du Sud-Est marocain. - Trav. Inst. scient. chérif., Série géol. et Géogr. Phys., 10, Rabat.

JOLY, F. & RAYNAL, R. 1961: Originalité des phénomènes periglaciaires au sud de la Méditerrannée. - Biuletyn Peryglacjalny, 10: 31-33, Lodz.

LAVILLE, E., LESAGE, J.-L. & SEGURET, M. 1977: Géometrie cinématique (dynamique) de la tectonique atlasique sur le versant sud du Haut Atlas marocain. Apercu sur les tectoniques hercyniennes et tardi-hercyniennes. - Bull. Soc. géol. France VII, 19: 527-539, Paris.

MARGAT, J. 1953: Plissement d'une terrasse quaternaire dans le Haut Atlas (Maroc). - C.R. Acad. Sci., 237: 1731-2.

MENSCHING, H. 1953: Morphologische Studien im Hohen Atlas von Marokko. - Würzburger Geogr. Arb., 1: 1 104, Würzburg.

MENSCHING, H. 1954: Fußflächen in Marokko. - Pet. Geogr. Mitt., 98: 171-176, Gotha.

MENSCHING, H. 1955: Das Quartär in den Gebirgen Marokkos. - Pet. Geogr. Mitt., Erg. 256: 1-79, Gotha.

MENSCHING, H. & RAYNAL, R. 1954: Fußflächen in Marokko. - Pet. Geogr. Mitt., 98: 171-176, Gotha.

MENSCHING, H.G. 1984: Waditerrassen als paläoklimatische Zeugen in Nordafrika. - Mitt. Geol.-Paläont. Inst. Univ. Hamburg (Festschrift Knetsch), 56: 133-141, Hamburg.

MICHARD, A. 1976: Eléments de géologie Marocaine. - Notes et Mémoires du Service Géologique, 252: 1-408, Rabat.

MÖLLER, K., STÄBLEIN, G., WAGNER, P. & ZILLBACH, K. 1983: Georelief, Abtragung und Gefügemuster an einem aktiven Kontinentalrand; Bericht zum Forschungsprojekt in Süd-Marokko. - DIE ERDE, 114: 309-331, Berlin.

MÜLLER-HOHENSTEIN, K. 1978: Nordafrikanische Trockensteppengesellschaften; zur ökologischen Erklärung der räumlichen Differenzierung der Vegetation zwischen Mittelmeer und Sahara. - Erdkunde, 32: 28-39, Bonn.

RAYNAL, R. 1965: Morphologie de piedmonts et tectonique quarternaire au Maroc oriental. - Notes Serv. Géol. Maroc, 25: 87-90, Rabat.

RISER, J. 1973: Le barrage de Mansour Eddahbi et les aménagements agricoles du Dra Moyen. - Rev. Géogr. Maroc Rabat, 23/24 (1973): 167-177, Rabat.

RISER, J. 1978: Le gisement de l'Oued Neffid dans le Tinzouline (vallée moyenne du Dra): un example de l'Acheuléen du Sud-Est marocain. - Bulletin de la Société Préhistorique Francaise, 75 (10): 291-302.

RISER, J. 1979: La Periode Charniere entre Pliocene et Quaternaire. - Bull. Assoc. Géogr. Franc., 462: 224-229, Paris.

RISER, J. 1981: Formes et processus géomorphologiques dans le Haut Atlas Marocain. - Revue de Geographie Alpine, 69: 561-582.

SABELBERG, U. & ROHDENBURG, H. (Hg) 1979: Relief- und Bodenentwicklung im Mediterrangebiet, Kurzberichte eines Symposiums (Braunschweig 1979). - Landschaftsgenese und Landschaftsökologie, 5: 1-120, Cremlingen-Destedt.

SAUNDERS, I. & YOUNG, A. 1983: Rates of surface processes on slopes, slope retreat and denudation. - Earth Surface Processes and Landforms, 8: 473-501, Bristol.

SCHMIDT, K.H. 1986: Strukturbestimmte Relieftypen und Tektonik im Grenzbereich zwischen Hohem Atlas und Anti-Atlas. - Berliner Geowiss. Abh., A 66: 503-514, Berlin.

SERVICE GEOLOGIQUE DU MAROC 1959/1978: Ouarzazate. Carte Géologique du Maroc 1/500 000: feuille No. 4. - Notes et Mémoires du Service Géologique, 70: Rabat.

SERVICE GEOLOGIQUE DU MAROC 1966/1970: Carte Géologique de l'Anti-Atlas Central et la Zone Synclinale de Ouarzazate 1/200 000. - Notes et Mémoires du Service Géologique, 138: Rabat.

SERVICE GEOLOGIQUE DU MAROC 1974-1977: Carte Géologique du Maroc, Jbel Saghro-Dades 1/200 000. - Notes et Mémoires du Service Géologique, 161: Rabat.

STÄBLEIN, G. 1968: Reliefgenerationen der Vorderpfalz, geomorphologische Untersuchungen im Oberrheingraben zwischen Rhein und Pfälzer Wald. - Würzburger Geogr. Arb., 23: 1-191, Würzburg.

STÄBLEIN, G. 1970: Grobsediment-Analyse als Arbeitsmehtode der genetischen Geomorphologie. - Würzburger Geogr. Arb., 27: 1-203, Würzburg.

STÄBLEIN, G. 1984: Neotektonische Verhältnisse und rezente Abtragungsbedingungen im Becken von Ouarzazate/Süd-Marokko. - Berliner Geogr. Arb., 38: 4-6, Berlin.

STETS, J. & WURSTER, P. 1981: Zur Strukturgeschichte des Hohen Atlas in Marokko. - Geologische Rundschau Stuttgart, 70 (1981): 801-841, Stuttgart.

WAGNER, P. 1984: Rezente Abtragung und geomorphologische Bedinguzngen im Becken von Ouarzazate (Süd-Marokko). - Berliner Geogr. Abh., 38: 1-112, Berlin.

WICHE, K. 1953: Klimamorphologische und talgeschichtliche Studien im M'Goungebiet (Hoher Atlas). - Mitt. Geogr. Ges. Wien, 95: 4-41, Wien.

WICHE, K. 1955: Fußflächen im Hohen Atlas. - Sitz.Ber. Österr. Akad. Wiss., Math.Phys.Kl. I/164: 389-417, Wien.

WURSTER, P. 1981: Der Hohe Atlas in Marokko, ein mißglückter Seitensprung, der steckengebliebene Versuch einer Ozeanbildung. - Forschung (DFG), 1/81: 6-9, Bonn-Bad Godesberg.

ZILLBACH, K. 1984: Geoökologische Gefügemuster in Süd-Marokko. - Berliner Geogr. Abh., 37: 1-95, Berlin.

ZYLKA, R. & JACOBSHAGEN, V. 1986: Zur strukturellen Entwicklung am Südrand des zentralen Hohen Atlas. - Berliner Geowiss. Abh., A 66: 415-432, Berlin.

RATES OF SCARP RETREAT: A MEANS OF DATING NEOTECTONIC ACTIVITY

Karl-Heinz Schmidt
Institut für Physische Geographie, Freie Universität Berlin
Altensteinstr. 19
D-1000 Berlin 33

Summary: Tectonically induced scarps have attracted much geomorphic and geologic interest in recent years. The distance of a cuesta scarp from a fault line or a domal structure is an age indicator of tectonic movements, if information on the rate of scarp retreat can be obtained. In Southern Morocco rates of retreat were determined by using the information given (1) by dated volcanic or sedimentary material in the foreland of scarps, (2) by gravels that were deposited on the cuesta backslope before the formation of the scarp, and (3) especially by talus relics or sequences of talus relics in the foreland of scarps. The methods were applied to a number of different cuesta scarps in Mio-Pliocene conglomerates, in Paleogene and Cretaceous limestones. The rates of retreat avarage 1,3 km/Mio. years for the thin conglomeratic caprocks, and 0,5 km/Mio. years for the more resistant and thicker limestone caprocks. Sequences of talus relics proved to be most valuable for determining rates of retreat, because the talus and pediment flatirons represent former scarp slope positions. As caprock lithology does not differ very much in the area under investigation the recession rates calculated so far yield a reliable basis for estimating the time of activation of scarps in geodynamically relevant positions. The transition from depositional environments to subaerial erosion in the Pre-Saharan depression happened much earlier between Goulmima and Errachidia (Middle Tertiary) than in the western part (Late Miocene to Early Pliocene).

1. Introduction

The interest of geological and geomorphological research in tectonically induced escarpments has grown in recent years. Attempts have been made to date tectonic activity by using indices of scarp morphometry or rates of slope retreat. The former approach implies that fault scarps in a given climate experience a regular degradation chronology and that morphometric indices may provide a quantitative measure for determining their age (WALLACE 1977; PETERSEN 1985; STERR 1985). This method is usually applied for late Quaternary fault scarps in homogeneous and non-consolidated material. For these kinds of slopes, which follow a slope decline evolution, escarpment form is a means for age assignment. Scarps, however, which are characterized by parallel retreat keep a constant, time-independent form. Parallel retreat, especially in a semiarid climate, is typical for heterogeneous, bipartite cuesta-type scarps (SCHUMM & CHORLEY 1966; CARSON & KIRKBY 1972; BLUME 1971), which generally consist of a resistant caprock above a less resistant layer. When dealing with these kinds of scarps inquiry into rates of retreat may yield important information for dating tectonic events. Cuesta scarp development and its bearing on tectonism and geodynamic evolution was investigated in Southern Morocco.

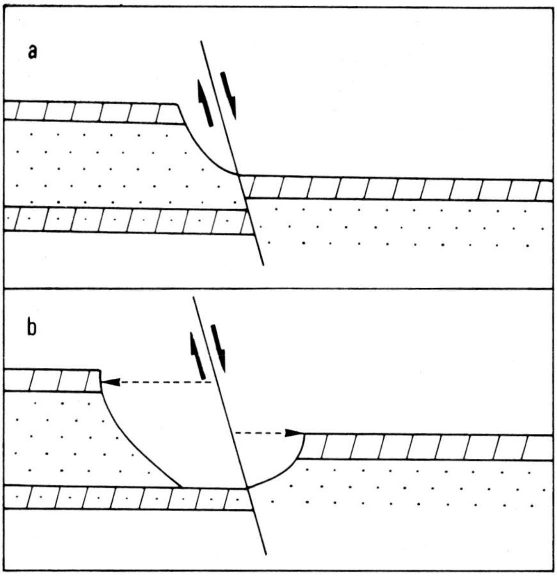

Fig. 1:

Backward migration of scarps from a fault

After vertical displacement in sedimentary rocks of different resistance cuesta scarps are formed, which retreat from the fault line.

2. The problem

Cuesta scarps come into existence when a soft sedimentary layer beneath a resistant caprock is exposed to erosion. This exposure is generally due to fluvial incision but may also be caused by faulting or regional uplift. Faulting results in fault scarp development. Fault scarps in sedimentary rocks of different resistance are subject to the geomorphodynamics of cuesta scarps, they migrate back from the fault parallel to themselves (Fig. 1).

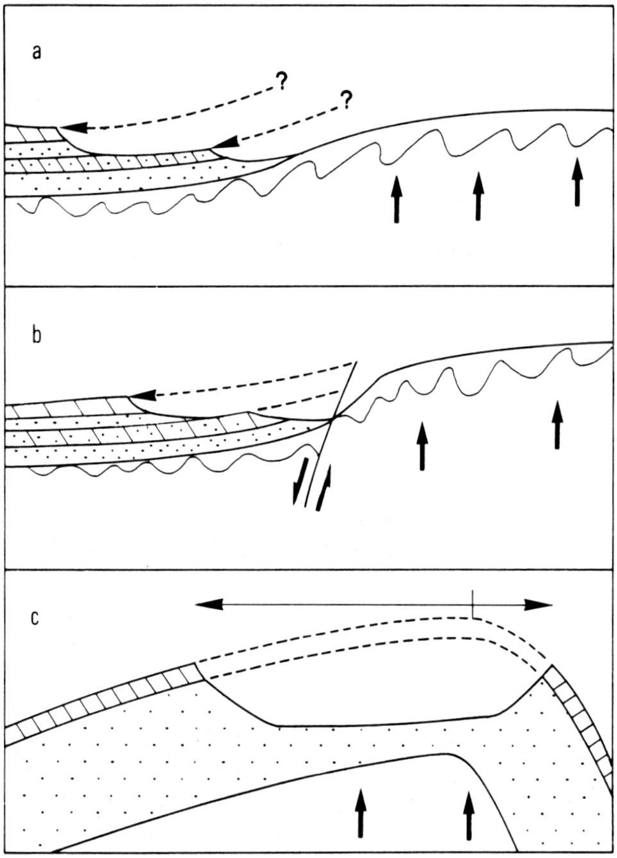

Fig. 2:

Scarp retreat from uplifts
a) retreat from a basement dome (former extension of sedimentary cover can only be inferred from facies and fossil evidence)
b) retreat from a basement dome with boundary fault
c) retreat from crest of an anticlinal uplift, here asymmetrical

Renewed uplift of a structural dome causes the removal of its sedimentary cover by scarp retreat. A number of different initiations are possible:

- uplift of a basement dome with sedimentary cover on its flanks (Fig. 2a),
- uplift of a basement dome with boundary fault (Fig. 2b),
- anticlinal uplift of sedimentary cover (Fig. 2c).

The distance of a cuesta scarp from a fault line or a domal structure is an age indicator for faulting or updoming, if some information on the velocity of scarp retreat can be obtained. There is, however, only a limited number of possibilities to determine the rate of scarp retreat. Sources of information lie in:
- dated sedimentary or volcanic material that has been shed against a former position of the scarp foot or slope (LUCCHITTA 1975; LUCCHITTA & YOUNG 1986),
- gravels that were deposited on the cuesta backslope before the formation of the cuesta scarp,
- talus relics and sequences of talus relics of different age in the foreland of scarps (GERSON 1982; SCHMIDT 1986, in press),
- beheaded valleys of known age on the scarp front (SCHMIDT 1980),
- known rates of scarp retreat from cuestas with similar characteristics in regions with similar climatic conditions.

All of these methods have been tested in the area under investigation. Some were successfully applied and yield estimates in the right (and same) order of magnitude. To illustrate the theoretical considerations in more detail some examples are presented.

3. Areal and geologic setting

The investigations were regionally concentrated on the Pre-Saharan depression between Ouarzazate and Errachidia and the transitional zones to the mountain systems of the High Atlas and Anti-Atlas (Fig. 3). This region especially in its western and eastern parts is dominated by structural landforms because of the mostly moderate to gentle dip of the strata and the interstratification of resistant and less resistant layers in the Cretaceous and Paleogene sequence.

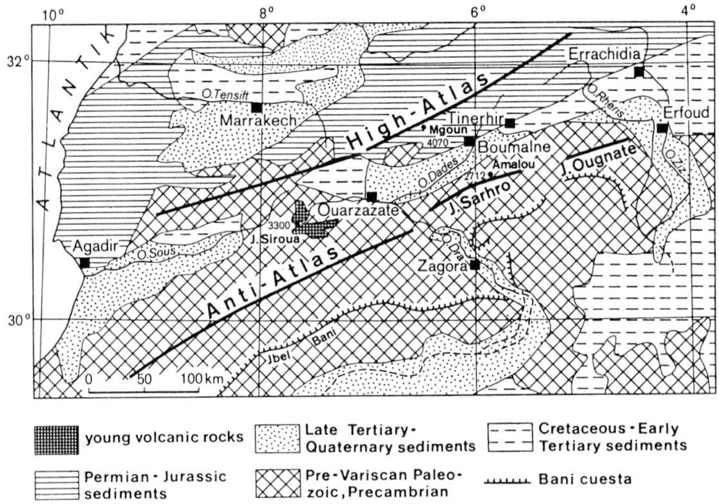

Fig. 3:

Geologic sketch map of Southern Morocco

The principal caprocks are limestones of Cenomanian-Turonian age as well as Paleocene and Eocene limestone series. The Early Tertiary limestones attain a mean thickness of about 10 m, values greater than 20 m are rare. The Cretaceous limestone caprocks have a general thickness of about 20-30 m; these limestone formations thicken to the east (CHAMAYOU & RUHARD 1977). The limestones are underlain by marls, siltstones and poorly consolidated sandstones. As the lithological composition of the cuesta scarps in the Cretaceous and Paleogene rocks does not differ very much, these landforms show morphologically similar features. An erosional outlier (Gara Aferdou) west of Errachidia with a Tertiary (Eocene? Aquitanian?, HERBIG, in press) limestone caprock also belongs to this group of cuesta scarps (Fig. 8).

Between Ouarzazate and Boumalne in the late Tertiary and Quaternary sedimentary basin of Ouarzazate conglomerates with a maximum thickness of about 10 m in the Mio-Pliocene terrestrial sequence are another type of caprock. The conglomerates are underlain by soft sandstones and marls. Hogbacks on the northern margin of the Anti-Atlas are formed in Ordovician quartzites of the Bani series.

The backslopes of the cuestas are stripped surfaces and no truncating plains, differences in rock resistance are well pronounced due to the selective work of arid weathering and erosion. A sharp crest (Trauf) marks the transition between cuesta scarp and backslope. In the gently dipping strata there are a great number of beheaded valleys on the crest of the scarp as a consequence of active scarp recession. The upper scarp slope in the resistant caprock usually forms a cliff, whereas the lower scarp slope is moderately inclined with about 25 to 30°. The lower slope is concave, steeper towards the upper slope and with gently inclined pediments at the lower end. Presently the lower scarp slopes are being dissected under the prevailing semiarid to arid climate and bedrock is exposed below the debris cover.

The rainfall regime is highly variable with mean annual precipitation values in the Pre-Saharan depression between 100 and 175 mm. Episodic high intensity rainstorms of short duration are typical for the regime. Hourly intensities of 16 mm have been measured during a field trip (WAGNER 1984).

4. Determination of rates of scarp retreat

In the following section the use of different sources of information for the determination of scarp retreat will be demonstrated and discussed for a number of cuesta scarps in the Pre-Saharan depression.

4.1 Scarps in Mio-Pliocene conglomerates in the basin of Ouarzazate (source of information: talus relics)

The talus relics described in this chapter are found not only on the Mio-Pliocene conglomerate scarps, but also on the limestone scarps (Aquitanian to Cretaceous). The conglomerate scarps are chosen as an example for introducing the talus and pediment flatirons as landforms, which can be used for estimating scarp retreat. The specific example is preceded by a more general description and interpretation of these landforms.

Talus relics on cuesta scarps are often preserved as talus flatirons or pediment flatirons. These landforms have been described by a number of authors (KOONS 1955; BLUME & BARTH 1972; DONGUS 1980; GERSON 1982; SCHMIDT 1986). They are remnants of formerly active smooth slopes covered by a thin debris cover. These flatirons have a triangular to trapezoidal shape with their top directed towards the cuesta scarp (Fig. 4). Talus flatirons, which rest on the scarp slope, and pediment flatirons in the transition zone between slope and scarp foreland develop as a consequence of slope dissection and scarp retreat.

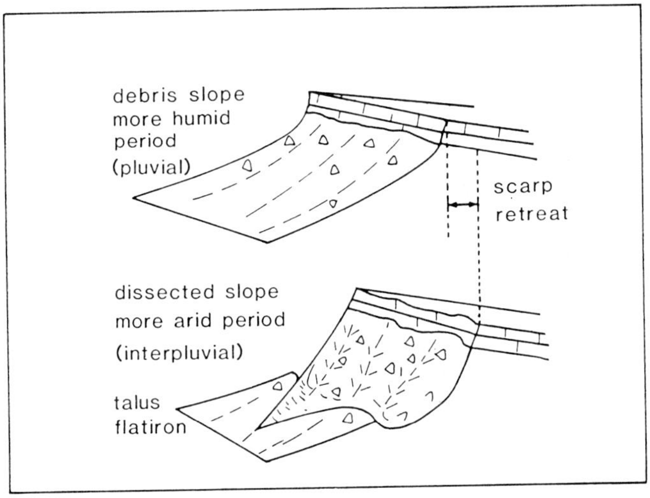

Fig. 4:

Development of talus flatirons

The smooth debris covered slope of the humid period (pluvial) is dissected and steepened in the more arid period (interpluvial). The scarp retreats and flatirons develop.

The smooth debris-covered slopes are formed in more humid periods and dissected in more arid periods (JOLY 1962; ERGENZINGER 1972; GERSON 1982; SCHMIDT, in press). The distance of the most recent talus flatirons from the present slope shows the extent of backwearing after the last pluvial. Scarp retreat takes place both in the debris slope period and the dissection period (DONGUS 1980; GERSON 1982). In some slope profiles sequences of talus and pediment flatirons of different age are found, the surfaces of the flatirons representing the latest stage of the debris slope of the respective pluvial. The distance between two flatirons is an indicator for scarp retreat in a complete interpluvial/pluvial cycle, it can be used to estimate rates of retreat (Fig. 5). To obtain a value of total retreat for a cycle, one cannot simply measure the distance from the top of a talus flatiron to the next, because the front slope of the flatiron itself after its formation will migrate away from the scarp slope (SCHMIDT 1986).

Flatirons with different backslope inclinations represent different positions of the formerly continuous pluvial scarp slope. To determine actual retreat it is necessary to measure the distance from the top of the older flatiron to the point of equal inclination on the next younger flatiron (Fig. 5). The interpluvial/pluvial cycle from Tensiftien (Saale) to Soltanien (Weichsel) lasted about 10^5 years, the cycle from Amirien (Elster) to Tensiftien about 2 x 10^5 years. With the length of the intervals between times of debris slope formation known, rates of retreat can be calculated.

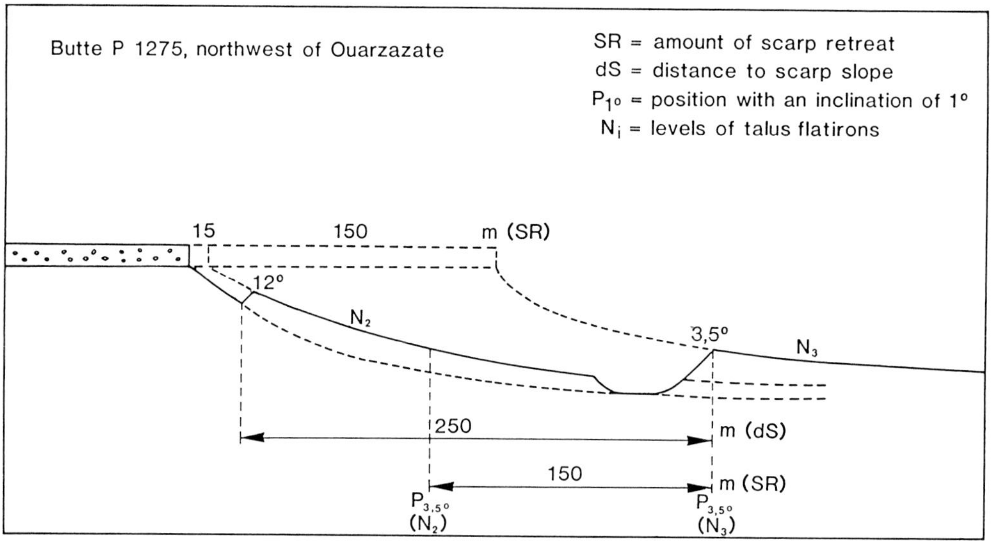

Fig. 5:

Method of determining the amount of scarp retreat (SR) with sequences of talus and pediment flatirons in the scarp foreland

N_2 and N_3 are levels of talus (pediment) flatirons. N_1 is the youngest (Holocene) terrace. For determining the amount of scarp retreat for an interpluvial/pluvial cycle the distance from the top of the older flatiron (e.g. $P_{3,5}$o; N_3) to the point of equal inclination on the next younger flatiron backslope (e.g. $P_{3,5}$o; N_2) is measured. The amount of retreat in the example is 150 m. As the cycle from Tensiftien (N_3) to Soltanien (N_2) lasted about 10^5 years, a rate of retreat of about 1,5 km/Mio. years is obtained.

Two isolated buttes covered by a 5 m thick Mio-Pliocene conglomerate are situated about 9 km northwest of Ouarzazate. They are surrounded by a number of talus and pediment flatirons. A profile from the western flank of the southeastern butte is shown in Figure 5. The distance from the top of the younger talus flatiron (N_2) (N_1 is a Holocene terrace) to the scarp slope is 17 m, the distance from the older pediment flatiron (N_3) to the scarp slope is about 250 m. The respective backslope inclinations are 12 and 3,5°.

Using the procedure described above, an amount of retreat between the Tensiftien (N_3) and Soltanien (N_2) of 150 m and of 15 m for the time after the latest pluvial is obtained (Fig. 5). This corresponds to a rate of retreat of about 1,5 km/Mio. years. The average rate for the Mio-Pliocene scarps investigated in the basin of Ouarzazate is about 1,3 km/Mio. years (SCHMIDT, in press). This means that the scarps with the thin, well jointed conglomeratic caprock retreated much faster than the limestone cuestas, for which different values were calculated with the same procedure (cf. chapter 4.2, 4.3 and SCHMIDT, in press).

4.2 The cuesta scarps west of Tinerhir
(source of information: gravels on cuesta backslope, volcanic material in scarp foreland, talus relics)

The cuesta scarps consisting of Paleogene and Cretaceous rocks, facing the about 3 to 4 km wide strike valley of the Assif n' Targuit west of Tinerhir are of particular interest for the visualization of the tectonic history of the transition zone between the Pre-Saharan depression and the Djebl Sarhro. The Lower Cretaceous redbeds unconformably overlie the folded Paleozoic rocks of the Anti-Atlas zone. The Cretaceous and Paleogene sequence has been stripped away from the basement rocks in a northward direction (Fig. 6).

The activation of scarp erosion must be younger than the sedimentation of Mio-Pliocene material, which we find in the syncline of Timadriouine north of the profile as the easternmost outcrop of Neogene sediments in the basin of Ouarzazate. There is another maximum age indicator for the formation of the scarp. On the crest of the scarp and on its backslope on the Early Tertiary rocks there are gravels, the gravels of Timadriouine, which contain large quantities of Anti-Atlas material. Clearly, the foreland valley cannot have been in existence, when the gravels were deposited. The formation of the scarp, which in all probability was initiated by the uplift of the southern highlands and the transition zone, interrupted the ancient northward drainage. The gravels so far were only indirectly dated and thought to be of Villafranchian age (GAUTHIER 1957) because of their high elevation above the regional base level.

According to latest results an older age of the gravels must be assumed, because they are younger than the Neogene sediments in the syncline of Timadriouine, and older than the volcano Foum el Kous, which erupted into the foreland of the scarp, which, as was shown, was not yet formed when the gravels from the Anti-Atlas were accumulated. There are preliminary K-Ar data for the ankaratrite of the Foum el Kous which centre around 2,9 Mio. years (SCHERMERHORN 1987, oral communication). As quite a long time interval is needed for the formation of the wide foreland strike valley, the Mio-Pliocene sedimentation must have ended earlier than in the western parts of the basin of Ouarzazate (cf. GÖRLER & ZUCHT 1986) and the gravels are possibly a lithostratigraphical equivalent of the conglomerates in the Neogene sequence and are of latest Miocene to Early Pliocene age.

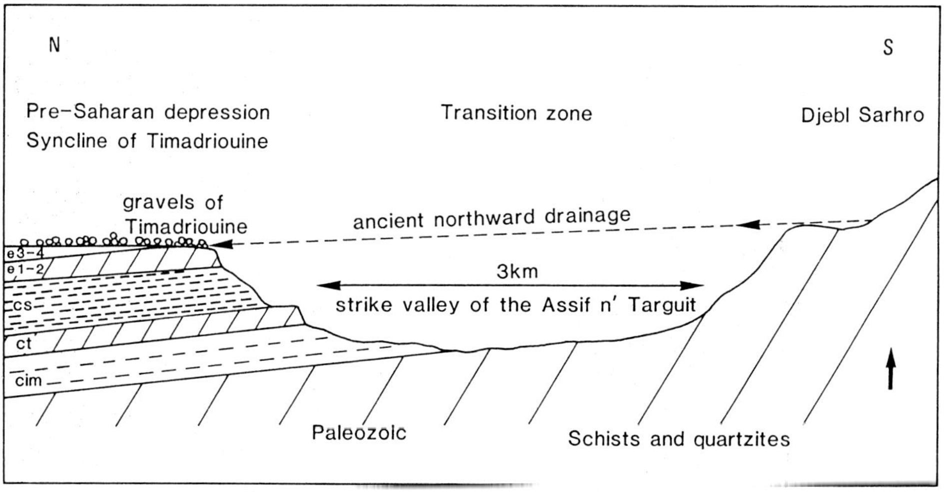

Fig. 6:

The transition zone between the Pre-Saharan depression and the Djebl Sarhro

The gravels of Timadriouine, which contain large proportions of Djebl Sarhro material, lie on the backslope of the Paleogene scarp and must have been deposited before the formation of the scarp. The gravels are probable of Latest Miocene to Early Pliocene age.

Stratigraphic abbreviations: e_{3-4}: *Early Eocene*

e_{1-2}: *Paleocene*

cs : Senonian

ct : Cenomanian-Turonian

cim : Infracenomanian

The Paleogene limestones exposed in the scarp west of Tinerhir were formed in a near-shore environment with a shore located very near the present northern rim of the Anti-Atlas (HERBIG 1986). Thus a rate of retreat for the Early Tertiary limestones of about 0,4 to 0,7 km/Mio. years can be estimated. As erosion reached the Cretaceous limestones later than the overlying rocks a somewhat faster rate of retreat, but in the same order of magnitude is assumed. The rates are lower than previously calculated (1 km/Mio. years, SCHMIDT 1986) because the Villafranchian age assignment of the gravels (GAUTHIER 1957) was used. Today the Cenomanian-Turonian limestone caprock crops out at the foot of the Paleogene scarp about 1 km in front of the upper scarp's crest. East of the Foum el Kous the distance becomes larger and the Cretaceous limestone forms a separate cuesta.

The Foum el Kous volcano erupted in the immediate foreland of the Paleogene escarpment (Fig. 7). Lava flowed against the scarp slope and into reentrants of the scarp. In these parts the slope became stabilized. The highest point of the volcano has an altitude of 1558 m, the crest of the Paleogene scarp only an altitude of about 1530 m, yet with one exception, no lava flow is found north of the crest of the scarp. This exception is a south-trending valley which originated as an obsequent valley on the slope of the Eocene scarp above the main scarp. In the other parts there must have been some obstacle at the time of eruption. This obstacle was formed by this secondary cuesta scarp in Eocene limestones and marls, which, in the meantime, has moved to the north 100 to 200 meters (not shown in Fig. 7). The greatest part of the volcanic material flowed to the south, where it covers and stabilizes the slope and the plateau on the Cretaceous rocks and the respective escarpment.

In the area influenced by the ankaratrite extrusion the scarps remained almost stationary and vertical erosion through the flows and the underlying sedimentary material was dominant. So the Eocene scarp which had prevented the northern extension of the lava flows retreated with a rate of only 0,05 km/Mio. years. But as in the case of the main Paleogene scarp an unknown time needed for detachment from the lava has to be taken into account. The Paleogene scarp which is in direct contact with the centre of eruption has remained in the same position for almost 3 million years. To the west the distance between scarp and lava flow increases up to 500 m. So the retreat of the Paleogene scarp in this area ranges from 0 to 0,2 km/Mio. years.

Another indicator for scarp retreat in this section are talus relics. Here the Tensiftien talus flatirons (N_3) have distances from the Soltanien slope (N_2) of about 20 to 40 m, which means that the scarp has retreated at a velocity of 0,2 to 0,4 km/Mio. years. The rate derived from the radiometric dating is lower because of the unknown period of scarp stabilization, which is included in the time interval from eruption to present.

West of the Foum el Kous, where it is not stabilized by the lava cover, the scarp formed by the Cenomanian-Turonian limestones has retreated a greater distance, about 500 m, than the corresponding lava-protected scarp, which makes a minimum rate of retreat of 0,2 km/Mio. years. To the east the Cretaceous scarp is much dissected by long obsequent valleys which reach down from the Paleogene scarp slope. An estimate of retreat is difficult to obtain.

In an earlier article (SCHMIDT 1986) there was not yet a radiometric dating available for the Foum el Kous, and a Middle Pleistocene age was estimated from geomorphological inference, because the distal parts of the flow (Fig. 7) are located on Paleozoic rocks on the southern side of the Assif n'Targuit as erosional outliers in a relative elevation of only 35 m above the wash, an elevation that fluvial erosion in other river systems in this area reached as late as the Middle Pleistocene (Amirien, Tensiftien, cf. JOLY 1962; MÖLLER et al. 1983). As this part of the volcanic material is geochemically younger than the northern lava (cf. SCHERMERHORN & GROTHE 1986, 339), a Pleistocene age of the southern flow is to be expected.

4.3 Scarps in Cretaceous limestones between Goulmima and Errachidia (sources of information: sedimentary material in scarp foreland, beheaded valleys, talus relics)

In a profile from the Jebel Ougnate in the south to Jebel Timidrout there are two cuesta scarps, both of which have Cretaceous limestone as a caprock (Fig. 8). The southern scarp is highly embayed, whereas the northern one runs very straight. The southern scarp has had a long erosional history since its initiation caused by the uplift of the Jebel Ougnate in Middle Tertiary time.

The distance from the cuesta scarp to the basement rocks of the Jebel Ougnate increases from 15 km near Goulmima to about 20 km near Tarda.

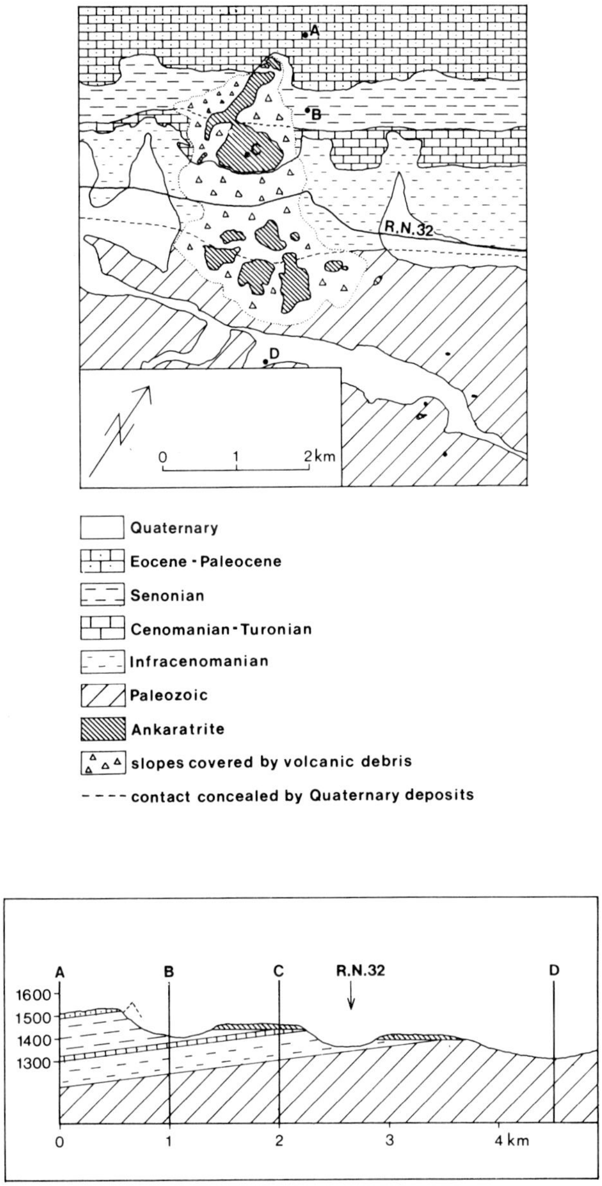

Fig. 7:

Geologic setting of the Foum el Kous volcano with profile

The approximate position of the volcano cone is inserted in the profile. The southern band of Quaternary sediments marks the location of the Assif n' Targuit.

The Cretaceous rocks must have covered the northern margin of the Jebel Ougnate, because the Oued Ferkla (eastern projection of Oued Todrha) crosses a spur of resistant basement rocks in a superimposed valley. The distance from this valley to the present scarp near Goulmima is about 18 km.

The straight course of the northern scarp supports the assumption that it has been induced tectonically in recent earth history. In the foreland of the scarp on the southern end of the anticline there is a flexure with steeply inclined strata in an average distance of about 3 km. In this zone a hogback has developed. The backward migration of the cuesta scarp started from the steeply inclined southern flank of the anticline (cf. Fig. 2 c). In the east the southern steep flank of the anticline disappears below the Senonian cover and a 30 m thick sequence of well rounded gravels (Ikhf n' Tfounast, Fig. 8), which were fluvially deposited in a strike valley at the foot of a former position of the scarp, which since then has retreated about 3 km. The gravels are thought to be of late Tertiary age because of their elevation. From this age assignment a rate of retreat of about 0,5 km/ Mio. years would follow.

The southern scarp is characterized by a great number of beheaded valleys. But they could not successfully be used for estimating rates of retreat for the scarps in the Cretaceous limestones (cf. SCHMIDT 1980), as sediments which could be used for the determination of the age of the valleys, are missing. But information on retreat of the scarps was provided by talus and pediment flatirons. The sequences of talus relics in front of the scarp of the Jebel Rheris east of Goulmima (Fig. 8), when using the procedure described above (cf. 4.1, Fig. 5), yield rates of retreat between 0,3 and 0,9 km/Mio. years, averaging 0,55 km/Mio. years. The formation of the anticline from this line of evidence and from the figure given above must have begun at least 6-8 million years ago and is still active today as can be deduced from the deformation of young pediments in the foreland of the scarp (SCHMIDT 1986). The rate of retreat calculated means that backwearing from the Jebel Ougnate must have been initiated in Middle Tertiary time. Stronger uplift of the southern highlands must have begun at that time. It is also indicated that the area between Goulmima and Errachidia was a structural high in the Pre-Saharan depression in late Tertiary time, because there was continuous subaerial erosion. No Mio-Pliocene sediments are found in this part of the depression and

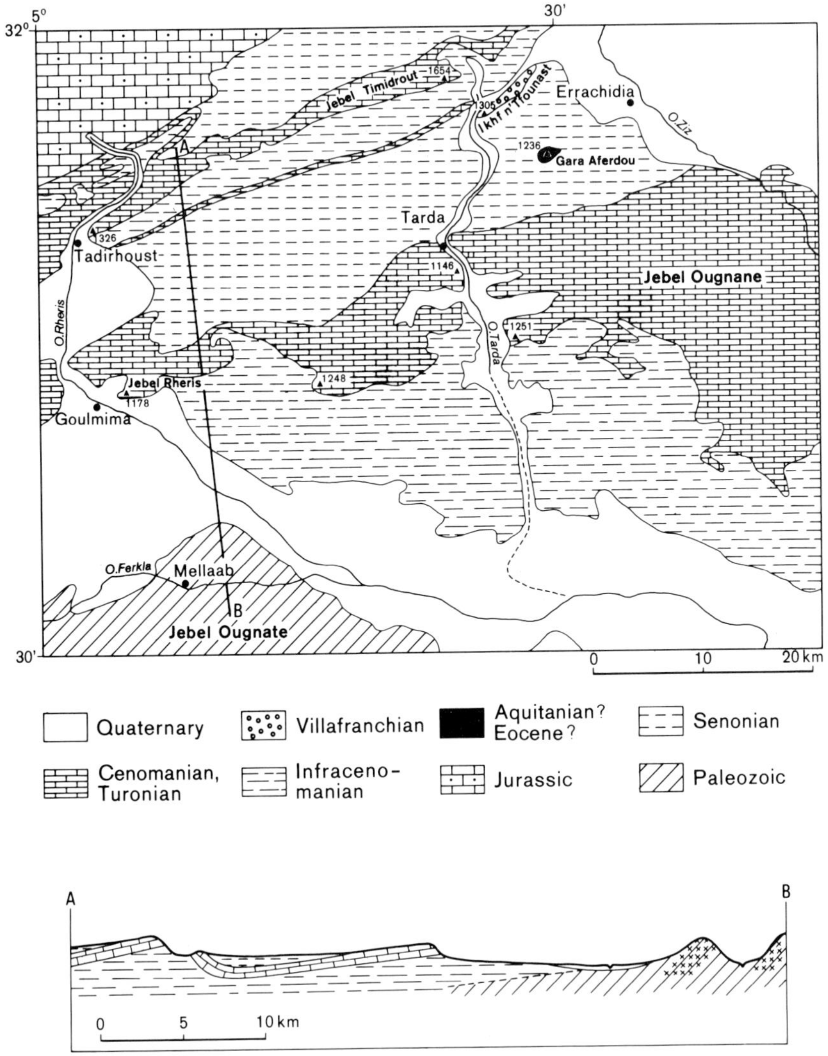

Fig. 8:

Geologic map of the eastern part of the Pre-Saharan depression between Goulmima and Errachidia with a schematic profile (generalized map after SAADI et al. 1978)

Note that scales of map and profile are not identical. Dip is only shown in the Cenomanian-Turonian limestone caprocks. Crosses mark the positions of more resistant rocks in the Jebel Ougnate. The southern end of the profile shows the superimposed valley of Oued Ferkla.

the area is also free of the young Hammada sediments which are found east of Errachidia. The youngest sediment in this area, with the exception of alluvial sediments, is the limestone cap of an erosional remnant (Gara Aferdou) which is of probable Aquitanian, if not at all Eocene age (HERBIG, in press). So here scarp retreat began earlier than in the western part, where it started in late Miocene to Early Pliocene time (cf. 4.2).

5. Conclusion

The rates of retreat for the Mio-Pliocene conglomerates average 1,3 km/Mio. years and the rates for the limestone scarps 0,5 km/Mio. years. This is in good agreement with values of 0,2 to 0,6 km/Mio. years obtained by GERSON (1982) for limestone scarps in Israel. Values calculated for sandstone scarps in the Southwestern United States are somewhat higher (3 km/Mio. years; SCHMIDT 1980), but scarps in resistant limestones also migrate backwards at a slower rate (0,5 km/ Mio. years). Most valuable for the determination of rates of scarp retreat are sequences of talus relics, because the talus and pediment flatirons represent former slope positions. The dating of these talus relics needs more precision, their association with pluvial periods, however, is sufficiently ascertained (SCHMIDT, in press).

The sedimentary material on the cuesta backslopes and in the scarp forelands can only be dated relatively and is therefore of limited value for calculations of rates of retreat. Resistant volcanic material, that has been shed against a scarp slope, stabilizes the scarp for an unknown time interval and allows only minimum estimates of retreat, though the most precise dating of the material is available. The combination of different approaches, however, yields satisfactory estimades of scarp retreat and allows a better understanding of the structural history and geodynamic evolution of the Pre-Saharan depression and its vicinity.

As the number of caprocks of different lithologies is quite limited in the Atlas System, the recession rates calculated so far offer a reliable basis for this region for estimating the beginning of scarp retreat for tectonically induced scarps in key positions.

References:

BLUME, H. (1971): Probleme der Schichtstufenlandschaft. - Erträge der Forschung, 5: 1-117, Darmstadt.

BLUME, H. & BARTH, H.K. (1972): Rampenstufen und Schuttrampen als Abtragungsformen in ariden Schichtstufenlandschaften. - Erdkunde, 26: 108-116, Bonn.

CARSON, M.A. & KIRKBY, M.J. (1972): Hillslope - Form and Process. - 1-475, Cambridge (Cambridge University Press).

CHAMAYOU, J. & RUHARD, J.P. (1977): Sillon préafricain à l'est du Siroua: les bassins de Ouarzazate et de Errachidia (Ksar es Souk) - Boùdenib. - Notes et Mémoires du Service Géologique, 231: 224-242, Rabat.

DONGUS, H. (1980): Rampenstufen und Fußflächenrampen. - Tübinger Geogr. Studien, 80: 73-78, Tübingen.

ERGENZINGER, P. (1972): Reliefentwicklung an der Schichtstufe des Massif d'Abo (Nordwest-Tibesti). - Z.Geomorph., N.F., Suppl. Bd. 15: 93-112, Berlin-Stuttgart.

GAUTHIER, H. (1957): Contribution à l'étude géologique des formations post-liassique des bassins du Dadès et du Haut Todra (Maroc méridional). - Notes et Mémoires du Service Géologique, 119: 1-212, Rabat.

GERSON, R. (1982): Talus relicts in deserts: a key to major climatic fluctuations. - Israel Journ. of Earth Sci., 31: 123-132, Jerusalem.

GÖRLER, K. & ZUCHT, M. (1986): Stratigraphie und Tektonik des kontinentalen Neogens im Süden des zentralen Hohen Atlas, Provinz Ouarzazate (Marokko). - Berliner geowiss. Abh., A 66: 471-494, Berlin.

HERBIG, H.-G. (1986): Lithostratigraphisch-fazielle Untersuchungen im marinen Alttertiär südlich des zentralen Hohen Atlas (Marokko). - Berliner geowiss. Abh., A 66: 343-380, Berlin.

HERBIG, H.-G. (in press): The Upper Cretaceous to Tertiary Hammada west of Errachidia (SE Morocco): A continental sequence involving paleosol development. - N. Jb. Geol. Paläont. Abh., Stuttgart.

JOLY, F. (1962): Etudes sur le relief du Sud-Est marocain. - Trav. Inst. Sci. Chér., Sér. Géol. et Géogr. Phys., 10: 1-578, Rabat.

KOONS, D. (1955): Cliff retreat in the Southwestern United States. - Am. J. Sci., 253: 44-52, Washington.

LUCCHITTA, I. (1975): Application of ERTS images and image processing to regional geologic problems and geologic mapping in Northern Arizona-Part IV B, The Shivwits Plateau. - National Aeronautic and Space Administration Techn. Rep., 32-1597: 41-72.

LUCCHITTA, I. & YOUNG, R.A. (1986): Structure and geomorphic character of western Colorado Plateau in the Grand Canyon - Lake Mead region. - In: NATIONS, J.D. (Ed.): Geology of central and northern Arizona. - 159-176, Flagstaff.

MÖLLER, K., STÄBLEIN, G., WAGNER, P. & ZILLBACH, K. (1983): Georelief, Abtragung und Gefügemuster an einem aktiven Kontinentalrand. - Die Erde, 114: 309-331, Berlin.

PETERSEN, J.F. (1985): Equilibrium tendency in piedmont scarp denudation, Wasatch Front, Utah. - In: MORISAWA, M. & HACK, J.T. (Eds.): Tectonic Geomorphology: 261-278, London.

SAADI, S.E.M., HILALI, E.A. & BOUDDA, A. (1978): Carte géologique du Maroc 1 : 500 000, Ouarzazate. - Notes et Mémoires du Service Géologique, 70 (1959), reprint, Rabat.

SCHERMERHORN, L.J.G. & GROTHE, C. (1986): Vulkanismus im Anti-Atlas (Marokko). - Berliner geowiss. Abh., A 66: 333-342, Berlin.

SCHMIDT, K.-H. (1980): Eine neue Methode zur Ermittlung von Stufenrückwanderungsraten, dargestellt am Beispiel der Black Mesa, Colorado Plateau, USA. - Z. Geomorph., N.F., 24: 180-191, Berlin, Stuttgart.

SCHMIDT, K.-H. (1986): Strukturbestimmte Relieftypen und Tektonik im Grenzbereich zwischen Hohem Atlas und Anti-Atlas. - Berliner geowiss. Abh., A 66: 503-514, Berlin.

SCHMIDT, K.-H. (in press): Das Schichtstufenrelief der präsaharischen Senke, Süd-Marokko. - Z. Geomorph., N.F., Suppl. Bd, Berlin, Stuttgart.

SCHUMM, S.A. & CHORLEY, R.J. (1966): Talus weathering and scarp recession in the Colorado Plateau. - Z. Geomorph., N.F., 10, 11-36, Berlin, Stuttgart.

STERR, H.M. (1986): Erdbebenrisiko im Westen der USA. - Geowissenschaften in unserer Zeit, 4: 131-138, Weinheim.

WAGNER, P. (1984): Rezente Abtragung und geomorphologische Bedingungen im Becken von Ouarzazate (Süd-Marokko). - Berliner geogr. Abh., 38: 1-110, Berlin.

WALLACE, R.E. (1977): Profiles and ages of young fault scarps, North-central Nevada. - Geol. Soc. Am. Bull., 88: 1267-1281, Boulder.

Syntheses

EVOLUTION OF THE ALBORAN BLOCK MARGIN
FROM EARLY MESOZOIC TO EARLY MIOCENE TIME

M. Durand-Delga and Ph. Olivier.
Laboratoire de Géologie Méditerranéenne (U.A. 145),
Université Paul-Sabatier - 38, rue des Trente-Six Ponts; F-31400 Toulouse (France)

ABSTRACT

The evolution of the Alboran (or Betic-Rifian) margin can be divided into two main periods :

- from early Mesozoic to Oligocene it was a passive continental margin, with a shelf (Ghomaride-Malaguide and Dorsalian domains), a slope (Predorsalian domain) and a partly oceanic, deep basin (Mauretanian and Massylian domains). The shelf appeared during late Triassic time, the slope as early as Liassic time and the flysch basin during late Jurassic-early Cretaceous time. During Dogger, Malm and Neocomian time this margin coincided with the large left lateral transcurrent fault which separated Europe from Africa. By the end of Cretaceous time the Betic-Rifian domain became independent from Europe with which it was bound until then.

- the collision of the internal Betic-Rifian zones with the external Rifian zones (African Plate) and with the external Betic zones (European Plate) took place during late Oligocene and early Miocene time. This led to the formation of thrust sheets with an external vergence at the boundary of the internal and external zones and to the formation of the tectonic curvature of Gibraltar.

INTRODUCTION

The Alpine Maghrebides trend E-W from Calabria (South Italy) to the Atlantic Ocean. Like the other parts of this large range, the Rif Mountains consist of tectonic units which were overthrust with a southward and westward vergence mainly during the Miocene. The strata in these units represent three main zones :

a) the northern margin of the African continent (confined to the external zones of the Rif).

b) the Mograbin Basin (Bouillin,1986), with clastic rocks of Cretaceous to early Miocene age which now form the allochthonous flysch nappes.

c) the southern margin of a northern continent (confined to the internal zones of the chain) which is the subject of this paper.

The internal zones of the Rif and of the Betic Cordillera are geologically homologous (Didon *et al.*, 1973) and are linked by the Arc of Gibraltar. This Betic-Rifian or Alboran (Andrieux *et al.*, 1971) internal block is the westernmost part of the Alboran-Kabylian-Peloritan-Calabrian ensemble (or "ALKAPECA", Bouillin *et al.*, 1986). It is possible to consider (Leblanc and Olivier, 1984; Bouillin *et al.*, 1986) that, at the beginning of the Mesozoic, this ensemble belonged to southern Europe but in a more eastern location than now. During Dogger, Malm and Neocomian time the southern boundary of ALKAPECA coincided with the large left lateral strike-slip fault which separated Europe from Africa. Probably as early as the late Cretaceous, a large right lateral strike-slip fault (North Betic fault) appeared between Iberia and western ALKAPECA; this caused ALKAPECA to behave differently from the surrounding large plates. In Paleogene time, the N-S convergence of Europe and Africa gave rise, under ductile conditions, to thrust sheets of ENE vergence in the ALKAPECA basement (Cuevas *et al.*,1986; Balanyá et García-Dueñas, in press).

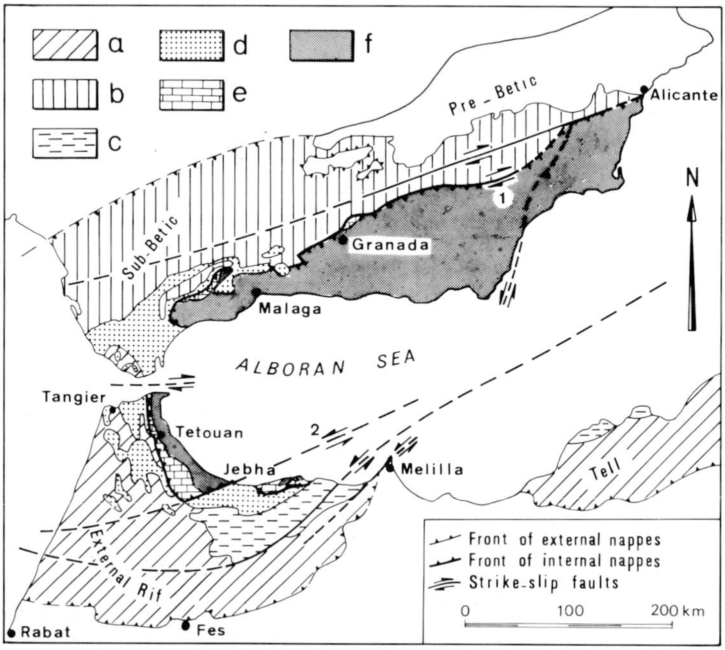

Fig. 1.- The Betic-Rifian Orogen . (a) Maghrebide external non-schist zones. (b) Betic external non-schist zones. (c) Maghrebide external schist zones. (d) Flysch and Predorsalian area (e) "Dorsale calcaire". (f) Ghomarides-Malaguides/ Sebtides-Alpujarrides/ Nevado-Filabrides. 1- North Betic fault. 2- Jebha fault.

Then, as internal strain was not able to take up all the regional stress, ALKAPECA broke up, leading so to the formation of the Alboran Block (in Oligocene to early Miocene time). This block, bound to the north by the North Betic fault and to the south by the Jebha fault, moved relatively toward the WSW and progressively collided with the external Rifian zones (Africa) and the external Betic zones (Europe). This movement gave rise to thrust sheets, this time with external vergence, first at the margin of the Alboran Block and then in more and more external domains. Formed at the same time was the Arc of Gibraltar (fig. 2) whose curvature is thus mainly tectonic in origin.

Emplacement of the main tectonic units in the northern Rif was long and complicated. To determine the relative paleogeographical locations of the depositional domains during the Mesozoic and Paleogene , it is necessary, first to separate the tectonic units in a SW-NE or N-S direction and then to displace them eastward such that the more internal the unit, the greater the displacement.

This means that it is impossible to make direct palinspastic comparisons between, for instance, the Mesozoic formations of the Alboran Block, and those of the Moroccan Atlas because at this time these domains were widely separated.

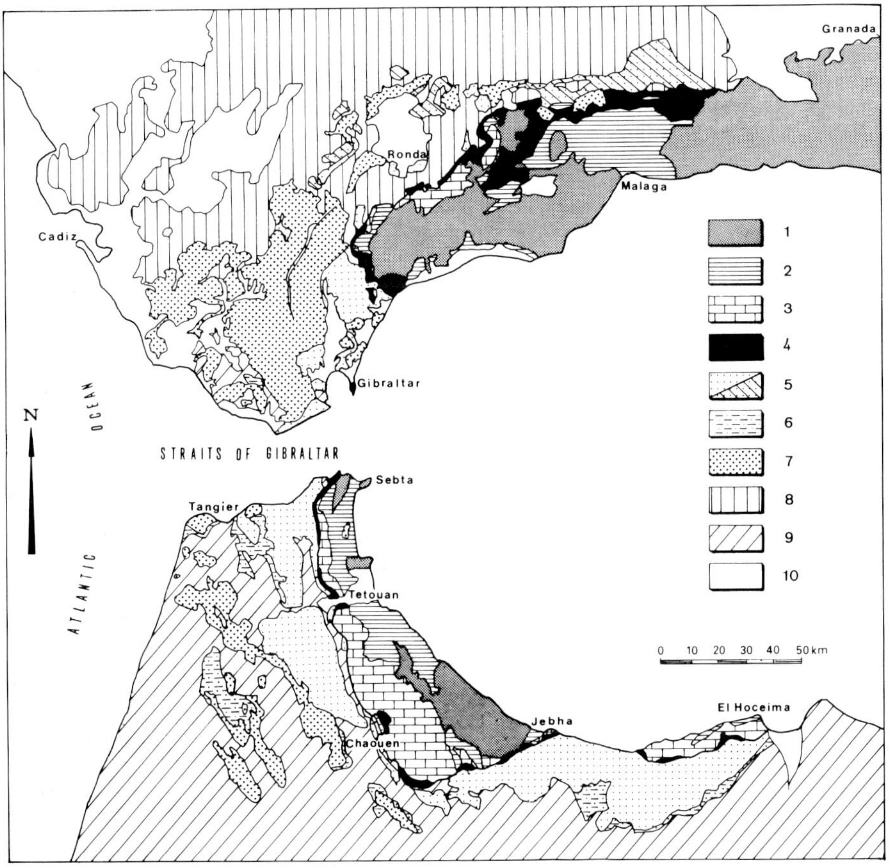

Fig. 2.- The Arc of Gibraltar. 1- Sebtides-Alpujarrides. 2- Ghomarides-Malaguides. 3- Dorsale calcaire. 4- Predorsalian zone. 5- Mauretanian flysch and its possible substratum. 6- Massylian flysch. 7- Numidian nappe. 8- External Betic zones. 9- External Rifian zones. 10- Post-orogenic formations.

A. -THE MAIN STRUCTURAL UNITS OF THE ALBORAN MARGIN.

1. *The Sebtides* . - These are the Rifian equivalents of the Betic Alpujarrides which, like them, contain mantle peridotites (whose conditions and age of emplacement are still uncertain), overlain

by a thick Precambrian(?)-Paleozoic series affected by several important Alpine and olders periods of metamorphism. This is overlain, with apparent stratigraphic continuity, by a cover of "Permo-Triassic" formations, exhibiting greenschist facies metamorphism and lying in nappes of Alpine age.

2. *The Ghomarides* . - Equivalent to the Betic Malaguides, the Ghomarides generally lie on the Sebtides. This relationship, likely due to overthrusting of the Ghomarides from S-W to N-E, becomes more complex towards the south-western and southern borders of the internal Rifian zones. The Ghomaride terranes, occurring in three main Alpine nappes, consist essentially of Paleozoic slates overlain disconformably by Triassic red sandstones and by a thin and discontinuous carbonate cover (fig. 3; fig. 4), mainly late Triassic and Liassic in age (Maaté, 1985). The youngest formations incorporated in the Ghomaride nappes are late Eocene in age (Olivier, 1978).

3. *The "Dorsale calcaire"* . - This borders the Ghomarides- Malaguides to the S, to the W and partly to the N. It is split into many subvertical slices (N of Tetouan) or into thrust sheets with an external vergence (S of Tetouan and in the Bokoyas; in the Betic Serrania de Ronda). The largest part is made up of Mesozoic carbonates locally overlain by Paleogene (up to Upper Oligocene) detrital formations (fig. 3; fig.4; fig. 5). It is generally detached at the level of the Triassic from its unknown but probably Ghomaride-type basement. Indeed, SE of Tetouan, the Triassic and Liassic carbonates of the most internal unit of the Dorsale (J. Gorgues unit) are a normal continuation of middle Triassic red sandstones (Baudelot *et al*, 1984) and of Ghomaride Carboniferous formations (Koudiat Tizian nappe). Thus we conclude that the Dorsale and the Ghomarides constitute a single large Alpine tectonic unit.

When going from the more internal to the more external parts of the Dorsale , a marked facies change can be observed, particularly in strata of Triassic-early Liassic, Malm and Eocene age : a shallow environment generally prevails in the " Dorsale interne", while a deeper and more pelagic environment characterizes the "Dorsale externe". Intermediate facies are known from the "Dorsale médiane".

4. *The "Predorsalian" units.* - The name Predorsalian was used (Durand-Delga, 1972; Didon *et al.*, 1973) for some tectonic slices with unusual stratigraphy (fig. 3; fig. 4; fig. 5), which are generally located at the contact between the internal and external zones, but which may also be thrust on more external units or back-thrust on the Ghomarides-Malaguides. The Predorsalian units form four groups (Olivier, 1984), that differ mainly in the facies of their Liassic and Cretaceous formations but also in the nature of their Oligo-Miocene sequences : going towards the external zones, these are Beni Derkoul, Jbel Moussa, Camarote and Argüelles groups. This subdivision has a paleogeographic significance in that the internal groups have the closest affinities to the calcareous Dorsale and the external groups to the flysch. Nevertheless these groups are not regularly distributed all along the Arc of Gibraltar and the complexity increases from the Rifian

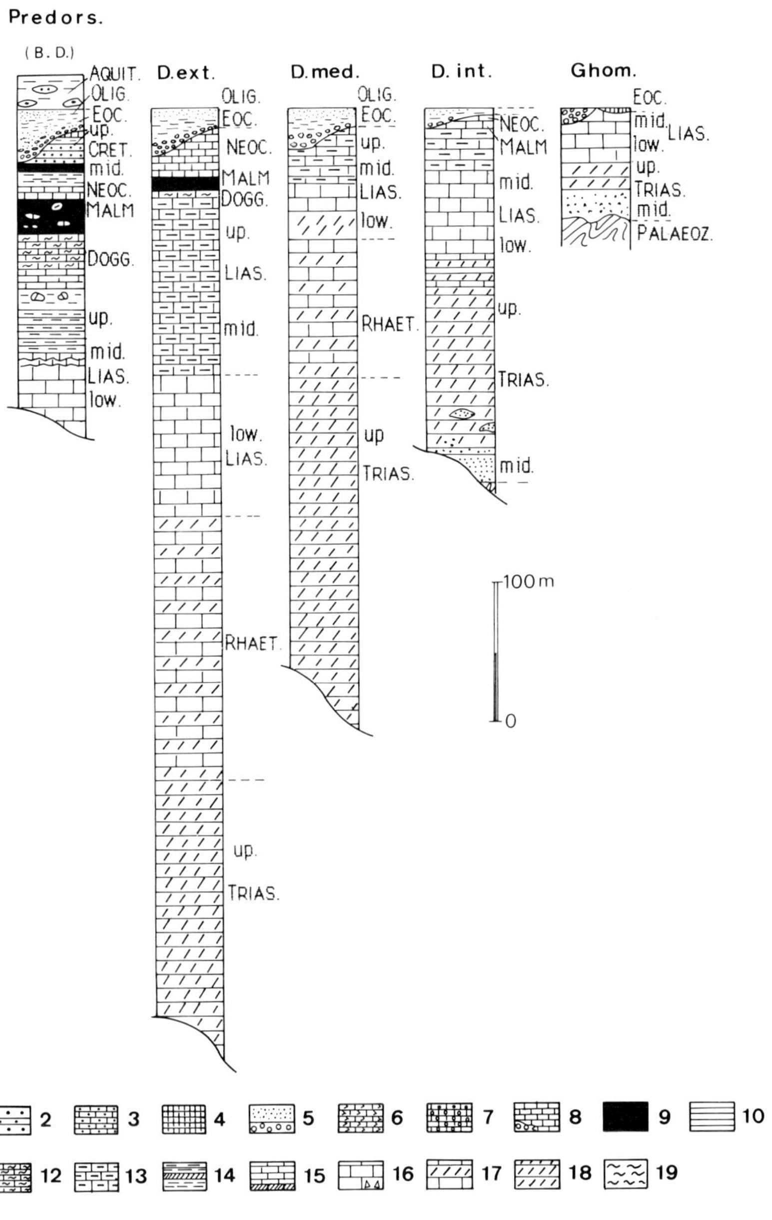

Fig. 3.- Stratigraphic columns at the boundary of the internal and external zones (south of Tetouan, Rif). Ghom. : Ghomarides; D. int. : Dorsale interne; D. méd. : Dorsale médiane; D. ext. : Dorsale externe; Predors. (B.D.) : Predorsalian (Beni Derkoul southern unit).

1. "Silexites"; 2. Numidian sandstones; 3. Gritty limestones; 4. *Microcodium* -bearing limestones; 5. Sandstones and conglomerates; 6. Calcarenites; 7. Oolithic limestones; 8. "Biancone"-type limestones; 9. Radiolarites and phtanites; 10. Sublithographic limestones; 11. Pseudo-knobly limestones; 12. "Filaments"-bearing limestones; 13. Cherty limestones; 14. Marls; 15. Marly limestones; 16. Massive limestones; 17. Alternation limestones/dolomites; 18. Dolomites; 19. Evaporitic rocks.

side of the Arc (one or two units on a single cross-section) to the Betic one (up to four units on a single cross-section).

In the Alboran margin the Predorsalian zone represents the slope, which was probably progressively wider to the west (before the formation of the Arc of Gibraltar) and which was located between the continental shelf (Dorsale and Ghomarides-Malaguides) and the flysch basin that existed from late Jurassic time onwards.

The Predorsalian series are highly deformed and commonly the argillaceous and marly terranes envelope slices of competent rocks, mainly limestones. Some authors interprete this as a large olisthostrome "éo-oligo-(burdigalien?)" (De Wever *et al.*, 1985) or Middle Burdigalian in age (Bourgois, 1978). In fact, the argillaceous and marly formations with intercalations of Numidian-type sandstones which overlie the Predorsalian series, nowhere contain incontestable olistholiths and are Aquitanian in age, i.e. coeval with the Numidian flysch s.s (Olivier, 1984).

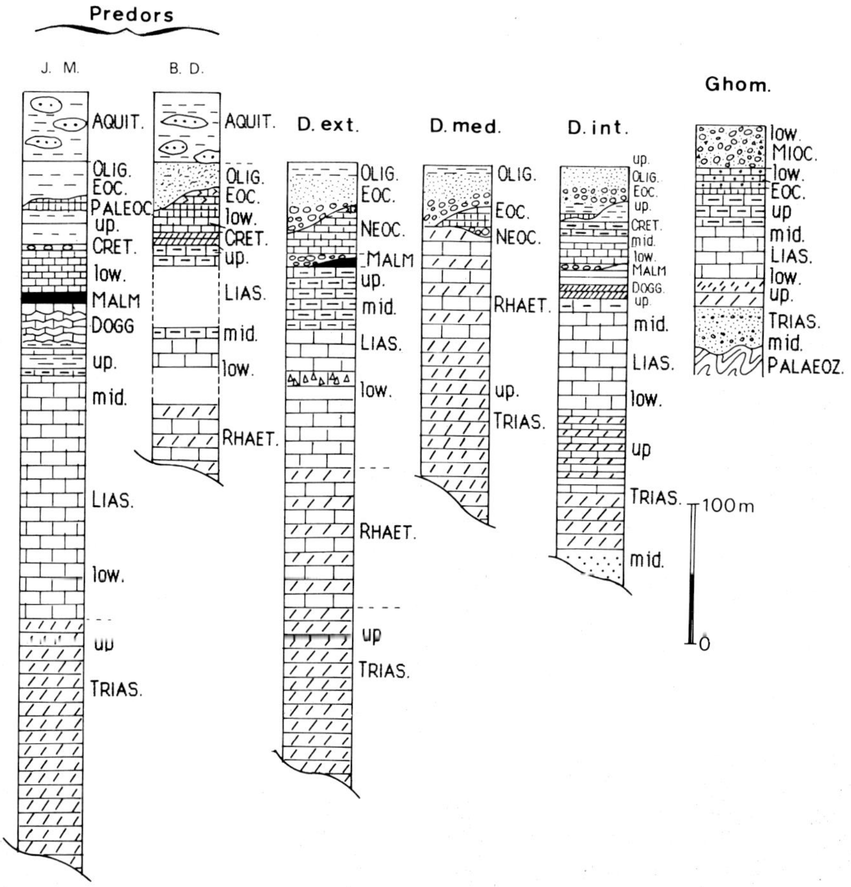

Fig. 4.- Stratigraphic columns at the boundary of the internal and external zones (north of Tetouan, Rif). See fig. 3; J.M. : Jbel Moussa unit.

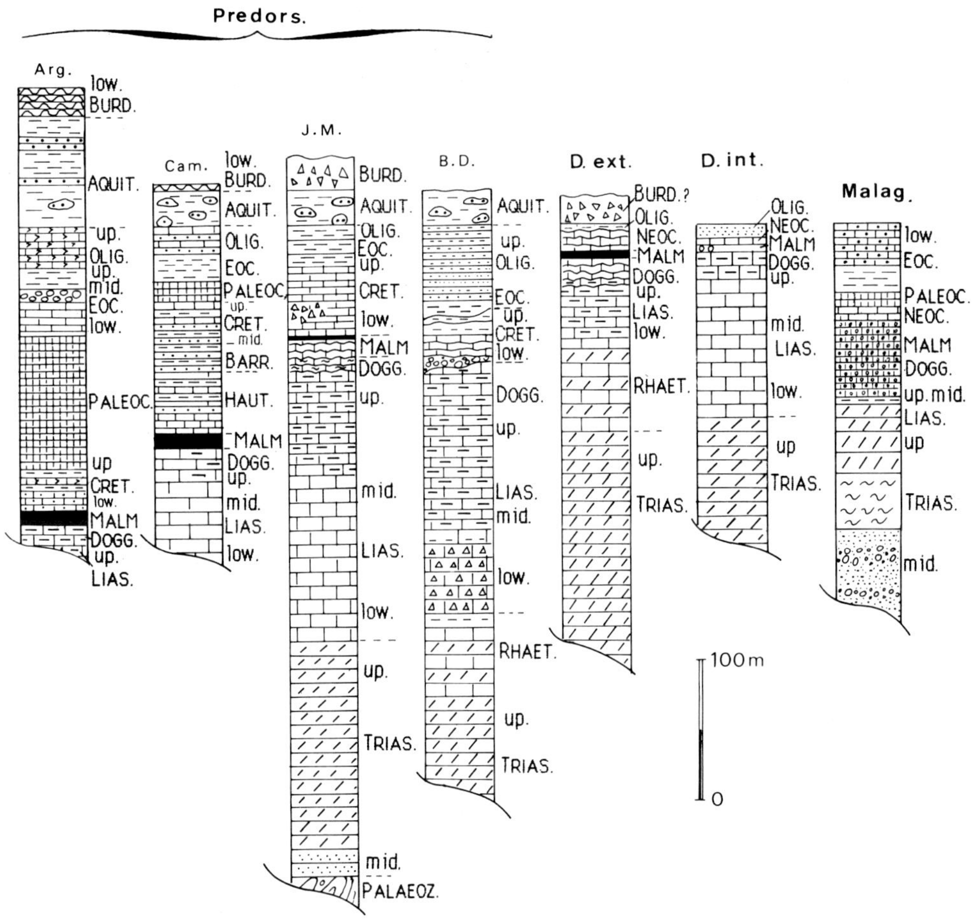

Fig. 5.- Stratigraphic columns at the boundary of the internal and external zones (in the Ronda area, Betic Cordillera). See fig. 3; Malag. : Malaguides; B.D. : Beni Derkoul group (Cañada Perella unit,....); J.M. : Jbel Moussa group (Tajo de Cabrilla, Los Enamorados, Robledo units); Cam. : Camarote group (Camarote and Espildora units); Arg. : Argüelles group (Argüelles unit, ...).

5. *Flysch* . - The structure of the Rifian flysch nappes changes when crossing the N70°E trending left lateral strike-slip Jebha fault. In the S.E., the Jbel Tisirene nappe, with several sub-units, is made up of thick Lower Cretaceous strata (Neocomian calcareous preflysch, Barremian to Albian sandstone-like and pelitic flysch, Gübeli, 1982; Thurow, 1987), locally overlain by a Cenomanian-Turonian carbon-rich siliceous formation (Olivier, 1978; Besson *et al.*, 1984). The youngest strata dip northward under the Predorsalian and Dorsalian units. This large nappe overlies the "Jbel Chouamat nappe", (made up of Aptian-Albian quartzitic-pelitic flysch, Cenomanian-Turonian phtanites and Upper Cretaceous calcareous flysch, Thurow, 1987) and both nappes overlie the external zones s.s., i.e. the "Ketama unit" in which Upper Cretaceous marls

succeed a thick Aptian-Albian pelitic and sandstone-like sequence.

North-west of the Jebha fault, where the curvature of Gibraltar begins, the flysch nappes are more complex. The Beni Ider nappe is characterized by its calcareous and marly flysch (Upper Cretaceous to Middle Eocene, Thurow, 1987) and by its thick (1 km) "gréso-micacé" (gritty-micaceous) flysch, mainly late Oligocene in age. It constitutes the stratigraphical complement of the mainly Lower Cretaceous Tisirene-type flysch (fig. 6).

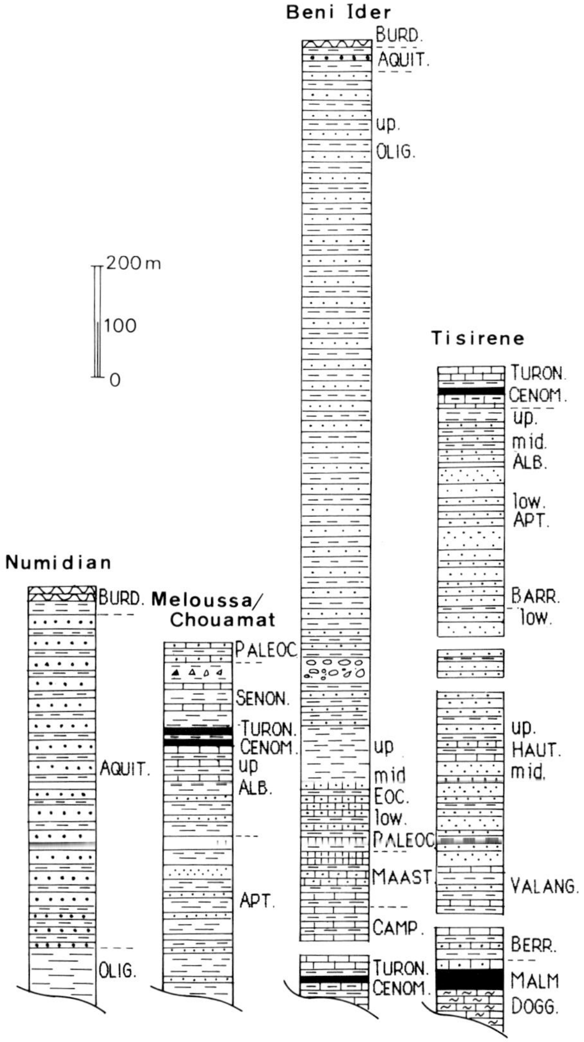

Fig. 6.- Stratigraphical columns of the Rifian flysch. (see fig. 3).

Under the Beni Ider nappe, east of Tangier, the Meloussa nappe is formed mainly by pelitic and quartzitic flysch, related to the J. Chouamat flysch but with fewer sandstones. To the east, near the Dorsale, Upper Cretaceous pelitic and calcareous flysch with a black phtanitic layer of Cenomanian to Turonian age appears as a window under the Beni Ider nappe : this flysch, regarded for a long time (Didon *et al.*, 1973) as the eastward equivalent ("Tanger oriental") of the coeval marly facies of the western part of the external Tangier unit, must in fact constitute a distinct unit which we shall name the "Beni Hassan nappe" (after the name of the ancient market of Souk el Arbâa des Beni Hassan, N. of Chaouen); this nappe probably lies directly on the parautochthonous Tangier unit.

From a paleogeographical point of view, using the terminology of equivalent units in Algeria (Bouillin *et al.*, 1970), we may group the J. Tisirene and Beni Ider nappes into an internal, (Upper Jurassic to Lower Burdigalian) Mauretanian ensemble (near the "Dorsale calcaire"), tectonically high in the flysch pile, and the J. Chouamat, Meloussa and Beni Hassan nappes into a more external (Lower Cretaceous to Paleocene), Massylian ensemble, that is in a lower tectonic position.

This second ensemble may originally have included the sequence of the Numidian nappe (Paleocene to Lower Burdigalian). Mainly formed by the Numidian sandstones , essentially Aquitanian in age, this nappe overlies the external zones s.s. principally the Tangier unit.

Lastly, a recent revision of the Tangier region and of the Campo de Gibraltar (Durand-Delga et Esteras, *in litt.*) confirmed our idea that these flysch units constitute true nappes (their incompetent horizons have commonly experienced superficial flow, thereby complicating their original structure) and are not olistholiths or blocks inserted in a Miocene matrix (Bourgois, 1978; Martín-Algarra, 1987) for which we have vainly searched.

B. - PRE-OROGENIC EVENTS FROM MESOZOIC TO PALEOGENE TIME.

As we have said before the assumptions underlying our palinspastic reconstruction are the following :

a) the Cretaceous and Paleogene flysch basins were located south of the continental ensemble we have named ALKAPECA.

b) as evidenced by the gradual evolution of the facies during the early Mesozoic, the Ghomaride-Malaguide domain was in a more internal (northern) position than the Dorsale; consequently, the Sebtide-Alpujarride domain was still more internal in the Betic-Rifian ensemble.

c) the Betic-Rifian ensemble was probably shifted a large distance to the west, in which case the "austro-alpine" type formations of the Sebtides- Alpujarrides and of the Dorsale calcaire were deposited in a region near present-day Tunisia.

We propose the following evolution for the Rifian and Betic internal zones.

1. *Triassic* - Conglomerates and continental red sandstones ("Verrucano", at least partly mid-Triassic in age) covered the ALKAPECA domain, especially the future Alboran Block. After a

short evaporitic event, the sea transgressed during the Middle Triassic (Anisian) in the Sebtide-Alpujarride domain and during the late Triassic in the Ghomaride-Malaguide, Dorsalian and Predorsalian domains . Thick dolomites and subsidiary limestones were then deposited in the Sebtide-Alpujarride domain due to a strong subsidence. One can interprete this as resulting from initiation of the south-European continental margin which ceased shortly afterwards since this large Sebtide-Alpujarride domain seems to have been emergent from the end of the Triassic until the Miocene (when post-nappe deposits were laid down).

Not long after, south of the Ghomaride-Malaguide and inner Dorsalian domain, emergent or at no great depth and hardly subsident, the domain of the Dorsale médiane and of the Dorsale externe began to subside as proved by the kilometric thickness of the Upper Triassic (?Norian-Rhetian) dolomites : from this time onwards , it is here that the Alboran part of the European continental margin evolved.

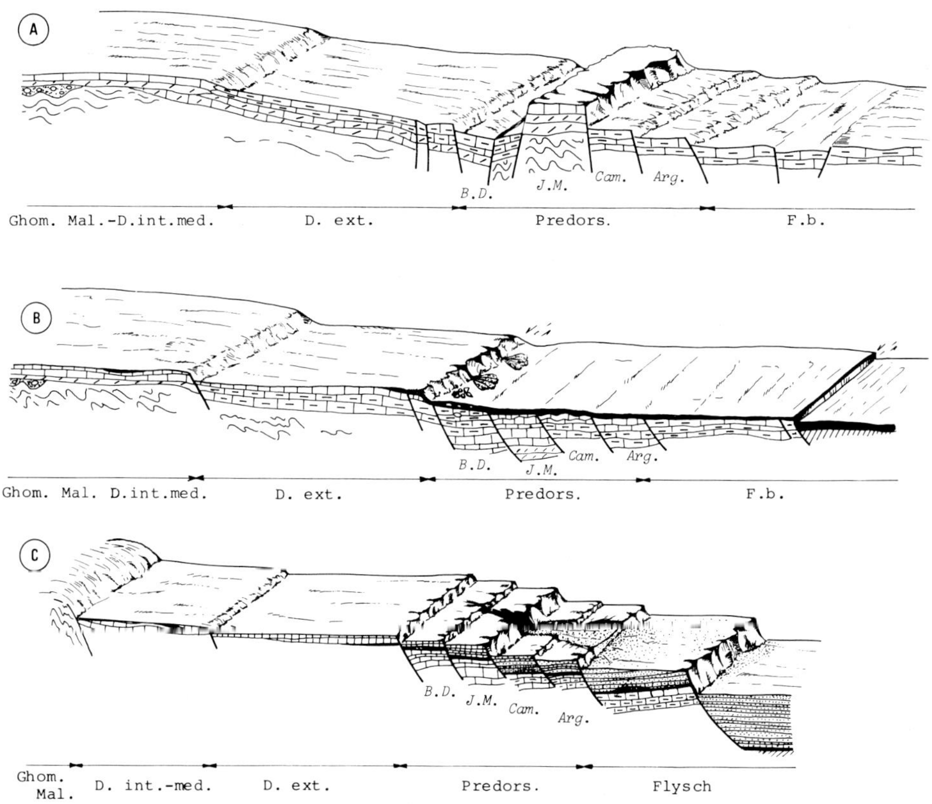

Fig. 7.- Paleogeographic reconstruction of the Alboran Block margin. (A): early to Middle Lias. (B) : Malm. (C) : early Cretaceous (Barremian). Ghom.-Mal. : Ghomaride-Malaguide domain; D. int. : Dorsale interne; D. méd. : Dorsale médiane; D. ext. : Dorsale externe; Predors. : Predorsalian domain (B.D. : Beni Derkoul group; J.M. : Jbel Moussa group; Cam. : Camarote group; Arg. : Argüelles group); F.b. : Flysch basement; Flysch : Flysch domain.

2. *Lower(-Middle) Lias* . - The environments became more truly marine. One can distinguish : a) a northern area (Ghomarides, Dorsale interne and Dorsale médiane) with inter- to supra-tidal, often brecciated, carbonates. b) a southern area (Dorsale externe and Predorsalian zone p.p.) with thicker inter- to sub-tidal limestones; the Sinemurian is characterized by pelagic influences (Ammonites) and by synsedimentary slides. Whithin the Predorsalian zone, the Jbel Moussa group is conspicuous with its inter- to supra-tidal well bedded white limestones, which probably correspond to a ridge developed mainly in the Betic part of the arc (fig. 7-A).

3. *Middle Lias (Domerian?) to latest Dogger* - Deepening continued and the sediments became really pelagic. Upper Lias and younger strata of the Dorsale contain many gaps and are not very thick whereas in the Predorsalian zone the marly limestones, with local intercalations of collapse breccias and blocks of Triassic and Liassic carbonates (Beni Derkoul), are generally thicker and more continuous.

4. *Latest Dogger to Kimmeridgian.* - Radiolarites, which seem to express a renewed deepening, appeared in most of the parts of the Dorsale externe, in all of the Predorsalian zone and what was to become the flysch zone (fig. 7-B), probably related to the local appearance of oceanic crust in latter which was just beginning to develop. Contemporeanously, in the inner part of the Dorsale, condensed series continued to accumulate. As regards the Ghomaride-Malaguide zone it seems it was only episodically and partially invaded by the sea.

5. *Late Malm-early Cretaceous.* - Sedimentation of "Biancone"- type micrites occurred in a large part of the Dorsale, giving way to a more marly calcareous facies in the most external Dorsale and in the internal Predorsalian domain (Beni Derkoul and J. Moussa groups). In the external part of the Predorsalian zone (especially Camarote), a coloured flyschoid facies ("Complexe à Aptychus") appeared during the Hauterivian but perhaps earlier, evoking the thick marly-calcareous pre-flysch (Berriasian to Hauterivian) at the base of the Mauretanian nappe of the J. Tisirene. This "Complex à Aptychus" crops out essentially in Andalusia but some remnants of it can be found in the Predorsalian zone of the Bokoyas (Rif). The post-Neocomian and pre-Cenomanian strata are scarce in the Dorsalian and internal Predorsalian domain where they are represented locally by thin dark marls (Upper Aptian, Upper Albian) that were perhaps deposited in perched basins (Olivier et Magné, 1985). Thus, this part of the margin may have been crossed by the detritus of probable internal origin that accumulated in the Mauretanian basin (about 1 km of thick-bedded, sub-arkosic sandstone-like turbidites (Chiocchini *et al.*, 1980; Gübeli, 1982; Thurow, 1987)(fig. 7-C). In the external part of the Predorsalian domain (Camarote) , above the "Complexe à Aptychus", were deposited some gritty-pelitic flysch-type strata of Barremian age.

6. *Latest early Cretaceous to late Cretaceous.* - During late Albian, Cenomanian and Turonian, sedimentation in the flysch zone was affected by a significant decrease in the amount of quartzous input. This "Cenomanian-Turonian boundary event" is characterized especially by

carbonate-detrital turbidites, siliceous layers and bituminous black-shales. Such formations are known from the Massylian domain (Thurow and Kuhnt, 1986), the Mauretanian domain (J. Tisirene nappe : Olivier, 1978; Besson *et al.*, 1984 / Beni Ider nappe : Thurow and Kuhnt, 1986) and probably in the Beni Derkoul unit of the Predorsalian zone (Nold *et al.*, 1981; Olivier, 1984).

During the Senonian, calcareous, marly or pelitic, coloured flysch was deposited in the Mauretanian and Predorsalian domains. On the other hand, in the Dorsalian domain, the Cenomanian to Senonian period is characterized by very scattered limestones and pelagic marls which do not differ significantly from the Lower Cretaceous strata in this domain. Nevertheless, especially in the Betic part of the chain, red marly limestones, closely resembling the coeval red beds of the external Betic zone (Penibetic), occur locally. This probably means that these realms, once widely separated, became much closer together at that time.

7. *Paleocene.* - The most striking facies is a calcareous turbidite with fragments of *Microcodium.* It is widely developed in the most internal (Mauretanian) part of the flysch zone and in the Predorsalian zone, where it overlies upper Cretaceous flysch. Thinner *Microcodium* -bearing limestones also appear in some parts of the Dorsale externe where they may directly overlie the lower Mesozoic strata. These *Microcodium* fragments probably originated in the Ghomaride-Malaguide domain, then emergent, where in situ *Microcodium* are known (Maaté, 1984).

8. *Early to Middle Eocene.* - Three facies zones may be distinguished. In the Ghomaride and internal Dorsalian domains, the sea transgressed beyond the limit of Mesozoic strata, and series of more or less diachronous limestones, rich in algae, large foraminifera and detritus from the surrounding regions were deposited. In the more external Dorsale domain, variegated marls with pelagic foraminifera and intercalations of large foraminifera-rich, turbiditic biocalcarenites and conglomerates of local or more internal origin were then deposited.

Finally, in the Mauretanian zone, a true marly-calcareous, coloured flysch, with many conglomeratic intercalations derived from the Dorsale domain, was laid down.

9. *Late Eocene and Oligocene.* - Extensional deformation characterizes this period; no important compressional events have been recorded. In the Dorsalian domain, the tilting of blocks created considerable topographic relief, perhaps as early as the Paleocene but essentially during the middle and late Eocene : the scarps of Triassic and Liassic carbonates produced thick conglomerates. These are intercalated in pink or yellowish planktonic marls. Light brown calcareous sandstones, sometimes with olistholiths in the Dorsale interne, and giving rise to thick turbiditic accumulations in the external Dorsalian and Predorsalian domains, were then deposited. In the Predorsalian group of Argüelles, the upper Oligocene is characterized by a calcarenitic flysch rich in large foraminifera.

Detritus in Oligocene strata came essentially from the emergent Ghomaride-Malaguide zone. In the Mauretanian basin, this detritus gave rise to the thick "gréso-micacée" formation of the Beni

Ider-Algeciras nappe. Erosion of the internal basement followed its gradual uplift that was related to emplacement of nappes in the most internal zones, especially in the Ghomaride-Malaguide domain, after late Eocene (Priabonian) and before latest Oligocene time.

C. - THE SYN-OROGENIC EVENTS OF THE EARLY MIOCENE.

1. *Late Oligocene and Aquitanian* . - At the end of the Oligocene, the ALKAPECA ensemble broke up into several blocks. The westernmost one was the Alboran Block (or Betic-Rifian Block) which was then bound to the North by the right lateral North Betic fault and to the south by the left lateral Jebha fault (Leblanc and Olivier, 1984).

At the same time nappes of external (southward and southwestward) vergence developed in the Dorsalian and Ghomaride domains. Related to this, stretching of continental crust occurred in the area of the present Alboran sea,and oceanic crust appeared north of the Kabylia (Rehault *et al.*, 1984), both events being associated with sinking leading to the first Mediterranean Sea. A progressively deepening sea then invaded the internal parts of ALKAPECA, covering the Rifian Ghomaride units with Chattian to Aquitanian deposits whose diversity (conglomerates, sandstones, marls) was related to the topographic relief of the domain invaded by the sea. These deposits formed the "Fnidek formation" , equivalent to the Kabylian "Oligo-Miocene" and to the "Stilo-Capo d'Orlando formation" of the Calabrian-Sicilian Arc. The Dorsalian domain, after it has been deformed, seems to have been almost completely emergent in the Arc of Gibraltar area. This relatively narrow and probably discontinuous ridge separated the new Mediterranean Sea from the basin which continued to exist in the more external parts of the Maghrebides (fig. 8-A) and to link the Atlantic ocean to the peri-Apennine sea.

In the Rifian part of this domain, south and south-west of the Dorsale, the Numidian flysch was deposited throughout the Aquitanian . The very mature quartzous detritus of this thick flysch cannot have originated from erosion of the exposed parts of the internal zones because at the same time these were receiving the very immature and polygenic strata discussed above (Durand-Delga, 1980).

The paleogeographic position of the Numidian, now completely allochthonous, is uncertain : we shall assume that it constituted the upper part of the Massylian flysch, detached at the level of the infra-Numidian coloured shales which are as old as the Paleocene or Eocene in the Campo de Gibraltar (Spain) and Oligocene in the Rif region.

Unlike North Algeria, where the Numidian was defined, there is in the Arc of Gibraltar a domain with peculiar Numidian-type strata which occur in the Predorsalian zone still active at this epoch. They are mainly brown, more or less marly shales with intercalations of Numidian sandstones the thickness of which increases towards the external zones (Argüelles). The area where these formations were deposited was connected with the Numidian s.s. basin but probably in rather a complex way. To understand the nature of this connection, more must be known about the evolution during Aquitanian time of the intervening Mauretanian domain . We know only, that in the external part of this domain the "Merinide" series, intercalations of Numidian sandstones in the "gréso-micacée" series (Didon et Hoyez, 1978) probably of latest Oligocene to earliest Aquitanian age, were deposited. In a more internal position, S.W. of Ksar Es Srhir, above the "gréso-micacée" Mauretanian sequence, there is a thin argillaceous and pelitic series with Numidian-type beds which have recently been dated as Aquitanian (unpublished). This attenuated series may have accumulated in a region between the Predorsalian and Massylian- Numidian (s.s.) domains.

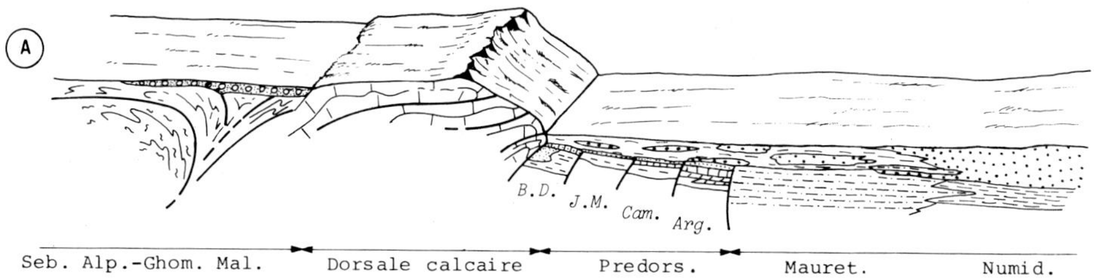

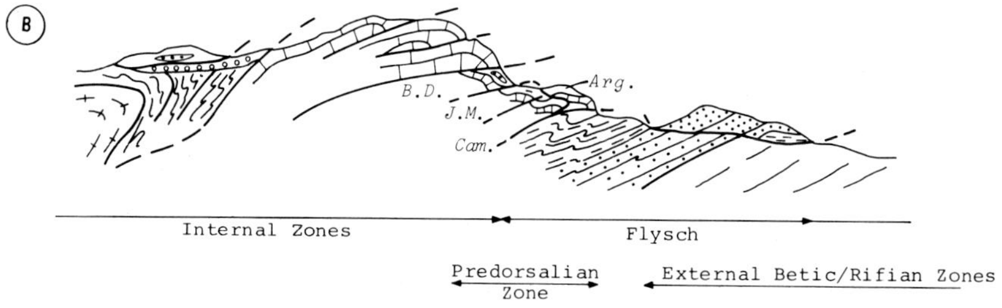

Fig. 8.- Paleogeographic reconstruction of the Alboran Block margin. (A) Aquitanian; (B) Middle to late Burdigalian. (see fig. 7).

2. *Burdigalian*. - A stoppage of the detrital inputs and a decrease of the subsidence then took place which were followed by the deposition in several domains of siliceous clays, generally associated with marls or radiolaria-rich clear shales . The total thickness is not over some tens of meters. This "silexites"-bearing formation occurs in scattered outcrops on one hand in the Rif at the top of the Numidian s.s. sequence and at the top of the Mauretanian succession S.W. of Ksar Es Srhir (Durand-Delga et Didon, 1984) and on the other hand in the Betic Cordillera at the top of Predorsalian series (Argüelles) and above the Malaguide Aquitanian sequence.

These layers have been dated as early to Middle Burdigalian. They represent the youngest, pre-thrusting Miocene strata known from the Betic-Rifian margin.

Immediately after the deposition of these beds, during Middle to late Burdigalian time, westward thrusting of the internal zones occurred. The thrust plane was located at the level of the

Predorsalian zone which was especially weak. Some Predorsalian strata and the Mauretanian and Massylian ensembles were in this way thrust toward the external domains of the Arc of Gibraltar and some elements of these units experienced back-thrusting onto the "Oligo-Miocene" cover of the Ghomarides and Malaguides (fig. 8-B).

By this time, the Betic-Rifian margin had acquired most of its structural features and after that, during the middle and upper Miocene, tectogenesis shifted towards the more external parts of the chain .

ACKNOWLEDGMENTS

We are very grateful to H. Charlesworth for reviewing the english version of this paper and to C. Cavaré-Hester for the drawings.

REFERENCES

Andrieux, J., Fontboté, J.-M., et Mattauer, M., 1971. Sur un modèle explicatif de l'Arc de Gibraltar. Earth Planet. Sci. Lett. , 12 (2) : 191-198.

Balanyá, J.C. et García-Dueñas, V., (in press): Les directions structurales dans le Domaine d'Alborán de part et d'autre du Détroit de Gibraltar. C. R. Acad. Sc. Paris.

Baudelot, S., Bouhdadi, S. et Durand-Delga, M., 1984 . Datation palynologique du Trias moyen au sein des Grès rouges "permo-triasiques" des environs de Tétouan (Rif septentrional, Maroc). C. R. Acad. Sc. Paris , 299-II (15) : 1061-1068.

Ben Yaich, A., 1981 . Etude géologique de la Dorsale calcaire entre Tlata Taghramt et Ben Younis (Haouz, Rif, Maroc). Thèse doct. 3e cycle , Univ. Rabat, 202 p.

Besson, F., Müller, C., Sigal, J., Faure-Muret, A., Choubert, G. et Frizon de Lamotte, D., 1984 . L' âge du flysch pélito-gréseux de l'édifice oriental des nappes du Rif (Maroc). C. R. Acad. Sc. Paris , 298-II (17) : 745-750.

Bouillin, J.-P., 1986 . Le "bassin maghrébin" : une ancienne limite entre l'Europe et l'Afrique à l' Ouest des Alpes. Bull. Soc. géol. France , 8-II (4) : 547-558.

Bouillin, J.-P., Durand-Delga, M., Gélard, J.P., Leikine, M., Raoult, J.F., Raymond, D., Tefiani, M. et Vila, J.M., 1970. Définition d'un flysch massylien et d'un flysch maurétanien au sein des flyschs allochtones de l'Algérie. C. R. Acad. Sc. Paris, 270-D : 2249-2252.

Bouillin, J.P., Durand-Delga, M. and Olivier, Ph., 1986 . Betic-Rifian and Tyrrhenian Arcs : distinctive features, genesis and development stages. *in* : Wezel F.-C., Edit., The Origin of Arcs - *Developments in tectonics*, 21, Elsevier, Amsterdam : 281-304.

Bourgois, J., 1978 . La transversale de Ronda (Cordillères bétiques, Espagne). Données géologiques pour un modèle d'évolution de l'Arc de Gibraltar. Ann. Sci. Univ. Besançon, 30, 445 p.

Chiocchini, U., Franchi, R., Guerrera, F., Ryan, W.B.F. e Vannucci, S., 1978. Geologia di alcune successioni torbiditiche cretaceo-terziarie appartenenti ai "Flysch maurétaniens" e alla "nappe numidienne" del Rif settentrionale (Marocco). Studi geol. Camerti, IV : 37-66.

Cuevas, J., Aldaya, F., Navarro-Vilá, F. et Tubía, J.M.,1986. Caractérisation de deux étapes de charriage principales dans les nappes Alpujarrides centrales (Cordillères bétiques, Espagne). C. R. Acad. Sc. Paris, 302-II (18): 1177-1180.

De Wever, P., Duée, G. et El Kadiri K., 1985. Les séries stratigraphiques des klippes de Chrafate (Rif septentrional, Maroc) témoins d'une marge continentale subsidente au cours du Jurassique-Crétacé. Bull. Soc. géol. France. 8-I (3) : 363-379.

Didon, J., 1969. Etude géologique du Campo de Gibraltar (Espagne méridionale). Thèse Sci., Univ. Paris VI, 3t., 539 p.

Didon, J., Durand-Delga, M. et Kornprobst, J., 1973. Homologies géologiques entre les deux rives du détroit de Gibraltar. Bull. Soc. géol. France. 7-XV (2) : 77-105.

Didon, J. et Hoyez, B., 1978. Les séries à faciès mixte, numidien et gréso-micacé, dans le Rif occidental (Maroc). C. R. somm. Soc. géol. France., 6 : 304-307.

Didon, J., Durand-Delga, M., Esteras, M., Feinberg, H., Magné J. et Suter, G., 1984. La formation des Grès numidiens de l'arc de Gibraltar s'intercale stratigraphiquement entre des argiles oligocènes et des marnes burdigaliennes. C. R. Acad. Sc. Paris, 299-II (3) : 121-128.

Durand-Delga, M., 1972. La courbure de Gibraltar, extrémité occidentale des chaînes alpines, unit l'Europe et l'Afrique. Eclog. Geol. Helv., 65 (2) : 267-278.

Durand-Delga, M., 1980. La Méditerranée occidentale : étapes de sa genèse et problèmes structuraux liés à celle-ci. Soc. géol. France, Mém. h.-sér. (10) : 203-224.

Durand-Delga, M. et Didon, J., 1984. Carte géologique au 1/50 000e de Ksar es Srhir (Rif, Maroc). Notes et Mém. Serv. géol. Maroc, 295.

El Kadiri, K., 1984. Les radiolarites jurassiques des klippes de Chrafate (Rif septentrional-Maroc) : stratigraphie, taxonomie. Thèse 3e cycle, Univ. Pau, 2 t., 460 p.

Feinberg, H. et Olivier, Ph., 1983. Datation de termes aquitaniens et burdigaliens dans la zone prédorsalienne bético-rifaine et ses conséquences. C. R. Acad. Sc. Paris, 296-II : 473-476.

Gübeli, A.A., 1982. Stratigraphische und sedimentologische Untersuchung der detritischen Unterkreide-Serien des zentralen Rif (Marokko). Inaug.-Dissertation, Univ. Zürich, 190 p.

Kornprobst, J., Raoult, J.F., Griffon, J.C. et Mouterde, R., 1966. Etudes géologiques sur la chaîne du Rif.(I). Notes et Mém. Serv. géol. Maroc, 184, 243 p.

Kuhnt, W., 1985. Biostratigraphie und Paläoenvironment der externen Kreideserien des westlichen Rif und Betikum : Ein Ansatz zur Rekonstruktion der Kreide-Paläogeographie des Gibraltarbogens. Dissertation, Univ. Tübingen. (unpublished).

Leblanc, D. and Olivier, Ph., 1984. Role of strike-slip faults in the betic-rifian orogeny. Tectonophysics, 101 : 345-355.

Leikine, M., Gutnic, M. et Mégard, F., 1969. Etudes géologiques sur la chaine du Rif.(II). Notes et Mém. Serv. géol. Maroc, 194, 198 p.

Maaté, A., 1984. Etude géologique de la couverture mésozoique et cénozoique des unités ghomarides au Nord de Tétouan (Rif interne, Maroc). Thèse 3e cycle, Univ. Toulouse, 161 p.

Martín-Algarra, A., 1987. Evolución geológica alpina del contacto entre las Zonas Internas y las Zonas Externas de la Cordillera bética . Tesis, Univ. Granada, 1271 p.

Nold, M., Uttinger, J. et Wildi, W., 1981. Géologie de la Dorsale calcaire entre Tétouan et Assifane (Rif interne, Maroc). Notes et Mém. Serv. géol. Maroc, 300, 233 p.

Olivier, Ph., 1978. Etude géologique et structurale de la région de Jebha (Rif, Maroc). La terminaison N.E. de l'Accident de Jebha-Chrafate. Thèse 3e cycle, Univ. Toulouse, 129 p.

Olivier, Ph., 1984. Evolution de la limite entre zones internes et zones externes dans l'Arc de Gibraltar (Maroc-Espagne). Thèse Sci., Univ. Toulouse, 229 p.

Olivier, Ph. et Magné, J., 1985. Témoins de la sédimentation aptienne et albienne à la bordure des Zones internes bético-rifaines (Arc de Gibraltar). C. R. Acad. Sc. Paris, 301-II (5) : 329-334.

Rehault, J.P., Boillot, G. and Mauffret, A., 1984. The western Mediterranean basin geological evolution. Marine Geology, 55 : 447-477.

Thurow, J. , 1987. Die kretazischen Turbiditserien im Gibraltarbogen : Bindeglied zwischen atlantischer und tethyaler Entwicklung. Dissertation, Univ. Tübingen, 495 p.

Thurow, J. and Kuhnt, W., 1986. Mid-Cretaceous of the Gibraltar Arch area. *in* : Summerhayes, C.P. & Shackleton, N.J., edit.- North Atlantic Palaeoceanography, Geol. Soc. spec. pub. 22 : 423-445.

Wildi, W., 1983. La Chaîne tello-rifaine (Algérie, Maroc, Tunisie) : structure, stratigraphie et évolution du Trias au Miocène. Rev. Géol. dyn. Géogr. phys., 24 (3) : 201-297.

GEODYNAMIC EVOLUTION OF THE ATLAS SYSTEM (MOROCCO) IN POST-PALAEOZOIC TIMES

Volker Jacobshagen[+], Konrad Görler[+] and Peter Giese[++]
Freie Universität Berlin, D-1000 Berlin 33
[+]Institut für Geologie, Altensteinstr. 34 A
[++]Institut für Geophysikalische Wissenschaften, Rheinbabenallee 49

ABSTRACT

The Atlas system of Morocco comprises the Mediterranean Rif orogen and three foreland intracontinental ranges: Middle, High and Anti-Atlas, which behave as mobile zones within the stable platform. For this region, a review is given on the major structural elements, the sedimentary and volcanic evolution from Triassic to recent, the crustal and upper mantle structure and the seismicity. Based on these data the following geodynamic model has been derived: The break-up of Pangaea and the subsequent formation of the North Atlantic and the western Tethys affected to some extent the northwestern margin of the African continent. During the Triassic-Dogger interval, rift grabens, pull-apart basins and accompanying volcanism are features of these moderate mobilizations. From Upper Jurassic onward the rifting processes slowed down and ceased during the Eocene. The accompanying magmatism changed from basaltic effusions to alcaline intrusions, thus indicating a change from a tensional to a compressional stress-field. The third period of the evolution, from Oligocene to recent, was governed by compressional movements between Africa and Europe, culminating in collisional structures of the Betic Cordilleras and the Rif. Compression and uplift of the Rif Atlas and of the three intracontinental belts took place within the same time intervals, thus implying a common relationship to certain plate tectonic events. As evidence by magnetotelluric and seismic studies intracrustal shearing and detachment have been facilitated by the existence of intracrustal high conductivity and low velocity layers, caused by Mesozoic and Tertiary thermal activities. A mechanism of transferring compressional stress from the active margin to its foreland is provided by the BALLY model of thin-skinned thrusting. The combination of these principles with the structural situation in northern Africa suggests the formation of an inclined thrust plain structure, which has penetrated the sedimentary cover and deeper crustal complexes.

A stronger compression had taken place within the fractured and weakened former rift zones, causing the observed young uplift of the ranges. A final stage of collision is marked by the NE/SW striking Transalboran fault zone, extending over a length of 1100 km from southeastern Spain across the Alboran Sea and along the Middle Atlas to Agadir.

1. Introduction

The Atlas system of Morocco consists of four mountain belts: Rif, Middle Atlas, High Atlas and Anti-Atlas. The geotectonic situation, structural evolution and the sequences of deformed rocks, show considerable differences. But all of them have developed to high mountains since the middle of the Tertiary. Therefore most geologists believe that there are close connections in their structural history. Geodynamic interpretations of the Atlas system differ, however, widely due to many gaps in our geological knowledge and, even more, due to the present state of geophysical investigations.

The Rif Atlas and its eastern prolongation, the Tell Atlas, are parts of a normal orogen of the Mediterranean type called the Maghrebides, that displays nappe structures facing towards the south. The structure and geological evolution of the Rif have been summarized by FAURE-MURET and CHOUBERT (1971), CHOUBERT and FAURE-MURET (1974), MICHARD (1976), DURAND-DELGA and FONTBOTE (1980) and by DURAND-DELGA and OLIVIER (this vol.). BOUILLIN et al. (1986) were of the opinion that the Palaeozoic and Mesozoic rocks of the Rif originated on a continental fragment called ALKAPECA which was originated situated far to the northeast. During the Lower Tertiary this fragment was transported to its present position by transcurrent movements and, from the Oligocene onward, involved in West Mediterranean collisions. The age of the overthrusts within the Rif differs according to the stacking order of the nappes: The upper nappes, now occupying an internal position along the Mediterranean coasts, were overthrust during the Oligocene, whereas the lower ones of the external zones developed during Miocene phases. The northern margin of the African plate is buried beneath the Rif orogen.

Middle and High Atlas are similar in structure and geological history. Recent overviews on the High Atlas have been given by CHOUBERT and FAURE-MURET (1962), FAURE-MURET and CHOUBERT (1971) and MICHARD (1976). The post-Palaeozoic history has been summarized e.g. by STETS and WURSTER (1981), JACOBSHAGEN (1986) and JACOBSHAGEN et al. (this vol.). For the Middle Atlas, DRESNAY presents a first review in this volume. Both ranges seem to have developed from Mesozoic rift grabens and pull-apart basins in a mobile zone at the northern margin of the African plate, which was the foreland of the Rif. Fracturation and subsidence of these grabens and basins was accompanied by magmatic activity, mainly by the effusion of basaltic rocks, up to the Middle Jurassic. A second period of calm subsidence in Cretaceous times was followed by inversion which may have started locally in the Upper Cretaceous (FROITZHEIM 1984, FROITZHEIM et al., this vol.), but generally took place from Oligocene or even Lower Miocene onward (GÖRLER et al., this vol.).

Geological literature on the Anti-Atlas is more or less focussed on its Precambrian and Palaeozoic history (CHOUBERT and FAURE-MURET 1971, MICHARD 1976, WALLBRECHER, this vol.). Its post-Palaeozoic evolution is poorly known. Early Mesozoic fracturation is testified by the presence of basaltic dykes (e.g. grand dyke de Foum Zguid, LEBLANC 1973). Further evidence on fractural events has been given by HEINITZ (1984). Finally, we can conclude from the influxe of Anti-Atlas detritus within the sedimentary cover at the High Atlas and within the adjacent Upper Cenozoic basins, that several important periods, uplift took place during the Mesozoic and Tertiary. The total thickness of rocks eroded on top of the crystalline domes of the Central Anti-Atlas is estimated to 8-10 km. The present height of the mountain range originated, however, by simple updoming of the Precambrian and Lower Palaeozoic rocks of the Sahara craton. This happened from the uppermost Tertiary onward, as documented by relics of a Neogene basin now being elevated more than 2000 m N.N. (GÖRLER et al. 1987 and this vol.).

In recent years, several geodynamic models have been proposed to explain structure and origin of the High and the Middle Atlas (MICHARD et al. 1975, MICHARD 1976, MATTAUER et al. 1977, STETS and WURSTER 1981, COURBOULEIX et al. 1982). The uplift of the Anti-Atlas, however, has not been included in these considerations. All hypotheses are based on the idea that the post-Palaeozoic structural development of these ranges was strongly influenced by reactivation of major pre-existing Palaeozoic or even Proterozoic fractures within the basement. Most of these authors agree, furthermore, that lateral displacement along these faults played a key role in the Mesozoic and Cenozoic evolution of Middle and High Atlas.

Looking for a general geotectonic model to explain the origin of the whole ensemble of the Atlas system, MATTAUER et al. (1977) have envisaged that the regional stress field of NW Africa had changed its orientation in the course of Mesozoic and Cenozoic plate movements. Thus, Mesozoic opening of the Atlantic Ocean would have caused the formation of rift grabens and pull-apart basins in the mobile zone north of the Sahara craton. These structures were compressed and uplifted in connection with West Mediterranean collisions in Cenozoic times. Contrary to this view, the model of STETS and WURSTER (1981) referred only to Atlantic opening, but not to Mediterranean influences. An explanation for the inversion, of the High Atlas has not been offered. Recently, however, the same authors have also considered a transpression model (FROITZHEIM et al., this vol.) which necessarily implies Mediterranean influence.

The aim of this paper is, to present a new model of the geodynamic evolution of the Atlas system of Morocco in Mesozoic and Cenozoic times. This model has to be based upon a discussion of the recent geodynamic situation of the Atlas system's palaeotectonic development since the early Mesozoic.

2. Recent geotectonic situation

From a geotectonic point of view, Morocco south of the Rif (which will not be discussed here) can be subdivided into three stable regimes (fig. 1):

- The Moroccan Meseta, including parts of the western High Atlas;
- the High Plateau of Oran;
- the Sahara craton with the Anti-Atlas as its updomed northern margin.

These regimes are separated by two fractural zones which predominantly consist of faults striking to the NE and to the ENE, respectively. A major zone, slightly sigmoidal-shaped, crosses Morocco in a northeasterly direction from the Souss basin south of Agadir through the "massif ancien" and the western parts of the central High Atlas and along the Middle Atlas to the Mediterranean coast in the vicinity of Melilla. The second zone runs in an easterly direction from Jebilet and along the central High Atlas. In the eastern High Atlas it branches to the NE and ends within the High Plateau of Oran.

Although one has to, where tectonic character and activity periods of faults are concerned, envisage these fractural zones as somewhat heterogeneous, it is evident that the NE striking zone is, on the whole, still active. This is due to the fact that a linear arrangement of recent earthquake-hypocenters coincides with this fault zone (fig. 2, see also SAADI 1982, map 21, SCHWARZ and WIGGER, this vol.). Several

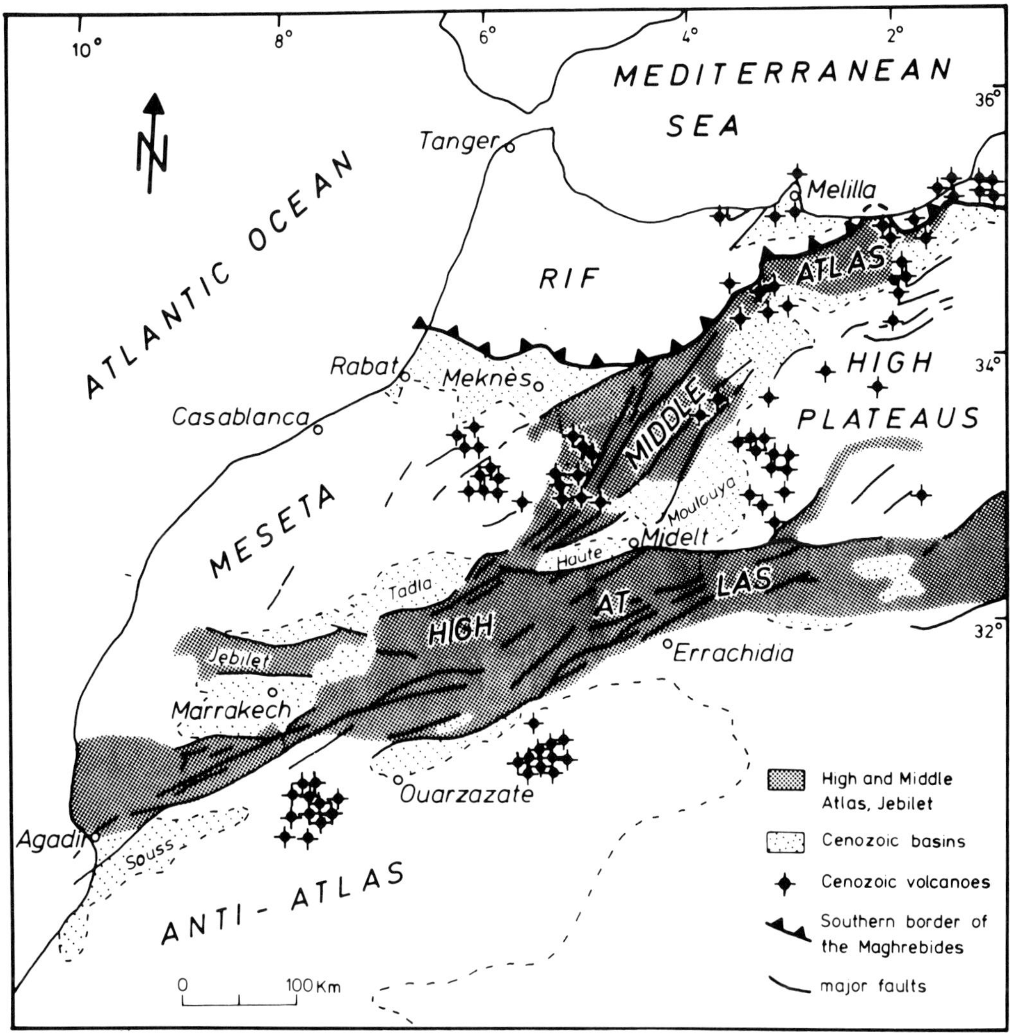

Fig. 1: Geotectonic sketch-map of the Atlas system, based upon MICHARD (1976), JENNY (1983), DRESNAY (this vol.) and JACOBSHAGEN et al. (this vol.).

of them were located in a depth of 100-200 km. Along the E/W-striking zone, however, recent earthquakes happened only within the central High Atlas, near the crossing of both zones.

Another region of high seismic activity is situated in the north, comprising the Alboran Sea, the Rif and the Upper Cenozoic molasse basin south of the Rif.

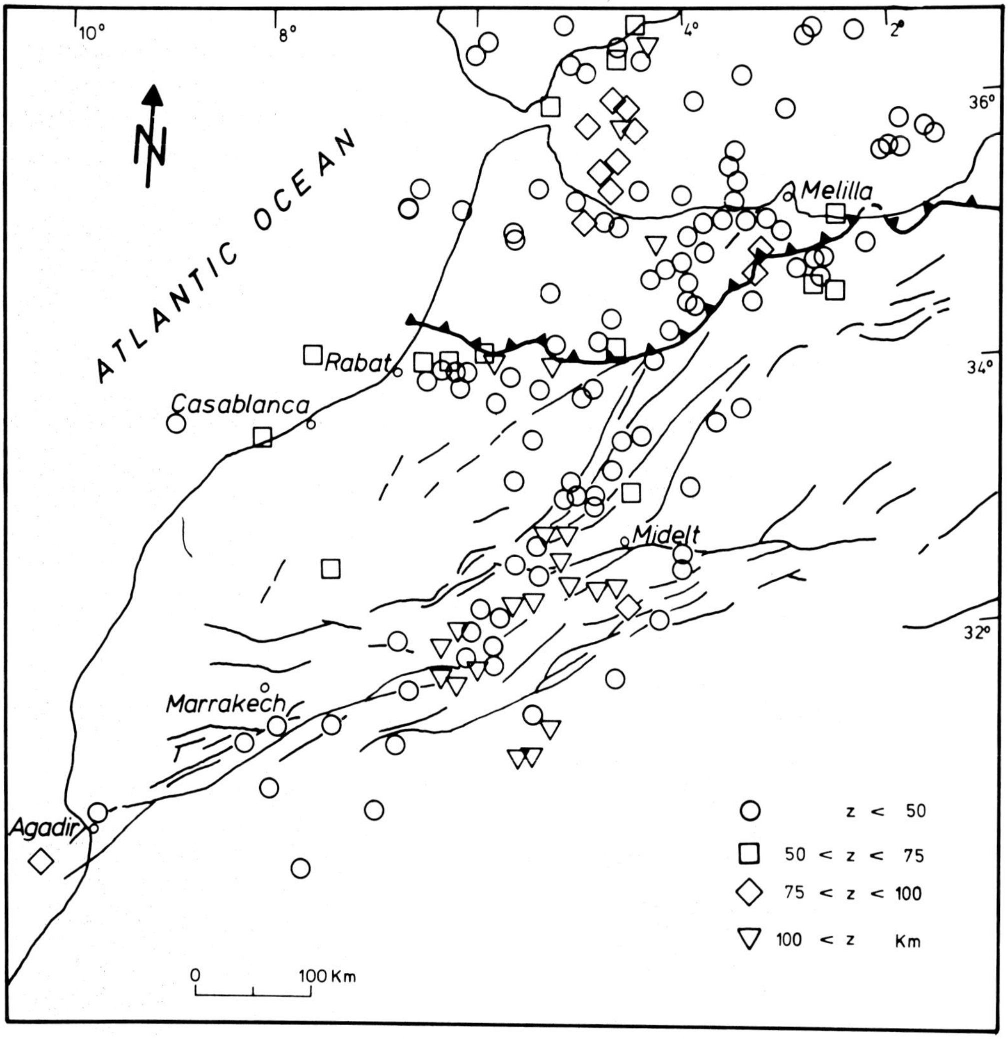

Fig. 2: Distribution of earthquake hypocenters in relation to the fractural pattern of the Atlas system (data from HATZFELD and FROGNEUX 1981).

The ranges of the High and the Middle Atlas coincide roughly, but not exactly with the fractural zones. Modern authors agree that these mountain ranges developed from Mesozoic grabens and basins by inversion. Both are accompanied by Upper Cenozoic basins, and this may give a hint to the age of inversion - but not to its cause.

To the major features of the young geodynamic evolution of Morocco belongs a wide-spread volcanic activity within the last 14 m.y. Recent publications have presented geochemical and geochronological data mainly for the northeastern part of Morocco (BELLON et al. 1984, HERNANDEZ and BELLON 1985, HARMAND and MOUKADIRI 1986, HERNANDEZ et al. 1987). Up to now, the results do not allow a general interpretation. In the eastern Rif and in the northern Middle Atlas, calcalcaline eruptions occurred from 14-8 m.y. They were followed by shoshonites and alkalibasalts, the latter reaching up to 2 m.y. In the southern part of the Middle Atlas two periods of volcanic activity can be distinguished (14-10 m.y. and 1,8-0,5 m.y.), both being of alcaline character. The volcanic centers of the Anti-Atlas (J. Sirwa, J. Sarrho, Foum el Kouss) are still under investigation; publications of J. Schermerhorn and Ch. Grothe (Berlin) will be published in the near future. From the geochemical character we may deduce an intracontinental volcanism which was caused by thermal activation of the upper mantle: only the calcalcaline eruptions in the north point to a subduction-related magmatism.

The regional distribution of the young volcanic centers is not strictly bound to distinct faults, but they are situated within or adjacent to the NE/SW-striking fault zone (fig. 1). This fits exactly with the ideas of TORRES-ROLDANI et al.(1986) and HERNANDEZ et al. (1987), who have found the Upper Cenozoic volcanism to be connected with a SW striking Transalboran shear zone they could follow from southeastern Spain across the Alboran Sea to the eastern Rif. This shear zone can be prolongated, now, to the Atlantic coast of Africa at Agadir, i.e. to an extension of about 1100 km.

3. Structure of crust and upper mantle

SCHWARZ and WIGGER (this vol.), have presented an excellent review of the crustal and upper mantle structure between Rif and Anti-Atlas.

Where the Bouguer gravity is concerned two extended negative anomalies with values down to -150 mgal have been observed beneath the Atlas system, one being correlated to the Rif mountains and the other one to the Middle and the central High Atlas (BOSCH 1971). From gravity and crustal seismic measurements an average crustal thickness of about 30 km could be derived for Morocco. Only beneath the High and the Middle Atlas may it increase to about 35 km, and such values seem to be valid also for the Rif (DURAND-DELGA and FONTBOTE 1980). Moreover, WIGGER and HARDER (1986) have detected a low velocity zone in the lower crust of the Middle and the High Atlas. The same authors also have found reduced velocities of 7.5-8.0 km/s for the upper mantle which points again to high thermal activity.

Magnetotelluric and geomagnetic depth soundings have led to the detection of a zone of high electric conductivity at mid crustal depths from the High Atlas towards the Middle Atlas (SCHWARZ et al. 1986). Comparable anomalies have been found at the southern border of the High Atlas (MENVIELLE and LE MOUEL 1985) and at the southern border of the Rif, where it dips to the N (MENVIELLE and ROSSIGNOL 1982). SCHWARZ and WIGGER have speculated that these anomalies could be caused by fluids which might have been activated by deep-seated basic intrusions. Therefore, the origin of the induction anomalies as well as of the low velocity zone is explained by small amounts of melt in the middle and lower crust.

4. Tectonic and magmatic evolution of the post-Palaeozoic history

As mentioned above, the pre-Oligocene rocks of the Rif had originated far from their present position, and the post-Palaeozoic development of the Anti-Atlas is poorly documented. Thus, the key areas in deciphering the geodynamic evolution are the sedimentary cover of the High and the Middle Atlas and the Neogene basins adjacent to these ranges. We shall, therefore, now examine their tectonic and magmatic history in Mesozoic and Cenozoic times.

During the Triassic, the northwestern edge of the later African plate was strongly influenced by two geotectonic events: the formation of the Mid-Atlantic rift zone and the expansion of the Tethys to central and western Mediterranean areas. Both events caused active fault zones and fault-bounded basins in central and eastern Morocco (MANSPEIZER 1982; LORENZ, this vol.). Consequently, the prevailing fault directions were called Atlantic (about NE/SW) or Mediterranean (about E/W) by some authors (KANES et al. 1974, MICHARD 1976). During the Upper Triassic, localized subsidence of the basins decreased, but a more regional subsidence began causing transgressions from the expanding Tethys ocean and from the just originating Atlantic (LORENZ, this vol.). In early Jurassic times, large areas were already flooded by seas. Fig. 3 shows a large Tethys embayment ("Atlas Gulf") which covered the Middle Atlas with the northeastern Meseta, the central and eastern

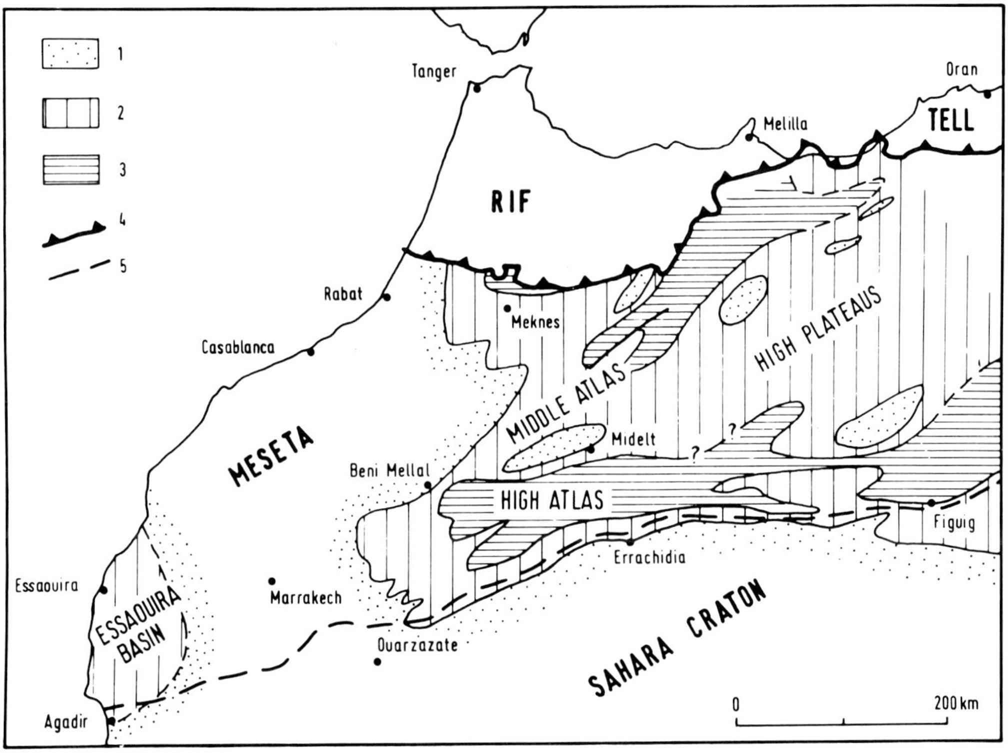

Fig. 3: Major features of Middle Liassic palaeogeography of Morocco (simplified according to DRESNAY 1971). 1 continental, 2 shallow marine, 3 deeper basins, 4 southern border of the Maghrebides, 5 southern border of the High Atlas.

Fig. 4: Mesozoic transgressions and regressions a) of the world (according to HAQ et al. 1986), b) of the Essaouira basin (data from HOSSNER 1985) and c) of the Atlas Gulf (Middle Atlas and central High Atlas).

The two Moroccan regions show a different evolution from the Dogger onwards.

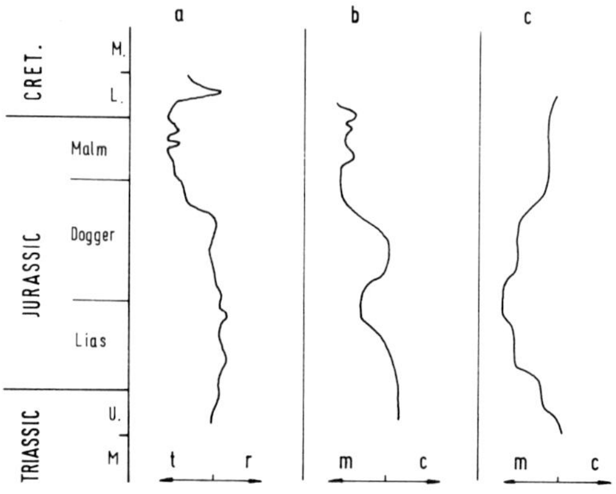

High Atlas and the High Plateaus. Facies distribution and local highs without sedimentation reflect, however, the inherited structures of the Triassic. The relatively small Essaouira basin, which now is part of the western High Atlas, was open to the Atlantic. According to the majority of recent publications, both basins were separated by a small land bridge between the Meseta and the Sahara craton in the region of the "massif ancien". But an interpretation also exists which claims a connection between the "Atlas Gulf" and the Essaouira basin throughout the whole Mesozoic period (STETS and WURSTER 1981, FROITZHEIM et al., this vol.). This opinion contradicts, however, the fracies distribution of the marine Jurassic in the central High Atlas (see e.g. BRECHBÜHLER et al., JACOBSHAGEN et al. and WARME, this vol.) as well as biogeographic arguments. We may, in addition, consider the different tectonic development of both basins from the Middle Jurassic onward (fig. 4): The Essaouira basin shows, in accordance with the VAIL curve, an Upper Jurassic to Lower Cretaceous transgression, whereas in the Atlas Gulf continental influence increased from the Bathonian upward (JENNY et al. 1981, DRESNAY, this vol., and JACOBSHAGEN et al., this vol.). In reference to the recent geodynamic pattern described above, most of the Essaouira basin is today part of the Meseta block. Perhaps, this may also explain the very low degree of compressional deformation in the western High Atlas.

It has been demonstrated by many authors that Lower and Middle Jurassic subsidence and sedimentation in the Atlas Gulf were accompanied by fracturing and by vertical and lateral displacement along inherited faults (reviews in DRESNAY, this vol., and JACOBSHAGEN et al., this vol.). Compressional structures of Upper Liassic or Dogger age are confined to narrow zones and may, consequently, be explained by local block rotations within a regime of distention and general subsidence. However, this tectonic period ceased in the Atlas Gulf during the Upper Dogger, and was followed by wide-spanned, but moderate subsidence throughout Upper Jurassic and Cretaceous times up to the end of the Eocene.

Mesozoic and Lower Tertiary distension and subsidence were associated with magmatic activity, but up to now we still have uncomplete knowledge of the number, the age and the geochemical character of the magmatic events. A Late Triassic dolerite formation is known from the whole region south of the Rif (MANSPEIZER 1981, SALVAN 1984, LORENZ, this vol.) and further basaltic eruptions or dykes, probably being of Upper Liassic, Dogger and Lower Cretaceous age, were reported from the High Atlas (see JACOBSHAGEN et al., this vol.). The frequency of these eruptions decreased with time. All authors agree that they were relics of an aborted Mesozoic rifting. Besides that, intrusions of different petrological and geochemical character and with ages between Dogger and Eocene are known in the central and eastern High Atlas. They show an alcaline trend (e.g. CHEVREMONT 1972, TISSERANT et al. 1976), thus indicating a change to a more continental development. At present, however, these intrusions can not be definitely connected with certain geotectonic events, due to the lack of reliable geochronological and geochemical data.

The second period of the post-Palaeozoic evolution is marked by the uplift of the mountain ranges of the Atlas system. This means an inversion of the previous graben structures for the Middle and the High Atlas.

According to most authors, these processes were, to a certain degree, connected with compressional deformation, probably caused by Mediterranean collisions. At present, there is a vivid discussion about the age and the geodynamic conditions which led to uplift and inversion. On account of the reasons mentioned above, we preferably consider the High and the Middle Atlas, again.

In the central and eastern High Atlas and in the Middle Atlas, inversion did not start earlier than the Oligocene. This is testified by a great number of sections that show continuous sedimentation from the Cenomanian upwards. Therefore, we do not agree with FROITZHEIM et al. (this vol.), who, due to an angular discontinuity south of Marrakech, have placed the beginning of inversion in the Upper Cretaceous. These outcrops seem to be of a limited importance (JACOBSHAGEN et al., this vol.).

The history of uplift of the High and the Middle Atlas is clearly reflected by the sedimentary filling of the adjacent Cenozoic basins, which largely consist of psephitic deposits (GÖRLER and ZUCHT 1986, GÖRLER et al. 1987 and this vol., FRAISSINET et al., this vol.). GÖRLER et al. succeeded in separating microfaunas in the Ouarzazate and the Ait Kandoula basins south of the High Atlas which delimitate the age of several phases of uplift: The first influxe of High Atlas detritus was observed within fanglomerates at the base of the Ait Kandoula formation (Lower Miocene - ?Pleistocene). They are followed by very thick conglomerates which consist almost completely of pebbles of High Atlas rocks. Strong Lower Miocene uplift ist, moreover, demonstrated by Mesozoic limestone slabs which slid into the basins by the force of gravity. The largest of them was the "nappe of Toundoute" of approximately 300 km^2 in size. This second phase was followed by at least 10 m.y. of relative tectonic quiescence from the Serravallian to the Upper Pliocene. During this intervall the uplift of the High Atlas went on with diminished velocity. The top of the Ait Kandoula formation is made up of another conglomeratic sequence of Upper Pliocene or Lower Pleistocene age, that,owing to slumping structures, intercalation of megabreccias and glide-slabs of Jurassic limestones, marks a third stage of uplift. FRAISSINET et al. (this vol.) have even discerned four stages of uplift which are, however, based on lithostratigraphic data only. The final Quaternary stages of High Atlas uplift are documentated by a series of glacis levels south of the High Atlas (see STÄBLEIN, this vol.):

Gravity sliding seems to be confined to areas with high differences in vertical displacement. In the Toundoute region this rate has climbed up to at least 3-4000 m between the bottom of the Ouarzazate basin and the summits of the central High Atlas. GÖRLER & ZUCHT (1986) have estimated even 5000 m referring to the base of the Eocene. A similar rate seems correct for the northern slope of the High Atlas between the Tichka region and the Haouz basin, where large-scale gravity sliding has been described by FERRANDINI and LE MARREC (1982). East of the Ouarzazate basin, gravity sliding could not be recognized along the southern rim of the central High Atlas. The total amount of vertical displacement is reduced, there, to 2000 m.

There is good evidence that the stages of uplift coincided with compressional deformation in the central and eastern High Atlas; we are not sure whether this applies to the Oligocene uplift. The main compressional phase of the central High Atlas has been dated as Lower Miocene, since its structures include the Oligocene Hadida formation, which is disconformably overlain by the Ait Kandoula formation. Upper Pliocene and Pleistocene uplift was associated, again, with compression which produced tilting and folding of the Ait Kandoula formation together with the "nappe of Toundoute". STÄBLEIN (this vol.) has described an upthrust in the Ouarzazate basin which even tilted Middle Pleistocene beds. Furthermore, DUTOUR and FERRANDINI (1985) have concluded from deformation studies that the Haouz basin south of Marrakech has been under compression from Early to Late Quaternary. It is, moreover, important to mention that the southern and northern borders of the central High Atlas show differences. In the south, an alternation of paroxystic phases of folding and of thick sedimentation during the Oligocene - Pliocene interval is evident, whereas along the northern border the main deformations are younger than the Mio-Pliocene (FRAISSINET et al., this vol.).

The Cenozoic history of the Middle Atlas is not known in detail. The tectonic structure of the southeastern strip known as Moyen-Atlas plissé indicates a relatively weak compressional deformation, whereas the northwestern zone (Moyen-Atlas tabulaire) is only built up of flat-lying strata cut by a few fractures and wrench-faults (DRESNAY, this vol.). Compared with the High Atlas, the Middle Atlas shows a smaller degree of compression (see sections of MARTIN 1980/81, ROBILLARD 1981), but was apparently more influenced by lateral displacement. CHOUBERT and FAURE-MURET (1962) have distinguished three orogenic phases culminating in Late Eocene, Late Oligocene and Middle Miocene times, respectively. The main uplift happened in the early Pleistocene, according to these authors. Bearing in mind the latest stratigraphic determinations for the Tertiary formations of the High Atlas' southern margin (GÖRLER et al., this vol.) we presume, similarly for the Middle Atlas a change in chronology of the tectonic phases. We are sure that the deformational processes of Middle and High Atlas were closely connected.

5. Geodynamic model

When looking for a geodynamic model for the development of the intra-continental Atlas mountain system, one has to distinguish between two periods: A Mesozoic period in which rift grabens and basins formed due to a tensional and shearing stress-field between the North American, African and European plates. Crustal thinning in the future graben zones, subsidence of the basement there and sedimentary filling resulted from this geodynamic situation. A geothermally activated upper mantle beneath Northwest Africa is seen as the primary source for such features. In Upper Oligocene and Miocene, a new period has commenced with the change from a tensional to a compressional stress-field. In our opinion, this is related to the double collision of the westernmost ALKAPECA-fragment (Alboran block) with Iberia and the

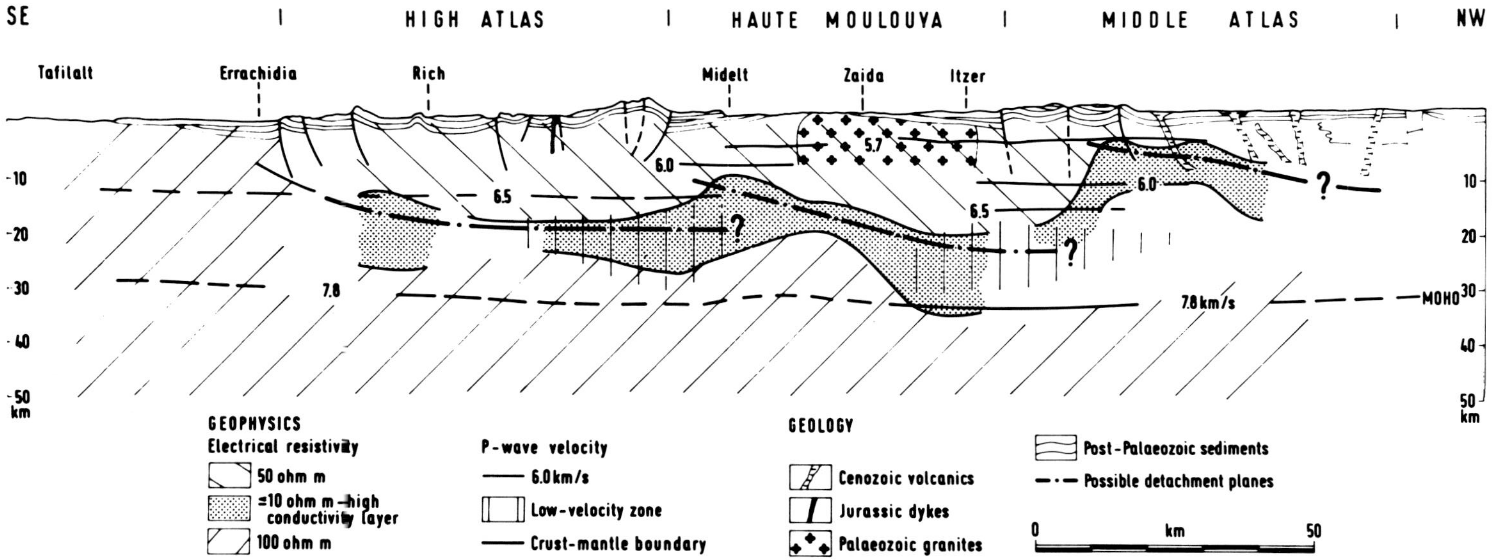

Fig. 5: Geotectonic interpretation of the crustal section through the Middle and High Atlas published by SCHWARZ and WIGGER (this vol.).

northwestern edge of Africa. This process caused not only the folding and thrusting of the Betic Cordilleras and the Rif, but also the uplift and folding of the Middle and High Atlas.

As evidenced by seismic and magnetotelluric measurements, there is a zone of low velocity and high electrical conductivity in the crust beneath the High Atlas, the High Plateaus and the Middle Atlas which descends like a ramp within the crust from S to N (HAAK and HUTTON 1986). Most likely it extends northwards beneath the Rif Atlas (SCHWARZ and WIGGER, this vol.). Especially the zone of high electrically conductivity may be associated with a zone of reduced shear strength, thus indicating the existence of detachment horizons. The model of thin-skinned tectonics as proposed by BALLY (1981, 1984) offers a possibility how the intracrustal mountain ranges might have been formed. Due to the assumed intracrustal detachment zones as evidence by geophysical results, the compressional stress could have been transferred far to the south. The sedimentary filling of the former rift grabens has been fractured, whereas the High Plateaus remained more or less uneffected, intermediate blocks. No typical mountain root, like in the Alps, could develope by this type of tectonics.

The type of tectonics in the graben zone depends on the angle between the confining stress-field and the graben axis. Though nearer to the Rif, the Middle Atlas shows a lower degree of compressional deformation than the central High Atlas. Running diagonal to the direction of maximum compressional strain, it was mainly deformed by left-lateral strike-slip faults; the folds and upthrusts observed are due to transpression (HARLAND 1971, SANDERSON and MARCHINI 1984). The Upper Pliocene "back-thrusting" of the High Atlas originated during a late stage of compression.

We have, however, to remember that lateral shortening does not surmount a degree of 20% in the central High Atlas. It is, without any doubt, lower not only in the Middle Atlas, but also in the western High Atlas and in the Anti-Atlas. Looking for further conditions to produce uplift, we may take into account that the thermal processes mentioned may have caused density inversions within the upper mantle and the lower crust. This view seems to be supported, too, by the distribution of Upper Cenozoic volcanic centers in both ranges (fig. 1), which are missing in the High Atlas.

The origin of the Transalboran fault is interpreted as an event subsequent to collision and orogeny. Being a prolongation of an old Northwest African fault zone, this lineament crosses the whole ensemble of the West Mediterranean orogens as well as the mobile zone in the foreland of the Rif. Deep-seated earthquakes along that zone demonstrate that the process of collision between Africa and Iberia is not yet fully extinct.

Acknowledgement: The authors are obliged to Dr. P. Wigger for scientific discussion and to R. Cousino, S. Huhn, St. Köhler, and D. Reich for technical assistance. We also thank G. Lindemann for helping with the English.

References

BALLY, A.W. (1981): Thoughts on the tectonics of folded belts.- in K.R. McCLAY and N.J. PRICE (eds.): Thrust and nappe tectonics.- Spec. Publs. geol. Soc. London, 9: 13-32, London.

-- (1984): Tectogenèse et séismique réflexion.- Bull.Soc. géol. France, (VII), 26: 279-285.

BELLON, H.; GUARDIA, P. and MAGNE, J. (1984): Les associations volcaniques du Miocène supérieur de la région Oranaise (Algérie occidentale). Conséquences géodynamiques.- Géol. médit., 11: 255-264, Nice.

BOSCH, J.W.H. van den (1971): Carte gravimétrique du Maroc, au 1:500.000.- Notes Mém. Service géol. Maroc, 234: 7 feuilles, Rabat.

BOUILLIN, J.-P., DURAND-DELGA, M. and OLIVIER, Ph. (1986): Betic, Rifian and Tyrrhenian arcs: Distinctive features, genesis and development stages.- in F.A. WEZEL (ed.): The origin of arcs, 281-304, Amsterdam (Elsevier).

CHEVREMONT, P. (1977): Les roches éruptives basiques des boutonnières de Tassent et Tasraft et leurs indices metallifères (Haut Atlas central, Maroc).- C.R. Acad.Sci. Paris, (D), 284: 1493-1496, Paris.

CHOUBERT, G. and FAURE-MURET, A. (1962): Evolution du domaine atlasique marocain depuis les temps paléozoiques.- Livre Mém. P. Fallot, 1960, 1: 447-527, Paris.

-- (1971): Anti-Atlas (Maroc).- in UNESCO, Tectonique de l'Afrique,. Sci. de la Terre, 6: 163-175, Paris.

-- (1974): Maroccan Rif.- in A.M. SPENCER (ed.): Mesozoic-Cenozoic orogenic belts. Data for orogenic studies.- Geol. Soc. London spec. Publ., 4: 37-46, Edinburgh.

COURBOULEIX, S.; DELPONT, G. and DESTEUCQ, C.H. (1981): Un grand décrochement est-ouest au nord du Maroc à l'origine des structures plissées atlasiques. Arguments géologiques et expérimentaux.- Bull.Soc. géol. France, (VII), 23: 33-43, Paris.

DRESNAY, R. du (1971): Extension et développement des phénomènes récifaux jurassiques dans le domaine atlasique marocain, particulièrement au Lias moyen.- Bull.Soc. géol. France, (VII), 13: 46-56, Paris.

DURAND-DELGA, M. and FONTBOTE, J. (1980): Le cadre structurale de la Méditerranée occidentale.- 26e Congr. géol. internat. Paris 1980, Colloque 5 "Les chaînes alpines issues de la Tethys": 67-85, Paris.

DUTOUR, A. and FERRANDINI, J. (1986): Nouvelles observations néotectoniques dans le Haut Atlas de Marrakech et le Haouz central (Maroc). Apports sur l'évolution récente d'un segment du bâti atlasique.- Rev. Géol. dyn. Géogr. phys., 26(5): 285-298, Paris.

FAURE-MURET, A. and CHOUBERT, G. (1971): Le Maroc. Domaine rifain et atlasique.- in Unesco (ed.): Tectonique de l'Afrique: 17-45. Sciences de la Terre, 6, Paris.

FERRANDINI, J. and LE MARREC, A. (1982): La couverture jurassique à paléogène du Haut Atlas de Marrakech est allochthone dans la "zone des cuvettes" d'Ait Ourir (Maroc).- C.R. Acad. Sci. Paris, (II), 295: 813-816, Paris.

FROITZHEIM, N. (1984): Oberkretazische Vertikaltektonik im Hohen Atlas SW von Marrakech/Marokko - Rekonstruktion eines Bewegungsablaufs im Frühstadium der Atlas-Orogenese.- N.Jb. Geol. Paläont. Mh., 1984: 463-471, Stuttgart.

GÖRLER, K. and ZUCHT, M. (1986): Stratigraphie und Tektonik des kontinentalen Neogens im Süden des zentralen Hohen Atlas, Provinz Ouarzazate (Marokko).- Berliner geowiss. Abh., (A), 66, 471-495, Berlin.

GÖRLER, K.; HELMDACH, F.-F. and ZUCHT, M. (1987): Continental Neogene south of the central High Atlas Mountains.- in G. MATHEIS and H. SCHANDELMEIER (eds.): Current research in African earth sciences: 133-136, Rotterdam (Balkema).

HAAK, V. and HUTTON, R. (1986): Electrical conductivity in continental lower crust.- in J.B. DAWSON (ed.): The nature of the lower continental crust.- Geol. Soc. spec. Publ., 25: 35-49.

HAQ, B.U.; HARDENBOHL, J. and VAIL, P.R. (1987): Chronology of fluctuating sea levels since the Triassic.- Science, 235: 1156-1167, 5 Abb., Washington, D.C.

HARLAND, W.B. (1971): Tectonic transpression in Caledonian Spitzbergen.- Geol. Mag., 108: 27-42.

HARMAND, CH. and MOUKADIRI, A. (1986): Synchronisme entre tectonique compressive et volcanisme alcalin: exemple de la provence quaternaire du Moyen Atlas (Maroc).- Bull.Soc. géol. France, (VIII), 2: 595-603, Paris.

HATZFELD, D. and FROGNEUX, M. (1981): Intermediate depth seismicity in the Western Mediterranean unrelated to subduction of oceanic lithosphere.- Nature, 292: 443-445.

HEINITZ, W. (1984): Die Deformation des sedimentären Deckgebirges im zentralen Anti-Atlas (Südmarokko).- Berliner geowiss. Abh., (A), 55, 84 S., Berlin.

HERNANDEZ, J. and BELLON, H. (1985): Chronologie K-Ar du volcanisme miocène du Rif oriental (Maroc): implications tectoniques magmatologiques.- Revue Géol. dyn. Géogr. phys., 26: 85-94, Paris.

HERNANDEZ, J.; LAROUZIERE, F.D. de; BOLZE, J. and BORDET, P. (1987): Le magmatisme néogène bético-rifain et le couloir de décrochement trans-Alboran.- Bull. Soc. géol. France, (VIII), 3: 257-267, Paris.

HÜSSNER, H. (1985): Jurassische Karbonate des westlichen Hohen Atlas (Marokko): Mikrofaziesanalyse und plattentektonischer Rahmen.- Facies, 12: 141-218, Erlangen.

JACOBSHAGEN, V. (1986): Zur Baugeschichte des zentralen Hohen Atlas (Marokko).- Berliner geowiss. Abh.,(A), 66: 433-454, Berlin.

JENNY, J. (1983): Les décrochements de l'Atlas de Demnat (Haut Atlas central, Maroc): prolongation orientale de la zone de décrochement du Tizi-n-Test et clef de la compréhension de la tectonique atlasique.- Eclogae geol. Helv., 76: 243-251, Basel.

JENNY, J.; LE MARREC, A. and MONBARON, M. (1981): Les couches rouges du Jurassique moyen du Haut Atlas central (Maroc): correlations lithostratigraphiques, éléments de datation et cadre tectonosédimentaire.- Bull. Soc. géol. France, (VII), 23: 627-639, Paris.

KANES, W.H.; SAADI, M.; EHRLICH, E. and ALEM, A. (1974): Moroccan crustal response to continental drift.- Science, 180: 950-952.

LEBLANC, M. (1973): Le grand dyke de dolérite de l'Anti-Atlas et le magmatisme jurassique du Sud marocaine.- C.R. Acad. Sci. Paris, (D), 276: 2943-2946, Paris.

MANSPEIZER, W. (1982): Triassic-Liassic basins and climate of the Atlantic passive margins.- Geol. Rdsch., 71: 895-917, Stuttgart.

MARTIN, J. (1980/81): Le Moyen Atlas central. Etude géomorphologique.- Notes Mém. Service géol. Maroc, 258 (1980): 10-91, and 258 bis: 268-416, Rabat.

MATTAUER, M.; TAPPONIER, P. and PROUST, F. (1977): Sur le mécanismes de formation des chaines intracontinentales. L'exemple des chaînes atlasiques du Maroc.- Bull.Soc. géol. France, (7), 19: 521-526, Paris.

MENVIELLE, M. and LE MOUEL, J.L. (1985): Existence d'une anomalie de conductivité dans le Haut Atlas marocain et concentration des courants telluriques à l'échelle régionale.- Bull. Soc. géol. France, (VIII), 1: 553-558, Paris.

MENVIELLE, M. and ROSSIGNOL, J.C. (1982): Conséquences tectoniques de l'éxistence d'une anomalie de conductivité électrique au Nord du Maroc.- Canad. J. Earth Sci., 19: 1507-1517.

MICHARD, A. (1976): Eléments de géologie marocaine.- Notes Mém. Serv. géol., 252: 1-408, Rabat.

MICHARD, A.; WESTPHAL, M.; BOSSERT, A. and HAMZEH, A. (1975): Tectonique de blocs dans le socle atlaso-mésétien du Maroc: une nouvelle interprétation des données géologiques et paléomagnétiques. Earth planet. Sci. Lett., 24: 363-368.

ROBILLARD, D. (1981): Etude stratigraphique et structurale du Moyen Atlas septentrional (région de Taza, Maroc).- Notes Serv. géol. Maroc, 42: 101-193, Rabat.

SALVAN, H.M. (1984): Les formations évaporitiques du Trias marocain. Problèmes stratigraphiques, paléogéographiques et paléoclimatologiques. Quelques réflexions.- Rev. Géol. dyn. Géogr. phys., 25: 187-203, Paris.

SANDERSON, D.J. and MARCHINI, W.R.D. (1984): Transpression.- J. struct. Geol., 6: 449-458.

SCHWARZ, G.; RAT, V. and HAAK, V. (1986): Magnetotellurische Messungen und erdmagnetische Tiefensondierungen im marokkanischen Atlas.- Berliner geowiss. Abh., (A), 66: 289-300, Berlin.

STETS, J. and WURSTER, P. (1981): Zur Strukturgeschichte des Hohen Atlas in Marokko.- Geol. Rdsch., 70: 801-841, Stuttgart.

TISSERANT, D.; THUIZAT, R. and AGARD, J. (1976): Données géochronologiques sur le complexe de roches alcalines du Tamazeght (Haut Atlas de Midelt), Maroc.- Bull. B.R.G.M., 1976(3)): 279-283, Paris.

TORRES-ROLDAN, R.L.; POLI, G. and PECCERILLO, A. (1986): An early Miocene arc - tholeiitic magmatic dike event from the Alboran Sea - evidence for precollisional subduction and back-arc crustal extension in the westernmost Mediterranean.- Geol. Rdsch., 75: 219-234, Stuttgart.

WIGGER, P. and HARDER, S. (1986): Seismologische und krustenseismische Untersuchungen im Atlas-System, Marokko.- Berliner geowiss. Abh., (A), 66: 273-288, Berlin.